# Lecture Notes in Statistics 115

Edited by P. Bickel, P. Diggle, S. Fienberg, K. Krickeberg,
I. Olkin, N. Wermuth, S. Zeger

**Springer**
*New York*
*Berlin*
*Heidelberg*
*Barcelona*
*Budapest*
*Hong Kong*
*London*
*Milan*
*Paris*
*Santa Clara*
*Singapore*
*Tokyo*

P.M. Robinson
M. Rosenblatt
(Editors)

# Athens Conference on Applied Probability and Time Series Analysis

## Volume II: Time Series Analysis
## In Memory of E.J. Hannan

Springer

P.M. Robinson
London School of Economics
Department of Economics
Houghton Street
London WC2A 2AE
Great Britain

Murray Rosenblatt
University of California, San Diego
Department of Mathematics
La Jolla, CA 92093
and
Colorado State University
Department of Statistics
Ft. Collins, CO 80523

CIP data available.
Printed on acid-free paper.

Camera ready copy provided by the author.
Printed and bound by Braun-Brumfield, Ann Arbor, MI.
Printed in the United States of America.

9 8 7 6 5 4 3 2 1

ISBN 0-387-94787-6 Springer-Verlag New York Berlin Heidelberg   SPIN 10538542

# Preface

A conference was held March 22 to 26, 1995 at the Hotel Titania in Athens, Greece partly in memory of Ted Hannan. Hannan was one of the very important figures in the area of time series analysis from the 1960s until his death in January 1994. Some of the participants in the meeting as well as colleagues who were not able to attend contributed papers.

The initial paper of Robinson is a memorial article that discusses Ted Hannan's researches and their influence on current work in time series analysis. The later papers are then arranged in alphabetical order according to the names of the authors. The papers cover a broad range of topics in time series analysis. A brief description and rough grouping by topic follows.

A classical part of time series analysis is centered about methods for finite parameter Gaussian models (introduced on a heuristic basis by Peter Whittle and partially anticipated by others like Mann and Wald). Such methods are all the more important in that they typically have some usefulness under much broader circumstances than the Gaussian property, as Hannan himself incisively demonstrated. Especially during the latter part of his career, Hannan contributed immensely to statistical theory concerning time series with rational spectrum and Gaussian identification and estimation. A number of the papers pertain to these topics. Bauer and Deistler examine the topological properties of parametrizations of state space models with application to identification. Taniguchi presents a number of results on higher order asymptotic theory for hypothesis tests and studentized estimates in Gaussian ARMA processes. Brockwell discusses the interpolation problems for ARMA sequences. Rosenblatt considers nonGaussian autoregressive models where roots of the characteristic polynomial may have any absolute value other than one. El Matouat and Hallin present a stochastic complexity justification and an information-theoretic derivation of the Hannan-Quinn autoregressive order selection criterion. Camarinopoulos, Zioutas and Bora-Senta compute, via a quadratic programming algorithm, certain robust estimates of autoregressions. Subba Rao and Chandler consider Gaussian estimation for point process data, in the spirit of Brillinger, Franke and Gründer discuss interpolation for a spatial-temporal process with the work motivated by measurements of gas concentration and a number of meterological parameters.

Early work on time series with infinite variance or stable marginal distribution was done by Kanter and Hannan. Bhansali's paper is a survey of results on estimation for finite parameter stable nonGaussian models (second moments infinite) where rates of convergence are faster than in the case of classical Gaussian models. Klüppelberg and Mikosch consider spectral methods for linear processes generated by infinite variance symmetric $\alpha$-stable noise. Chobanov, Mateev, Mittnik and Rachev investigate the performance of stable distributions in modelling exchange rate data.

Hannan's extensive frequency domain work is reflected in some of the papers. Kavalieris is concerned with estimating the number of spectral lines and the model order for the absolutely continuous component in a mixed spectrum. Priestley discusses the estimation of harmonics. Quinn describes various techniques for the estimation and tracking of frequency in a sinusoidal model in the presence of additive noise.

Several of the papers focus on long range dependent processes. Though this was not an explicit interest of Hannan's, some of his asymptotic statistical theory in fact admits such processes. Hosoya obtains a number of limit theorems for stationary sequences with long range dependence. Koul establishes asymptotic properties of nonlinear regression estimates in the presence of long range dependent errors. For a nonparametric regression, Ray and Tsay consider how the choice of bandwidth in a kernel estimate is affected by the long range dependence. Bandwidth selection is also discussed by Henry and Robinson, but in the context of a particular method of semiparametrically estimating the long memory parameter. Smith and Chen suggest an extension of this method to linear regression models. Taqqu and Teverovsky review the same method, and others, for semiparametrically estimating the long memory parameter and propose a graphically-based approach to implementation.

Nonstationary processes of various types are emphasized in several papers. For processes with a local stationarity property, Dahlhaus establishes asymptotic theory for nonparametric estimates of time-varying autocovariance and spectral density. Phillips and Lee show how the Grenander-Rosenblatt equivalence of least squares and generalized least squares regression estimates can break down in the presence of integrated or near-integrated errors. Davis, Chen, and Dunsmuir in their paper deal with estimation (and testing) of a root on or near the unit circle in a seasonal moving average model. Granger, Hyung, and Jeon study a class of fractionally integrated processes with a unit root. Gray and Thomson discuss the design of moving average local trend estimation filters for nonseasonal time series. An's paper considers the deterministic sequence of a number of chaotic maps. Cheng and Tong provide a wavelet representation and decomposition theory for a general class of stochastic process.

With Boston, Hannan wrote an early paper on the modelling of nonlinear time series. Some of the papers deal with problems in which nonlinearity or the nonGaussian property is dealt with explicitly. Brillinger considers the analysis of a finite-valued time series. Bossaerts, Hafner and Härdle nonparametrically estimate stochastic volatility models for exchange rate data. Skaug and Tjostheim review a number of nonparametric tests for serial dependence that use a measure of distance between estimated probability densities.

We hope that this volume will serve as a proper memorial for Ted Hannan who made such a strong contribution as a researcher and as a person to this important field in statistics, applied mathematics, engineering and other areas of scientific application.

We wish to thank, in alphabetical order, Caroline Cook of Colorado State University, Judy Gregg of the University of California, San Diego and Sue Kirkbride of the London School of Economics for helping to assemble the manuscript and complete the book.

P. M. Robinson                                                          M. Rosenblatt
London School of Economics                     University of California, San Diego
and Political Science                                      and Colorado State University

# Contents

Memorial Article:

## EDWARD J. HANNAN, 1921–1994

By P. M. Robinson

*London School of Economics*

Statistical time series analysis has lost a unique creative force and brilliant technician through the death of Ted Hannan on 7 January 1994, of a heart attack. During some forty years, he profoundly influenced the development of many areas of time series. This period has seen an immense growth in the breadth, sophistication and technical standards of the subject, and a great deal of this is due to the originality, usefulness and depth of Hannan's ideas.

It is remarkable that so resourceful a mathematician as Hannan was largely self-taught in mathematics, and that his statistical research, which led to four books and over one hundred and thirty published articles, did not begin until he was in his thirties. On leaving school, in Melbourne, at the age of 15, he went to work as a bank clerk, until joining the Australian Army in 1941. He saw active service in New Guinea. In 1946, he took advantage of an assistance scheme for ex-servicemen and enrolled for a commerce degree at the University of Melbourne. There he studied mostly economics, and on graduation took a job as a junior economic researcher in the Reserve Bank of Australia, in Sydney, becoming involved in statistical and econometric analysis of data.

In 1953 the Bank sent him on a year's leave to the newly-founded Australian National University in Canberra, ostensibly to work in economics. One day he was reading mathematics in the library when P.A.P. Moran, who had recently taken up the Chair of Statistics at the ANU, looked over his shoulder and, impressed, asked Hannan to come and see him. Moran arranged a research fellowship in statistics. Hannan quickly completed a Ph.D. under Moran's supervision, and was to remain in Canberra for the rest of his life. He was appointed Professor and Head of the Department of Statistics in the ANU's School of General Studies in 1959, and in 1971 moved over to the ANU's Institute of Advanced Studies, where he took up a full-time research position as Professor. He served as Head of the Statistics Department there from 1982 to 1985. He retired in 1986.

0143-9782/94/06 563–576          JOURNAL OF TIME SERIES ANALYSIS Vol. 15, No. 6
© 1994 Basil Blackwell Ltd., 108 Cowley Road, Oxford OX4 1JF, UK and 238 Main Street, Cambridge, MA 02142, USA.

Despite his extensive exploration of theoretical issues much of Hannan's research had its roots in real-world problems, especially ones in geophysics and economics. Some of his first time series research, published in [1], was an analysis of rainfall data, in which he tested for the presence of periods of heavy rainfall, by comparing daily means, with correction for seasonality and first order autoregressive autocorrelation. Following earlier work of Anderson, Durbin, Watson and others, testing for lack of autocorrelation against autoregressive alternatives was at the centre of attention during the 1950's, and it was the main focus of Hannan's early research ([2]–[5], [7] and [11]). Much of his effort, like others', was directed to establishing exact tests, for scalar and vector series, and for regression errors. However, Hannan realized the limitations of finite-sample theory in time series analysis, and also made asymptotic Pitman efficiency comparisons. Another notable paper from the period was [6] (with G. Watson), which investigated the effects on regression-based inference of misspecification of the error spectral density, and with [8], marks the beginning of his interest in the frequency domain. Nonparametric weighted autocovariance and periodogram estimates of the spectral density, and their basic statistical properties, had by then been developed, by Bartlett, Tukey, Parzen, Priestley and others, but Hannan saw how they could play a crucial role in inference on other features of interest. In [8] he modified a procedure of Jowett for studentizing the sample mean of a stationary series and suggested an asymptotic approximation using a spectrum estimate at zero frequency. This idea has led to comparatively recent research in econometrics. Hannan was not blinded by the simplicity of the asymptotic formula, pointing out the conflict in choosing the spectral bandwidth to avoid both strong bias and dependence on the sample mean. In a related paper [10], he discussed bias-correction in spectrum estimates computed from detrended data, while [15] modified Fisher's test for a jump in the spectral distribution function by allowing for a nonparametric spectrum under the alternative.

Contact with the oceanographer B. Hamon led to even more interesting use of nonparametric spectrum estimates. Hamon had been computing narrow-band regressions and wondered how to optimally combine them. In 1963 Hamon and Hannan [19] proposed a frequency-domain weighted least square estimate, weighting inversely with smoothed estimates of the nonparametric error spectrum, while in a truly seminal paper [20], Hannan showed that it is asymptotically as efficient as the generalized least squares estimate, in the presence of nonparametric error autocorrelation. This was a delicate task because of the slow convergence of the spectrum estimates, and Hannan brought to it a technical virtuosity which was unusual in time series analysis at the time; his regularity conditions still seem quite mild. [20] is remarkable also as an early example of the demonstration of adaptive estimation, in the sense that the same first-order efficiency can be achieved in semiparametric models by using a smoothed nonparametric estimate of an infinite-dimensional nuisance function as by using a correct parametric one. Over a decade passed before the ability to adapt for unknown distributional form in the errors in location and regression models with independent observations was proved, while adaptivity to regression error heteroscedasticity of unknown form with independent observations (a problem analogous to Hannan's after Fourier transformation of his time series regression) was not shown until even later. This line of research proved very fruitful; in several subsequent papers (e.g. [22], [26], [31]; and [53] with R.D. Terrell) Hannan employed the same approach in more elaborate models, in particular ones relevant to econometrics, while [32] with the oceanographer G. Groves is a detailed application to data on sea level and weather.

Much of Hannan's work of the period, and subsequently, illustrates in other ways how the frequency domain provides an elegant setting for parametric and semiparametric inference, as well as for the nonparametric problems with which it is traditionally associated. Many statistics are most neatly expressed in the frequency domain. An example is his estimate [20, 31] of the coefficients in a lagged regression; it approximates Fourier coefficients of nonparametric estimates of the frequency response function, and has the orthogonality property that further lags can be included without recomputing the previous coefficients. Also in 1963, Hannan introduced the idea of omitting frequencies to alleviate the errors-in-variables problem. He argued that the signal-to-noise ratio for a regressor observed with error might vary significantly with frequency, and suggested leaving out from a broad-band frequency domain regression estimate, non-degenerate bands of frequencies for which the ratio is believed small, such as high frequencies or neighbourhoods of seasonal frequencies. He rigorously justified the method in [59], and it was subsequently developed in the econometric literature.

In his work for the Reserve Bank, Hannan had come across the problem of seasonality and gained considerable insight into the nature of seasonality, and the general properties that seasonal models should exhibit. Three of his papers published in the early 1960's introduced new ideas. In [13] and [21] he used operators in estimating seasonality in the presence of trend and stationary noise, noting how the effects of a moving average operator to eliminate nonseasonal effects can be easily reversed. In [24] he modelled the seasonal component by a cosinusoid whose coefficients are stationary processes, so the seasonal pattern changes over time. Thus the seasonal component can be modelled to have spectrum with smooth peaks, not singularities. The same model was studied in [29], but here the coefficients had roots on the unit circle; this was published in 1967, many years before unit roots became the major preoccupation of many theoretical and applied workers which they are today, and in fact the model in [29] is similar to random component models of seasonal patterns which are currently popular.

Meanwhile Hannan had also started to indulge his interest in abstract mathematics. In [16] he derived results for canonical correlation from general theories concerning the spectral decomposition of operators, allowing for infinite-dimensionality and with applications to stationary Gaussian processes. His monograph [25], *Group Representations and Applied Probability* appeared in 1965; in part through teaching experimental design, Hannan noticed the analogy with spectral analysis, while he had also become interested in the connection between Fourier analysis and representations of groups. In [28] Hannan obtained a deep description of filters for a class of processes that includes covariance stationary ones. In other circumstances he could well have ended up as a pure mathematician rather than a statistician.

The appearance in 1970 of *Multiple Time Series* [40] was a major landmark. It considerably developed his 1960 volume *Time Series Analysis* [12], which had provided a concise and readable account of both the time and frequency domain in the scalar case. [40] embodied also most of Hannan's research in the intervening years, mostly in a multivariate setting, and is notable for its rigorous and detailed treatment of both continuous and discrete time processes. The, often difficult, exercises completing each chapter actually posed some new research problems (Hannan provided no solutions!). It has proved a stimulating reference to many researchers.

By this time Hannan was equipped with the techniques to solve, in a distinctive fashion, a wide range of complex problems in frequency domain time series analysis. A sequence of papers written with P. Thomson in the early 1970's indicates some of the themes covered. [44] established the central limit theorem for the discrete Fourier transform of vector, tapered time series at finitely many adjacent harmonic frequencies, under both $\alpha$-mixing and linear process assumptions. In [46] and [55], improved estimates of coherence and group delay were proposed, in the presence of a rapidly changing phase which can seriously bias the usual coherence estimates (see also [61] with Hamon). Inference on frequency-dependent time-lags (modelling "echo times") was developed in [60]. Later work with Thomson [95], [130] on delay estimation had a significant impact on the signal processing community. In [45] and [56] Hannan studied a simple cosinusoidal regression with unknown real-valued frequency $\theta$. The estimates of $\theta$ were shown to be $n^{3/2}$-consistent, where $n$ is sample size, while when $\theta = 0$ the estimate actually equals 0 for all sufficiently large $n$. The first journal to which [56] was submitted found this result too surprising, and Hannan experienced a very rare, possibly unique, rejection. Several papers entailed a spatial dimension, discussing the analysis of noise-corrupted measurements, in particular of a plane wave, recorded over time by several recorders at distinct geographical locations, when also the plane wave can be a transient (see [61] with Hamon, [66], [67] with J. Goncz, and [75] and [82] with M. Cameron). Others studied refined properties of the periodogram: [91] with Z. Chen; [108] with H-Z. An and Chen.

A great deal of Hannan's work during his last quarter-century was on linear time series models. The literature on this topic is enormous but he dealt definitively with some of the most difficult and important issues. Linear ARMA and transfer function models had come to the fore around 1970 due in part to the book of Box and Jenkins. A proper study of such models requires an understanding of the identification problem. This is difficult due to the presence of lags in the innovations and possible inputs, as well as the output variables, especially when these are vector-valued. It was an unexplored problem when, in 1969 and 1971, Hannan [36], [43] gave conditions for identification of stationary vector ARMA and ARMAX systems of known order. He used techniques he had learnt years earlier from L. MacDuffee's book *Theory of Matrices*. Soon after, the relevance of the state space representation to ARMAs was pointed out by Akaike. Then Hannan noticed work of the engineer J. Clark who discussed the manifold of ARMA transfer functions corresponding to the state space form. This led Hannan to link up these representations in his study of identification [70], and to an extremely profound and extensive study of their algebraic and topological properties via the concept of McMillan degree, the dimension of the minimal state vector (see e.g. [81], [92]; [71] with W. Dunsmuir; [93] with M. Deistler; [77] and [90] with Dunsmuir and Deistler; and [109] with L. Kavalieris).

A most important practical issue is the choice of ARMA and ARMAX orders. Following earlier suggestions of Akaike, various rules had been proposed for estimating the order of $p$ of a scalar autoregression; $p$ is chosen to minimize $\log \hat{\sigma}_p^2 + pC_n$ where $\hat{\sigma}_p^2$ is an estimate of the innovations variance. In [83], Hannan and B. Quinn showed via the law of the iterated logarithm that use of $C_n \sim 2(\log \log n)/n$, which decreases faster than had earlier been conjectured in thew literature, provides a consistent estimate of $p$. Order determination is considerably more difficult to study in case of ARMA models because of the consequences of overspecifying both $p$ and $q$. But in [89], Hannan established consistency of scalar ARMA order deterministic procedures. In [92], he showed how to estimate McMillan degree. Another major paper [99] discussed Akaike's methods in relation to ARMA order determination. [116] with Kavalieris considered ARMAX models. A very good account of the topic is Hannan's review paper [121], published with discussion.

Relatively few of Hannan's papers contain empirical applications, and when they did these were typically carried out by co-authors. However, he had an excellent grasp of what is required in practice, and intuition in to what is likely to be used by practitioners. He saw the importance of efficient computations and brought considerable ingenuity and common-sense to the task of developing new algorithms. Many of his methods were based on Fourier transformed data and he reminded readers of the economies due to use of the fast Fourier transform. In the early 1970's he invented a clever use of the FFT for the spectral analysis of series with missing observations, but it turned out that R.H. Jones had already had the same idea and was in process of publishing it. For the estimation of scalar ARMA models, Hannan [38] published in 1969 a method which once again makes elegant use of the frequency domain, and is asymptotically efficient and avoids the necessity of iteration and factorisation. ARMAX models were similarly studied in a paper with D. Nicholls [48]. Several papers developed ARMA estimates that are recursive, in the sense of being updated as new observations come to hand (see e.g. [88]). The method of scalar ARMA estimation in [98] with Rissanen (modified in [111] with Kavalieris) made a significant impact. It involves order-determination and estimation of innovations from a long autoregression, followed by economical recursive calculation of a sequence of regressions. Analogous methods for multivariate ARMA and ARMAX models were studied in [109] with Kavalieris and [115] with Kavalieris and M. Mackisack.

A feature of nearly all of Hannan's research of the last thirty years was the superb quality of his treatment of asymptotic statistical theory. Exact theory for time series is often intractable, and asymptotics can provide approximate justification under broad conditions. Hannan achieved conditions which are weak, sometimes minimal, yet also primitive and comprehensible. He brought a formidable array of mathematical techniques to this work, such as Fourier analysis: he moved elegantly between the time and frequency domains in expressing regularity conditions on stationary time series. Although, as has been seen, much of his work focused on rational spectral densities, he established many results under very weak summability conditions on the Wold decomposition weights. While he nearly always used Gaussian procedures, he often stressed the achievement of mild moment conditions. He was at the forefront of applying martingale theory to time series problems from the early 1970's, noting that martingale difference assumptions on innovations are natural because then the best linear predictor is the best predictor. These features are all illustrated in his account of the asymptotic theory of scalar linear time series models [54], published in 1973, in which he improved upon results of Whittle and Walker. This paper was also notable as a relatively early rigorous treatment of implicitly-defined extremum estimates in time series; prompted by R. Jennrich's 1969 work on nonlinear least squares for independent observations, Hannan established almost sure convergence of the estimates, rather than the convergence in probability which was then standard in the time series literature, as well as the central limit theorem (see also [45]). Extensions to more general models were given in [71], [81], [90], [109], for example. Fundamental results on the asymptotic properties of sample autocovariances were established in [62] and [69], and also [51] with Heyde, and [101] with An and Chen. For example, [69] showed that under mild regularity conditions square integrability of the spectrum is both necessary and sufficient for asymptotic normality with $\sqrt{n}$-norming. Over the years he revised his work on the central limit theorem and invariance principle for linear regression [14], [58], [84]. Like Eicker had done in 1967, Hannan required only square summability of the Wold decompositon weights (thereby allowing for long range dependence) in some of these results, but he relaxed Eicker's conditions in important respects and ended up in [84] with very definitive results, again using martingale techniques.

Several pieces of Hannan's work defy any of the above categories. He did little work on nonlinear time series, yet [49], with R. Boston, preceded much of the activity in this area, employing a nonlinear transformation of a moving average. In [64] he extended Grenander and Rosenblatt's work on efficiency of least squares to continuous time. A paper "Aliasing", with new results on the discrete sampling of continuous processes, was never published. Outside of time series analysis, [30] described the connection between canonical correlation and limited information estimates in multiple equation econometric models. (Hannan's contribution to econometrics was immense, and would have been perceived as even greater had he not ceased publishing in econometric journals in the 1970's.)

"Retirement", in 1986, at the age of 65 saw no apparent diminution in the quality, variety or volume of Hannan's output. In 1988 he had seven publications, including a fourth book *The Statistical Theory of Linear Systems* [125], with Deistler, collecting together much of their work on linear time series models. The remaining years continued to bring diverse and fascinating contributions, for example the application of stochastic complexity to spectral bandwidth ([129] with Rissanen), estimation of cosinusoidal regression with frequencies close together relative to $n^{-1}$ ([132] with Quinn) and a generalization of autoregressive modelling involving discrete Laguerre filters ([138] with B. Wahlberg). At the time of his death, approaching his 73rd birthday, he had several papers submitted and was working on a further book, [146] with D. Huang and Quinn.

Hannan's Ph.D. students included (in approximate chronological order) R. D. Terrell, D. Nicholls, P. Thomson, Y. Hosoya, W. Dunsmuir, J. Henstridge, M. Cameron, V. Solo, K. Tanaka, B. Quinn, L. Kavalieris, M. Mackisack, A. MacDougall and D. Huang. Working with Hannan was a unique learning experience, and at times an awe-inspiring and even intimidating one. Hannan had immense affection for his students, and they became his life-long friends.

Hannan served with distinction on the editorial boards of several journals: *The Annals of Statistics*, *Econometrica*, *The Journal of Multivariate Analysis*, *The Journal of Applied Probability*, *The Advances in Applied Probability*, *The Journal of Time Series Analysis*, *International Economic Review* and the *Journal of Forecasting* (he was associated with the *Applied Probability* journals and the *Journal of Time Series Analysis* from their inception until his death). As an associate editor and referee he was consistently penetrating, fair-minded and extraordinarily quick.

Hannan was elected a Fellow of the Australian Academy of Science, a Fellow of the Australian Academy in the Social Sciences, a Fellow of the Econometric Society, an Honorary Fellow of the Royal Statistical Society, and a Member of the International Statistical Institute. He was awarded the Lyle medal for research in mathematics and physics by the Australian Academy of Sciences in 1979, and the Pitman medal by the Statistical Society of Australia in 1986. In 1986 a Festschrift volume in his honour "Essays in Time Series Analysis and Allied Processes" (edited by J. Gani and M. B. Priestley) was published. Also in 1986, his 65th birthday was marked by a conference in his honour, in Canberra, sponsored by the Australian mathematical Society.

Hannan would most likely have received even greater recognition had he not chosen to remain in the relative isolation of Australia. However, he travelled a lot, with several extended visits to the United States, although family ties kept him from a permanent move. He was extremely close to his wife of 45 years, Irene, and family. As well as Irene, he leaves four children and six grandchildren. Hannan's intellectual interests embraced much more than mathematics. He was very widely read—literature, politics, biography, and especially the poetry of Yeats—and his acute and original obserations on any number of topics enlivened many a letter or coffee-break conversation. The sharpness and rigour of his mind, his enthusiasm, excitability and humour left an indelible impression. He could lose his temper but would quickly end any quarrel with generous and often-unwarranted apologies. The integrity which shines through his writing was matched in his personal life. The loss of his vitality and humanity is felt by many friends. The influence on time series analysis of his magnificent life's work will continue.

ACKNOWLEDGEMENT

I am extremely grateful to Joe Gani, Peter Thomson and Des Nicholls for invaluable help and advice in the preparation of this article.

## THE PUBLICATIONS OF EDWARD J. HANNAN

### 1955

[1] A test for singlularities in Sydney rainfall. *Austral. J. Phys.* 8, 289–297.
[2] Exact tests for serial correlation. *Biometrika* 42, 133–142.
[3] An exact test for correlation between time series. *Biometrika* 42, 316–326.

### 1956

[4] The asymptotic powers of certain tests based on multiple correlations. *J. R. Statist. Soc.* B, 18, 227–233.
[5] Exact tests for serial correlation in vector processes. *Proc. Camb. Phil. Soc.* **52**, 482–487.
[6] With G. S. Watson. Serial correlation in regression analysis. *Biometrika* 43, 436–448.

### 1957

[7] Testing for serial correlation in least squares regression. *Biometrika* 44, 57–66.
[8] The variance of the mean of a stationary process. *J. R. Statist. Soc.* B, 19, 282–285.
[9] Some recent advances in statistics. *Economic Record* 33, 337–352.

### 1958

[10] The estimation of the spectral density after trend removal. *J. R. Statist. Soc.* B, 20, 323–333.
[11] The asymptotic powers of certain tests of goodness of fit for time series. *J. R. Statist. Soc.* B, 20, 143–151.

## 1960

[12] *Time Series Analysis*. Methuen: London. (Published in Russian with an Appendix by Yu. B. Rozanov in 1964; published in Japanese.)
[13] The estimation of seasonal variation. *Austral. J. Statist.* 2, 1–15.

## 1961

[14] A central limit theorem for systems of regressions. *Proc. Camb. Phil. Soc.* 57, 583–588.
[15] Testing for a jump in the spectral function. *J. R. Statist. Soc.* B, 23, 394–404.
[16] The general theory of canonical correlation and its relation to functional analysis. *J. Austral. Math. Soc.* 11, 229–242.

## 1962

[17] Systematic sampling. *Biometrika* 49, 218–283.
[18] Rainfall singularities. *J. Appl. Math.* 1, 426–429.

## 1963

[19] With B. V. Hamon. Estimating relations between time series. *J. Geophys. Res.* 68, 6033–6041.
[20] Regression for time series. In *Time Series Analysis*, edited by M. Rosenblatt, Wiley: New York, 17–37.
[21] The estimation of seasonal variation in economic time series. *J. Amer. Statist. Assoc.* 18, 31–44.
[22] Regression for time series with errors of measurement. *Biometrika* 50, 293–302.

## 1964

[23] The statistical analysis of hydrological time series. In *Proceedings of the Symposium on Water Resources, Use and Management*, Melbourne University Press: Melbourne, 233–243.
[24] The estimation of a changing seasonal pattern. *J. Amer. Statist. Assoc.* 59, 1063–1077.

## 1965

[25] *Group Representations and Applied Probability*. Methuen: London. (Published in Russian with introduction by A. M. Yaglom in 1970.)
[26] The estimation of relations involving distributed lags. *Econometrica* 33, 206–224.

## 1966

[27] Spectral analysis for geophysical data. *Geophys. J.R. Astronom. Soc.* 11, 225–236.

## 1967

[28] The concept of a filter. *Proc. Camb. Phil. Soc.* 63, 221–227.
[29] The measurement of a wandering signal amid noise. *J. Appl. Prob.* 4, 90–102.
[30] Canonical correlation and multiple equation systems in economics. *Econometrica* 35, 123–138.
[31] The estimation of a lagged regression relation. *Biometrika* 54, 315–324.

## 1968

[32] With G. W. Groves. Time series regression of sea level on weather. *Rev. Geophys.* 6, 129–174.
[33] With R.D. Terrell. Testing for serial corelation after least squares regression. *Econometrica* 36, 133–150.
[34] Least squares efficiency for vector time series. *J. R. Statist. Soc.* B, 30, 490–499.

## 1969

[35] Fourier methods and random processes. *Bull. Internat. Statist. Inst.* 42, 475–496.
[36] The identification of vector, mixed autoregressive-moving average, systems. *Biometrika* 56, 223–225.
[37] A note on an exact test for trend and serial correlation. *Econometrica* 37, 485–489.
[38] The estimation of mixed moving average autoregressive systems. *Biometrika* 56, 579–593.
[39] Fourier methods and linear models. *Austral. J. Sci.* 32, 171–175.

## 1970

[40] *Multiple Time Series*. Wiley: New York. (Published in Russian in 1974.)
[41] With R. D. Terrell and N. E. Tuckwell. The seasonal adjustment of economic time series. *Internat. Econ. Rev.* 11, 24–52.
[42] Data smoothing, in *Data Representation*, edited by R. S. Anderssen and M. R. Osborne, *Univ. of Queensland Press*: Brisbane, 34–42.

## 1971

[43] The identification problem for multiple equation systems with moving averge errors. *Econometrica* 39, 751–765.
[44] With P. J. Thomson. Spectral inference over narrow bands. *J. Appl. Prob.* 8, 157–169.
[45] Non-linear time series regression. *J. Appl. Prob.* 8, 767–780.
[46] With P. J. Thomson. The estimation of coherence and group delay. *Biometrika* 58, 469–482.

## 1972

[47] Spectra changing over narrow bands. In *Statistical Models and Turbulence*, edited by M. Rosenblatt, Springer: New York, Wiley, 400–469.
[48] With D. F. Nicholls. The estimation of mixed regression, autoregression, moving average and distributed lag models. *Econometrica* 40, 529–547.
[49] With R. C. Boston. The estimation of a non-linear system. In *Proceedings of Conference on Optimisation*, edited by R. S. Anderssen, E. S. Jennings, D. M. Ryan, Univ. of Queensland Press: Brisbane, 69–85.
[50] With R. D. Terrell. Time series regression with linear constraints. *Internat. Econ. Rev.* 13, 189–200.
[51] With C. C. Heyde. On limit theorems for quadratic functions of discrete time series. *Ann. Math. Statist.* 43, 2058–2066.

## 1973

[52] Multivariate time series anlysis. *J. Multivariate Anal.* 3, 395–407.
[53] With R. D. Terrell. Multiple equation systems with stationary errors. *Econometrica* 41, 299–320.
[54] The asymptotic theory of linear time series models. *J. Appl. Prob.* 10, 130–145.
[55] With P. J. Thomson. Estimating group delay. *Biometrika* 60, 241–253.

[56] The estimation of frequency. *J. Appl. Prob.* 10, 510–519.
[57] With P. A. P. Moran. The effects of serial correlation on a randomised rainmaking experiment. *Austral. J. Statist.* 15, 256–261.
[58] Central limit theorems for time series regression. *Z. Wahrsch. verw. Gebiete* 26, 157–70.
[59] With P. M. Robinson. Lagged regression with unknown lags. *J. R. Statist. Soc.* B, 35, 252–267.


## 1974


[60] With P. Thomson. Estimating echo times. *Technometrics* 16, 77–84.
[61] With B.V. Hamon. Spectral estimation of time delay for dispersive and non-dispersive systems. *Appl. Statist.* 23, 134–142.
[62] The uniform convergence of autocovariances. *Ann. Statist.* 2, 803–806.
[63] Time series analysis. *IEEE Trans. Automatic Control* AC-19, 706–715.


## 1975


[64] Linear regression in continuous time. *J. Austral. Math. Soc.* 19A, 146–159.
[65] The estimation of ARMA models. *Ann. Statist.* 3, 975–981.
[66] Measuring the velocity of a signal. In *Perspectives in Probability and Statistics, Papers in honour of M. S. Bartlett*, edited by J. M. Gani, Appl. Prob. Trust: Sheffield, 227–238.
[67] With J. Goncz. New methods of estimating dispersion from stacks of surface waves. *Bull Seismological Soc. Amer.* 65, 1519–1529.


## 1976


[68] The convergence of some recursions. *Ann. Statist.* 4, 1258–1270.
[69] The asymptotic distribution of serial covariances. *Ann. Statist.* 4, 396–399.
[70] The identification and parameterization of ARMAX and state space forms. *Econometrica* 44, 713–724.
[71] With W. Dunsmuir. Vector linear time series models. *Adv. Appl. Prob.* 8, 339–364.
[72] ARMAX and state space systems and recursive calculations. In *Frontiers in Quantitative Economics*, edited by M. Intriligator, North Holland: Amsterdam, 321–336.


## 1977


[73] With M. Kanter. Autoregressive processes with infinite variances. *J. Appl. Prob.* 14, 411–415.
[74] With D. F. Nicholls. The estimation of the prediction error variance. *J. Amer. Statist. Assoc.* 72, 834–840.


## 1978


[75] With M. A. Cameron. Measuring the properties of plane waves. *Math. Geol.* 10, 1–22.
[76] A note on a central limit theorem. *Econometrica* 46, 451–453.
[77] With M. Deistler and W. Dunsmuir. Vector linear time series models, corrections and extensions. *Adv. Appl. Prob.* 10, 360–372.
[78] Multivariate ARMA Theory. In *Stability and Inflation*, edited by A. R. Bergstrom, A. J. L. Catt, M. H. Preston, and D. J. Silverstone, Wiley: New York, 201–212.
[79] With K. Tanaka. ARMAX models and recursive calculations. In *Systems Dynamics and Control in Quantitative Economics*, edited by H. Myoken, Bushindo: Tokyo, 173–198.
[80] Rates of convergence for time series regression. *Adv. Appl. Prob.* 10, 740–743.

## 1979

[81] Statistical theory of linear systems. In *Development in Statistics* Vol.2, edited by P. Krishnaiah. Academic Press: New York, 83–121.
[82] With M. Cameron. Transient signals. *Biometrika* 66, 243–258.
[83] With B. G. Quinn. The determination of the order of an autoregression. *J. R. Statist. Soc. B*, 41, 190–195.
[84] The central limit theorem for time series regression. *Stoch. Proc. Appl.* 9, 281–289.
[85] A note on autoregressive-moving average identification. *Biometrika* 66, 672–674.
[86] "Saisonale Variation" and "Zeitreihenanalyse". In *Handwörterbuch du Mathematischen Wirtschafswissenschaften* 2, Gabler, Wiesbaden, 157–159, 299–306.

## 1980

[87] Time Series. *Math. Chronicle* 9, 101–119.
[88] Recursive estimation based on ARMA models. *Ann. Statist.* 8, 762–777.
[89] The estimation of the order of an ARMA process. *Ann. Statist.* 8, 1071–1081.
[90] With W. Dunsmuir and M. Deistler. Estimation of vector ARMAX models. *J. Multivariate Anal.* 10, 275–295.
[91] With Chen, Zhao-Guo. The distribution of periodogram ordinates. *J. Time Series Anal.* 1, 73–82.

## 1981

[92] Estimating the dimension of a linear system. *J. Multivariate Anal.* 11, 459–473.
[93] With M. Deistler. Some properties of the parametrization of ARMA systems with unknown order. *J. Multivariate Anal.* 11, 474–497.
[94] System identification. In *Stochastic Systems: The Mathematics of Filtering and Identification and Applications*, edited by M. Hazewinkel and J. C. Willems, Reidel: Dordrecht, 221–226.
[95] With P. J. Thomson. Delay estimation and the estimation of coherence and phase. *IEEE Trans. Acoust. Speech Signal Processing* ASSP-29, 485–490.

## 1982

[96] Fitting multivariate ARMA models. In *Statistics and Probability: Essays in Honour of C. R. Rao*, edited by G. Kallianpur, P. R. Krishnaiah and J. K. Ghosh, North-Holland: Amsterdam, 307–316.
[97] A note on bilinear time series models. *Stoch. Proc. Appl.* 12, 221–224.
[98] With J. Rissanen. Recursive estimation of mixed autoregressive-moving average order. *Biometrika* 69, 81–94.
[99] Testing for autocorrelation and Akaike's criterion. In *Essays in Statistical Science, Papers in Honour of P.A.P. Moran*, edited by J.M. Gani and E.J. Hannan, Appl. Prob. Trust: Sheffield, 403–412.
[100] Edited with J. M. Gani *Essays in Statistical Science, Papers in Honour of P. A. P. Moran*. Appl. Prob. Trust: Sheffield.
[101] With An, Hong-Zhi and Chen, Zhao-Guo. Autocorrelation, autoregression and autoregressive approximation. *Ann. Statist.* 10, 926–936.
[102] With V. Wertz and M. Gevers. The determination of optimum structures for the state space representation of multivariate stochastic processes. *IEEE Trans. Automatic Control* AC-27, 1200–1211.

## 1983

[103] With An, Hong-Zhi and Chen, Zhao-Guo. A note on ARMA estimation. *J. Time Series Anal.* 4, 9–17.

[104] With L. Kavalieris. Linear estimation of ARMA processes. *Automatica* 19, 447–448.
[105] With L. Kavalieris. The convergence of autocorrelations and autoregressions. *Austral. J. Statist.* 25, 287–297.
[106] Signal estimation. In *Handbook of Statistics* Vol. 3, edited by D. R. Brillinger and P. R. Krishnaiah, North Holland: Amsterdam, 111–123.
[107] Limit theorems for autocovariances and Fourier coefficients, in *Recent Trends in Statistics*, edited by S. Heiler, Vandenhouk and Ruprecht, 132–142.
[108] With An, Hong-Zhi and Chen, Zhao-Guo. The maximum of the periodogram. *J. Multivariate Anal.* 13, 383–400.

### 1984

[109] With L. Kavalieris. Multivariate linear time series models. *Adv. Appl Prob.* 16, 492–561.
[110] Multivariate ARMA systems and practicable calculations. *Qüestio* 8, 1–8.
[111] With L. Kavalieris. A method for autoregressive-moving average estimation. *Biometrika* 72, 273–280.
[112] The estimation of ARMA processes. In *Robust and Nonlinear Time Series Analysis*, edited by J. Franke, W. Hardle and D. Martin. Lecture Notes in Statistics No.26, Springer-Verlag: New York, 146–162

### 1985

[113] Linear systems, statistical theory. In *Encyclopedia of Statistical Sciences*, Vol.5, edited by S. Kotz and N.L. Johnson, Wiley, New York, 65–70.

### 1986

[114] Lectures on time series. *Advances in Mathematics* (Beijing), 15, 283–314.
[115] With L. Kavalieris and M. Mackisack. Recursive estimation of linear systems. *Biometrika* 74, 119–134.
[116] With L. Kavalieris. Regression, autoregression models. *J. Time Series Anal.* 7, 27–50.
[117] With M. Mackisack. A law of the iterated logarithm for an estimate of frequency. *Stoch. Proc. Appl.* 22, 103–109.
[118] With J. Franke. Comments on "Influence functionals for time series" by R.D. Martin and V.J. Yohai. *Ann. Statist.* 14, 822–824.
[119] Remembrance of things past. In *The Craft of Probabilistic Modelling*, edited by J. Gani, Springer-Verlag: New York, 31–42.
[120] Time series and stochastic models. In *Time Series and Linear Systems*, edited by Sergio Bittanti, Springer-Verlag: New York, 1–36.

### 1987

[121] Rational transfer function approximation. *Statist. Sci.* 2, 135–161.
[122] With C.H. Hesse. Comments on "What is an analysis of variance" by T. P. Speed, *Ann. Statist.* 15, 923–924.
[123] The statistical theory of linear systems. In *The Second International Tampere Conference in Statistics*, edited by T. Pukkila and S. Puntanen, Department of Mathematical Statistics, Tampere, Finland, 53–72.
[124] Approximation of linear systems. In *Time Series and Econometric Modelling*, edited by I. B. MacNeill and G. J. Umphrey, Reidel: Dordrecht, 1–12.

### 1988

[125] With M. Deistler. *The Statistical Theory of Linear Systems*. Wiley: New York.

[126] With D. Poskitt. Unit canonical correlation between future and past. *Ann. Statist.* 16, 784–790.
[127] With A. McDougall. Regression procedures for ARMA estimation. *J. Amer. Statist. Assoc.* 83, 490–498.
[128] With C. H. Hesse. Rates of convergence for the quantile function of a linear process. In *Studies in Modelling and Statistical Science*, Festschrift for J. M. Gani Special Volume 30A, *Austral. J. Statist.*, edited by C. C. Heyde, 283–295.
[129] With J. Rissanen. The width of a spectral window. *J. Appl. Prob.* 25A, 301–307.
[130] With P. J. Thomson. Time delay estimation. *J. Time Series Anal.* 9, 21–33.
[131] With P. G. Hall. On stochastic complexity and non-parametric density estimation *Biometrika* 75, 705–714.

*1989*

[132] With B. G. Quinn. The resolution of closely adjacent spectral lines. *J. Time Series Anal.* 10, 13–31.
[133] With A. McDougall and D. Poskitt. Recursive estimation of autoregressions. *J. R. Statist. Soc.* B, 51, 217–234.
[134] With B. Wahlberg. Convergence rates for inverse Toeplitz matrix forms. *J. Multivariate Anal.* 31, 127–135.

*1990*

[135] With L. Guo and D. W. Huang. On ARX(∞) approximation. *J. Multivariate Anal.* 92, 17–47.

*1992*

[136] Missed opportunities. In *The Art of Statistical Science*, edited by K.V. Mardia, Wiley: New York, 33–44.
[137] Determining the number of jumps in a spectrum. In *Developments in Time Series Analysis*, edited by T. Subba Rao. Chapman & Hall, London, 127–138.

*1993*

[138] With B. Wahlberg. Parametric signal modelling using Laguerre filters. *Ann. Appl. Prob.* 3, 467–496.
[139] With D. Huang. On line frequency estimation. *J. Time Series Anal.* 14, 147–161.
[140] The Whittle likelihood and frequency estimation. In *Probability, Statistics and Optimization: a Tribute to Peter Whittle*, edited by F.P Kelly, John Wiley, Chichester, 205–212.

———

Papers shortly to appear in Conference Proceedings:
[141] With D. W. Huang and B. G. Quinn. Frequency estimation. *Proc. IFAC Conference*, Sydney, May 1993.

———

Papers submitted during 1993:
[142] With D. W. Huang. Estimating time varying frequency.
[143] With L. Kavalieris. Determining the number of terms in a triogonometric regression.
[144] With L. Kavalieris and M. Salau. The estimation of ARMA models: Some comments.
[145] With L. Kavalieris. A note on ARMA parametrization.

———

Book in progress during 1993–4, to be completed by colleagues.
[146] With D. W. Huang and B. G. Quinn. Title unknown—on frequency estimation.

# A note on chaotic maps and time series

Hong Zhi An*

Institute of Applied Mathematics, Academia Sinica, China

## Abstract

A time series in this paper can be a random sequence, or a deterministic sequence of iterates of a chaotic map. The random behavior of such time series will be classified into three types: both sides random, one-sided random and pseudo-random. To illustrate the pseudo-random time series we investigate a chaotic map by showing that a time series of iterates of such map is pseudo-random, and the empirical distribution of such time series converges to the invariant probability distribution of the map for all initial values. The result can be applied for the computer generation of quasi-random numbers.

**Key words and phases**: Time series, Random behavior, Chaotic map, Empirical Distribution, Ergodic theory.

**AMS 1991 Subject Classifications:** Primary 60G10. 62M10.

---

* The work was done, when the author visited the Department of Statistics, the Chinese University of Hong Kong

# 1. Introduction

In probability and statistics, randomness plays an important role. For example, a stationary time series can display the characteristics typically associated with randomness. One of the characteristics of the randomness is the unpredictable behavior.

In the chaos literature, a time series is defined by iterates of a deterministic chaotic map with initial values. These sequences are intriguing in that the iterates corresponding to different, although extremely close, initial values typically diverge. The practical implication of this phenomenon is that, despite the underlying determinism, we cannot predict with any reasonable precision. For this reason a sequence of iterates of a deterministic chaotic map may look like a random time series, it may not be a real one in probability.

However, the relationship between the random behavior of random time series in probability and deterministic sequence in chaos is important. It has been investigated in the chaos literature. For instance, Berliner(1992) gave a good review entitled Statistics, Probability and Chaos, in which ergodic theory was emphasized. Bartlett(1990), Lawrance(1992), and Tong and Cheng(1992) have discussed a link between some deterministic chaos and stochastic models by reversing the time axis. Roughly speaking, a random time series under time reversal can be a deterministic sequence, that is, the iterates of a chaotic map. On the other hand, some iterates of a chaotic map can be a sample sequence from a random time series under time reversal. These links may be helpful in understanding the relationship between the random behavior of time series in probability and in chaos.

In this paper we shall consider a classification of time series by random behavior based on the above link. In Section 2 we shall define three kinds of time series: both sides random, one-sided random and pseudo-random. In Section 3, to illustrate the pseudo-random time series, we investigate a chaotic map without period by showing the invariant probability measure, and the convergence of the empirical distributions of iterates of such chaotic map for all initial values. Hence, such chaotic map can be applied for the generation of quasi-random numbers on the computer.

# 2. Classification of randomness

A random time series is a stochastic process in discrete time index. The characteristic behavior of a stochastic process is indicated in a family of distributions of all possible finite collections of the process. Hence we propose the following definition.

**Definition 2.1.** A time series $\{x_t : t = 0, 1, \cdots\}$ is said to be both sides random, if neither of the following two statements is true:

(1) For all positive integers $k$ and some positive $j$, and some function $\xi$,

$$x_{k+j} = \xi(x_{k+j-1}, x_{k+j-2}, \cdots, x_k), \qquad a.s.(P), \tag{2.1}$$

where "a.s.(P)" denotes "almost surely with respect to the joint probability measure P of the random variables $x_{k+j-1}, x_{k+j-2}, \cdots x_k$".

(2) For all positive integers $k$ and some positive $j$, and some function $\eta$,

$$x_k = \eta(x_{k+1}, x_{k+2}, \cdots x_{k+j}), \qquad a.s.(P). \tag{2.2}$$

**Example 2.1:** Let $\{x_1, x_2, \cdots\}$ be independent identically distributed ( i.i.d. ) Bernoulli, where $P(x_t = 1) = p = 1 - P(x_t = 0)$, and $p \neq 0$ *or* 1. Because $x_{k+j}$ and $\xi(x_{k+j-1}, x_{k+j-2}, \cdots, x_k)$ are independent for $k$, $j \geq 1$ and any function $\xi$, and the distribution of $x_{k+j}$ is not degenerate, then (2.1) cannot be true. Similarly (2.2) cannot be true neither. So the time series is both sides random.

**Example 2.2:** Consider the following stationary time series

$$x_t = \sum_{k=0}^{\infty} (0.5)^{k+1} a_{t-k} \tag{2.3}$$

where $\{a_t : t = 0, \pm 1, \cdots\}$ is i.i.d. with a common uniform distribution on $[0, 1]$. It is easy to check that the time series (2.3) satisfies the following linear autogeressive( AR ) model

$$x_t = 0.5 x_{t-1} + 0.5 a_t \tag{2.4}$$

and $a_t$ is independent of $\{x_k : k < t\}$. Hence

$$x_{k+j} = 0.5 x_{k+j-1} + 0.5 a_{k+j} = \xi(x_{k+j-1}, x_{k+j-2}, \cdots, x_k), \qquad a.s.(P),$$

or equivalently

$$a_{k+j} = 2\xi(x_{k+j-1}, x_{k+j-2}, \cdots, x_k) - x_{k+j-1}, \qquad a.s.(P),$$

cannot be true by the same arguments used in Example 2.1, and then (2.1) is not true. Consider (2.2) for $j = 1$, i.e.

$$x_k = \eta(x_{k+1}) = \eta(0.5x_k + 0.5a_{k+1}) \qquad a.s.(P) \qquad (2.5)$$

Because $x_k$ and $a_{k+1}$ are independent, and $a_{k+1}$ has a uniform distribution on $[0, 1]$, it is easy to show that the probability measure P in (2.5) is continuous with respect to the Lebesgue measure on $[0, 1] \times [0, 1]$. Thus (2.5) is not true. Similarly, but perhaps in although complicated way, we can show that (2.2) is not true for $j \geq 1$ neither. We omit the details. Thus the time series (2.3) is both sides random.

If we consider $\{a_t : t = 0, \pm 1, \cdots\}$ to be i.i.d. Bernoulli instead of uniform distribution on $[0, 1]$, then we may find that

$$x_{t-1} = 2x_t mod(1), \qquad t \geq 1, \qquad (2.6)$$

which implies (2.2) with $k = t - 1$ and $j = 1$. Hence the time series (2.3) is not both sides random. For this case we need the following definition.

**Definition 2.2.** A time series is said to be one-sided random if either (2.1) or (2.2) is true.

According to this definition, the time series (2.3) under the Bernoulli assumption on the distribution of $a_t$ is one-sided random.

**Example 2.3:** Consider a multiplicative congruence map

$$f(y) = 2ymod(1), \qquad 0 \leq y \leq 1. \qquad (2.7)$$

Let $\{y_t : t = 0, 1, \cdots\}$ denote the iterates of the map in (2.7) with initial value $y_0$, then

$$y_{k+1} = 2y_k mod(1), \qquad k = 0, 1, 2, \cdots \qquad (2.8)$$

which implies (2.1) with $j = 1$ and arbitrary $k \geq 0$. On the other hand, because of the non-invertibility of the map in (2.7), the value of $y_k$ cannot be uniquely determined by $y_{k+1}$. In fact, there are two choices

$$y_k = 0.5y_{k+1} \quad \text{or} \quad y_k = 0.5y_{k+1} + 0.5 \qquad (2.9)$$

If we determine the value of $y_k$ in (2.9) by Bernoulli trial, i.e. $y_k = 0.5y_{k+1} + 0.5a_k$, where $a_k$ is a Bernoulli random variable, then we obtain the model of (2.4) again, i.e.

$$y_k = 0.5y_{k+1} + 0.5a_k$$

where $\{a_t : t = 0, \pm 1, \cdots\}$ is i.i.d. Bernoulli again, but the time is now reversed. In view of Example 2.2 the time series of (2.8) is one-sided random. In particular, if $p = 0.5$ in Bernoulli $(p)$, the uniform distribution on $[0, 1]$ is ergodic both for the map of (2.7) and the model of (2.4).(see Tong and Cheng,1992)

**Remark:** When (2.1) (or (2.2)) holds for $j = 1$ and some $k$, then (2.1)(or (2.2)) holds for all $k$. In general, (2.1) or (2.2) may be true for some positive integers $k$ and $j$. On the other hand, for a one-sided random time series we may have a linear or non-linear AR model to link with it.(see Tong and Cheng, 1992)

If a time series is neither both sides random nor one-sided random, a third kind of time series should be introduced. Because we are interested in chaos, such time series must be the iterates of some chaotic map. Then a definition of chaos in terms of ergodic theory is necessary for the definition of the third type of time series. In order to make use of ergodic theory in deterministic chaos, we consider X as a random variable. If X and $f(X)$ have a common distribution measure P, $f$ is said to be measure invariant with respect to P. In particular, When P is ergodic and absolutely continuous with respect to the Lebesgue measure, we say that the function $f$ is chaotic. Although there are different definitions of chaos in the literature(e.g. see Berliner, 1992), the above definition of chaos is used in this paper.

**Definition 2.3.** A time series is said to be pseudo-random, if both of (2.1) and (2.2) are satisfied for $j = 1$ and all $k \geq 0$, and the function $\xi$ in (2.1) is chaotic.

In particular, a pseudo-random time series $\{x_t : t = 0, 1, \cdots\}$ satisfying (2.1) and(2.2) with $j = 1$, as mentioned in the above remark, is deterministic both under forward time and backward time. For simplicity of discussions in the next section, we only consider $j = 1$ in Definition 2.3.

To illustrate the pseudo-random time series, a chaotic map will be shown in the following section.

## 3. A chaotic map without period

This section is a special case of empirical distributions of time series. First we introduce some notations with a brief discussion about the empirical distribution of a one-sided random time series. The empirical distribution of time series is defined in the usual way by

$$F_n(x) = (1/n) \sum_{t=1}^{n} I(x_t < x), \qquad n = 1, 2, \cdots \qquad (3.1)$$

where $I(x_k < x)$ denotes the indicator function. When a time series $\{x_t : t = 0, 1, \cdots\}$ consists of iterates of a chaotic map $f$ with initial value $x_0$, then $x_t$ can be denoted by

$$x_t = f^t(x_0), \qquad \text{for} \quad t = 0, 1, 2, \cdots \qquad (3.2)$$

where $f^0(x_0) = x_0, f^1(x_0) = f(x_0), f^2(x_0) = f(x_1) = f(f(x_0))$ and so on, and then

$$F_n(x) = (1/n) \sum_{t=1}^{n} I(f^t(x_0) < x). \qquad (3.3)$$

If the function $f$ in (3.3) is ergodic with respect to the invariant probability measure P, from the ergodic theory(e.g., see Berliner, 1992) it follows that

$$F_n(x) \rightarrow F(x) = P(X < x), \qquad \text{for} \quad a.s. \quad x_0(P) \qquad \text{as} \quad n \rightarrow \infty \qquad (3.4)$$

where X is a random variable, and has distribution $F(x) = P(X < x)$.

As we know, most of the methods to generate quasi-random numbers are based on the result of (3.4). However, even if $f$ in (3.3) is chaotic, (3.4) may not be satisfied because of " a.s. " in (3.4). For example, if $f$ is defined by (2.7), it is easy to see that for a rational initial value $x_0$

$$F_n(x) \rightarrow J(x), \qquad \text{as} \quad n \rightarrow \infty,$$

where $J(x)$ is some stepwise function with finite jumps on $[0,1]$. Since all the numbers running on the computers are rational, it is not true that

$$F_n(x) \rightarrow U(x), \quad \text{as} \quad n \rightarrow \infty, \tag{3.5}$$

where $U(x)$ is the uniform distribution on $[0,1]$, and is the ergodic and invariant distribution of $f$ of (2.7). In fact, (3.5) is required to generate quasi-random numbers with asymptotically a uniform distribution $U(x)$. For other multiplicative congruence methods to generate quasi-random numbers, the same problem may also arise. This problem is caused by periodicity when starting with a rational. In order to overcome this trouble, a chaotic map without period is given below.

Consider the following two-piece linear function

$$f(x) = \begin{cases} bx + a(b-1)/(b-a), & \text{if } 0 \le x < (1-a)/(b-a), \\ ax - a(1-a)/(b-a), & \text{if } (1-a)/(b-a) \le x \le 1, \end{cases} \tag{3.6}$$

where $a$ and $b$ are some constant, $0 < a < 1 < b$, and $(\log a)/(\log b)$ is irrational. In the remainder of this section we will show that $f$ is chaotic, find the ergodic and invariant distribution F for $f$ in (3.6), and then prove that

$$F_n(x) \rightarrow F(x), \quad \text{for all initial values, as} \quad n \rightarrow \infty. \tag{3.7}$$

To do this, we start with the following map from $[0,\infty)$ to $[0,\infty)$, i.e.

$$\phi(u) = \begin{cases} bu, & \text{if } 0 \le u < 1, \\ au, & \text{if } 1 \le u < \infty, \end{cases} \tag{3.8}$$

where $a$ and $b$ are the same as in (3.6). It is easy to see that the attractor of $\phi$ of (3.8) is the interval $[a,b]$. From the ergodic point of view we can restrict our study on $[a,b]$. Then (3.8) becomes

$$\phi(u) = \begin{cases} bu, & \text{if } a \le u < 1, \\ au, & \text{if } 1 \le u \le b. \end{cases} \tag{3.9}$$

Consider an invertible transform from $[a,b]$ to $[0,1]$ defined by

$$x = G(u) = (\log u - \log a)/(\log b - \log a), \quad a \le u \le b \tag{3.10}$$

Then corresponding to $\phi$ of (3.9), we have

$$h(x) = \begin{cases} x + \alpha, & \text{if } 0 \leq x < (1 - \alpha), \\ x - (1 - \alpha), & \text{if } (1 - \alpha) \leq x \leq 1, \end{cases} \tag{3.11}$$

where $\alpha = \log b/(\log b - \log a)$. Because $0 < a < 1 < b$, and $(\log a/\log b)$ is irrational, then $\alpha$ is irrational, and $0 < \alpha < 1$.

Then consider another transform from $[0, 1]$ to the nuit circle $C = \{z : |z| \leq 1\}$, given by

$$z = T(x) = e^{i2\pi x}, \qquad \text{for } 0 \leq x \leq 1. \tag{3.12}$$

Notice that $e^{i2\pi(x+m)} = e^{i2\pi x}$ for any integer m, then corresponding to $h$ of (3.11) we have

$$\psi(z) = ze^{i2\pi\alpha}, \qquad \text{for } z \in C, \tag{3.13}$$

where $\psi$ is a rotation of the unit circle C.

By Theorem 1.8 of Walters(1982) we know that, the normalised circular Lebesgue measure is ergodic for map $\psi$ of (3.13), if and only if $\alpha$ is irrational. Note that $\alpha$ in (3.13) is irrational by the assumption in (3.6). Consequently $\psi$ of (3.13), then $h$ of (3.11) and $\phi$ of (3.9) are ergodic and chaotic.

Notice that the n-th iterate of $\phi$ of (3.9) with initial $u_0$ can be denoted as

$$u_n = \phi^n(u_0) = a^q b^{n-q} u_0,$$

for some integer $q$ depending on $u_0$ and n. Because $(\log a/\log b)$ is irrational, then $u_n$'s for all $n \geq 1$ never come back to initial $u_0$. Thus the iterates of $\phi$ of (3.9) for any initial value $u_0$ have no periodic tail.

Now we consider a transform from $[a, b]$ to $[0, 1]$ defined by

$$x = Q(u) = (u - a)/(b - a), \qquad \text{for} \quad u \in [a, b]. \tag{3.14}$$

Then corresponding to $\phi$ of (3.9) we obtain the map of (3.6).Thus $f$ of (3.6) is ergodic and chaotic without period.

By the result of ergodic theorem mentioned above and by (3.13), the invariant measure of $h$ of (3.11) is the Lebesgue measure on $[0, 1]$. Hence by the transform G of (3.10) we know that the invariant probability distribution of $\phi$ of (3.9) is

$$F_\phi(x) = (\log x - \log a)/(\log b - \log a), \qquad x \in [a, b] \tag{3.15}$$

Similarly by (3.14) we know that the invariant probability measure of $f$ of (3.6) is

$$F(x) = \{\log[x(b - a) + a] - \log a\}/(\log b - \log a), \qquad x \in [0, 1] \tag{3.16}$$

Now we prove (3.7). First consider iterates of $\psi$ of (3.13). Let $z_0 = e^{i2\pi\theta}$, and then

$$z_t = \psi^t(z_0) = z_0 e^{i2\pi\alpha t} = e^{i2\pi(t\alpha + \theta)}. \tag{3.17}$$

For $0 \leq c < d < 1$, define

$$F_n(c, d; \theta) = (1/n) \sum_{t=1}^{n} I(c \leq \{\arg(e^{i2\pi(t\alpha + \theta)})/2\pi\} < d), \tag{3.18}$$

where $\arg(z)$ denotes the argument of complex z, and $\{w\}$ denotes the fractional part of real number $w$. By the ergodic theorem again,

$$F_n(c, d; \theta) \to (d - c), \qquad a.s.(L) \quad \theta \in [0, 1] \qquad \text{as} \quad n \to \infty \tag{3.19}$$

where L denotes the Lebesgue measure on $[0, 1]$. By (3.19) there exists at least one $\theta \in [0, 1)$ such that (3.19) holds for all $0 \leq c < d < 1$. Let $z_0 = e^{i2\pi\beta}$ be another initial value. Because $\psi$ of (3.13) is a rotation of the unit circle C, it is easy to see that for the case that $\theta > \beta$, and $d - \beta + \theta < 1$,

$$F_n(c, d; \beta) = F_n(c - \beta + \theta, d - \beta + \theta; \theta) \to (d - c), \quad \text{as} \quad n \to \infty \tag{3.20}$$

For another case that $\theta > \beta$, and $c - \beta + \theta < 1 < d - \beta + \theta$, then

$$F_n(c, d; \beta) = F_n(c - \beta + \theta, 1; \theta) + F_n(1, d - \beta + \theta; \theta) \to (d - c), \quad \text{as} \quad n \to \infty$$

Of course there are other cases for which (3.20) can be obtained similarly. We omit the details. Because of the arbitrariness of $\beta$, (3.20) holds for all initial values.

From the invertible transform G of (3.10) and Q of (3.14), and the results of (3.20) and (3.16), it follows that

$$F_n(x) \to \{\log[x(b-a)+a]-\log a\}/(\log b-\log a), \text{for any initial } x_0, \text{ as } n \to \infty, \tag{3.21}$$

which is just (3.7).

It is obvious that, if $f$ of (3.6) is restricted on $[0,1)$, then $f$ is invertible, and $f$ is also chaotic and ergodic with respct to the same F of (3.16). Let $x_0 \in [0,1)$ be an arbitrary number, and

$$x_t = f(x_{t-1}) = f^t(x_0), \qquad t = 1, 2, \cdots$$

and let $f^{-1}$ be the inverse of the function $f$ in (3.6), then

$$x_{t-1} = f^{-1}(x_t) = \begin{cases} x_t/a + (1-a)/(b-a), & \text{if } 0 \le x_t < (ab-a)/(b-a), \\ x_t/b - a(b-1)/b(b-a), & \text{if } (ab-a)/(b-a) \le x_t < 1, \end{cases} \quad (3.22)$$

which with $x_t = f(x_{t-1})$ implies that the time series $\{x_t = f^t(x_0) : t = 1, 2, \cdots\}$ is pseudo-random. Consequently, such time series have no periodicity, at least from the theoretical point of view, and therefore can be regarded as quasi-random numbers of the non-uniform distribution F of (3.16), while the iterates of $h$ of (3.11) can be regarded as quasi-random numbers of the uniform distribution on $[0,1]$. Here we should point out that the calculations of the iterates of $f$ of (3.6) involve only the rational numbers, while $h$ of (3.11) involves an irrational number $\alpha$. From the computational point of view, to use $f$ of (3.6) in generating quasi-random numbers is more convenient than to use $h$ of (3.11). For example we show the following special case of the function $f$ in (3.6).

Consider $a = 1/2$ and $b = 3/2$ in ( 3.6). Then

$$f(x) = \begin{cases} (3/2)x + 1/4, & \text{if } 0 \le x < 1/2, \\ (1/2)x - 1/4, & \text{if } 1/2 \le x < 1. \end{cases} \quad (3.23)$$

Let $x_k$ be the k-th iterates of $f$ of (3.23) with initial value $x_0$, and put $u_k = x_k + 1/2$, then by (3.23)

$$u_k = (3/2)u_{k-1}, \quad \text{if } u_{k-1} < 1; \quad = (1/2)u_{k-1}, \quad \text{if } u_{k-1} \ge 1. \quad (3.24)$$

Because $x_0$ is rational, $u_k$ is rational as well. Denote $u_k$ by $u_k = p_k/q_k$, where $p_k$ and $q_k$ are some positive integers, which can be obtained by the following recursive forms

$$q_k = 2^k, \qquad k = 0, 1, 2, \cdots$$

$$p_k = 3p_{k-1}, \quad \text{if } p_{k-1} < q_{k-1}; \quad = p_{k-1}, \quad \text{if } p_{k-1} \ge q_{k-1}, \quad k = 1, 2, \cdots$$

where $u_0 = x_0 + 1/2 = p_0/q_0$, hence $p_0$ and $q_0$ are determined by $x_0$. Finally the iterates of $f$ of (3.23) are given by

$$x_k = p_k/q_k - 1/2, \quad k = 1, 2, \cdots \tag{3.25}$$

By (3.21), we know that the empirical distribution of the time series in (3.25) satisfy

$$F_n(x) \to F(x) = \{\log(x + 0.5) + \log 2\}/\log 3, \quad \text{as} \quad n \to \infty \tag{3.26}$$

for any initial value $x_0 \in [0, 1)$ without exception.


Accordingly, this kind of chaotic maps can be applied in the generation of quasi-random numbers on the computers.

## References

Berliner, L.M.(1992). Statistics, Probability and Chaos. *Statist. Sci.* 7(1), 69–122.

Bartlett, M. S. (1990). Chance or chaos ? *J. R. Statist. Soc.* **A 153**, 321–348.

Lawrance, A. J.(1992). Uniformly distributed first-order autoregressive stochastic models and multiplicative congruential random number generators. *J. Appl. Prob.* **29**(4), 891–903.

Tong, H. and Cheng, B.(1992). A note on one-dimensional chaotic maps under time reversal. *Adv. Appl. Prob.* **24**, 219–220.

Walters, P.(1982). *An Introduction to Ergodic Theory* . Springer–Verlag, New York.

# Balanced Parametrizations: A Structure Theory for Identification

## D.Bauer [1]
## M. Deistler [1]

ABSTRACT  The paper deals with balanced state space realizations and balanced forms. A structure theory treating the topological properties of the parametrization and the parameter space is developed. These results are important for identification. Finally balanced parametrizations are compared with parametrizations corresponding to Echelon forms.

## 1   Introduction

State space systems are important representations for linear dynamic time-constant finite-dimensional causal systems. For the sake of simplicity, we here restrict ourselves to the case of unobserved white noise inputs $\varepsilon_t$. However all results remain valid for the case, where also observed inputs are present. A state space model then is written in the form:

$$
\begin{aligned}
x_{t+1} &= Fx_t + G\varepsilon_t \\
y_t &= Hx_t + \varepsilon_t
\end{aligned}
\tag{1.1}
$$

where $y_t$ is the observed $s$-dimensional output, $\varepsilon_t$ is $s$-dimensional white noise with covariance matrix $\Omega$, $x_t$ is the $n$-dimensional state and $F \in \mathbb{R}^{n \times n}, G \in \mathbb{R}^{n \times s}, H \in \mathbb{R}^{s \times n}$ are parameter matrices.

Throughout the paper we will impose the stability condition, i.e. that $|\lambda_{max}(F)| < 1$, where $\lambda_{max}(.)$ denotes the eigenvalue of maximum modulus. For the sake of notational simplicity, the miniphase condition $|\lambda_{max}(F - GH)| \leq 1$ will not be imposed (for the implications of the miniphase condition on the parameter spaces see Hannan & Deistler [7]). We are only interested in the steady state solution $y_t = \sum_{j=0}^{\infty} K(j)\varepsilon_{t-j}$, which is described by the transferfunction $k(z)$:

$$
k(z) = \sum_{j=0}^{\infty} K(j)z^j = zH(I - zF)^{-1}G + I
\tag{1.2}
$$

[1]Institut für Ökonometrie, Operations Research und Systemtheorie, TU Wien, Argentinierstr.8, A- 1040 Vienna, Austria

Here $z$ is used both as a complex variable and as the backward shift on the integers. Note that every steady state solution of a state space system is a stationary process with a rational spectral density:

$$f(\lambda) = k(e^{-i\lambda})\Omega k^*(e^{-i\lambda}) \tag{1.3}$$

where $*$ denotes the conjugate transpose. Conversely every stationary process with a rational spectral density can be represented by a state space system satisfying our assumptions. For a given rational spectral density $f$ the transfer function $k$ and the covariance matrix $\Omega$ in (1.3) are unique under the assumptions:

1. $k$ has no poles for $|z| \leq 1$ and no zeros for $|z| < 1$

2. $k(0) = I$

3. $\Omega > 0$

To repeat, we will not impose a miniphase assumption (which is implied in the second part of 1.), but only the stability assumption. Let $S$ denote the set of all state space systems $(F, G, H)$ for given $s$ and $n$, satisfying the stability assumption, i.e. $S = \{(F, G, H) \in \mathbb{R}^{n \times n + n \times s + s \times n} : |\lambda_{max}(F)| < 1\}$. Note that $S$ is an open subset of $\mathbb{R}^{n \times n + n \times s + s \times n}$, since $\lambda_{max}(.)$ is a continuous function of $F$. Let $U$ denote the set of all transfer functions corresponding to $S$ and let $\pi : S \to U$ denote the mapping defined by (1.2) attaching to every $(F, G, H)$ a transfer function $k$.

The set of all $(F, G, H)$ that satisfy our assumptions giving rise to the same transfer function $k$, i.e. $\pi^{-1}(k) \subset S$, is called the class of all observationally equivalent state space systems. In many cases in addition we assume, that $(F, G, H)$ is minimal (i.e. for a given transfer function $k$ the state dimension $n$ cannot be reduced). The integer $n$ then is called the order of $(F, G, H)$ or of the transfer function. By $M(n) \subset U$ we denote the set of all transfer functions of order $n$. The set $S_n$ of all minimal $(F, G, H) \in S$ is an open and dense subset of $S$ (see Hannan & Deistler [7]). By a well known result (see e.g. Hannan & Deistler [7]), two minimal state space systems $(F_1, G_1, H_1)$ and $(F_2, G_2, H_2)$ are equivalent if and only if there exists a nonsingular matrix $T$ such that:

$$\begin{aligned} F_1 &= TF_2T^{-1} \\ G_1 &= TG_2 \\ H_1 &= H_2T^{-1} \end{aligned} \tag{1.4}$$

We endow $U$ with the so called pointwise topology as follows. Identify $k$ with the sequence of its power series coefficients $\{K(j)\} \in (\mathbb{R}^{s \times s})^N$. Then $(\mathbb{R}^{s \times s})^N$ is endowed with the product topology of the Euklidean spaces, which is metrizable and $U$ is endowed with the corresponding relative topology. Note that the closure of $M(n)$, $\bar{M}(n)$, is equal to $U = \bigcup_{i \leq n} M(i)$ (see

Hannan & Deistler,[7]).

Consider an ARMA system $a(z)y_t = b(z)\varepsilon_t$, where $a(z) = \sum_{j=0}^{p} A(j)z^j$, $b(z) = \sum_{j=0}^{p} B(j)z^j$, $A(j), B(j) \in \mathbb{R}^{s \times s}$. Again we impose the stability assumption, which is in this context of the form: $\det(a(z)) \neq 0, \forall |z| \leq 1$. Note that every causal rational transfer function can be represented either by a state space or an ARMA system. There is a one-one relation between state space and causal ARMA equivalence classes. However the corresponding equivalence classes themselves are different. The state space equivalence classes can be imbedded in $\mathbb{R}^{n^2 + 2ns}$, whereas the ARMA equivalence classes may be imbedded in $\mathbb{R}^{2s^2 p}$. Consider for example the case $s = 1$: In this case typically we have $n = p$ and (if $(a, b)$ has degree $p$ and no common factors) the ARMA equivalence classes are singletons in $\mathbb{R}^{2p}$. On the other hand for state space systems the transformation $T \in \mathbb{R}^{n \times n}$ is an arbitrary nonsingular matrix, therefore the equivalence classes are no singletons. If we want to choose representatives from the equivalence classes, besides general aspects like continuity of this choice, one might be interested in additional properties of the representatives. In this paper we consider the selection of representatives for state space systems, which are in the so called balanced form. There is no ARMA analogon to balancing. Balanced representations have a property of minimum sensitivity e.g. with respect to roundoff errors (see Mullis & Roberts [13]).

The paper is organized as follows: in section 2, balancing (Moore,[12]) is introduced and some properties of balanced realizations are presented. In section 3 we describe the canonical balanced forms introduced by Ober [14]. In section 4 we present the main results of the paper concerning the continuity of the parametrization and the properties of the parameter spaces and their boundary points. Finally in section 5 we discuss the advantages and the disadvantages of using balanced canonical forms compared to Echelon forms for identification.

# 2  Balancing

For a state space system $(F, G, H)$ let $\mathcal{O} = [H^T | F^T H^T | (F^2)^T | \cdots]^T$ denote the observability matrix and $\mathcal{C} = [G | FG | F^2 G | \cdots]$ denote the controllability matrix. $W_o = \mathcal{O}^T \mathcal{O}$ is called *observability Gramian*, $W_c = \mathcal{C}\mathcal{C}^T$ is called *controllability Gramian*. Balancing has been introduced by Moore [12] as follows:

**Definition:** A minimal state space representation is called *balanced*, if $W_o = W_c = \Sigma = diag(\sigma_1, \cdots, \sigma_n)$, where in addition $\sigma_1 \geq \cdots \geq \sigma_n > 0$. The $\sigma_i$ are called *second order modes*.

In order to form the Gramians the stability assumption is needed. Let us repeat, that there is no ARMA analogon to balancing. In the following, we repeat a few facts about balancing. First note that balanced realizations always exist. For two minimal observationally equivalent systems the Gramians are related by:

- $\mathcal{O}_1^T \mathcal{O}_1 = T^{-T} \mathcal{O}_2^T \mathcal{O}_2 T^{-1}$
- $\mathcal{C}_1 \mathcal{C}_1^T = T \mathcal{C}_2 \mathcal{C}_2^T T^T$

where $T$ is the nonsingular matrix in (1.4). The idea to show the existence of a balanced representation is to construct a transformation $T = T_2 T_1$, that balances an arbitrary minimal representation, in two steps. In the first step choose $T_1$ as any square root of the observability Gramian. This results in a new observability Gramian equal to unity. Now decompose the new controllability Gramian as $T_1 \mathcal{C}_2 \mathcal{C}_2^T T_1^T = U \Sigma^2 U^T$, where $U$ is a unitary matrix and $\Sigma$ the diagonal matrix of the second order modes. Use as the next transformation $T_2 = \Sigma^{-1/2} U^T$. Then the both Gramians are equal to $\Sigma$. For an efficient implementation of a balancing algorithm see Laub et al. [9].

As easily can be seen, balanced realizations are not unique (there is more than one balanced realization in each equivalence class): commencing from a balanced realization every basis change (1.4) with an orthogonal matrix $T$, which fulfills $T \Sigma T^T = \Sigma$ results in an observationally equivalent balanced realization. In particular for minimal $(F, G, H)$, if all entries of $\Sigma$ are different, the set of all feasible matrices $T$ relating balanced realizations consists of all diagonal matrices having only $+1$ or $-1$ as diagonal entries. The other extreme is, when all second order modes are equal and nonzero. Then the set consists of all orthogonal matrices. Note that $\Sigma$ is an invariant in the class of all observationally equivalent balanced realizations. More generally we define $\Sigma^2$ as the matrix of eigenvalues (with elements ordered in size) of $\mathcal{O}^T \mathcal{O} \mathcal{C} \mathcal{C}^T$. Then $\Sigma$ is an invariant in the whole equivalence class.

Let us remark, that the notion of balancing is more general than the notion used in our paper. For alternative definitions see for example Arun & Kung [1] or Desai & Pal [5].

The Lyapunov equations for the system (1.1) are of the form:

$$\begin{aligned} F W_c F^T - W_c &= -G G^T \\ F^T W_o F - W_o &= -H^T H \end{aligned} \qquad (2.1)$$

These equations are an immediate consequence from the definition of $\mathcal{O}$ and $\mathcal{C}$. The Lyapunov equations play an important role for the construction of balanced forms. Note that for given $W_c$ and $G$ ($W_o$ and $H$ respectively) these equations are quadratic in $F$, which causes serious complications. This is the reason for introducing a homeomorphism $i$ defined by:

$$\begin{aligned}
i(F_c, G_c, H_c) &= ((I - F_c)^{-1}(I + F_c), \sqrt{2}(I - F_c)^{-1}G_c, \\
&\quad \sqrt{2}H_c(I - F_c)^{-1}) & (2.2) \\
i^{-1}(F_d, G_d, H_d) &= ((I + F_d)^{-1}(F_d - 1), \sqrt{2}(I + F_d)^{-1}G_d, \\
&\quad \sqrt{2}H_d(I + F_d)^{-1}) & (2.3)
\end{aligned}$$

This homeomorphism attaches to the discrete time system (1.1) a continuous time system of the form:

$$\begin{aligned}
\dot{x}(t) &= F_c x(t) + G_c \varepsilon(t) & (2.4) \\
y(t) &= H_c x(t) + D_c \varepsilon(t)
\end{aligned}$$

where $D_c$ is defined as $D_c = I - H(I + F)^{-1}G$. Here the transfer function of the discrete time system $k(z)$ and of the continuous time system $g(s) = D_c + H_c(sI - F_c)^{-1}G_c$ (where $s$ denotes the differential operator) are related by:

$$g(s) = k(z(s)) \qquad (2.5)$$

where

$$z(s) = \frac{1 - s}{1 + s} \qquad (2.6)$$

Note that the bilinear transformation (2.6) is a bijection on the compactified complex plane, which maps the open left halfplane of the complex plane onto the complement of the closed unit disc and gives a one-one relation between the poles and the zeros of $k$ and $g$. The supremum norms are preserved under the homeomorphism defined above, i.e.

$$\sup_{|z|=1} k(z) = \sup_{s=iw, w \in \mathbb{R}} g(s) \qquad (2.7)$$

where (2.5) holds. Clearly the homeomorphism $i$ preserves system equivalence. Define the observability Gramian and the controllability Gramian for continuous time systems (2.4) as:

$$\begin{aligned}
W_c &= \int_0^\infty e^{tF} G G^T e^{tF^T} dt \\
W_o &= \int_0^\infty e^{tF^T} H^T H e^{tF} dt
\end{aligned} \qquad (2.8)$$

Note that $W_c$ and $W_o$ respectively are not changed by the homeomorphism. Now balancing can be defined completely analogously to the discrete time case (note that Moore [12] originally defined balancing for continuous time systems). By considering the corresponding matrix differential equation (2.4) it is easily verified that the following continuous time Lyapunov equations hold:

$$\begin{aligned}
F_c W_c + W_c F_c^T &= -G_c G_c^T \\
F_c^T W_o + W_o F_c &= -H_c^T H_c
\end{aligned} \tag{2.9}$$

Note that these equations are linear in $F_c$ for given $W_c$ and $G_c$ (given $W_o$ and $H_c$ respectively), which is an essential advantage compared to the discrete time versions.

The homeomorphism (2.2) also preserves balancing, which can be seen as follows: Substituting the homeomorphism in the controllability part of the continuous time Lyapunov equation gives:

$$-2(I + F)^{-1}(\Sigma - F\Sigma F^T - GG^T)(I + F)^{-1} = 0 \tag{2.10}$$

and thus $(F_c, G_c, H_c)$ satisfies the controllability part of the Lyapunov equations (2.9) if and only if $(F, G, H)$ satisfies the discrete time counterpart and the same holds true for the observability part. Since for given $(F, G, H)$ and $(F_c, G_c, H_c)$ the Lyapunov equations give unique positive semidefinite symmetric matrices $W_c$ and $W_s$ (see e.g Aoki [2]), we see, that balancing is preserved by the homeomorphism (2.2). The stability condition for the continuous time systems (2.4) is that all eigenvalues of the matrix $F_c$ have real part strictly less than zero (The miniphase condition restricts the eigenvalues of $F_c - G_c D_c^{-1} H_c$ to the closed left halfplane in the complex plane.) Clearly since the poles (and the zeros) of $k(z)$ and $g(s)$ are related by the bilinear transformation (2.6), the stability (and the miniphase) property are preserved by the homeomorphism.

One of the advantages of the balanced forms are their model reduction properties. By simply truncating the matrices $(F, G, H)$, a lower dimensional system is attained. The quality of the lower order approximation is determined by the second order modes. Consider an approximation $(F_{11}, G_1, H_1)$ to the system $(F, G, H) = (\begin{bmatrix} F_{11} & F_{12} \\ F_{21} & F_{22} \end{bmatrix}, [G_1 G_2], \begin{bmatrix} H_1 \\ H_2 \end{bmatrix})$, where $F_{11} \in \mathbb{R}^{n_1 \times n_1}$ and where we always assume $\sigma_{n_1} > \sigma_{n_1+1}$. Then in continuous time we have error bounds for the corresponding transfer functions (cf. Glover[6]). In the discrete time case, we do not have such bounds on the error of the lower order approximation.

In the continuous time case, balancing is preserved for truncated systems. This is a consequence of the linearity of the equations (2.9) as is straightforward to see. In the discrete time case, balancing is not preserved by truncation in general.

# 3    Ober's Balanced Canonical Form

As has been stated already, balancing does not uniquely describe a representative from the class of all observationally equivalent systems. Therefore in order to obtain a canonical form (i.e. a map that picks a unique representative for each equivalence class), additional restrictions have to be imposed. In this section we give an outline of the approach given by Ober [14]. As in Ober [14] we define the canonical form for the continuous time case. The discrete time canonical forms then are defined by $k \to g \to (F_c, G_c, H_c) \to i^{-1}(F_c, G_c, H_c)$, where $k \to g$ is defined by (2.5) and (2.6), $g \to (F_c, G_c, H_c)$ is the Ober form and $i$ is defined by (2.2). For notational simplicity we will omit the subscript $_c$ in this section. We use the word *Ober form* for the continuous time $(F_c, G_c, H_c)$ and the discrete time form $(F, G, H)$.

First consider the case $\Sigma = \sigma I$, where the class of all observationally equivalent balanced systems is defined by the class of all orthogonal basis transformations (1.4). A unique orthogonal matrix can be obtained as follows: Commencing from an arbitrary balanced realization $(F_1, G_1, H_1)$ we apply an orthogonal basis transformation, which makes $G_1$ 'positive upper triangular', i.e. there exist integers $1 \leq t(1) < t(2) < \cdots < t(r) \leq s$, where $r$ is the rank of $G_1$, such that ($x$ denotes arbitrary entries):

$$G = TG_1 = \begin{bmatrix} 0 & \cdots & 0 & g_{1,t(1)} & x & \cdots & \cdots & x \\ 0 & \cdots & \cdots & 0 & g_{2,t(2)} & x & \cdots & x \\ & & & & \cdots & & & \\ 0 & & \cdots & & 0 & g_{r,t(r)} & x & x \\ & & & & \mathbf{0} & & & \end{bmatrix} \quad (3.1)$$

where $g_{i,t(i)} > 0, \forall i$. This fixes the orthogonal transformation in the first $r$ columns. For the special case $\Sigma = \sigma I$, the Lyapunov equations are of the form:

$$\begin{aligned} \sigma(F + F^T) &= -GG^T \\ \sigma(F^T + F) &= -H^T H \end{aligned} \quad (3.2)$$

This shows, that the symmetric part $F_s = \frac{1}{2}(F + F^T)$ of $F$ is uniquely determined from $G$ and that $G$ and $H^T$ are square roots of the same positive semidefinite matrix. Therefore we have $H = [U0](G_r G_r^T)^{1/2}$, where $G_r$ is the matrix of the first $r$ rows of $G$ and $(G_r G_r^T)^{1/2}$ denotes the uniquely defined positive definite symmetric square root of $(G_r G_r^T)$ and the $r$ columns of $U$ are orthonormal vectors.

Write $F$ as $F = F_s + F_{sk}$, where $F_{sk}$ is the skewsymmetric part of $F$ $(F_{sk} + F_{sk}^T = 0)$. In the special case $r = n$, also the matrix $F_{sk}$ is unique. If $r < n$ then the orthogonal transformation is not uniquely fixed in the last $n - r$ columns. A way of uniquely defining these columns by additional restrictions is described in Ober [14]. For the case $\Sigma = \sigma I$ the free parameters for given integers $r, t(1), \cdots, t(r)$ are:

- $\sigma > 0$

- the elements in $G$, which are not a-priori zeros (where $g_{i,t(i)} > 0, 1 \leq i \leq r$)

- the entries in the orthonormal vectors of $U$

- certain entries in the skewsymmetric part of $F$ (see Ober [14] for details)

The form of the parametrization assures, that the corresponding $F$ matrix is stable (see e.g. Pernebo & Silverman,[16]). Also minimality is guaranteed for every choice of parameter values, as for systems satisfying the Lyapunov equations (2.9) with nonsingular $\Sigma$ stability is equivalent to minimality, see for example Pernebo & Silverman [16] (The miniphase condition restricts the set of feasible parameter values.).

The next case, we consider, is the case, where all $\sigma_i$'s are different. In this case a matrix is a feasible orthogonal basis transformation preserving balancing if and only if it is diagonal with $+1$ or $-1$ as diagonal entries. This in particular implies that the set of observationally equivalent balanced realizations is finite containing $2^n$ elements. Thus if we take the first nonzero element in every row of $G$ as positive (Note that for the continuous time case by minimality $G$ can have no zero row, since $(F, G, H)$ is balanced and all multiplicities of the second order modes are 1. This can easily be seen from (2.9)), the feasible transformation is uniquely defined and thus we have a unique balanced realization by this prescription. For the $i$-th row of $G$ define the integer $s(i)$ as the position of the first nonzero element, further let $n_i$ denote the multiplicity of $\sigma_i$. Then for given integers $n, n_1 = \cdots = n_n = 1$ and $s(1), \cdots, s(n)$ the free parameters are:

- $\sigma_1 > \cdots > \sigma_n > 0$

- The entries in $G$, which are not a priori restricted to zero ($g_{i,s(i)} > 0, 1 \leq i \leq n$)

- $n$ orthogonal vectors $u_1, \cdots, u_n$ determining the directions of the columns in the $H$ matrix.

The $F$ matrix is uniquely determined by:

$$F_{ii} = -\frac{G_i G_i^T}{2\sigma_i} \tag{3.3}$$

$$F_{ij} = \frac{\sigma_j G_i G_j^T - \sigma_i H_i^T H_j}{\sigma_i^2 - \sigma_j^2} \tag{3.4}$$

where $G_i$ denotes the $i$-th row of $G$ and $H_i$ the $i$-th column of $H$. These equations follow immediately from the continuous time Lyapunov equations (2.9). The stability assumption can be shown to hold for arbitrary parameter values, since the diagonal entries of the $F$ matrix are negative, see Kabamba [8]. (The miniphase condition introduces an additional restriction on the set of feasible parameter values.)

The discussion above can easily be generalized to the case of arbitrary multiplicities of the second order modes: if we partition the matrices $(F, G, H)$ according to the multiplicities of the modes and consider the diagonal block in the continuous time Lyapunov equations (2.9), we see that the subsystems $(F_{ii}, G_i, H_i)$ are balanced systems with Gramian equal to $\Sigma_i = \sigma_i I$. This case was discussed in the first paragraph of this section. Now the off-diagonal blocks of $F$ can be calculated using the equations (3.4). This gives a complete description of the canonical form.

# 4    Parametrization and Topological Properties

As other canonical forms, Ober's (canonical) form is characterized by integer and real parameters. By fixing the integer parameters, we obtain a welldefined subset (piece) of $U$. As can be seen from the last section for the Ober form in order to cover the set $M(n)$ a great number of pieces has to be used. This means that in identification a great number of integers (defining dynamic specification) has to be estimated prior to e.g. maximum likelihood estimation of the real parameters. For this reason we restrict ourselves to the case where the multiplicities of all second order modes are equal to 1. Let $U_n$ denote the set of all $k \in M(n)$ such that the corresponding second order modes are distinct. The next theorem shows, that this case is generic:

**Theorem 4.1** $U_n$ *is open and dense in* $\bar{M}(n)$.

**Proof:** The proof is along the lines indicated by Kabamba [8]. Remember that $S_n \subset S$ denotes the set of all minimal $(F, G, H) \in S$ (the matrix triple $(F, G, H)$ here denotes a discrete time system) and that $S_n$ is an open and dense subset of $S$. The second order modes are the solutions to the equation $\det(\lambda I - W_o W_c) = 0$. If the multiplicity of such a solution $\lambda_1$ is greater than one, also the equation $\frac{d \det(\lambda I - W_o W_c)}{d\lambda}|_{\lambda=\lambda_1} = 0$ holds and thus the matrix polynomials $a(\lambda) = \sum_{j=0}^{n} a_j \lambda^j = \det(\lambda I - W_o W_c)$ and $b(\lambda) = \sum_{j=0}^{n-1} b_j \lambda^j = \frac{da(\lambda)}{d\lambda}$ have a common zero in this case. Therefore the determinant of the Sylvester matrix

$$R = \begin{bmatrix} a_0 & a_1 & \cdots & \cdots & a_n & 0 & \cdots & 0 \\ 0 & a_0 & a_1 & \cdots & & \cdots & a_n & \\ & & \ddots & & & & \ddots & 0 \\ 0 & \cdots & 0 & a_0 & a_1 & \cdots & \cdots & a_n \\ b_0 & \cdots & \cdots & \cdots & b_{n-1} & 0 & \cdots & 0 \\ 0 & \ddots & & & & \ddots & & \\ & & \ddots & & & & \ddots & 0 \\ 0 & \cdots & 0 & b_0 & \cdots & \cdots & \cdots & b_{n-1} \end{bmatrix}$$

is equal to zero then. Note that the function, $f : S \to \mathbb{R}$ say, attaching to every triple $(F, G, H)$ the determinant of the Sylvester matrix $R$ can be represented by a convergent power series in the entries of the triple $(F, G, H)$ and thus in particular is continuous. Now consider a series $k_T$ of transfer functions in $M(n)$ converging to $k \in M(n)$. Taking an overlapping parametrization (see Hannan & Deistler [7]) we have a continuous function $\varphi$ (defined in the neighborhood of $k$) attaching a triple $(F, G, H)$ to $k$. Thus $k_T \to k$ implies $f(\varphi(k_T)) \to f(\varphi(k))$ from a certain $T_0$ onwards, i.e. the function $h$ attaching the determinant of the Sylvester matrix $R$ to $k$ is continuous. Therefore $U_n = h^{-1}(\mathbb{R} - \{0\})$ is open in $M(n)$ and since $M(n)$ is open in $\bar{M}(n)$ (cf. Hannan & Deistler [7]) also in $\bar{M}(n)$.
In the next step we show, that $U_n$ is dense in $\bar{M}(n)$. Let $D_n \subset S_n$ denote the set of all triples $(F, G, H)$ (discrete time) which in addition correspond to distinct second order modes. Assume that $D_n$ is not dense in $S_n$, i.e. $S_n - D_n$ contains a nontrivial open set. But then the power series $f$ must be zero on $S_n$, which cannot the case as easily can be seen. Thus $D_n$ is dense in $S_n$ and by the continuity of the function $\pi$ attaching transfer functions to the triples $(F, G, H)$, the set $U_n$ is dense in $\bar{M}(n)$. This completes the proof.

However $U_n$ cannot be parametrized in a continuous way using our canonical forms, i.e. the function $\varphi$ attaching to every $k \in U_n$ the balanced Ober form is not continuous. The discontinuities occur at transfer functions, where the corresponding continuous time canonical forms have zeros at certain entries in the $G_c$ matrix (note that we have restricted some elements to be positive with the choice of certain integers). Consider for example the case, where $g_{1,1}$ is zero, again suppressing the subscript $_c$, whereas $g_{1,2}$ has an arbitrary positive value. Now choose a sequence of balanced realizations (which are not necessarily in the balanced Ober form), where we change only the entry $g_{1,1,t}$ such that $g_{1,1,t} \to 0$ and $g_{1,1,t}$ is positive for even $t$ and negative for odd $t$. Then in the corresponding Ober form (using the same symbols) $g_{1,2,t}$ is equal to $g_{1,2}$ for even $t$ and equal to $-g_{1,2}$ for odd $t$. Thus the parametrization cannot be continuous at such a point.
In order to make the parametrization continuous on a generic set, we have

to exclude points of discontinuity described above. Let us define $D_n^+$ as the set of all $(F, G, H) \in D_n$, where in addition all elements in the first column of $G_c$ (continuous time) are strictly positive. Let $U_n^+$ denote the set of the corresponding transfer functions, $\pi(D_n^+)$. In the next step we show, that the parametrization $\varphi$ restricted to $U_n^+$ is continuous. This follows from the fact, that the parameters can be obtained from an SVD of the Hankel matrix $\mathcal{H} = \mathcal{O}\mathcal{C}$: If we decompose $\mathcal{H} = (U\Sigma^{1/2})(\Sigma^{1/2}V^T)$ and choose $\mathcal{O} = U\Sigma^{1/2}, \mathcal{C} = \Sigma^{1/2}V^T$, then the balanced realizations can be calculated as follows: $G$ consists of the first $s$ columns of $\mathcal{C}$ and $H$ of the first $s$ rows of $\mathcal{O}$. $F$ can be calculated from the equation $F\mathcal{C} = \vec{\mathcal{C}}$, where $\vec{\mathcal{C}}$ denotes the matrix obtained by omitting the first $s$ columns of $\mathcal{C}$. Since the SVD is continuous at points, where the singular values are distinct and the signs of the singular vectors are fixed, the mapping attaching the observability and the controllability matrix to the transfer function is continuous. Then clearly also $G$ and $H$ depend continuously on the transfer function $k$. $F$ depends on $k$ in a continuous way, since the controllability matrix $\mathcal{C}$ is of full rank.

As easily can be seen, $U_n^+$ is a generic subset of $U_n$ and thus of $M(n)$. Denseness is evident, since the mapping attaching transfer functions to triples $(F, G, H)$ is continuous. In order to show that $U_n^+$ is open in $U_n$, we note that $\varphi$ restricted to $U_n^+$ is a homeomorphism by what was said above. Since for every $(F, G, H) \in D_n^+$ exists an open neighborhood contained in $D_n^+$, the same is true for the corresponding transfer functions. Thus we have shown:

**Theorem 4.2** $U_n^+$ *is open and dense in* $\bar{M}(n)$. $\varphi : U_n^+ \rightarrow D_n^+$ *is a homeomorphism.*

The next theorem shows, that in the closure of $D_n^+$ the same transfer functions are described as in the closure of $U_n^+$. This is a difference to the case of e.g. Echelon forms. If $U_\alpha$ denotes the generic neighborhood for the Echelon forms and $T_\alpha$ the corresponding parameter space, then, for $s > 1$, $\bar{U}_\alpha - \pi(\bar{T}_\alpha)$ is not empty (cf. Hannan & Deistler [7]), i.e. in the Echelon case there are transfer functions in $\bar{M}(n)$ which can only be obtained as a limit of transfer functions, where the corresponding parameters diverge to infinity.

**Theorem 4.3** $\bigcup_{j \leq n} M(i) = \bar{U}_n^+ = \pi(\bar{D}_n^+)$

**Proof:** From the continuity of $\pi$ we see that $\bar{U}_n^+ \supset \pi(\bar{D}_n^+)$. In order to prove that $\bar{U}_n^+ \subset \pi(\bar{D}_n^+)$ we proceed as follows: Let $k \in \bar{U}_n^+$. Then there exists a triple $(F, G, H) \in \mathbb{R}^{n \times n + n \times s + s \times n}$ (which of course is not uniquely defined) such that $\pi(F, G, H) = k$. From the proof of Theorem 4.1 we see, that then there exists a sequence $(F_t, G_t, H_t)$ converging to $(F, G, H)$, such that $\pi(F_t, G_t, H_t) \in U_n^+$. We now transform $(F_t, G_t, H_t)$ to the uniquely defined Ober form $(F_t^b, G_t^b, H_t^b)$ in discrete time, which defines a sequence of transformations $T_t$ (comp. (1.4)). From section 2 we see that $T_t$ is of the form $T_t = P_t \Sigma^{-1/2} U_t^T$, where we choose $P_t$ as the Cholesky factor of the observability Gramian $W_{o,t}$ and where $P_t W_{c,t} P_t^T = U_t \Sigma_t^2 U_t^T$ corresponds to the eigenvalue decomposition. Note that $P_t$ and $\Sigma_t$ continuously depend on $W_o$ and $W_c$ (for the latter see e.g. Deif [4]). Since $U_t$ are orthogonal matrices, the sequence $T_t$ is bounded. Thus there exists a subsequence of $T_t$ which is convergent. From $\det T_t = \det P_t \det \Sigma^{-1/2}$ we conclude that the limiting element of the subsequence is nonsingular. Thus there exists a sequence $(F_t^b, G_t^b, H_t^b) \in D_n^+$ (using $t$ as the index for the subsequence) which is convergent and where $\pi(F_t^b, G_t^b, H_t^b) \to k$. This completes the proof.

Next we consider the relation between the $(F, G, H) \in D_n^+$ and their free parameters. Let us denote $i^{-1}(D_n^+)$ by $C_n^+$. Remember that $i$ defines a homeomorphism from $C_n^+$ onto $D_n^+$. As has been stated in section 3, the free parameters $\tau$ for the $(F_c, G_c, H_c)$ consist of $\sigma_1 > \cdots > \sigma_n > 0$, the entries in $G_c$ where $g_{j1,c} > 0$ and the suitably chosen free parameters of $n$ $\mathbb{R}^s$-vectors of length 1. Clearly the set of all $\mathbb{R}^s$ vectors of length 1 is the unit sphere in $\mathbb{R}^s$ and thus a manifold, which cannot be described by one coordinate system. However removing one point (e.g. the 'southpole') from the unit sphere, we have a homeomorphism to $\mathbb{R}^{s-1}$. Thus $\tau$ consists of $n + ns + (s-1)n = 2ns$ components. Note that $2ns$ is the dimension of $M(n)$ (see Hannan & Deistler [7]). Let $F_n^{++} \subset \mathbb{R}^{2ns}$ denote the set of free parameters corresponding to $C_n^+$, where in addition the southpoles have been removed and let $D_n^{++}$ denote the set of the corresponding $(F, G, H) \in D_n^+$. It is straightforward to see that the mapping $\psi : D_n^{++} \to F_n^{++}$ is a homeomorphism and that $D_n^{++}$ is open and dense in $D_n^+$ and that $U_n^{++} = \pi(D_n^{++})$ is open and dense in $U_n^+$.

## 5   Implications for Identification

The usual approach for identification of linear dynamical systems can be decomposed into first estimating integer parameters such as e.g. the order $n$ e.g. with information criteria such as AIC or BIC and then to estimate the real valued parameters e.g. by maximum likelihood methods. This is a consequence of the fact, that the set of all rational stable transfer functions has no continuous parametrization and thus we have to decompose this set into pieces (subclasses) such that each piece allows for a continuous parametrization. The pieces are characterized by integers. The likelihood function $L_T$ then is optimized for a given piece. Under rather general conditions such a maximum likelihood estimator $\hat{k}_T$ of the transfer function $k_0$ can be shown to be consistent (Hannan & Deistler [7]), i.e. if $U_\alpha$ denotes such a piece, where $\alpha$ denotes a multiindex of integers, then $\hat{k}_T = argmax_{k \in U_\alpha} L_T(k)$ will converge a.s. to the true transfer function $k_0 \in U_\alpha$. The properties of the maximum likelihood estimates for $(F, G, H)$ then depend on the properties of the parametrization chosen for the set $U_\alpha$. Note that in optimizing the likelihood, boundary points of $U_\alpha$ cannot be avoided (i.e. in a certain sense we are optimizing over $\bar{U}_\alpha$). However if $\hat{k}_T \to k_0 \in U_\alpha$ and if $U_\alpha$ is open in its closure $\bar{U}_\alpha$, then $\hat{k}_T$ is contained in $U_\alpha$ from a certain $T_0$ onwards. If there exists a continuous parametrization $\varphi_\alpha : U_\alpha \to D_\alpha$, then $\varphi(\hat{k}_T) \to \varphi(k_0)$, i.e. we also have consistency for $(F, G, H)$.

Recently balanced parametrizations have become popular in systems engineering because of certain numerical properties, see for example Mullis & Roberts [13]. For identification, balancing has been used e.g. by Maciejowski [10] and McGinnie [11]. In the previous section we were concerned with deriving results on balanced canonical forms, which are important for identification. In this section we discuss the strengths and the weaknesses of using balanced parametrizations for identification rather than Echelon forms.

In both cases, the integer parameter $n$ (the order) has to be estimated. In both cases there exists an open and dense subset of $M(n)$, which allows for continuous parametrization: for Ober forms this subset is $U_n^+$ and for Echelon forms this is the generic neighborhood $U_\alpha$, $\alpha = (n_1, \cdots, n_s)$ where for the Kronecker indices we have $n_1 = \cdots = n_i = n_{i+1}+1 = \cdots = n_s+1$ for a suitable chosen $1 \leq i \leq s$ (Hannan & Deistler [7]). Typically in order to cover $M(n)$ more pieces have to be used in the case of Ober forms compared to Echelon forms. Consider for example the case $n = 4, s = 2$. Then in the Echelon case $M(n)$ is covered by $\binom{n+s-1}{s-1} = 5$ pieces (each of which allows for a continuous parametrization), whereas for the Ober forms we have 82 pieces (see Bauer [3]). This example shows, that estimation of all pieces is intractable from a practical point of view for the Ober forms. Even in the case $s = 1$, where for the Echelon form $M(n)$ is covered by one piece, for the Ober form we have to use several pieces. So the advice would be for this case only to estimate the integer $n$ and then use the parametrization $\varphi : U_n^+ \to D_n^+$ with a danger of missing (nongeneric) transfer functions.

One advantage of Ober forms is that the second order modes are estimated as well as the elements of $G$ as real parameters and these parameters convey information about the integer parameters $n$, the multiplicities of the second order modes and the positions of the a priori zeros in $G_c$. In addition if the order $n$ initially has been chosen too large, truncation provides us with reasonable initial estimates in searching over a class of lower order systems.

Another advantage of Ober forms compared to Echelon forms is stated in Theorem 4.3, which says that every transfer function from $\bar{M}(n)$ can be represented either in $D_n^+$ or as an equivalence class at the boundary of $D_n^+$, whereas for the Echelon forms the transfer functions contained in $M(n) - \pi(T_\alpha)$ can neither be described in $T_\alpha$ nor by an equivalence class at the boundary of $T_\alpha$.

# 6   References

[1] K.S.Arun, S.Y. Kung: Balanced Approximation of Stochastic Systems, SIAM J. Matrix Anal. Appl., Jan. 1990, Vol. 11, No.1, p.42-68

[2] M.Aoki: State Space Modelling of Time Series, Springer-Verlag Berlin, 1987

[3] D. Bauer: Strucutre of Balanced Realizations of Discrete Time State Space Systems, Master Thesis, TU Wien, 1994

[4] A. S. Deif: Advanced Matrix Theory for Scientists and Engineers, Abacus Press, Gordon and Breach Science Publishers, 1991

[5] U. B. Desai, D. Pal: A Transformation Approach to Stochastic Model Reduction, IEEE Transactions on Automatic Control, Dec. 1984, Vol.29, p.1097-1100

[6] K. Glover: All Optimal Hankel-norm Approximations of Linear Multivariable Systems and Their $L^\infty$-error bounds, Int. J. Control, 1984, Vol. 39, No. 6, p.1115-1193

[7] E. J. Hannan, M. Deistler: The Statistical Theory of Linear Systems, Wiley Series, 1988

[8] P. T. Kabamba: Balanced Forms: Canonicity and Parametrization, IEEE Transactions on Automatic Control, Nov. 1985, Vol. 30, p.1106-1109

[9] A. J. Laub, M. T. Heath, C. C. Paige, R. C. Ward: Computation of System Balancing Transformations and Other Applications of Simultaneous Diagonalization Algorithms, IEEE Transactions on Automatic Control, Feb. 1987, Vol.32, No.2, p.115-121

[10] J. M. Maciejowski: Balanced Realizations in System Identification, Proc. 7th IFAC Symposium on Identification and Parameter Estimation, York, U.K., 1985

[11] B. P. McGinnie: A balanced view of System Identification, PhD Thesis, University of Cambridge, 1993

[12] B.C.Moore: Principal Component Analysis in Linear Systems: Controllability, Observability, and Model Reduction, IEEE Transactions on Automatic Control, Feb. 1981, Vol.26, p.17-31

[13] C. T. Mullis, R. A. Roberts: Synthesis of Minimum Roundoff Noise Fixed Point Digital Filters, IEEE Transactions on Circuits and Systems, Sept 1976, Vol.23, No.9, p.551-561

[14] R. Ober: Balanced Realizations: Canonical Forms, Parametrization, Model Reduction, Int. J. Control, 1987, Vol.46, No.2, p.643-670

[15] R. Ober: Balanced Parametrization of Classes of Linear Systems, SIAM J. Control and Optimization, Nov. 1991, Vol. 29, No.6, p.1251-1287

[16] L. Pernebo, L. Silverman: Model Reduction Via Balanced State-Space Systems, IEEE Transactions on Automatic Control, April 1982, Vol.27, p.382-387

# Recent Developments in Analysis of Time Series with Infinite Variance: A Review

R. J. Bhansali

University of Liverpool, U. K.

ABSTRACT The class of linear stable processes is considered and a review of the literature focusing on the behaviour of the standard techniques for model selection, spectral analysis and parameter estimation for this class of processes is given.

## 1   Introduction

Consider the autoregressive-moving average model of order $(m, h)$, ARMA $(m, h)$, for a discrete-time stationary process, $\{x_t\}$:

$$\sum_{j=0}^{m} \tau_{mh}(j) x_{t-j} = \sum_{j=0}^{h} \beta_{mh}(j) \epsilon_{t-j}, \quad \tau_{mh}(0) = \beta_{mh}(0) = 1, \qquad (1.1)$$

where the coefficients $\tau_{mh}(j)$ and $\beta_{mh}(j)$ are such that

$$\Phi_{mh}(z) = \sum_{j=0}^{m} \tau_{mh}(j) z^j \neq 0, \quad \Theta_{mh}(z) = \sum_{j=0}^{h} \beta_{mh}(j) z^j \neq 0, \quad |z| \leq 1,$$

and that $\Phi_{mh}(z)$ and $\Theta_{mh}(z)$ do not have a common zero.

Having observed $x_1, \ldots, x_T$, it is common to assume that $\{\epsilon_t\}$ is a sequence of independent Gaussian variates, each with mean 0 and variance

$\sigma^2$, the assumption being made for obtaining the likelihood function of the parameters and their maximum likelihood estimates and although the sampling properties of these estimates may be studied without the Gaussian assumption, for non-Gaussian time series they may be interpreted only as least-squares estimates; moreover, since the version of AIC used for time series model selection, see Findley (1985), is based on the Gaussian likelihood, the normality assumption is also central to a number of time series model selection procedures.

Many observed time series, however, display characteristics that flatly contradict the Gaussian assumption. Mandelbrot (1960) observed that certain economic time series such as stock and commodity price changes have highly leptokurtic distributions with heavy tails, meaning that as compared with a normal density, their frequency functions have too much mass near the origin and in the tails and not enough mass between one and two standard deviations; see Du Mouchel (1983) for examples. Since the work of Mandelbrot, the use of stable distributions as models for such data has become popular. There has also been discussion of several different alternative approaches and other heavy-tailed distributions, such as the $t$, Log-Normal, Compound Normal and the Pareto family, have been suggested, see Tucker (1992), and also the family of ARCH models, see Harvey (1993). We do not intend to examine the relative merits of the various approaches for modelling financial data. Rather, we review the literature which shows that even under the stable law hypothesis some of the standard procedures for parameter estimation, model selection and spectral analysis remain robust.

## 2   Properties of Stable Laws

A detailed discussion of the stable laws may be found in Samorodnitsky
and Taqqu (1994). We only consider the class of standardized symmetric
stable laws with the location and scale parameters equal to zero and one
respectively, and with the characteristic function given by

$$c(t) = \exp(-|t|^\delta), \tag{1.2}$$

where $0 < \delta \le 2$ is called the characteristic exponent of the distribution.
The Normal distribution corresponds to taking $\delta = 2$ but it is the only
stable law with finite variance since for $\delta < 2$ all absolute moments of order
greater than or equal to $\delta$ are infinite. Some of the important properties of
the stable laws, such as the only class of distributions admitting domains of
attraction under convolution and the occurrence of large values, are rather
well-known, e.g., see Mittnik and Rachev (1993). It should be emphasised,
however, that as we consider time series whose innovations follow an infinite
variance stable law, the large values of $x_t$ may not be termed outliers as
they are due to the heavy-tails in the probability distribution of $\{x_t\}$.

## 3   Linear Stable Processes

Suppose that $\{x_t\}$ satisfying (1.1) is an ARMA$(m, h)$ process, but $\{\epsilon_t\}$
is a sequence of independent identically distributed random variates each
following a standardized symmetric stable law with characteristic function
(1.2). A straightforward argument, similar to the case $E\left(\epsilon_t^2\right) < \infty$, shows
that $\{x_t\}$, almost surely, has an infinite moving average representation,

$$x_t = \sum_{j=0}^{\infty} b(j)\epsilon_{t-j}, \quad b(0) = 1, \tag{1.3}$$

and an infinite autoregressive representation,

$$\sum_{j=0}^{\infty} a(j)x_{t-j} = \epsilon_t, \quad a(0) = 1, \tag{1.4}$$

in which the $b(j)$ and $a(j)$ are related to the $\tau_j$ and $\beta_j$ in the usual way, that is, their characteristic polynomials are given by:

$$B(z) = \sum_{j=0}^{\infty} b(j)z^j = [\Theta(z)/\Phi(z)], \quad A(z) = \sum_{j=0}^{\infty} a(j)z^j = [B(z)^{-1}], \tag{1.5}$$

see Cline and Brockwell (1985), who consider a general ARMA process; Yohai and Maronna (1977) and Smith (1983) obtain this result for a pure autoregression with $h = 0$ and Bhansali (1988) for a finite moving average process with $m = 0$.

In view of these representations and to ensure greater generality we consider the class of linear stable processes satisfying (1.3) in which $\{\epsilon_t\}$ is a sequence of independent variates, each having a normalized symmetric stable distribution with characteristic function (1.2), and the $b(j)$ are real coefficients such that $B(z) \neq 0, |z| \leq 1$, and for $d = \min(\delta, 1)$,

$$s_{bd} = \sum_{j=0}^{\infty} |b(j)|^d < \infty. \tag{1.6}$$

An extension of the Wiener-Levy theorem due to Zelazko (1965) ensures that, under these assumptions, $\{x_t\}$ also has an infinite autoregressive representation (1.4) with $A(z) \neq 0, |z| \leq 1$, and $s_{ad} < \infty$, where $s_{ad}$ is given by (1.6) but with $a(j)$ replacing $b(j)$.

For the class of processes defined above, which we call the $M_\delta$ class of stationary stable processes, the marginal distribution of $x_t$ for each $t$ is symmetric stable about 0 with characteristic exponent $\delta$ and scale $s_{bd}$, and, if $\delta < 2, \{x_t\}$ is strictly stationary but not covariance-stationary. Moreover, this class of stable processes is distinct from and does not overlap with the

class of harmonizable stable processes, see Cambanis and Soltani (1984).
For the $M_\delta$ class of processes, Cline and Brockwell (1985) consider the
question of linear prediction of the future, $\{x_h, h > 0\}$, of $\{x_t\}$ from the
complete past, $\{x_t, t \leq 0\}$, and show that if the optimality of the predictor is judged by the criterion of minimizing the error 'dispersion' then
the optimal predictor coincides with the classical Wiener-Kolmogorov linear
least-squares predictor. Kanter (1979) considers the question of the non-
linear prediction for a general class of moving averages of the form (1.3)
and obtains a 'painless' lower bound for the one-step mean squared error of
prediction. This author also makes some remarks on non-linear prediction
for infinite-variance linear processes, though the actual class of processes
considered by him does not necessarily coincide with that considered here.

The basic tools used for model identification of a covariance-stationary
process, namely the correlation, partial correlation and spectral density
functions, do not exist for the class of $M_\delta$ processes with $\delta < 2$. Nevertheless, having observed $x_1, \ldots, x_T$, the sample estimates of these quantities
may be computed. Rosenfeld (1976), Granger and Orr (1972) and Nyquist
(1980) observe empirically that for this class of processes, the Box and Jenkins (1970) model identification technique continues to be reliable. Next, we
review the literature which gives theoretical underpinning for this empirical finding. We observe, first, that although for $0 < \delta < 2$, the correlation
function does not exist, a pseudo-correlation function may be defined in
terms of the coefficients $b(j)$ in (1.3). Let,

$$r(-u) = r(u) = \frac{\sum_{j=0}^{\infty} b(j)b(u+j)}{\sum_{j=0}^{\infty} b^2(j)} \qquad (u = 0, 1, \ldots). \qquad (1.7)$$

If $\delta = 2$, $\{r(u)\}$ defines the (auto) correlation function of $\{x_t\}$. For $\delta < 2$,
however, the $r(u)$ still exist even though they do not have a stochastic

interpretation. Similarly, the normalized pseudo-spectral density function of $\{x_t\}$ may be defined by

$$\tilde{f}(\mu) = (2\pi)^{-1} \sum_{u=-\infty}^{\infty} r(u) \exp(-iu\mu)$$

$$= \frac{(2\pi)^{-1} \left| \sum_{j=0}^{\infty} b(j) \exp(-ij\mu) \right|^2}{\sum_{j=0}^{\infty} b^2(j)}. \tag{1.8}$$

The related quantities, like the (pseudo) partial correlation, inverse correlation and inverse spectral density functions may be defined in terms of the $b(j)$ by analogy with their definitions for a covariance-stationary process.

Some of the results discussed below hold without requiring that $\{x_t\}$ belongs to the $M_\delta$ class of processes; however, to minimise the technical details and to ensure a unified treatment, we do not stress this aspect in detail.

# 4 Consistency of the Sample Correlation Function

Suppose the $\{x_t\}$ belongs to the $M_\delta$ class, $0 < \delta < 2$, of stable processes. Having observed $x_1, \ldots, x_T$, let

$$r^{(T)}(u) = r^{(T)}(-u) = \frac{\sum_{t=1}^{T-u} x_t x_{t+u}}{\sum_{t=1}^{T} x_t^2}, \quad u = 1, \ldots, T-1,$$

denote the standard sample correlation function, obtained under the assumption that $\delta = 2$, and let $r(u)$ be defined by (1.7).

We have, for all $\varphi > \delta$ and each fixed $u \neq 0$,

$$\lim_{T \to \infty} T^{\frac{1}{\varphi}} \left| r^{(T)}(u) - r(u) \right| = 0, \quad \text{in probability.} \tag{1.9}$$

Kanter and Steiger (1974) establish this result for a pure autoregressive process and Bhansali (1988) for a finite moving average process. The more general case is considered by Davis and Resnick (1986) who treat a slightly larger class of processes than the $M_\delta$ by only requiring that the distribution of $\epsilon_t$ has Pareto tails and $\{x_t\}$ admits a two-sided moving average representation and derive the asymptotic distribution of a finite collection of the $r^{(T)}(u)$ as a ratio of two independent stable random variables with exponents $\delta$ and $\delta/2$.

An implication of (1.9) is that the sample correlations, $r^{(T)}(u)$, are reasonably well-behaved for the $M_\delta$ class of stable processes and that now the rate of convergence is faster than in the Gaussian case; moreover, since the convergence rate is inversely related to the value of $\delta$, for sufficiently large values of $T$, the distribution of the estimate may be expected to 'concentrate' more and more around $r(u)$ as $\delta$ decreases. It should be noted, however, that their actual distribution may contain outliers and the confidence intervals for $r(u)$ constructed from the $r^{(T)}(u)$ can be large, see Brockwell and Davis (1991).

# 5   Consistency of the estimated Autoregressive Coefficients

Suppose now that $\{x_t\}$, belonging to the $M_\delta$ class, is an autoregressive process of order $m$, i.e, $h = 0$ in (1.1) and $a(j) = \tau_j, j = 1, \ldots, m, a(j) = 0, j > m$, in (1.4). Let $\hat{a}_{is} = [\hat{a}_{is}(1), \ldots, \hat{a}_{is}(s)]'(i = 1, 2)$ denote the $s$th order estimates of the autoregressive coefficients, $s = 1, 2, \ldots, L$, where $L \geq m$ denotes a known integer,

$$\hat{a}_{is} = -\hat{\Gamma}_{is}\hat{r}_{is}, \tag{1.10}$$

$$\hat{\Gamma}_{is} = [\hat{R}_i(u, j)](u, j = 1, \ldots, s), \quad \hat{r}_{is} = [\hat{R}_i(0, j)](j = 1, \ldots, s)$$

and

$$\hat{R}_1(u,j) = \sum_{t=1}^{T-|u-j|} x_t x_{t+|u-j|}, \quad \hat{R}_2(u,j) = \sum_{t=s+1}^{T} x_{t-u} x_{t-j}.$$

Note that, $\hat{a}_{1s}$ and $\hat{a}_{2s}$ provide the $s$th order Yule-Walker and least-squares estimates. The corresponding parameter vector, $a_s = [a_s(1), \ldots, a_s(s)]'$ is given by $a_s = -\Gamma(s)^{-1} r(s)$, where $\Gamma_s$ and $r_s$ have $r(u-j)$ and $r(j)$ as their typical elements and if $s > m, a_s(j) = 0, j > m, a_s(j) = a_m(j), 1 \leq j \leq m$.

Some of the early work on linear Stable processes concerns the consistency of the $m$th order least-squares estimates. Hannan and Kanter (1974), see also Kanter and Steiger (1974) and Yohai and Maronna (1977), show that for $i = 2, s = m, \varphi > \delta$ and each fixed $u = 1, \ldots, m$,

$$\lim_{T \to \infty} T^{\frac{1}{\varphi}} |\hat{a}_{is}(u) - a_s(u)| = 0, \quad \text{a.s.} \tag{1.11}$$

The corresponding result for the Yule-Walker estimate, $i = 1$, but in probability, has been given by Bhansali (1988), who also considers the situation in which $m$ is unknown and thus $s \neq m$. Davis and Resnick (1986) derive the asymptotic distribution of the $m$th order least-squares estimates, $\hat{a}_{2m}(u)$, as a ratio of two independent stable variates.

For explaining the faster convergence rate of the least-squares estimates in the Cauchy case, $\delta = 1$, Cox (1966) observes that the outliers in $\epsilon_t$ produce additional outliers in the observed time series and these provide 'leverage' points by depressing any correlation present in the remaining data points.

# 6    Order Determination

For a Gaussian autoregressive process of unknown order, $m$, the order to
be fitted may be selected by minimizing the criterion

$$FPE_\alpha(p) = \hat{\sigma}^2(p)(1 + \alpha p/T) \qquad (p = 0, 1, \ldots, L) \qquad (1.12)$$

in which $\alpha > 1$ is a constant, see Bhansali and Downham (1977), and

$$\hat{\sigma}^2(p) = \frac{\hat{R}_2(0,0) + \sum_{j=1}^{p} \hat{a}_{2p}(j)\hat{R}_2(0,j)}{\hat{R}_2(0,0)}.$$

Alternatively, the Yule-Walker autoregressive estimates may be used instead. As is well-known, this criterion includes as its special case, with
$\alpha = 2$, Akaike's (1973) information criterion, and, with $\alpha = \ln T$, Schwarz's
(1978) Bayesian criterion, and for a fixed value of $\alpha$ it fails to be consistent for $m$, but consistent criteria are obtained if $\alpha = \alpha(T)$, a function of
$T$, such that as $T \to \infty, \alpha(T) \to \infty$, but sufficiently slowly; see Bhansali
(1993b).

   In view of (1.11), however, for the $M_\delta$ class of processes with $\delta < 2$,
(1.12) is consistent for $m$ with any fixed $\alpha > 0$. We have, see Bhansali
(1988), if $\hat{m}_\alpha$ denotes the autoregressive order selected by minimizing the
criterion (1.12) with $\alpha > 0$ and $\{x_t\}$ is an autoregressive process of order
$m$ belonging to the $M_\delta$ class of processes with $0 < \delta < 2$,

$$\lim_{T \to \infty} P\{\hat{m}_\alpha = j\} = 1, \quad j = m, \quad \lim_{T \to \infty} P\{\hat{m}_\alpha = j\} = 0, \quad j \neq m. \quad (1.13)$$

   Bhansali (1988) also establishes the consistency of a computationally-
effective procedure, Bhansali (1983), for order selection of moving average
processes belonging to the $M_\delta$ class, $0 < \delta < 2$; the simulation evidence in
support of these asymptotic results is given by Bhansali (1984).

# 7  Estimation of the Impulse Response Coefficients

A parametric and a non-parametric approach is available for estimating the coefficients $b(j)$ occurring in (1.3). The former postulates an ARMA model, (1.1), for $\{x_t\}$ and estimates the $b(j)$ by first estimating the $\tau_j$ and $\beta_j$. Mikosch *et al.* (1995) show that the Gaussian maximum likelihood estimates of the ARMA coefficients are consistent for the $M_\delta$ class of processes, and go on to also derive the asymptotic distribution of these estimates.

Bhansali (1993a), by contrast, considers the non-parametric approach and constructs estimates of the $b(j)$ by two different methods from an initial $k$th order autoregression fitted to the data. He shows that if $k \to \infty$ as $T \to \infty$, but sufficiently slowly and certain additional regularity conditions hold then both these methods yield consistent estimates of the $b(j)$ and a result analogous to (1.11) also holds for both these estimation methods.

# 8  Consistency of the Spectral Estimates

If $E(\epsilon_t^2) < \infty$, then the normalized spectral density, $\tilde{f}(\mu)$, may be estimated either by the window method by smoothing the raw periodogram or by fitting an autoregression to the data.

Kluppelberg and Mikosh (1993) have shown that for the $M_\delta$ class of processes, under suitable regularity conditions, the normalized window spectral estimate is pointwise consistent for $\tilde{f}(\mu)$; these authors also derive the asymptotic distribution of the normalized periodogram for this situation.

Also, Bhansali (1995) has show that, under regularity conditions, the normalized autoregressive spectral estimate is uniformly consistent for $\tilde{f}(\mu)$ even when $\{x_t\}$ is a member of the $M_\delta$ class; the relative behaviour of the autoregressive and window estimates for this situation is also examined by repeating the simulations of Beamish and Priestley (1981) with several different values of $\delta$ in $(0.0, 2.0]$ and letting $T = 100$ and $500$.

## 9  REFERENCES

Akaike, H. (1973). Information theory and an extension of the maximum likelihood principle. In *2nd International Symposium on Information Theory*, B. N. Petrov and F. Csaki, editors, pages 267–281. Akademia Kiado.

Beamish, N. and Priestley, M. B. (1981). A study of autoregressive and window spectral estimation. *Applied Statistics*, **30**, 41–58.

Bhansali, R. J. (1983). Estimation of the order of a moving average model from autoregressive and window estimates of the inverse correlation function. *J. Time Series Anal.*, **4**, 137–162.

Bhansali, R. J. (1984). Order determination for processes with infinite variance. In *Robust and Nonlinear Time Series Analysis*, J. Franke, W. Hardle, and D. Martin, editors, pages 17–25. Springer Verlag.

Bhansali, R. J. (1988). Consistent order determination for processes with infinite variance. *J. Royal Statist. Soc.*, **B**, **50**, 46–60.

Bhansali, R. J. (1993a). Estimation of the impulse response coefficients of a linear process with infinite variance. *J. Mult. Anal.*, **45**, 274–290.

Bhansali, R. J. (1993b). Order selection for linear time series models: a review. In *Developments in Time Series Analysis*, T. S. Rao, editor, pages 50–66. Chapman & Hall.

Bhansali, R. J. (1995). Robustness of the autoregressive spectral estimate for linear processes with infinite variance. *J. Time Series Analysis*. to appear.

Bhansali, R. J. and Downham, D. Y. (1977). Some properties of the order of an autoregressive model selected by a generalization of Akaike's FPE criterion. *Biometrika*, **64**, 547–551.

Box, G. E. P. and Jenkins, G. M. (1970). *Time Series Analysis: Forecasting and Control*. Holden Day.

Brockwell, P. J. and Davis, R. A. (1991). *Time Series: Theory and Methods*. Springer Verlag, 2nd edition.

Cambanis, S. and Soltani, A. R. (1984). Prediction of stable processes: spectral and moving average representations. *Z. Wahrscheinlichkeits theorie*, **66**, 593–612.

Cline, D. B. H. and Brockwell, P. J. (1985). Linear prediction of ARMA processes with infinite variance. *Stochastic Proc. and Appl.*, **19**, 281–296.

Cox, D. R. (1966). The null distribution of the first serial correlation coefficient. *Biometrika*, **53**, 623–626.

Davis, R. and Resnick, S. (1986). Limit theory for the sample covariance and correlation functions of moving averages.

Du Mouchel, W. H. (1983). Estimating the stable index $\alpha$ in order to measure tail thickness: A critique. *Ann. Statist.*, **14**, 533–558. *Ann. Statist.*, **14**, 533–538.

Findley, D. F. (1985). On the unbiasedness property of AIC for exact or approximating linear stochastic time series models. *J. Time Series Anal.*, **6**, 229–252.

Granger, C. W. J. and Orr, D. (1972). "Infinite Variance" and research strategy in time series analysis. *J. Amer. Statist. Assoc.*, **67**, 273–285.

Hannan, E. J. and Kanter, M. (1974). Autoregressive processes with infinite variance. *J. Appl. Prob.*, **6**, 768–783.

Harvey, A. C. (1993). *Time Series Models*. Harvester Wheatsheaf, 2nd edition.

Kanter, M. (1979). Lower bounds for prediction error in moving average processes. *Ann. Prob.*, **7**, 128–138.

Kanter, M. and Steiger, W. L. (1974). Regression and autoregression with infinite variance. *Adv. Appl. Prob.*, **6**, 768–783.

Kluppelberg, C. and Mikosh, T. (1993). Some limit theory for the normalized periodogram of p-stable moving average processes. *Scand. J. Statist.* to appear.

Mandelbrot, B. (1960). The Pareto-Levy law and the distribution of income. *Internat. Eco. Rev.*, **1**, 79–106.

Mikosch, T., Gadrich, T., Kluppelberg, C., and Adler, R. J. (1995). Parameter estimation for ARMA models with infinite variance innovations. *Ann. Statist.*, **23**, 305–326.

Mittnik, S. and Rachev, S. T. (1993). Modeling asset returns with alternative stable distributions. *Econometric Reviews*, **12**, 261–389.

Nyquist, H. (1980). *Recent Studies on $L_p$-Norm Estimation*. Ph.D. thesis, University of Umea, Sweden.

Rosenfeld, G. (1976). Identification of a time series with infinite variance. *Appl. Statist.*, **25**, 147–153.

Samorodnitsky, G. and Taqqu, M. S. (1994). *Stable Non-Gaussian Random Processes*. Chapman & Hall.

Schwarz, G. (1978). Estimating the dimension of a model. *Ann. Statist.*, **6**, 461–464.

Smith, W. L. (1983). On positive autoregressive sequences. In *Contributions to Statistics: Essays in honour of N. L. Johnson*, P. K. Sen, editor, pages 391–405. Amsterdam: North Holland.

Tucker, A. L. (1992). A reexamination of finite- and infinite- variance distributions as models of daily stock returns. *J. Bus. & Eco. Statist.*, **10**, 73–81.

Yohai, V. and Maronna, R. (1977). Asymptotic behaviour of least-squares estimates for autoregressive processes with infinite variance. *Ann. Statist.*, **5**, 554–560.

Zelazko, W. (1965). *Metric Generalizations of Banach Algebras*. Warsaw: Panstwowe Wydawnictwo Naukowe.

# Foreign Exchange Rates Have Surprising Volatility

Peter Bossaerts        Wolfgang Härdle        Christian Hafner

January 1996 (First Draft: February 1995)[*]

### Abstract

Local Polynomial Estimation (LPE) is implemented on a dataset of high-frequency foreign exchange (FX) quotes. This nonparametric technique is meant to provide a flexible background against which to evaluate parametric time series models. Assuming a conditionally heteroscedastic nonlinear autoregressive (CHARN) model, estimates of the mean and volatility functions are reported. The mean function displays pronounced reversion. Surprisingly, the volatility function exhibits asymmetry. The CHARN model, however, captures only the short-run behavior of conditional volatility. Nevertheless, part of the evidence of persistent conditional volatility appears in reality to be the effect of conditional kurtosis. Stochastic volatility models are ideal to capture this time series feature.

*Keywords*: Local Polynomial Estimation, Conditional Volatility, Conditional Kurtosis, Nonlinear Autoregressive Models, Foreign Exchange Markets.

[*]First author's affiliation: California Institute of Technology and Tilburg University; Mailing address: CentER, Tilburg University, PO Box 90153, NL-5000 LE Tilburg, The Netherlands; Phone: +31.13.663.101; e-mail: pbs@rioja.caltech.edu. Second and third author's affiliation: Humboldt–Universität zu Berlin; Mailing address: Institut für Statistik und Ökonometrie, Humboldt–Universität zu Berlin, Spandauer Strasse 1, D-10178 Berlin, Germany; Phone +49.30.246.82.30; e-mail: SFB373@wiwi.hu-berlin.de. This research was financed through contributions from the Sonderforschungsbereich 373 "Quantifikation und Simulation Ökonomischer Prozesse." Comments from participants and discussants at the fifth $(EC)^2$ Conference as well as the first HFDF Conference are gratefully acknowledged. Michel Dacorogna and Christian Gouriéroux gave valuable criticism. Olsen Associates generously provided the data.

# 1 Introduction

Conditional volatility of asset prices in general and foreign exchange (FX) rates in particular have been the subject of intense investigation over the last few years. Virtually all estimation has been parametric. Most popular are the GARCH family (e.g., Baillie and Bollerslev [1990]), as well as stochastic volatility models (e.g., Mahieu and Schotman [1994]). In this paper, we report estimates from the application of a new, nonparametric technique to high-frequency FX quote data, namely, Local Polynomial Estimation (LPE). This leads to surprising insights on the dynamics of FX rates.

In their analysis of the shape of the volatility of foreign exchange rates and stock returns as a function of lagged, exogenous information, Pagan and Ullah [1988] and Pagan and Schwert [1990] have already pointed out that nonparametric modelling is urgent. While it is not meant to displace parametric modelling, it reveals important information with which to enhance the parametric estimation. For ease of analysis, the extant parametric models (the ARCH family, as well as models of stochastic volatility) are all linear after suitable transformation. Misspecifications are accomodated for by the addition of explanatory variables, at the cost of making the analysis less parsimonious. What we provide here is a simple nonparametric and nonlinear analysis of the volatility function. The result is not only a descriptive account of the data, but also a framework with which one can evaluate the suitability of the existing parametric models. A similar approach has recently been taken by e.g. Aït-Sahalia [1994] in continuous-time modelling, and by Härdle and Mammen [1993], in regression analysis.

One could argue that there are few theoretical reasons to expect nonlinearities in the volatility function of FX rates. In particular, there are no "leverage effects," unlike with common stock prices (see, e.g., Christie [1982], Pagan and Schwert [1990]). Yet, a closer inspection of the mechanics of the FX market reveals a potential for nonlinearities. Foremost, one should mention central bank intervention. Bossaerts and Hillion [1991], for instance, find a pronounced effect of eminent central bank intervention on bid-ask spreads. The nonlinearity of typical policy reaction functions (Neumann [1984]) is likely to be reflected in the process of FX rate changes.

Moreover, as far as FX rates are concerned, it is attractive to be able to capture the stochastic nature of persistence in conditional volatility. There appear to be two types of events in high-frequency FX rate changes: those that induce volatility and those that have no effect on subsequent volatility (see Bewley, Lowe and Trevor [1988], who analyzed intraday Australian dollar quotes). CHARN modelling allows for such phenomena. Of course, it would ultimately be desirable to identify the nature of these different events.

We explicitly fit the mean function as well, allowing it to be nonlinear. In some recent analyses, the mean change in the FX rate is even constrained to be zero (e.g., Mahieu and Schotman [1994]). The mean function is not unimportant, even if the purpose of the modelling of time-varying volatility is option pricing. As Lo and Wang [1994] illustrate, misestimation of mean changes in asset prices (let alone failure to account for a mean) leads to substantial biases in option prices and hedges. In continuous time, the mean function is irrelevant. But since estimation necessarily takes place in discrete time, biases enter whenever time-variation in volatility is estimated without accounting for the mean function.

We obtain estimates of the mean and volatility functions of the CHARN model by means of Local Polynomial Estimation (LPE). As its name indicates, LPE is based on locally fitting polynomials. This means that polynomials are estimated by weighted least squares, where the weights depend on the distance of an observation from the values of the arguments of the mean and volatility function at which an estimate is to be obtained. Kernel functions localize the weights in the space of predictor variables. The bandwidth of the kernel function determines the smoothness of the fit. For consistency, the bandwidth must be lowered appropriately as the number of observations increases.

LPE has a long tradition in regression estimation for cross-sectional data. Stone [1977] and Cleveland [1979] seem to have been the first to suggest the technique. Tsybakov [1986] proved asymptotic normality. For an application in finance, see Bossaerts and Hillion [1995], where LPE is used to obtain estimates of dynamic hedge portfolio weights. Recently, Härdle and Tsybakov have analyzed the properties of LPE in the context of CHARN models (Härdle and Tsybakov [1995]). Crucial in the analysis is the concept of geometric ergodicity, which ensures the existence of a time-invariant distribution and sufficiently strong mixing such that laws of large numbers and central limit theorems hold. The conditions for geometric ergodicity are familiar to option pricing theorists, where these are needed to prove existence of solutions to stochastic differential equations (e.g., Karatzas and Shreve [1983]). Duffie and Singleton [1993] also appealed to geometric ergodicity in order to show consistency and asymptotic normality of their simulation estimator of Markov models.

Other nonparametric procedures have been suggested for the analysis of time series, such as standard kernel estimation (see, e.g., Györfi, et al. [1989]). LPE has the advantage, however, of featuring improved smoothing bias, as well as being computationally straightforward (local least squares; also, derivatives are obtained in a straightforward way). Closely related to the nonparametric techniques is the (parametric) qualitative threshold ARCH models of Gouriéroux and Monfort [1992], which is in fact a nonparametric modelling procedure, whereby mean and volatility functions are approximated by step functions with a fixed number of steps.

The LPE technique is implemented on the Olsen high-frequency FX quote data. Substantial mean reversion is discovered, as well as asymmetry in the volatility function. Long-run autocorrelation in squared residuals, however, remained even after introducing additional conditioning variables, such as the bid-ask spread. These were meant to capture volatility persistence. The problem did not disppear after (admittedly timid) implementation of HARCH modeling (Müller, et al. [1995]). Undersmoothing, however, indirectly revealed pronounced evidence of *conditional kurtosis*. Stochastic volatility models (e.g., Mahieu and Schotman [1994]) are well-tailored to capture such time series properties.

The remainder of the paper is organized as follows. The next section briefly introduces LPE of CHARN models. Section 3 discusses the dataset. Section 4 reports the LPE results. Section 5 concludes.

## 2   LPE of CHARN models

Let $\{y_t\}$ be a Markov time series, satisfying the following stochastic difference equation:

$$y_t = f(y_{t-1}) + s(y_{t-1})\xi_t, \tag{1}$$

where $\xi_t$ are i.i.d. random variables with mean zero and unit variance. Here $f$ and $s$ are unknown mean and volatility functions, respectively, with $s(x) > 0$ for all values of $x$, and $y_0$ is a random variable independent of the series $\xi_t$. The model (1) is a heteroscedastic nonlinear autoregression (CHARN).

We estimate the volatility function $v(x)$ $(= s^2(x))$ from a sample $y_1, ..., y_T$ by means of Local Polynomial Estimation (LPE). The procedure will simultaneously generate an estimate of $f(x)$. Define $T$ vector functions $u_T(z)$, as follows:

$$u_T(z)' = [1 \ \ \frac{z}{h_T} \ \ (\frac{z}{h_T})^2 \ \ ... \ \ (\frac{z}{h_T})^{l-1}/(l-1)!], \tag{2}$$

where $h_T$ is a parameter to be referred to as the bandwidth parameter and $l$ denotes the degree of the polynomial. For consistency, the bandwidth parameter should decrease with the sample size $(T)$. Consider now the minimization problems:

$$\gamma_T(x) = \arg\min_{\gamma} \sum_{t=1}^{T} \left\{ y_t^2 - \gamma' u_T(y_{t-1} - x) \right\}^2 K\left(\frac{y_{t-1} - x}{h_T}\right), \tag{3}$$

$$\phi_T(x) = \arg\min_{\phi} \sum_{t=1}^{T} \left\{ y_t - \phi' u_T(y_{t-1} - x) \right\}^2 K\left(\frac{y_{t-1} - x}{h_T}\right), \tag{4}$$

where $K$ is a kernel function and $h_T$ the bandwidth. The estimate of $f$ at $x$, $\hat{f}(x)$, is then given by:

$$\hat{f}(x) = \phi_T(x)'u_T(0).$$ (5)

The estimate of $v$ at $x$, $\hat{v}(x)$, on the other hand, is given by:

$$\hat{v}(x) = \gamma_T(x)'u_T(0) - \{\phi_T(x)'u_T(0)\}^2.$$ (6)

From these equations, it is clear that the estimates are obtained as the intercepts of polynomials which are fit by weighted least squares, where the weight to be put on an observation is determined by its distance from the target value, $x$. In other words, our procedure can also be described as local weighted least squares.

There are plenty of valid kernel functions. The bandwidth parameter, however, should be chosen carefully, in order to avoid overfitting. Crossvalidation is a simple procedure, albeit computationally intensive. The "out of sample" average squared prediction error of the estimated model is minimized with respect to the bandwidth. The "out of sample" prediction error for an observation is obtained from estimates of the mean and volatility functions based on all the data except the observation at hand.[1]

Härdle and Tsybakov [1995] establish the theoretical properties of LPE estimation in the Markov model (1). They show that the estimates of the mean and volatility functions converge and are asymptotically normally distributed. The conditions stated there guarantee asymptotic stationarity. The effect from initial sampling from another distribution than the stationary one thereby dies out fast enough for central limit theorems to continue to hold even in the nonstationary case.

# 3  Data

The dataset was compiled by Olsen and Associates, and consists of bid and ask quotes from Reuter's FXFX page. The sample covers the period 1 October 1992 at 0:00:00 GMT till 30 September 1993 at 23:59:59 GMT. The data were filtered by Olsen to remove erroneous quotes and other outliers (less than 0.5% of the data). Quotes for two currencies are available: DEM/USD and YEN/USD. Obviously, this is a huge dataset: the DEM/USD file, for instance, contains 1,472,241 records. We focused on transactions ten and twenty minutes apart, so that we could safely use the average of the bid and ask quotes in our analysis. Over shorter intervals, temporary skewness in bid-ask spreads

---

[1]The estimations were performed with XploRe. For a description, see Härdle, et al. [1995].

because of inventory rebalancing by market making banks become important (see Guillaume, et al. [1994]), so that the average of the bid and the ask is devoid of economic meaning. Profit opportunities are obviously higher at shorter frequencies, because of the possibility to trade against banks' inventory rebalancing. But we set out to focus on the volatility effects net of such short-term inventory redressing. Incidentally, the impact of market making on short-term autocorrelations invalidates the analysis of FX quote changes as diffusion processes, for in that case, the importance of the mean function ought to decrease with the discretization mesh. Because of profit opportunities induced by market making, the mean function becomes actually more important with reductions in the length of the sampling interval.

Hence, if $b_t$ denotes the bid at $t$ and $a_t$ the ask, the price at $t$, $p_t$, is defined to be

$$p_t = [\log b_t + \log a_t]/2.$$

We model the time series behavior of the *change* in the price, i.e.,

$$y_t = p_t - p_{t-1}.$$

At one point, however, we also report results involving the bid-ask spread itself, defined as:

$$\log a_t - \log b_t.$$

We did not, however, sample quotes over intervals of ten or twenty minutes in calendar time. There is substantial seasonality in FX data, due to seasonalities in trading intensity. Volatility, for instance, is highest during the period of the 24 hour trading day when the European markets are open. In contrast, there is little activity, and, hence, little volatility, on Sunday mornings (measured relative to GMT). Therefore, time was first deformed, after which sampling took place in this newly-defined measure of time. Effectively, time intervals were shortened during busy periods, while the real-time equivalent of ten or twenty minutes was lenghtened over periods of low activity.

We used our own time deformation, where activity is measured as the kernel fit through the sample average number of quote revisions over twenty-minute intervals of the trading week. Quotes were obviously not always available at exactly twenty-minute marks. Whenever that happened, we took the first subsequent quote. Of course, sometimes no quote change is forthcoming even beyond the next time interval (this obviously occurs often during holidays which fell on an otherwise regular trading day). The empty intervals in deformed time are then skipped in order to match consistently the intervals in real and deformed time. The skipping of intervals in deformed time because of absence of quote changes meant that of the theoretical 26,280 quote changes, we sampled only 25,434 (DEM/USD) and 25,247 (YEN/USD).

We also used data sampled in Olsen's own redefinition of time ("theta time"; see, e.g., Dacorogna, et al. [1993]), which the Olsen Research Group graciously sent us. Theirs are quotes sampled over ten minutes in redefined time. Time deformation is not only based on a simple measure of activity (as ours was), but included other conditioning variables which are not directly available from the Reuters FXFX page (which provides only a limited picture of the state of the market). When comparing the CHARN estimation results, however, no differences could be found.[2] Because the dataset has become standard, we decided to report here only the results from LPE estimation on Olsen's series in "theta time."[3]

# 4 LPE Results

LPE was implemented on the samples discussed in the previous section. We fitted first-order polynomials (i.e., linear functions) locally, using a quartic kernel, with bandwidth selected by means of crossvalidation. Figure 1 displays our estimate of $v$, the variance function, for the DEM/USD, together with 95% confidence bounds. Most surprising is the asymmetric shape, as if there were a leverage effect (albeit inverted) in FX similar to that found in stock prices, with volatility increasing after increases in the DEM/USD rate.

We suspect that this "leverage" effect is caused by the asymmetric nature of central bank reaction policies, with more uncertainty about imminent intervention after increases in the DEM/USD rate. The asymmetry in the variance function is significant. To make this clear visually, Figure 2 plots the variance function against the absolute value of lagged spot quote changes.

Figure 3 plots the estimate of the mean function for the DEM/USD. It displays substantial mean reversion. The functional relationship is close to linear.

Mean reversion is even stronger for YEN/USD; asymmetry in the variance function is less pronounced for large lagged changes in the FX quote. The plots are available on request.

While CHARN modelling is able to capture interesting short-term mean and volatility patterns, it captures only part of the persistence in absolute values and squares of FX quote changes. Table 1 documents this. It lists serial correlations of the signed change in the exchange rate quotes ($y_t$), as

---

[2] Only when differencing over longer intervals could clear seasonalities in our own dataset be discovered. These seasonalities are absent in Olsen's data in "theta time".

[3] For some estimation results on the first dataset, see Bossaerts, Härdle and Hafner [1995].

well as of absolute and squared values. In comparison, it displays the same autocorrelations for $\xi_t$, the residual in the CHARN model. This residual is white noise, and, hence, should not be autocorrelated. Whereas the CHARN model is able to capture all the serial correlation in signed FX quote changes, it fails to account for most of the autocorrelation of absolute and squared values.

We tried to accomodate this persistence by adding conditioning variables to the volatility function, namely the lagged bid-ask spread (a persistent variable as well, which theory would claim changes systematically with conditional volatility) and more lags of quote changes. This had little effect on the higher-order autocorrelation of the noise. It does indicate that the autocorrelation in conditional volatility is mostly deterministic, reducing the importance of one of our conjectures, namely that large FX quote changes which fail to generate subsequent spells of high volatility are important.

The addition of the bid-ask spread as conditioning variable did generate the expected effects. In particular, a higher bid-ask spread predicted higher conditional volatility. The effect was nonlinear, however: small increases in the bid-ask spread are associated with minimal changes in conditional volatility.[4]

We also implemented Müller, et al. [1995]'s idea of HARCH modelling, whereby the sum of FX quote changes over several lags is used as conditioning variable in the variance function. Unlike Müller, et al. [1995], we only went up to twenty-four lags (four hours in redefined time). This failed to address satisfactorily even low-order serial correlation in absolute and squared residuals of the CHARN model.

In the LPE estimation, bandwidths were selected by means of crossvalidation. This provides optimal smoothness, in the sense that it balances bias against variance. When reducing the bandwidth, a better fit is obviously obtained. Surprisingly, however, it not only reduced first-order serial correlation in absolute and squared values of the residuals, *but also higher-order autocorrelations*. Table 1 illustrates this. The bandwidth was set equal to 0.0001 (about 1/60th of the optimal bandwidth).

In an attempt to discover the cause of this effect, we plotted the estimated volatility function for the reduced bandwidths. The estimates were extremely erratic, indicating that the improved higher-order serial correlation of the residuals was generated by mixing the residual of the original model with a random variable whose distribution depends on the lagged value of the FX quote change. Formally, the resulting model can be written in terms of the original CHARN model, as follows:

$$y_t = f(y_{t-1}) + s(y_{t-1})\eta_t\xi_t, \tag{7}$$

where $\eta_t$ is a mixing variable with mean one and variance depending on $y_{t-1}$. Of course, $s(y_{t-1})$ and

---

[4]See Bossaerts, Härdle and Hafner [1995] for estimation results.

$\eta_t$ could be merged to one random variable, thus obtaining a stochastic volatility model (see, e.g., Mahieu and Schotman [1994]).

The main effect of the use of a mixing variable is that it introduces explicitly *conditional kurtosis*. Conditional kurtosis is also present in the original model (Equation (1)), provided the unconditional distribution of $\xi_t$ is not normal. But it changes as a function of $y_{t-1}$ only indirectly, through the effect on the conditional volatility. Equation (7) has the potential of disentangling the impact of $y_{t-1}$ onto future volatility and future kurtosis.

There is evidence in the data that the impact of lagged FX quote changes on future volatility and kurtosis are different. In particular, they are inverted. Whereas higher conditional volatility is associated with large changes in exchange rate quotes (see Figure 1), conditional kurtosis is higher for small FX quote changes. Figures 4 and 5 show this. They display plots of conditional kurtosis as a function of $y_{t-1}$. The plots were generated as follows. FX quote changes are allocated to "bins". The bins are formed after sorting the data with respect to the lagged FX quote change. Each bin contained 100 observations, in ascending order. The sample kurtosis was computed per bin, and plotted in Figures 4 and 5 against the sample mean lagged FX quote change. One should smooth the estimates across bins, but we wanted to provide the reader with the raw results. No matter how one smooths, the effect would be the same: conditional kurtosis is far higher for small changes in exchange rates.[5]

# 5 Conclusion

We presented results from local polynomial estimation of the mean and volatility function in a conditionally heteroscedastic nonlinear autoregressive (CHARN) model of foreign exchange quote changes. The data were sampled over intervals of time that were deformed to remove seasonalities. To estimate the mean and volatility functions, linear functions were fit locally by least squares, using a kernel function to determine the weights to be put on each observation. The resulting estimates show clear (i) mean reversion, (ii) nonlinearity and asymmetry in conditional volatility.

The biggest challenge to the model came from the high persistence in absolute and squared residuals. In part, this appeared to be the result of conditional kurtosis that could not be captured by changes in the volatility parameter. It appears that parametric modelling could be improved by dis-

---

[5]The estimates of conditional kurtosis are valid only if fourth moments really exist. Incidentally, the same applies to the estimates of the serial correlation of squared exchange rate changes. For a critical view on this, see Dacorogna, et al. [1994].

entangling conditional volatility and conditional kurtosis. Stochastic volatility models may be ideal to attain this goal.

# References

Aït-Sahalia, Y., 1994, "A Specification Test For Continuous-Time Stochastic Processes," University of Chicago Graduate School of Business Working Paper.

Baillie, R.T. and T. Bollerslev, 1990, "Intra-Day and Inter-Market Volatility in Foreign Exchange Rates," *Review of Economic Studies* 58, 565-85.

Bewley, R., P. Lowe and R. Trevor, 1988, "Exchange Rate Changes: Are They Distributed As Stochastic Mixtures of Normals?" University of New South Wales Discussion Paper.

Bossaerts, P., W. Härdle and C. Hafner, 1995, "A New Method For Volatility Estimation With Applications In Foreign Exchange Rate Series," *Proceedings of the 5. Karlsruher Ökonometrie-Workshop*, Universität Karlsruhe, forthcoming.

Bossaerts, P. and P. Hillion, 1991, "Market Microstructure Effects of Government Intervention in the Foreign Exchange Market," *Review of Financial Studies* 4, 513-541.

Bossaerts, P. and P. Hillion, 1995, "Local Parametric Analysis of Hedging in Discrete Time," *Journal of Econometrics*, forthcoming.

Christie, A., 1982, "The Stochastic Behavior of Common Stock Variances: Value, Leverage and Interest Rate Effects," *Journal of Financial Economics* 10, 407-432.

Cleveland, W., 1979, "Robust Locally Weighted Regression and Smoothing Scatterplots," *Journal of the American Statistical Association* 74, 823-36.

Dacorogna, M.M., O. Pictet, U.A. Müller and C.G. de Vries, 1994, "The Distribution of Extremal Foreign Exchange Rate Returns in Extermely Large Datasets," Olsen & Associates, working paper.

Dacorogna, M.M., U.A. Müller, R.J. Nagler, R.B. Olsen and O.V. Pictet, 1993, "A Geographical Model for the Daily and Weekly Seasonal Volatility in the FX Market," *Journal of International Money and Finance* 12, 413-438.

Duffie, D. and K. Singleton, 1993, "Simulated Moments Estimation of Markov Models of Asset Prices," *Econometrica* 61, 929-952.

Gouriéroux, C. and A. Monfort, 1992, "Qualitative Threshold ARCH Models," *Journal of Econometrics* 52, 159-99.

Guillaume, D.M., M.M. Dacorogna, R.R. Davé, U.A. Müller, R.B. Olsen and O.V. Pictet, 1994, "From the Bird's Eye to the Microscope: A Survey of New Stylized Facts of the Intra-Daily Foreign Exchange Market," Olsen Associates working paper.

Györfi, L., W. Härdle, P. Sorda and P. Vieu, 1989, *Nonparametric Curve Estimation From Time Series*, New York: Springer Verlag.

Härdle, W., S. Klinke and B. Turlach, 1995, "XploRe – an Interactive Statistical Computing Environment", New York: Springer Verlag.

Härdle, W. and A. Tsybakov, 1995, "Local Polynomial Estimation of the Volatility Function," *Journal of Econometrics*, forthcoming.

Härdle, W. and E. Mammen, 1993, "Comparing Nonparametric Versus Parametric Regression Fits," *Annals of Statistics* 21, 1926-47.

Karatzas, I. and S. Shreve, 1983, *Brownian Motion and Stochastic Calculus*, New York: Springer Verlag.

Lo, A. and J. Wang, 1994, "Implementing Option Pricing Models When Asset Returns Are Predictable," MIT Sloan School Discussion Paper.

Mahieu, R. and P. Schotman, 1994, "Stochastic Volatility and the Distribution of Exchange Rate News," LIFE Discussion Paper.

Müller, U.A., M.M. Dacorogna, R.D. Davé, R.B. Olsen, O.V. Pictet and J.E. von Weizsäcker, 1995, "Volatilities of Different Time Resolutions - Analyzing the Dynamics of Market Components," Olsen & Associates Research Group preprint.

Neumann, M.J.M., 1984, "Intervention in the mark/dollar market: The authorities' reaction function," *Journal of International Money and Finance* 3, 233-39.

Pagan, A. and A. Ullah, 1988 "The Econometric Analysis of Models with Risk Terms," *Journal of Applied Econometrics* 3, 87-105.

Pagan, A. and G.W. Schwert, 1990 "Alternative Models for Conditional Stock Volatility," *Journal of Econometrics* 45, 267-290.

Stone, C., 1977, "Consistent Nonparametric Regression," *Annals of Statistics* 5, 595-645.

Tsybakov, A., 1986, "Robust Reconstruction of Functions by the Local-Approximation Method," *Problems of Information Transmission* 22, 133-46.

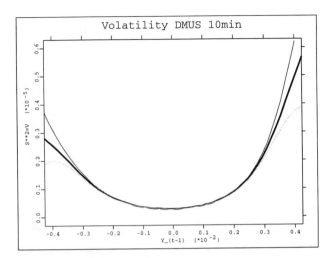

Figure 1: Estimated conditional variance function with 95% confidence bands in a conditionally het-eroscedastic nonlinearly autoregressive model of changes in DEM/USD quotes over ten-minute inter-vals in deformed time during the period 1 Oct 92/30 Sep 93. The estimates were obtained by locally fitting linear functions using a quartic kernel. Crossvalidation determined the bandwidth size.

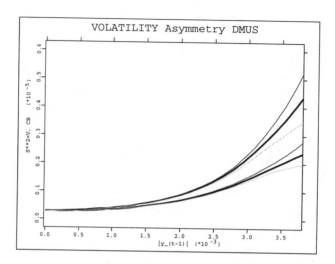

Figure 2: Estimated conditional variance function against absolute values of its argument, with 95% confidence bands in a conditionally heteroscedastic nonlinearly autoregressive model of changes in DEM/USD quotes over ten-minute intervals in deformed time during the period 1 Oct 92/30 Sep 93. The estimates were obtained by locally fitting linear functions using a quartic kernel. Crossvalidation determined the bandwidth size.

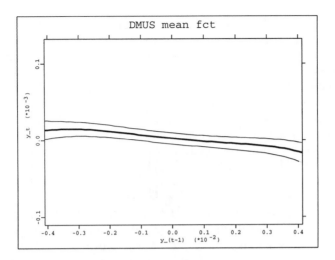

Figure 3: Estimated conditional mean function with 95% confidence bands in a conditionally heteroscedastic nonlinearly autoregressive model of changes in DEM/USD quotes over ten-minute intervals in deformed time during the period 1 Oct 92/30 Sep 93. The estimates were obtained by locally fitting linear functions using a quartic kernel. Crossvalidation determined the bandwidth size.

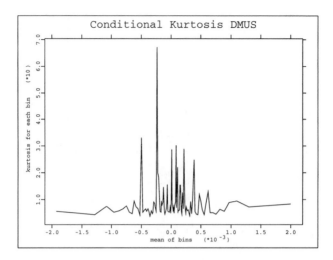

Figure 4: Estimated conditional kurtosis of changes in DEM/USD quotes over ten-minute intervals in deformed time during the period 1 Oct 92/30 Sep 93. The estimates were obtained by binning the data and computing the sample kurtosis per bin.

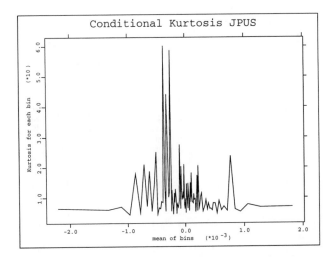

Figure 5: Estimated conditional kurtosis of changes in YEN/USD quotes over ten-minute intervals in deformed time during the period 1 Oct 92/30 Sep 93. The estimates were obtained by binning the data and computing the sample kurtosis per bin.

**Table 1**
**Serial correlations of FX quote changes**
**and residuals of the CHARN model**

| | order | FX quote changes | | | CHARN residuals (optimal bandwidth) | | | CHARN residuals (low bandwidth) | | |
|---|---|---|---|---|---|---|---|---|---|---|
| | | $y_t$ | $\|y_t\|$ | $(y_t)^2$ | $\xi_t$ | $\|\xi_t\|$ | $(\xi_t)^2$ | $\xi_t$ | $\|\xi_t\|$ | $(\xi_t)^2$ |
| DEM/USD | | | | | | | | | | |
| | 1 | -.007 | .220 | .204 | -.001 | .121 | .043 | -.003 | .005 | -.004 |
| | 2 | -.022 | .173 | .119 | -.015 | .145 | .075 | -.011 | .113 | .051 |
| | 3 | -.028 | .151 | .089 | -.025 | .132 | .075 | -.021 | .102 | .055 |
| | 6 | .003 | .121 | .071 | .001 | .111 | .068 | .002 | .088 | .049 |
| | 12 | -.007 | .107 | .059 | -.008 | .094 | .046 | -.008 | .079 | .038 |
| | 24 | .001 | .080 | .033 | .003 | .072 | .029 | .005 | .064 | .032 |
| | 48 | -.002 | .057 | .021 | -.002 | .054 | .024 | -.003 | .045 | .019 |
| YEN/USD | | | | | | | | | | |
| | 1 | -.062 | .295 | .238 | -.001 | .099 | .020 | -.001 | .016 | .007 |
| | 2 | -.010 | .206 | .146 | -.011 | .148 | .069 | -.010 | .115 | .050 |
| | 3 | -.015 | .164 | .074 | -.018 | .127 | .048 | -.014 | .105 | .042 |
| | 6 | -.006 | .133 | .061 | -.007 | .105 | .045 | -.008 | .085 | .039 |
| | 12 | .004 | .118 | .047 | -.000 | .097 | .037 | -.008 | .079 | .026 |
| | 24 | -.010 | .111 | .058 | -.008 | .091 | .044 | -.009 | .077 | .035 |
| | 48 | .004 | .084 | .049 | -.005 | .068 | .029 | .001 | .053 | .022 |

# An analysis of an ordinal-valued time series

**David R. Brillinger**

Department of Statistics, University of California, Berkeley, CA 94720

## 1 Abstract.

Time series and spatial processes are sometimes ordinal-valued. It can be convenient to handle such types of data via generalized linear model algorithms employing the complimentary *loglog* link function. This approach facilitates the use of standard statistical packages and leads to a convenient technique for handling serial dependence. Model fit is assessed by uniform residuals, amongst other tools. In this article an example of such an analysis is provided for a three-valued series corresponding to the possible results *loss, tie, win* of events involving a sports team.

> *"... - it is important to have in command the mathematics so you can solve the problem. Of course, the 64 dollar question is which mathematics to learn, because you can't learn all of it."*

> E.J. Hannan interviewed in [23]

## 2 Preamble.

Throughout my whole professional career Ted Hannan was there as a role model. The ever growing stack of his collected works was a constant research companion. In particular he was special for working on problems simultaneously from all sides: substantive, theoretical and computational. He always kept up with, indeed typically led, contemporary developments in time series. He has left us too soon, but his standards remain.

## 3 Introduction.

Ordinal data refers to quantities whose values are categories falling on a scale such that the order of the categories matters and is known. A characteristic is that adjacent categories may be sensibly merged with the ordinality remaining. One general reference is [21], Chapter 5. In the time

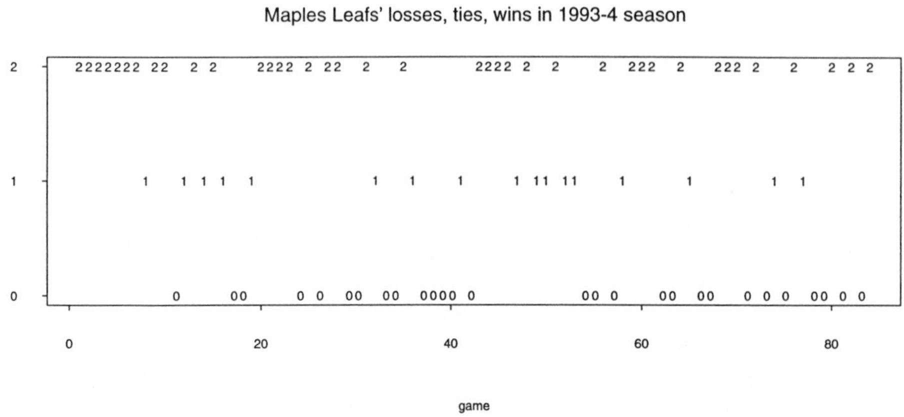

74

Maples Leafs' losses, ties, wins in 1993-4 season

FIGURE 1. Results of 84 games: 0, 1, 2 refer to loss, tie, win respectively.

series case the individual values are ordinal categories and questions of interest include: Is there serial dependence? Is a trend present? Are there useful explanatories? Is there change?

This work is concerned with a segment $Y(t)$, $t = 0, ..., T-1$ of an ordinal-valued time series taking values such that

$$Y(t) = 0, 1, \text{ or } 2$$

corresponding to the results *loss*, *tie*, *win* of the Toronto Maple Leafs Hockey team during the 1993-1994 season. (In assigning points in the standings 0, 1, 2 actually represent the points awarded.)

Figure 1 provides a graph of the results. There were 84 regular season games in all. The Toronto team began the season with a record setting string of 10 wins. In order that the results be more homogeneous for the analyses presented, the data graphed and analysed actually correspond to the state of the game after regulation time. (If the game is tied at the end of regulation time, an overtime is period is played, which may result in a win for one of the teams.) By this count Toronto had 28 losses, 17 ties and 39 wins in the course of the season.

Smooth 'trend' - classic wins

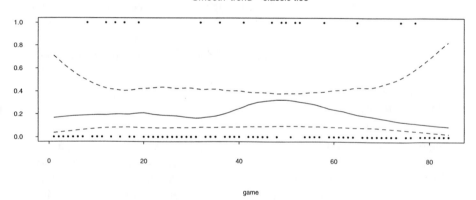

game

Smooth 'trend' - classic ties

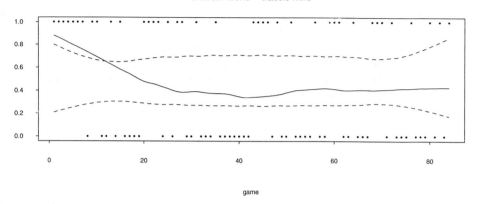

game

FIGURE 2. Smoothed rate for wins and ties respectively with uncertainty limits.

Figure 2 provides smoothed estimates of the probability of a classic (i.e. after regulation time) win and of a classic tie respectively. The approximate ±2 standard error limits are computed as if the successive games are statistically independent. Except for the early success, the win curve fluctuates about a constant mean level. In the case of the ties the curve fluctuates about the mean level throughout. These curves were produced employing the *cloglog* link and the functions *gam*() and *predict.gam*() of the statistical package $S$ (see[2, 8]).

The data are provided in an Appendix.

# 4   Ordinal Data.

A number of different models have been proposed for the analysis of ordinal data. These include: continuation ratio (see [12]), stereotype (see [1]) and the grouped continuous (see [20]).

The following presents an approach to building a stochastic model for ordinal data. Let $Y$ be 0, 1, 2 for a particular game, depending on whether the result is a *loss*, *tie* or *win*. Suppose that there exists a latent or state variable, $\Lambda$, whose value represents the strength of the Toronto team against a general opponent. Assume the existence of cutpoints $a$ and $b$ such that

$$Y = 0 \ if \ \Lambda < a, \quad Y = 1 \ if \ a < \Lambda < b \ and \quad Y = 2 \ if \ b < \Lambda$$

So for example

$$Prob\{Y = 1\} = F_\Lambda(b) - F_\Lambda(a) \tag{0.1}$$

where $F_\Lambda$ is the c.d.f. of $\Lambda$. Figure 3 presents an example of a graph of a possible density function for $\Lambda$ with the regions of *loss*, *tie*, *win* indicated. In the graph the term linear predictor refers to $\Lambda$. The approach involving a latent variable has the advantages of: easy interpretability, clear possibilities of merging adjacent categories and of flexibility.

Maximum likelihood is a natural method of estimating unknown parameters in many cases and will be employed in the present work. Goodness of fit may be assessed by procedures such as: deviance and chi-squared type statistics (see [21]), plots of estimated probability against the linear predictor ([5, 6]) or "uniform residuals" ([4]).

In a generalized linear model, the link function describes the relation between the mean of the basic variate and the natural parameter of its distribution. Its choice is sensibly based on the subject matter of the problem. The complimentary *loglog* corresponds to situations in which of an extremal variate crosses a threshold and an extreme value distribution, ([25]). In the present context this may be reasonable, with a win for the hockey team resulting from the team members putting out maximum efforts to exceed those of the opponent.

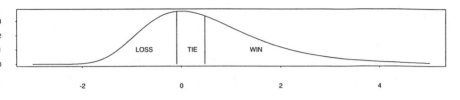

FIGURE 3. Areas of regions refer to probabilities of the respective events.

The extreme value distribution of the first type is given by

$$Prob\{\Lambda > \lambda\} = exp\{-e^\lambda\} \quad for \ \lambda > 0$$

The graph of Figure 3 is based on this distribution. One can write

$$log(-log(1 - Prob\{\Lambda \leq \lambda\})) = \lambda$$

and sees the appearance of the *cloglog* link. In the case of ordinal-valued $Y$ one writes

$$log(-log(1 - Prob\{Y \leq j\} = \theta_j$$

with $\theta_j > \theta_{j-1}$ and

$$log(-log(1 - Prob\{Y = j \mid Y \geq j\})) = \alpha_j \qquad (0.2)$$

for $j = 0, 1, 2$. Pregibon, [24], noted the fact that for the *cloglog* link the parametrization was of the same form and hence, by writing a probability as a product of conditional probabilities, one could employ standard statistical packages in analyses of such multinomial data. See also [17]. One can work with $Prob\{win\}$ and $Prob\{tie|not \ win\}$ in the present hockey game case.

Explanatory variables, $x$, may be introduced quite directly by writing

$$\Lambda = E + \beta'x$$

where E has the extreme value distribution. Now (0.1) becomes

$$F_E(b - \beta'x) - F_E(a - \beta'x)$$

# 5 The Time Series Case.

There is a massive literature concerning time series, that is sequences $Y(t)$, $t = 0, \pm 1, \pm 2, \ldots$ which are stochastic. The literature mainly refers to real-valued $Y$, some of it refers to count-valued ([3, 14, 18, 28]). What distiguishes the present circumstance are the values that $Y$ can take on. In this work the values correspond to ordinal categories. In the case of two categories the series are binary and there is a large existing literature ([5, 9]). There is further a literature for extensions to the case of the generalized linear models ([11, 10, 15, 16, 26, 27]). There are also approaches to categorical-valued time series based on Markov chains and on state space descriptions ([11, 10]). A distinction that arises in the literature concerns whether one realization of the time series is involved or several. The latter case is typically referred to as longitudinal data analysis ([7, 19, 22]).

Both parametric and nonparametric models can be considered. A direct parametric way to introduce temporal dependence is to set up an autoregressive-type model with past values of the series being employed as explanatories. In likelihood approaches it is then convenient to set up a likelihood as the product of a sequence of conditional mass or density functions, $f_Y$,

$$\prod_{t=0}^{T-1} f_{Y(t)}(y_t | H_{t-1})$$

with $H_t$ denoting the history $\{y_0, \ldots, y_t\}$. Taking this result together with the simplification resulting from the use of the complimentary *loglog* function, referred to in Section 4, means that parametric analyses can be carried out using standard functions such as *glm*() of S, [8]. The appearance of the conditional term (0.2) may be controlled by the use of the weight option.

# 6 Results

The graphs of Figure 1 may be considered a first-order analysis of the question of temporal dependence. What may be seen is a small indication of an increased probability of a win for the Toronto team at the beginning of the season. It is of further interest whether there is some clustering of the *losses*, *ties* or *wins* or if these perhaps alternate in some fashion.

A nonparametric second-order analysis may be developed by creating a bivariate time series. Define the two binary series $Y_1$ and $Y_2$ with $Y_1(t) = 1$ if the $t$-th game is a *win* and 0 otherwise, similarly define $Y_2(t) = 1$ if the game is a *tie* and 0 otherwise. To begin consider a frequency domain approach, the one so often taken by Ted Hannan [13]. In the case of a bivariate stationary white noise process, each of the second-order spectra are constant and the quadrature spectrum is identically 0. Figure 4 provides

Periodogram - classic wins

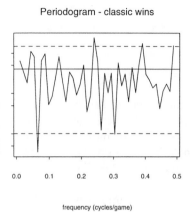

frequency (cycles/game)

Periodogram - classic ties

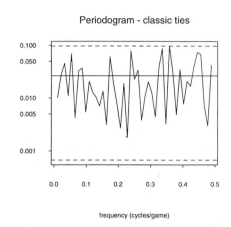

frequency (cycles/game)

Re(crossperiodogram)

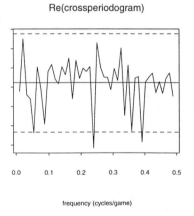

frequency (cycles/game)

Im(crossperiodogram)

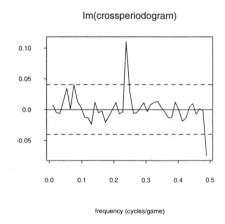

frequency (cycles/game)

FIGURE 4. Second-order periodograms of the data.

the periodograms and cross-periodograms of the data for the series $Y_1$ and $Y_2$. The solid lines are the estimated levels in the case that the successive observations are i.i.d. The dashed lines are approximate 95% marginal confidence limits. There is one unusual point in the crossperiodogram, but no substantial evidence for temporal dependence.

One type of parametric analysis involves fitting a process of autoregressive type. As an example consider the model

$$log(-log(1 - Prob\{Y(t) = j|H_{t-1}\})) = \theta_j + \phi_j \cdot y_{t-1} \qquad (0.3)$$

with the $\phi \cdot y$ term having the meaning that the value, $y_{t-1}$, of the series at the previous time point is to be viewed as a factor. The deviance change in fitting the model 0.3 with and without this term is 3.03 on 4 degrees of freedom with a corresponding probvalue of .553. There is no evidence for the postulated form of dependency on the previous time value. Earlier time values may be studied just as easily.

Various other explanatories may be considered, for example whether the game is home or away, goals scored and some measure of the strength of the opposing team. In the case of including whether the game was home or away, as an explanatory factor, the deviance change is only .012 on 1 degree of freedom. The corresponding probvalue is .911 . Again there is no evidence of an effect.

# 7   Goodness of Fit.

In any work with stochastic models, goodness of fit is a central issue. In work with generalized linear models the residual deviance is often employed; however its approximation by a chisquared variate in the null case is often poor. In [4] the idea of employing uniform residuals was introduced. One uses the probability integral transformation based on the fittted model. In the case that the parameter values are known, this will have a uniform distribution. These residuals may be plotted against explanatories, be used to construct probability plots and other such things.

The present approach acts as if the data are binary. Suppose that $X$ is a Bernoulli variate with $Prob\{X = 1\} = \pi$. Then a standard uniform variate, $V$, may be constructed by setting

$$V = uniform \; on \; (1 - \pi, 1) \; if \; X = 1$$

$$V = uniform \; on \; (0, 1 - \pi) \; if \; X = 0$$

This was done for the simplest model (of the $Y$ i.i.d.) and the observed data, based on the estimates of $Prob\{win\}$ and $Prob\{tie|not \; win\}$. Figure 5 gives plots of the $V$'s against game for the wins and conditional ties. In

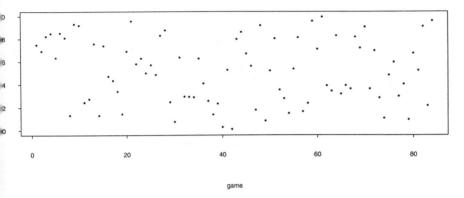

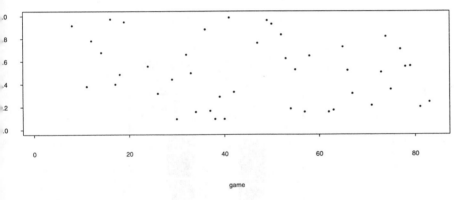

FIGURE 5. Variates created to be approximately standard uniform if the model holds.

## Wins and conditional ties

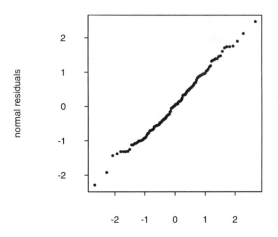

Quantiles of Standard Normal

## Uniform residuals vs. site

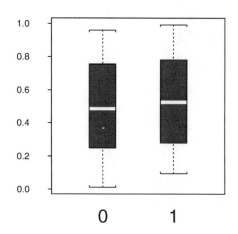

home - away

FIGURE 6. In top figure the uniform variates have been transformed to normals. In the lower boxplots of uniform residuals are plotted in the home and away cases.

the first case one sees some elevated values at the beginning, but randomness therafter. In the second case there is apparent randomness. Figure 6 gives a normal probability plot and a plot against the home-away variate respectively. There is no evidence to contradict the assumptions of the fitted model.

# 8   Summary.

The 1993-94 Toronto team began the season with a string of successes; however ultimately the results of the various games appear random. The analyses provide no real evidence for temporal dependence in rate or serial correlation. If temporal dependence had been noted there would have been the possibility of using the model for prediction.

# 9   Acknowledgements

This work was carried out with the partial support of the National Science Foundation Grant DMS-9300002 and the Office of Naval Research Grant N00014-94-1-0042. Mark Rizzardi made some helpful remarks on the manuscript.

# 10   Appendix

The site variable refers to whether the game was home or away, 0 is away. The overtime variable refers to whether the game ended in regulation time,

0 means it did.

| Goals for | Goals against | site | overtime |
|:---:|:---:|:---:|:---:|
| 6 | 3 | 0 | 0 |
| 2 | 1 | 0 | 0 |
| 5 | 4 | 1 | 0 |
| 7 | 1 | 0 | 0 |
| 6 | 3 | 0 | 0 |
| 2 | 1 | 1 | 0 |
| 7 | 2 | 0 | 0 |
| 4 | 3 | 1 | 1 |
| 2 | 0 | 1 | 0 |
| 4 | 2 | 1 | 0 |
| 2 | 5 | 1 | 0 |
| 3 | 3 | 1 | 1 |
| 6 | 3 | 0 | 0 |
| 3 | 3 | 1 | 1 |
| 5 | 3 | 0 | 0 |
| 2 | 2 | 1 | 1 |
| 2 | 3 | 1 | 0 |
| 2 | 3 | 0 | 0 |
| 5 | 5 | 0 | 1 |
| 4 | 3 | 1 | 0 |
| 3 | 2 | 1 | 0 |
| 3 | 2 | 1 | 0 |
| 5 | 2 | 1 | 0 |
| 3 | 5 | 1 | 0 |
| 4 | 2 | 0 | 0 |
| 0 | 3 | 0 | 0 |
| 4 | 2 | 0 | 0 |
| 5 | 4 | 1 | 0 |
| 3 | 4 | 0 | 0 |
| 4 | 5 | 0 | 0 |
| 3 | 1 | 0 | 0 |
| 3 | 3 | 1 | 1 |
| 0 | 1 | 0 | 0 |
| 2 | 6 | 1 | 0 |
| 4 | 1 | 0 | 0 |
| 2 | 2 | 0 | 1 |
| 2 | 3 | 1 | 0 |
| 2 | 5 | 1 | 0 |
| 0 | 4 | 1 | 0 |
| 4 | 7 | 0 | 0 |
| 3 | 3 | 1 | 1 |
| 0 | 1 | 0 | 0 |

| Goals for | Goals against | site | overtime |
|:---:|:---:|:---:|:---:|
| 6 | 3 | 0 | 0 |
| 5 | 3 | 0 | 0 |
| 3 | 0 | 1 | 0 |
| 2 | 1 | 1 | 0 |
| 4 | 3 | 0 | 1 |
| 5 | 1 | 1 | 0 |
| 3 | 3 | 0 | 1 |
| 3 | 3 | 1 | 1 |
| 4 | 3 | 0 | 0 |
| 4 | 4 | 0 | 1 |
| 4 | 4 | 1 | 1 |
| 3 | 4 | 0 | 0 |
| 1 | 2 | 0 | 0 |
| 3 | 1 | 1 | 0 |
| 2 | 3 | 1 | 0 |
| 5 | 4 | 0 | 1 |
| 2 | 1 | 0 | 0 |
| 3 | 2 | 0 | 0 |
| 6 | 4 | 1 | 0 |
| 3 | 6 | 1 | 0 |
| 0 | 3 | 0 | 0 |
| 4 | 1 | 1 | 0 |
| 6 | 5 | 1 | 1 |
| 1 | 4 | 1 | 0 |
| 2 | 3 | 0 | 0 |
| 4 | 2 | 0 | 0 |
| 4 | 2 | 1 | 0 |
| 3 | 1 | 0 | 0 |
| 1 | 4 | 0 | 0 |
| 4 | 2 | 0 | 0 |
| 3 | 6 | 0 | 0 |
| 1 | 1 | 1 | 1 |
| 1 | 2 | 0 | 0 |
| 6 | 3 | 0 | 0 |
| 2 | 3 | 1 | 1 |
| 3 | 5 | 1 | 0 |
| 1 | 3 | 1 | 0 |
| 6 | 4 | 1 | 0 |
| 3 | 5 | 1 | 0 |
| 7 | 0 | 0 | 0 |
| 3 | 4 | 0 | 0 |
| 6 | 4 | 1 | 0 |

# 11 References

[1] J.A. Anderson. Regression and ordered categorical variates. *J. Royal Statist. Soc. B*, 46:19–35, 1984.

[2] R.A. Becker, J.M. Chambers, and A.R. Wilks. *The New S Language.* Wadsworth, Pacific Grove, 1988.

[3] D.R. Brillinger. The natural variability of vital rates and associated statistics. *Biometrics*, 42:693–734, 1986.

[4] D.R. Brillinger and H.K. Preisler. Maximum likelihood estimation in a latent variable problem. In S. Karlin et al., editor, *Studies in Econometrics, Time Series and Multivariate Statistics*, pages 31–65, New York, 1983. Academic.

[5] D.R. Brillinger and J.P. Segundo. Empirical examination of the threshold model of neuron firing. *Biol. Cybernetics*, 35:213–220, 1979.

[6] D.R. Brillinger, A. Udias, and B.A. Bolt. A probability model for regional focal mechanism solutions. *Bull. Seismol. Soc. Amer.*, 70:149–170, 1980.

[7] G.J. Carr, K.B. Hafner, and G.G. Koch. Analysis of rank measures of association for ordinal data from longitudinal studies. *J. Amer. Statist. Assoc.*, 84:797–804, 1989.

[8] J.M. Chambers and T.J. Hastie. *Statistical Models in S.* Wadsworth, Pacific Grove, 1992.

[9] D.R. Cox. *Analysis of Binary Data.* Methuen, London, 1970.

[10] L. Fahrmeir. State space modelling and conditional mode estimation for categorical time series. In D.R. Brillinger et al., editor, *New Directions in Time Series*, pages 87–110, New York, 1992. Springer.

[11] L. Fahrmeir and G. Tutz. *Multivariate Statistical Modelling Based on Generalized Linear Models.* Springer-Verlag, New York, 1994.

[12] S.E. Fienberg. *The Analysis of Cross-Classified Categorical Data.* MIT Press, Cambridge, Mass., 1980.

[13] E.J. Hannan. *Multiple Time Series.* Wiley, New York, 1970.

[14] A.C. Harvey. *Forecasting, Structural Time Series Models and the Kalman Filter.* Cambridge Press, Cambridge, 1989.

[15] H. Kaufmann. Regression models for nonstationary categorical time series: asymptotic estimation theory. *Ann. Statist.*, 15:79–98, 1987.

[16] G. Kitagawa and W. Gersch. *Smoothness Priors Analysis of Time Series.* Springer Lecture Notes, 1995.

[17] E. Laara and J.N.S. Matthews. The equivalence of two models for ordinal data. *Biometrika*, 72:206–207, 1985.

[18] A. Latour. Existence and stochastic structure of a non-negative integer-valued autoregressive process. *Preprint*, 1995.

[19] K.Y. Liang and S. Zeger. Longitudinal data analysis using generalized linear models. *Biometrika*, 73:13–22, 1986.

[20] P. McCullagh. Regression model for ordinal data. *J. Roy. Statist. Soc. B*, 42:109–127, 1980.

[21] P. McCullagh and J.A. Nelder. *Generalized Linear Models, Second Edition.* Chapman and Hall, New York, 1989.

[22] R.D. Murison. Analysis of repeated measure of ordinal data. *Proc. Centre Math. Applications, Australian National University*, 28:109–118, 1991.

[23] A. Pagan. The ET interview: Professor E.J. Hannan. *Econometric Theory*, 1:263–289, 1985.

[24] D. Pregibon. Discussion of paper by P. McCullagh. *J. Royal Statist. Soc. B*, 42:139–139, 1980.

[25] H.K. Preisler. Analysis of a toxicological experiment using a generalized linear model with nested random effects. *Int. Statist. Review*, 57:145–159, 1989.

[26] M. West, P.J. Harrison, and H.S. Mignon. Dynamic generalized linear models and bayesian forecasting. *J. Amer. Statist. Assoc.*, 80:73–97, 1985.

[27] M. West, P.J. Harrison, and H.S. Migon. *Bayesian Forecasting and Dynamic Models.* Springer, New York, 1989.

[28] S.L. Zeger. A regression model for time series of counts. *Biometrika*, 75:621–629, 1988.

# On the Use of Continuous-time ARMA Models in Time Series Analysis

Peter J. Brockwell

Department of Statistics and Operations Research

Royal Melbourne Institute of Technology

ABSTRACT We review some applications of continuous-time ARMA processes in the modelling of time series observed at discrete times $t_1, \ldots, t_N$. The problem of finding a continuous-time ARMA process in which a given discrete-time ARMA process can be "embedded" is discussed and some of the properties of a family of continuous-time analogues of Tong's SETARMA models are considered. An approach to inference for such models is described and illustrated with reference to a variety of data sets.

## 1   Introduction

It is a great pleasure to participate in this meeting to honour the work of Joe Gani and Ted Hannan. They have had a profound effect not only on the fields of stochastic processes, time series and applied probability in general, but also on the many students, colleagues and friends fortunate enough to have benefitted from their work and guidance.

The purpose of this paper is to review some aspects of continuous-time ARMA models and particularly their use in modelling observations $Y(t_1), \ldots, Y(t_N)$ made at discrete times $t_1, t_2, \ldots, t_N$. Many such data sets *are* in fact observations of a continuously evolving process in which case the use of a continuous-time model is quite natural. Even when this is not the case however, continuous-time models provide a very convenient framework within which to analyze data observed at irregularly-spaced times (Jones(1981, 1985), Jones and Ackerson (1990)). Moreover certain non-linear models, in particular continuous-time threshold ARMA models, can be characterized in more explicit terms than the analogous discrete-time SETARMA (self-exciting threshold ARMA) processes of Tong(1983, 1990). The purpose of considering such models is primarily to obtain better prediction than is attainable with linear models. A further useful feature of non-linear models is the dependence on past history of the mean squared error of the minimum-mean-squared-error predictors (Pemberton (1989)). This means that it is possible to decide, on the basis of past observations, when prediction errors can be expected to be large or small.

In Section 2 we recall the definition of (linear) continuous-time ARMA (hereafter denoted CARMA) processes. Section 3 discusses the embedding problem, namely, given a discrete-time ARMA$(p, q)$ process, $\{X_t, t =$

$0, 1, 2, \ldots\}$, under what conditions can we find a CARMA process $\{Y(t), t \geq 0\}$ whose autocorrelation function at integer lags coincides with that of $\{X_t\}$? Given that such a process exists, how can we determine its defining parameters? By means of a counterexample we show that these problems have not yet been solved and indicate possible approaches to their solution. In Section 4 we define the continuous-time threshold ARMA (CTARMA) process and summarize results pertaining to its existence and uniqueness. Section 5 describes an approach to inference for CTARMA processes and Section 6 contains some illustrative examples.

## 2   The CARMA$(p, q)$ Process

A CARMA$(p, q)$ process (with $0 \leq q < p$) is defined to be a stationary solution of the $p$-th order linear differential equation,

$$Y^{(p)}(t) + a_1 Y^{(p-1)}(t) + \cdots + a_p Y(t) =$$

$$b_0 W^{(1)}(t) + b_1 W^{(2)}(t) + \cdots + b_q W^{(q+1)}(t) + c, \ t \geq 0, \qquad (2.1)$$

where the superscript $^{(j)}$ denotes $j$-fold differentiation with respect to $t$, $\{W(t)\}$ is standard Brownian motion and $a_1, \ldots, a_p, b_0, \ldots, b_q$ and $c$ are constants. It is assumed that $a_p \neq 0$ and $b_j \neq 0$ for some $j \leq q$, and we define $b_j := 0$ for $j > q$. Since the derivatives $W^{(j)}(t), j > 0$ do not exist in the usual sense, we interpret (2.1) as being equivalent to the *observation* and *state* equations,

$$Y(t) = \mathbf{b}' \mathbf{X}(t), \qquad (2.2)$$

and

$$d\mathbf{X}(t) = A\mathbf{X}(t)dt + \mathbf{e}(c\, dt + dW(t)), \ t \geq 0, \qquad (2.3)$$

where

$$A = \begin{bmatrix} 0 & 1 & 0 & \cdots & 0 \\ 0 & 0 & 1 & \cdots & 0 \\ \vdots & \vdots & \vdots & \ddots & \vdots \\ 0 & 0 & 0 & \cdots & 1 \\ -a_p & -a_{p-1} & -a_{p-2} & \cdots & -a_1 \end{bmatrix}, \ \mathbf{e} = \begin{bmatrix} 0 \\ 0 \\ \vdots \\ 0 \\ 1 \end{bmatrix}, \ \mathbf{b} = \begin{bmatrix} b_0 \\ b_1 \\ \vdots \\ b_{p-2} \\ b_{p-1} \end{bmatrix}$$

and (2.3) is an Ito differential equation for the state vector $\mathbf{X}(t)$. (We assume also that $\mathbf{X}(0)$ is independent of $\{W(t)\}$.) The state-vector $\mathbf{X(t)}$ is in fact the vector of derivatives,

$$\mathbf{X}(t) = \begin{bmatrix} X(t) \\ X^{(1)}(t) \\ \vdots \\ X^{(p-1)}(t) \end{bmatrix}, \qquad (2.4)$$

of the continuous-time $AR(p)$ process to which $\{Y(t)\}$ reduces when $b_0 = 1$ and $b_j = 0$, $j \geq 1$.

The process $\{Y(t), t \geq 0\}$ is thus said to be a $CARMA(p,q)$ process with parameters $(a_1, \ldots, a_p, b_0, \ldots, b_q, c)$ if $Y(t) = \begin{bmatrix} b_0 & b_1 & \cdots & b_{p-1} \end{bmatrix} \mathbf{X}(t)$ where $\{\mathbf{X}(t)\}$ is a stationary solution of (2.3). The solution of (2.3) can be written as

$$\mathbf{X}(t) = e^{At}\mathbf{X}(0) + \int_0^t e^{A(t-u)}\mathbf{e}\, dW(u) + c\int_0^t e^{A(t-u)}\mathbf{e}\, du,$$

which is strictly stationary if and only if

$$\mathbf{X}(0) \sim N(a_p^{-1}c\begin{bmatrix} 1 & 0 & \cdots & 0 \end{bmatrix}', \int_0^\infty e^{Ay}\mathbf{e}\,\mathbf{e}'e^{A'y}dy)$$

and all the eigenvalues of $A$ (i.e. the roots of $z^p + a_1 z^{p-1} + \cdots + a_p = 0$) have negative real parts. Under these conditions the autocovariance function of the process $\mathbf{X}(t)$ is easily found to be

$$E[\mathbf{X}^*(t+h)\mathbf{X}^*(t)'] = e^{Ah}\Sigma, \; h \geq 0,$$

where $\mathbf{X}^*(t)$ is the mean-corrected state vector

$$\mathbf{X}^*(t) = \begin{bmatrix} X(t) - c/a_p \\ X^{(1)}(t) \\ \vdots \\ X^{(p-1)}(t) \end{bmatrix},$$

and

$$\Sigma := \int_0^\infty e^{Ay}\mathbf{e}\,\mathbf{e}'e^{A'y}dy.$$

The mean and autocovariance function of the $CARMA(p,q)$ process $\{Y(t)\}$ are therefore given by

$$EY(t) = a_p^{-1}c, \tag{2.5}$$

and

$$\mathrm{Cov}(Y(t+h), Y(t)) = \mathbf{b}'\, e^{Ah}\Sigma\, \mathbf{b}. \tag{2.6}$$

Its spectral density is

$$f_Y(\omega) = \frac{1}{2\pi} \frac{|b_0 + i\omega b_1 + \cdots + (i\omega)^q b_q|^2}{|(i\omega)^p + (i\omega)^{p-1}a_1 + \cdots + a_p|^2}, \quad -\infty < \omega < \infty. \tag{2.7}$$

For the process to be *minimum phase* the roots of $b_0 + b_1 z + \cdots + b_q z^q = 0$ must have negative real parts. (This corresponds to *invertibility* for discrete time ARMA processes.)

**Example 1** The CARMA(2,1) process with zero mean and parameters $a_1, a_2, b_0, b_1$ is a stationary solution of the stochastic differential equation,

$$Y^{(2)}(t) + a_1 Y^{(1)}(t) + a_2 Y(t) = b_0 W^{(1)}(t) + b_1 W^{(2)}(t), \ t \geq 0. \qquad (2.8)$$

In order for such a solution to exist it is necessary that the roots of the equation,

$$\lambda^2 + a_1 \lambda + a_2 = 0, \qquad (2.9)$$

have negative real parts. In the case when (2.9) has two distinct complex conjugate roots,

$$\lambda_1 = \alpha + i\beta \ \text{ and } \ \lambda_2 = \alpha - i\beta, \ \alpha < 0, \ 0 < \beta, \qquad (2.10)$$

it follows at once from (2.6) that the autocorrelation function of $\{Y(t)\}$ is

$$\rho(h) = e^{\alpha h} \left[ \cos(\beta h) + \sin(\beta h) \frac{\alpha(b_1^2 a_2 - b_0^2)}{\beta(b_1^2 a_2 + b_0^2)} \right], \ h \geq 0. \qquad (2.11)$$

To conclude this section we represent the general CARMA$(p, q)$ process with $q < p$ as the first component of a continuous-time $p$-variate AR(1) process. This can be done by introducing a new state vector, $\mathbf{Y}(t) := B\mathbf{X}(t)$, where $\{\mathbf{X}(t)\}$, as before, is the stationary solution of (2.3) and

$$B = \begin{bmatrix} b_0 & b_1 & b_2 & \cdots & b_{p-1} \\ 0 & 1 & 0 & \cdots & 0 \\ 0 & 0 & 1 & \cdots & 0 \\ \vdots & \vdots & \vdots & & 0 \\ 0 & 0 & 0 & \cdots & 1 \end{bmatrix}$$

if $b_0 \neq 0$. If $b_0 = 0$, we select a value of $j$ such that $b_j \neq 0$ and then replace the $(j+1)^{\text{th}}$ row of $B$ by $\begin{bmatrix} 1 & 0 & \cdots & 0 \end{bmatrix}$. (This replacement ensures that $B$ is non-singular.) Then $Y(t)$ is the first component of $\mathbf{Y}(t)$, i.e.

$$Y(t) = \begin{bmatrix} 1 & 0 & \cdots & 0 \end{bmatrix} \mathbf{Y}(t), \ t \geq 0, \qquad (2.12)$$

where $\mathbf{Y}(t)$ satisfies the state equation,

$$d\mathbf{Y}(t) = BAB^{-1}\mathbf{Y}(t)dt + Be(dW(t) + c \, dt), \ t \geq 0. \qquad (2.13)$$

This representation shows that if $\{Y(t), t \geq 0\}$ is a CARMA$(p, q)$ process with $q < p$, then $\{Y(n), n = 0, 1, 2, \ldots\}$ is the first component of a discrete-time $p$-variate AR(1) process, and hence, by a result of Doob (1944), $\{Y(n)\}$ is a discrete-time ARMA$(p, q')$ process with $q' < p$. This result was established by Phillips (1959) in the case when $\{Y(t)\}$ is an AR$(p)$ process. The converse question, whether or not a given discrete-time ARMA process can be embedded in a CARMA process and if so in which ones, is more difficult and appears not yet to have been resolved.

# 3 Embedding an ARMA Process in a CARMA Process

When a CARMA$(p, p-1)$ model $\{Y(t), t \geq 0\}$ is fitted to observations of a time series $\{X_t, t = 1, 2, \ldots\}$, for example by maximizing the Gaussian likelihood, we know from the previous section that $\{Y(t), t = 1, 2, \ldots\}$ is an ARMA$(p, r)$ process with $r < p$. But does there exist some other ARMA$(p, q)$ model, $q < p$, with greater Gaussian likelihood? The answer to this question would be no if every ARMA$(p, q)$ process with $q < p$ were "embeddable" in a CARMA$(p, p-1)$ process, i.e. if for every ARMA$(p, q)$ process with $q < p$ there existed a CARMA$(p, p-1)$ process whose autocorrelation function at integer lags coincided with that of the ARMA$(p, q)$ process. This raises the general question of identifying those ARMA processes which are embeddable in CARMA processes and the related question of determining the parameters of a CARMA process in which a given ARMA process can be embedded.

It is clearly impossible to embed an AR(1) process satisfying $X_t = \phi X_{t-1} + Z_t$ with $\{Z_t\} \sim \text{WN}(0, \sigma^2)$ and $-1 < \phi < 0$ in a CAR(1) process, since the autocorrelation function of $\{X_t\}$ is $\rho(h) = \phi^{|h|}$ while the autocorrelation function of the general CAR(1) process satisfying $Y^{(1)}(t) + aY(t) = b_0 W^{(1)}(t)$ is $\rho(h) = \exp(-a|h|)$, which is non-negative. Chan and Tong (1988) showed however that it is possible to embed an AR(1) process with $-1 < \phi < 0$ in a suitably chosen CARMA(2,1) process. The general embedding problem was considered by He and Wang (1989). They concluded that every ARMA$(p, q)$ process with $q < p$ can be embedded in a CARMA$(p', q')$ process with $q' < p' = p + r$ where $r$ is the number of real negative zeroes of the autoregressive polynomial, $1 - \phi_1 z - \cdots - \phi_p z^p$, of the ARMA process. This result however is not correct, as the following example (Brockwell (1995)) shows.

**Example 2** Consider the causal AR(2) process defined by

$$X_t - \phi_1 X_{t-1} - \phi_2 X_{t-2} = Z_t,$$

where $\{Z_t\}$ is white noise with mean zero and variance $\sigma^2$ and the equation $z^2 - \phi_1 z - \phi_2 z^2 = 0$ has two distinct complex conjugate roots,

$$z_1 = \exp(\alpha + i\beta) \text{ and } z_2 = \exp(\alpha - i\beta), \ \alpha < 0, \ 0 < \beta < \pi.$$

It is easy to show (see for example Brockwell and Davis (1991), p.95) that the autocorrelation function of $\{X_t\}$ is, for $h = 0, \pm 1, \pm 2, \ldots$,

$$\rho(h) = e^{\alpha h} \left[ \cos(\beta h) - \sin(\beta h) \frac{\tanh(\alpha)}{\tan(\beta)} \right], \tag{3.1}$$

Now let $\rho_\beta(h)$, $-\infty < h < \infty$, denote the extension of $\rho(\cdot)$ defined by (3.1) for all real $h$. If $\rho_\beta(\cdot)$ is an autocorrelation function on $(-\infty, \infty)$, then $\{X_t\}$

is embeddable in a continuous-time stationary process with autocorrelation function $\rho_\beta(\cdot)$. This requires non-negativity, for all real $\omega$, of

$$f_\beta(\omega) = \frac{1}{2\pi} \int_{-\infty}^{\infty} \exp(-i\omega h)\rho_\beta(h) \, dh$$

$$= \frac{(\alpha^2 + \beta^2)(-\alpha - \beta\frac{\tanh\alpha}{\tan\beta}) + \omega^2(-\alpha + \beta\frac{\tanh\alpha}{\tan\beta})}{\pi(\alpha^2 + (\beta + \omega)^2)(\alpha^2 + (\beta - \omega)^2)}. \tag{3.2}$$

This non-negativity condition is clearly equivalent to the inequalities,

$$-1 \leq \frac{\beta\tanh(\alpha)}{\alpha\tan(\beta)} \leq 1. \tag{3.3}$$

Moreover if these conditions are satisfied, then comparison of (3.2) with (2.7) shows that $f_\beta(\cdot)$ is the spectral density of a CARMA(2,1) process and hence that $\{X_t\}$ is embeddable in such a process.

Direct comparison of the autocorrelation functions (2.11) and (3.1), taking into account the constraints $a_2 > 0$ and $-\infty < b < \infty$, shows that the inequalities (3.3) are also necessary for finding a CARMA(2,1) process with the same autocorrelation function at integer lags as the process $\{X_t\}$.

The parameter values $(\alpha, \beta)$ satisfying (3.3) constitute a proper subset of $(-\infty, 0) \times (0, \pi)$, showing that not all AR(2) processes with distinct complex conjugate roots are embeddable in a CARMA(2,1) process. If the values of $\alpha$ and $\beta$ (in $(0, \pi)$) do satisfy (3.3) then any CARMA(2,1) process with autoregressive coefficients,

$$a_1 = -2\alpha$$

and

$$a_2 = \alpha^2 + \beta^2$$

and ratio $b = b_1/b_0$ satisfying

$$\frac{\alpha(1 - b^2 a_2)}{\beta(1 + b^2 a_2)} = \frac{\tanh(\alpha)}{\tan(\beta)}, \tag{3.4}$$

has the same autocorrelations at integer lags as the AR(2) process $\{X_t\}$.

If $\alpha$ and $\beta$ do not satisfy the conditions (3.3), it can be shown (by calculating $f^*(\omega) = \frac{1}{2\pi} \int \exp(-i\omega h)\rho^*(h) \, dh$) that the function,

$$\rho^*(h) = \frac{1}{2}(\rho_\beta(h) + \rho_{\pi-\beta}(h)), \quad -\infty < h < \infty,$$

is the autocorrelation function of a CARMA(4,2) process, which coincides with the autocorrelation function of $\{X_t\}$ at integer lags, provided $\alpha$ and $\beta$ fall in a subset of the parameter space which overlaps and complements the region defined by (3.3).

Thus every AR(2) process with distinct complex conjugate autoregressive roots can be embedded in either a CARMA(2,1) process or a CARMA(4,2) process, or in some cases both.

## The General Case

The general causal discrete-time ARMA$(p, q)$ process with $q < p$, defined by

$$X_t - \phi_1 X_{t-1} - \cdots - \phi_p X_{t-p} = Z_t + \theta_1 Z_{t-1} + \cdots + \theta_q Z_{t-q},$$

where $\{Z_t\}$ is white noise with mean 0 and variance $\sigma^2$, has the state-space representation,

$$X_t = \boldsymbol{\theta}' \mathbf{X}_t = [\theta_{p-1} \ \cdots \ \theta_1 \ 1] \mathbf{X}_t,$$

where $\theta_j := 0$ for $j > q$ and

$$\mathbf{X}_t = \mathbf{\Phi} \mathbf{X}_{t-1} + \mathbf{e} Z_t,$$

with

$$\mathbf{\Phi} = \begin{bmatrix} 0 & 1 & 0 & \cdots & 0 \\ 0 & 0 & 1 & \cdots & 0 \\ \vdots & \vdots & \vdots & \ddots & \vdots \\ 0 & 0 & 0 & \cdots & 1 \\ \phi_p & \phi_{p-1} & \phi_{p-2} & \cdots & \phi_1 \end{bmatrix}, \quad \text{and } \mathbf{e} = \begin{bmatrix} 0 \\ 0 \\ \vdots \\ 0 \\ 1 \end{bmatrix}.$$

The autocovariance function of $\{X_t\}$ can therefore be written in the form,

$$\operatorname{Cov}(X_{t+h}, X_t) = \sigma^2 \boldsymbol{\theta}' \mathbf{\Phi}^h \sum_{j=0}^{\infty} (\mathbf{\Phi}^j \mathbf{e})(\mathbf{\Phi}^j \mathbf{e})' \boldsymbol{\theta}, \ h = 0, 1, 2, \ldots. \quad (3.5)$$

Comparing this expression with (2.6), we see that finding a CARMA process whose autocorrelations at integer lags coincide with those of $\{X_t\}$ reduces to finding real vectors $\mathbf{a} = [a_1 \ a_2 \cdots a_{p'}]'$ and $\mathbf{b} = [b_0 \ b_1 \ b_2 \cdots b_{q'}]$ with $q' < p'$ such that (3.5) and (2.6) are identical (when $A$ is replaced by the corresponding $p' \times p'$ matrix obtained from $\mathbf{a}$). In the case when $\mathbf{\Phi}$ has no negative eigenvalues, He and Wang claim that embedding of the ARMA$(p, q)$ process in a CARMA$(p, q')$ process is possible if there exists a real matrix $A$ such that $e^A = \mathbf{\Phi}$. However Example 2 shows that this is not the case and that the precise characterization of embeddable ARMA processes remains an open and apparently rather difficult problem.

## 4 Continuous-time Threshold ARMA Models

We define the CTARMA$(p, q)$ process with a single threshold at $r$ and boundary-width $2\delta > 0$ exactly as in (2.12) and (2.13), except that we allow the parameters $a_1, \ldots, a_p, b_0, \ldots, b_q$ and $c$ to depend on the first component $Y(t)$ of $\mathbf{Y}(t)$ in such a way that

$$a_i(Y(t)) = a_i^{(J)}, i = 1, \ldots, p; \quad b_i(Y(t)) = b_i^{(J)}, i = 1, \ldots, q; \quad c(Y(t)) = c^{(J)}, \tag{4.1}$$

where $J = 1$ or 2 according as $Y(t) \leq r - \delta$ or $Y(t) > r + \delta$. For $r - \delta < Y(t) \leq r + \delta$, we define the coefficients $a_i(Y(t)), b_i(Y(t)), \sigma(Y(t))$ and $c(Y(t))$ by linear interpolation. Thus $Y(t)$ is the first component of a $p$-dimensional diffusion process $\{\mathbf{Y}(t)\}$ whose drift and diffusion parameters are determined by (4.1). If $\mathbf{Y}(0) = \mathbf{y}$, then $\{\mathbf{Y}(t)\}$ is the unique strong solution of (2.13) subject to this initial condition. (We shall not require $\{\mathbf{Y}(t)\}$ to be stationary as in Section 2.) We shall consider only the case of a single threshold since the problems associated with more than one are quite analogous.

Note that *any* interpolated values $a_i(y), b_i(y)$ and $c(y), -\delta < y \leq \delta$, for which the functions $a_i(\cdot), b_i(\cdot)$ and $c(\cdot)$ are Lipschitz continuous, lead to a unique strong solution of the stochastic differential equation (2.13) defining $\{\mathbf{Y}(t)\}$. If however we allow $\delta$ to be zero, the coefficients may be discontinuous at the threshold $r$ and the standard construction of a strong solution of (2.13) no longer applies.

There are several classes of threshold models with $\delta = 0$ for which the problem of discontinuities can be handled by seeking weak rather than strong solutions of (2.13).

1. Processes for which the coefficients $b_i(y), i = 0, \ldots, q$, are independent of $y$. These processes (which include all CAR processes with constant scale parameter $b_0$) can be defined uniquely in terms of the weak solution $\{\mathbf{Y}(t)\}$ of (2.13). Furthermore the moments of the transition distribution of the Markov process $\{\mathbf{Y}(t)\}$ can be expressed in terms of expectations of functionals of standard Brownian motion (see Brockwell (1994)).

2. The general CTAR(1) process (in which $b_0(\cdot)$ may have a discontinuity at $r$). This case has been studied in detail by Stramer, Brockwell and Tweedie (1996) using results of Engelbert and Schmidt (1984) for a general class of first-order stochastic differential equations.

3. The general CTAR(2) process (in which $b_0(\cdot)$ may have a discontinuity at $r$). This case has been studied in detail by Brockwell and Williams (1997) using results of M. Nisio.

In each of these three cases it can be shown that, for any given distribution of $\mathbf{Y}(0)$ with finite second moments, there is a unique (in law) weak solution of equation (2.3) which uniquely defines the law of the CTARMA process. If $\{B(t)\}$ is a standard Brownian motion on $(\Omega, \mathcal{F}, P)$ then in case

1, $\{\mathbf{Y}(s), s \leq t\}$ coincides in law with an explicit functional of $\{B(s), s \leq t\}$ on $(\Omega, \mathcal{F}, \hat{P})$, where $\hat{P}$ is obtained from $P$ by a Cameron-Martin-Girsanov transformation. In cases 2 and 3, a random time change is also required. The existence and uniqueness problems for the general CTARMA process in which both sets of coefficients $a_i(\cdot)$ and $b_i(\cdot)$ are discontinuous at the threshold $r$ have yet to be solved.

The model-fitting procedure described in Section 5 requires the calculation of moments of the transition distribution of the state-vector $\{\mathbf{Y}(t)\}$. There are several possible methods for approximating these quantities. For numerical calculations in the purely autoregressive case, Brockwell and Hyndman (1992) used an approximating sequence of processes satisfying

$$Y_n(t) = [\, 1 \quad 0 \quad \cdots \quad 0\,]\, \mathbf{Y}_n(t), \; t = 0, 1/n, 2/n, \ldots, \qquad (4.2)$$

where

$$\mathbf{Y}_n\left(t + \frac{1}{n}\right) = [I + \frac{1}{n}A(Y_n(t))]\mathbf{Y}_n(t) + [b_0(Y_n(t))\frac{Z(t)}{\sqrt{n}} + \frac{1}{n}c(Y_n(t))]\, \mathbf{e}, \; (4.3)$$

and $A(Y_n(t))$ is the matrix defined in (2.3) with $a_1, \ldots, a_p$ defined as in (4.1). $\{Z(t)\}$ is an iid sequence of random variables with $P(Z = 1) = P(Z = -1) = 0.5$. For arbitrary $t \geq 0$, $\mathbf{Y}_n(t)$ is defined as $\mathbf{Y}_n(t) = \mathbf{Y}_n([nt]/n)$, where $[nt]$ denotes the integer part of $nt$. To deal with the CTARMA process defined by (2.12), (2.13) and (4.1), we simply replace $A(Y_n(t))$ in (4.3) by $C(Y_n(t)) = BAB^{-1}$ and $\mathbf{e}$ by $B\mathbf{e}$, where $a_1, \ldots, a_p$ $b_0, \ldots, b_q$ and $c$ are all dependent on $Y(t)$ as in (4.1). The convergence of these approximations for processes in Class 1 is discussed by Brockwell and Stramer (1995) and for Classes 2 and 3 by Stramer et al.(1996) and Brockwell and Williams (1997) respectively.

The process $\{\mathbf{Y}_n(t)\}$ in (4.3) with $A(Y_n(t))$ replaced by $C(Y_n(t)) = BAB^{-1}$ and $\mathbf{e}$ by $B\mathbf{e}$, is clearly Markovian. The conditional expectations,

$$\mathbf{m}_n(\mathbf{y}, t) = E(\mathbf{Y}_n(t)|\mathbf{Y}_n(0) = \mathbf{y}),$$

satisfy the backward Kolmogorov equations,

$$\mathbf{m}_n(\mathbf{y}, t + n^{-1}) = \frac{1}{2}\mathbf{m}_n(\mathbf{y} + n^{-1}(C(y)\mathbf{y} + c(y)B(y)\mathbf{e}) + n^{-1/2}\sigma(y)B(y)\mathbf{e}, t)$$

$$+ \frac{1}{2}\mathbf{m}_n(\mathbf{y} + n^{-1}(C(y)\mathbf{y} + c(y)B(y)\mathbf{e}) - n^{-1/2}\sigma(y)B(y)\mathbf{e}, t),$$
$$(4.4)$$

with the initial condition,

$$\mathbf{m}_n(\mathbf{y}, 0) = \mathbf{y}. \qquad (4.5)$$

These equations clearly determine the moments $\mathbf{m}_n(\mathbf{y}, t)$ uniquely. Higher order moments satisfy the same equation (4.4), and a modification of the initial condition (4.5) in which the right-hand side is appropriately replaced. For example

$$S_n(\mathbf{y}, t) := E[\mathbf{Y}_n(t)\mathbf{Y}_n(t)' | \mathbf{Y}_n(0) = \mathbf{y}],$$

satisfies (4.4) with the initial condition,

$$S_n(\mathbf{y}, 0) = \mathbf{yy}'.$$

The solution of the backward Kolmogorov equations is straightforward for orders of approximation $n$ up to about 10. For higher orders of approximation the moments $\mathbf{m}_n(\mathbf{y}, t)$ and $S_n(\mathbf{y}, t)$ can easily be estimated by simulation of the process $\{\mathbf{Y}_n(t)\}$.

An alternative approach to the determination of $\mathbf{m}_n(\mathbf{y}, t)$ and $S_n(\mathbf{y}, t)$ in the cases 1, 2, and 3 listed above is by simulation based on the representations of the processes in terms of Brownian motion. Although the results obtained in this way are consistent with those obtained by simulation of the approximating processes $\{\mathbf{Y}_n(t)\}$, there are difficulties associated with the large variance of the functionals whose expected values must be estimated. (Brockwell and Stramer (1995)).

## 5  Inference for CTARMA$(p, q)$ Processes

In this section we consider the problem of parameter estimation based on observations $Y(t_1), \ldots, Y(t_N)$ of the process,

$$Y(t) = [\, 1 \quad 0 \quad \cdots \quad 0 \,]\, \mathbf{Y}(t), \; t \geq 0, \tag{5.1}$$

where $\mathbf{Y}(t)$ satisfies the state equation,

$$d\mathbf{Y}(t) = BAB^{-1}\mathbf{Y}(t)dt + Be(dW(t) + c\, dt), \tag{5.2}$$

and $A$, $B$ and $c$ depend on $Y(t)$ as in (4.1) (with $\delta$ possibly zero under the conditions of Cases 1, 2 or 3 in Section 4).

We shall first establish recursion relations which determine the likelihood of the observations $\{Y(t_1), \ldots, Y(t_N)\}$ in terms of the probability density of $\mathbf{Y}(t_1)$ and the transition density of the Markov process $\{\mathbf{Y}(t)\}$, under the assumption that these densities exist. These can be used to determine the "Gaussian likelihood" of the data, defined as the likelihood computed under the assumption that the initial and transition distributions are Gaussian with the same first and second order moments as the true distributions. Provided the first and second order moments of the transition distributions are known, the conditional Gaussian likelihood of $\{Y(t_2), \ldots, Y(t_N)\}$ given $\mathbf{Y}(t_1)$ can also be determined from the recursions since it is the same as the Gaussian likelihood of $\{Y(t_2), \ldots, Y(t_N)\}$ with the initial distribution for $\mathbf{Y}(t_2)$ determined by the transition function.

To obtain the required recursions, we rewrite the state vector $\mathbf{Y}(t)$ as

$$\mathbf{Y}(t) = \left[ \begin{array}{c} Y(t) \\ \mathbf{V}(t) \end{array} \right],$$

where $\mathbf{V}(t)$ is the $(p-1) \times 1$ vector consisting of the last $p-1$ components of $\mathbf{Y}(t)$. Using the Markov property of $\{\mathbf{Y}(t)\}$, it is easy to check that the joint probability density, $f_r$, of $Y(t_r), \mathbf{V}(t_r), Y(t_{r-1}), Y(t_{r-2}), \ldots, Y(t_1)$. satisfies the recursions,

$$f_{r+1}(y_{r+1}, \mathbf{v}_{r+1}, y_r, y_{r-1}, \ldots, y_1) =$$

$$\int p(y_{r+1}, \mathbf{v}_{r+1}, t_{r+1} - t_r | y_r, \mathbf{v}_r) f_r(y_r, \mathbf{v}_r, y_{r-1}, \ldots, y_1) d\mathbf{v}_r, \quad (5.3)$$

where $p(y_{r+1}, \mathbf{v}_{r+1}, t_{r+1} - t_r | y_r, \mathbf{v}_r)$ is the density of $(Y(t_{r+1}), \mathbf{V}(t_{r+1})')'$, given $\mathbf{Y}(t_r) = (y_r, \mathbf{v}_r')'$. For a given set of observed values $y_1, \ldots, y_N$, at times $t_1, \ldots, t_N$, the functions $f_2, \ldots, f_N$ are functions of $\mathbf{v}_2, \ldots, \mathbf{v}_N$ respectively. These functions can easily be computed recursively from (5.3) in terms of $f_1$ and the functions $p(y_{r+1}, \cdot, t_{r+1} - t_r | y_r, \cdot)$. The likelihood of the observations $y_1, \ldots, y_N$, is then clearly

$$L(\boldsymbol{\theta}; y_1, \ldots, y_N) = \int_{\mathbf{v}_N} f_N(\mathbf{v}_n) d\mathbf{v}_n. \quad (5.4)$$

Note that the filtered value of the unobserved vector $\mathbf{V}(t_r)$, $r = 1, \ldots, N$, (i.e. the conditional expectation of $\mathbf{V}(t_r)$ given $Y(t_i) = y_i$, $i = 1, \ldots, r$) is readily obtained from the function $f_r$ as

$$\tilde{\mathbf{v}}_r = \frac{\int \mathbf{v} f_r(\mathbf{v}) d\mathbf{v}}{\int f_r(\mathbf{v}) d\mathbf{v}}. \quad (5.5)$$

On the other hand, the calculation of the expected value of $Y(t_{r+1})$ given $Y(t_i) = y_i$, $i = 1, \ldots, r$ involves a much more complicated higher dimensional multiple integration. An alternative natural predictor of $Y(t_{r+1})$ which is easy to compute can be found from

$$\tilde{y}_{r+1} = m((y_r, \tilde{\mathbf{v}}_r')', t_{r+1} - t_r), \quad (5.6)$$

where $m((y_r, \tilde{\mathbf{v}}_r')', t_{r+1} - t_r)$ is approximated by $m_n((y_r, \tilde{\mathbf{v}}_r')', t_{r+1} - t_r)$ as defined in Section 4.

Estimation of the parameter vector $\theta$ of the CTARMA process is based on maximizing the conditional Gaussian likelihood of $\{Y(t_2), \ldots, Y(t_N)\}$ given $\mathbf{Y}(t_1) = (y(t_1), 0, \ldots, 0)'$, where $y(t_1)$ is the observed value of $Y(t_1)$. The required moments of the transition distribution of $\{\mathbf{Y}(t)\}$ are computed numerically with the aid of the approximating process $\{\mathbf{Y}_n(t)\}$ defined in Section 4. We assume implicitly in these calculations that not only does the approximating sequence of processes $\{\mathbf{Y}_n\}$ converge to $\mathbf{Y}$ in distribution as is known, but that the first and second order moments of the transition distributions of $\mathbf{Y}_n$ also converge to the corresponding moments of the transition distributions of $\mathbf{Y}$. Although numerical calculations strongly indicate convergence of the moments of $\{\mathbf{Y}_n(t)\}$, this has yet to be established. The required moments can easily be computed exactly for $n = 10$ and by simulation for $n > 10$ (using 1000 replicates for each set of first and second moments). The value of $n$ is chosen large enough so that the Gaussian likelihood changes only slightly when $n$ is increased further (typically $n = 100$ is sufficient).

**Example 3** The above procedure was used by Brockwell and Hyndman (1992) to fit a CTAR(2) model to the annual sunspot numbers, $1770-1869$. Using $n = 10$ they found the model

$$D^2Y(t) + 8.74DY(t) + 0.33Y(t) = 43.3DW(t) + 31.6, \ Y(t) < 10.0,$$
$$D^2Y(t) + 0.55DY(t) + 0.46Y(t) = 28.4DW(t) + 23.0, \ Y(t) > 10.0.$$

Letting $GL$ denote the conditional Gaussian likelihood, computed as described above, it was found that $-2\ln(GL) = 796.6$, a substantial reduction (for the five additional parameters involved) on the maximum likelihood CAR(2) model, for which $-2\ln(GL) = 812.8$. Recomputing these likelihoods by simulation using $n = 100$ with 1000 replicates for each set of moments, gives very similar values for $-2\ln(GL)$. To check the forecasting performance of the non-linear model, one-step forecasts of the sunspot numbers for the following 20 years, $1870-1889$, were also computed as described above. The observed mean squared errors were 469.8 for the CAR(2) model and 450.9 for the CTAR(2) model. Another interesting feature of the non-linear model is the strong resemblance between the histogram of data generated from the model and the empirical histogram of the sunspot series itself.

**Example 4** A rather more interesting example is furnished by the Australian All-ordinaries Share Price Index $P(t)$ on 521 successive trading days terminating on July 18th, 1994. Brockwell and Williams (1997) consider the percentage price changes $U(t) = 100(P(t) - P(t-1))/P(t-1), t = 1,\ldots,520$. The efficient market hypothesis suggests that the series $\{U(t)\}$ should be an approximately uncorrelated sequence. Indeed the autocorrelation function is small, however the minimum-AIC CAR($p$) model for the data is of order 2. If we attempt to improve on the forecasting ability of this model, a natural model to try is a CTAR(2) process in order to allow for a possibly small deviation from linearity which may be useful for prediction. In this case the model found was

$$D^2U(t) + 0.98DU(t) + 4.91U(t) = 3.68DW(t) + .015, \ U(t) < -0.55,$$
$$D^2U(t) + 3.13DU(t) + 4.06U(t) = 3.88DW(t) - .185, \ U(t) > -0.55,$$

with $-2\ln(GL) = 1232.8$ as compared with 1249.4 for the maximum likelihood CAR(2) model. In spite of the large reduction in $-2\ln(GL)$, the empirical one-step mean squared error for the next 30 values of the series is slightly worse for the threshold model (0.430) than for the linear model (0.417). However if we look at the empirical mean squared errors *conditional* on larger values of $U(t-1)$, we find that the one-step predictors from the non-linear model are much superior. This is in accordance with theoretical calculations from the model and highlights the fact that a major reason for using non-linear models is precisely that they may be superior for certain identifiable past characteristics of the process.

**Example 5** An example of a CTARMA(2,1) process (with $b_0$ and $b_1$ constant) fitted to the logarithms to base 10 of the Canadian lynx data is given by Brockwell and Stramer (1995). Again the model is found to perform well compared with some of the other non-linear models which have been proposed.

Of course there are infinitely many potential non-linear models for time series analysis, including the SETARMA models already mentioned, bilinear models, random-coefficient autoregressive models. ARCH and GARCH models, all of which have been applied with success. The CTARMA processes have a number of appealing properties and provide a further class of non-linear models worthy of consideration when linear models are not satisfactory.

# 6   Acknowledgments

This paper draws on the work of many colleagues and coauthors to whom I am greatly indebted. I am also indebted for support of this work at various times from the National Science Foundation (under Grant 9100392), the Deutsche Forschungsgemeinschaft (at the University of Heidelberg) and the Japan Society for the Promotion of Science (at Tokyo Institute of Technology).

# 7   References

[1] Brockwell, P.J. and R.A. Davis, 1991, *Time Series: Theory and Methods*, 2nd edition (Springer-Verlag, New York).

[2] Brockwell, P.J. and R.J. Hyndman, 1992, "On continuous-time threshold autoregression", *Int. J. Forecasting* **8**, 157–173.

[3] Brockwell, P.J., 1994, "On continuous-time threshold ARMA processes", *J. Stat. Planning and Inference* **39**, 291–303.

[4] Brockwell, P.J., 1995, "A note on the embedding of discrete-time ARMA processes", *J. Time Ser. Anal.* **16**, 451–460.

[5] Brockwell, P.J. and O.Stramer, 1995, "On the approximation of continuous-time threshold ARMA processes", *Annals Inst. Stat. Mathematics* **47**, 1–20.

[6] Brockwell, P.J. and R.J. Williams, 1997, "On the existence and application of continuous-time threshold autoregressions of order two", *Adv. Appl. Prob.* **19**, to appear.

[7] Chan, K.S. and H. Tong, 1987, "A Note on embedding a discrete parameter ARMA model in a continuous parameter ARMA model", *J. Time Ser. Anal.* **8**, 277–281.

[8] Doob, J.L., 1944, "The elementary Gaussian processes", *Ann. Math. Statist.* **25**, 229–282.

[9] He, S.W. and J.G. Wang, 1989, "On embedding a discrete-parameter ARMA model in a continuous-parameter ARMA model", *J. Time Ser. Anal.* **10**, 315–323.

[10] Jones, R.H., 1981, "Fitting a continuous time autoregression to discrete data", *Applied Time Series Analysis II* ed. D.F. Findley (Academic Press, New York), 651–682.

[11] Jones, R.H., 1985, "Time Series Analysis with Unequally Spaced Data", in *Time Series in the Time Domain*, Handbook of Statistics **5**, eds. E.J. Hannan, P.R. Krishnaiah and M.M. Rao (North Holland, Amsterdam), 157–178.

[12] Jones, R.H. and L.M. Ackerson, 1990, "Serial correlation in unequally spaced longitudinal data", *Biometrika* **77**, 721–732.

[13] Pemberton, R.H., 1989, "Forecasting accuracy of non-linear time series models", Tech. Report, Dept. of Mathematics, University of Salford.

[14] Phillips, A.W., 1959, "The estimation of parameters in systems of stochastic differential equations", *Biometrika* **46**, 67–76.

[15] Stramer, O., Brockwell, P.J. and R.L. Tweedie, 1996, "Continuous-time threshold AR(1) processes", *J. Appl. Prob.* **33**, to appear.

[16] Stramer, O., Tweedie, R.L. and P.J. Brockwell, 1996, "Existence and stability of continuous-time threshold ARMA processes", *Statistica Sinica*, **6**, to appear.

[17] Tong, H., 1983, *Threshold Models in Non-linear Time Series Analysis*, Springer Lecture Notes in Statistics **21** (Springer-Verlag, New York).

[18] Tong, H., 1990, *Non-linear Time Series : A Dynamical System Approach* (Clarendon Press, Oxford).

# An Optimisation Technique For Robust Autoregressive Estimates

L. Camarinopoulos
G. Zioutas
E. Bora-Senta
University of Thessaloniki, Thessaloniki, Greece

ABSTRACT The robust estimation of the autoregressive parameters is formulated in terms of the quadratic programming problem. This article follows a proposal of Mallows, using simultaneously two weight functions. New robust estimates are yielded, by combining optimally the Huber-type " residual" weight function with a " position " weight function. The behavior of the estimators is studied numerically, under the additive and innovation outlier generating model. Monte Carlo results show that the proposed estimators compared favorably with respect to M-estimators and Bounded Influence estimators(GM-estimators). Based on these results we conclude that one can improve the robust properties of Ar(p) estimators using the proposed optimization technique.
Keywords : Time series, robust autoregression, outliers, quadratic programming, general M-estimates.

## 1 Introduction

For the last fifteen years, robust methods(M, GM-estimators) have been used to bound the influence of aberrant observations in autoregressive estimates. These observations, usually referred to as outliers in Statistics, cause serious bias in estimating the autoregressive (AR) and moving average (MA) parameters. Therefore, in many research papers [DM79], [Mar80b], [BY86], [dJdW85] and others the idea of M and GM regression outliers estimators has been used in order to estimate the parameters of time series models,in the presence of outliers.

Suppose that $y_t$ are observations corresponding to a stationary autoregressive process, with p autoregressive parameters, AR(p), obtained as

$$y_t = \mu + \phi_1\, y_{t-1} + \ldots\ldots + \phi_p\, y_{t-p} + u_t \tag{1.1}$$

where $u_t$ are independent identically distributed random variables and $\mu$ is the mean of $y_t$. When $u_t$ is distributed as $N(0, \sigma^2)$ the well known Box and

Jenkins estimation method[BJ76] is equivalent to least squares and both of them are efficient. In practice, however, normality never holds exactly, few outliers are often present in the data sets and the original methods are very sensitive to them.

We consider in this work the estimation of autoregressive parameters in the presence of two common types of outliers, innovation outliers and additive outliers. Formally, these two types of outliers, which have been studied for the first time by Denby and Martin[DM79], are as follows:

Innovation Outliers (IO): The model of time series $y_t$ is given by (1.1) but the $u_t$'s are distributed with a heavy tailed distribution F which may be considered as a Normal contaminated distribution. It is important to note that, although, the $u_t$'s have a heavy tailed distribution the dynamic equation (1.1) is always satisfied and the process $y_t$ is a perfectly observed AR(p) process,

$$x_t = \mu + \phi_1 x_{t-1} + \dots\dots + \phi_p x_{t-p} + u_t$$
$$y_t = x_t \tag{1.2}$$

Additive Outliers (AO): In this model , an outlier has an additive transient effect. The observed series is not itself an AR(p) process, but is the sum of an AR(p) process $x_t$ and an independent sequence of variables $v_t$, as follows:

$$x_t = \mu + \phi_1 x_{t-1} + \dots\dots + \phi_p x_{t-p} + u_t$$
$$y_t = x_t + v_t \tag{1.3}$$

The process $x_t$ follows the model (1.1) while the variable $v_t$ follows a contaminated distribution . This means that with high probability the AR(p) process $x_t$ itself is observed, and with lower probability the observation is the AR(p) process $x_t$ plus an error.

Bustos and Yohai[BY86] pointed out that in the presence of innovation outliers in an AR(p) process, the least squares estimate may be highly inefficient compared with robust estimates. Denby and Martin[DM79] and Lee and Martin[LM82] have proposed a successful class of robust estimates for ARMA models with innovation outliers. These estimates are Maximum Likelihood type (M-estimates) and are based mainly on the Huber's weight function, which is discussed in the next section.

The most influential are the additive outliers. In the case of time series contaminated with additive outliers Huber's proposal for AR estimates has an asymptotic bias which can be as catastrophic as that of least squares estimates.In other words, M-estimates are extremely sensitive to the presence of additive outliers.

Bustos and Yohai[BY86] presented two new families of resistant estimates with good robust properties. They followed the structure of AR(1) time series when downweighting observations in the presence of additive outliers. They concluded that further research is needed to determine the behavior of their estimators for higher-order models and other types of outliers, as innovation-type for example.

Generally, M-estimates are superior to the GM-estimates in the presense of innovation outliers, but GM-estimates are important in reducing bias in the event that outliers are of the additive effect variety. We use GM-estimates for reducing bias in AO model while suffering a rather small loss in efficiency at IO model.

[DM79] have suggested some useful motivations of GM-estimates for higher efficiencies at the IO model and smaller bias at the AO model. They have combined a "residual" weight function of Huber-type, for bounding the effect of innovation outliers, with a "position" weight function, for limiting the additive effect outlier appearing in the predictor variable $z_t^T = (y_{t-1}, y_{t-2}, ..., y_{t-p})$.

In this article we present a new approach to robust and resistant autoregressive estimation in the presence of innovation and additive outliers. We have transferred some of Huber's results and Mallows's idea of limiting the sensitivity ofregression estimators, in a robust estimation of autoregressive AR(p) model. The method is based on the optimal combination of Huber-type and bounded influence-type weight functions, using quadratic programming. These weight functions are suitable for innovation and additive outliers respectively.

The new estimator QPBI(Quadratic Programming Bounded Influence) is consistent and seems to be qualitatively robust for stationary autoregressive models. Monte Carlo results show that the QPBI estimates are compared favorably with GM-estimates. In the next section we define the new robust estimator for regressive and autoregressive model. In section 3 we implement the Mathematical Programming technique for defining the new estimator. The Monte Carlo design and results are presented in section 4. Finally the conclusions of this work are given in section 5.

# 2    Robust Estimates for Linear Models

Since this work is concerned primarily with autoregressive models of time series, a briefly review of the robust regression for the usual linear model is appropriate.

## 2.1    Regressive models

To motivate our robust estimators for the parameters of the AR(p) process, we first consider the most important M-estimators and GM-estimators of linear regression. Let

$$y_i = x_i^T \beta + u_i, \qquad i = 1, 2, .........., n \tag{2.1}$$

denote the linear model where $x_i^T = (x_{i1}, ..., x_{ip})$ and $\beta^T = (\beta_1, \beta_2, ..., \beta_p)$. For known scale parameter, an equivalent of Huber's M-estimator[Hub73], [Hub81] could be defined by minimizing

$$\sum_{i=1}^{n} \rho_c(\frac{u_i}{\sigma})\sigma^2 \qquad (2.2)$$

where $\rho_c$ is a less rapidly increasing function of the residuals than the corresponding one of ordinary least squares' estimator and its form is as follows:

$$\rho_c(\frac{u_i}{\sigma}) = \begin{cases} \frac{1}{2}(\frac{u_i}{\sigma})^2 & for \quad \frac{u_i}{\sigma} \leq c \\ c\frac{u_i}{\sigma} - c^2/2 & for \quad \frac{u_i}{\sigma} > c \end{cases} \qquad (2.3)$$

After a simple calculation on the right-hand side, and multiplying both sides by the fixed $2\sigma^2$ we obtain

$$2\sigma^2\rho_c(\frac{u_i}{\sigma}) = \begin{cases} u_i^2 & for \quad u_i \leq c\sigma \\ c^2\sigma^2 + 2c\sigma\varepsilon_i & for \quad u_i > c\sigma \end{cases} \qquad (2.4)$$

where $\varepsilon_i$ is the "pulling" distance of the $y_i$, when the residual $u_i$ is being downweighted and is given by

$$\varepsilon_i = \begin{cases} u_i - c\sigma & for \quad u_i > c\sigma \\ 0 & for \quad u_i \leq c\sigma \end{cases} \qquad (2.5)$$

Thus the M-estimator could be defined by solving an equivalent to the (2.2) problem

$$\sum_{i=1}^{n}(u_i^{*^2} + 2c\sigma\varepsilon_i) \qquad (2.6)$$

where $u_i^*$ are the weighted residuals

$$u_i^* = \begin{cases} u_i & for \ u_i \leq c\sigma \\ c\sigma & for \ u_i > c\sigma \end{cases} \qquad (2.7)$$

Thus, an M-estimator is defined by limiting the effect of large residuals due to $y$-outliers . Unfortunately, if there are outliers in the independent variables the influence of an observation $(y_i, x_i)$ may be arbitrarily large. When such deviations from the ideal model are a possibility the unbounded character of the influence curve can have unpleasant consequences. This is particularly relevant for time series autoregressions observed with additive outliers. Hampel[Ham78] has analyzed the total influence of an observation as the product of "residual" influence and the "position" influence.

A number of proposals have been made in order to bound the "position" influence. Mallows[Mal73],[Mal75] modified (2.2) by imposing certain weights $w(x_i)$ that may depend on the distance $d(x_i)$ of the vector

$x_i$ from the entire matrix $X$. The total influence function is limited, but the weight $w(x_i)$ causes a deficiency problem. Any observation with high distance $d(x_i)$ is not necessarily a real outlier when it is well fitted by the model, [LASR86].

Schweppe[EHF75] proposed a more efficient estimation method by scaling the residual $u$ by its own standard deviation $(1 - h_i)^{1/2}\sigma$ , and downweighting the high leverage points, where the high-leverage points are characterized the observations with large leverage

$$h_i = x_i^T \left[ X^T X \right]^{-1} x_i$$

For bounding the influence of "gross" errors in the independent variables Hampel[Ham78] and Krasker and Welsch[KW82] proposed a natural (and invariant) measure of the sensitivity of the parameter estimate. Then, they defined the robust estimators by limiting the sensitivity of the estimators in a efficient way. A more extreme downweighting of the outliers in the independent variables has been suggested by Welsch[Wel80] and Giltinan,Carroll and Ruppert[DMGR86]. They consider the problem of estimating the parameters by bounding the maximum influence a point can have on its own fitted value. For a sample of n observations the scaled change in fit is

$$DFFITS_i = \frac{h_i^{1/2} u_i}{s(i)(1 - h_i)} \tag{2.8}$$

where $s(i)$ is the standard deviation of the residual $u_i$ at the point i.

In all of the above approaches, the minimax estimate of $\beta$ will turn out to be the solution of the problem

$$\sum_{i=1}^{n} \rho_{c_i}(u_i/\sigma)\sigma^2 \tag{2.9}$$

or equivalently

$$\sum_{i=1}^{n} (u_i^{*2} + 2c_i\sigma\varepsilon_i) \tag{2.10}$$

The approaches differ only in the values of the trimming constants $c_i$. Roughly speaking, all the estimators can be defined from the solution of the problem (2.9) or (2.10) setting:
- $c_i = c$ for Huber estimator (H)
- $c_i = c(1 - h_i)^{1/2}$ for Schweppe estimator (SCH)
- $c_i = c((1 - h_i)/h_i^{1/2})$ for the bounded "self influence" sensitivity estimator (W).

## 2.2  New robust estimator for autoregressive models

Huber[Hub83] has recommended that in principle, $c_i$ should be cut down both for very low and for very high values of $h_i$. For small values of $h_i$ the $c_i$ should not be increased but we must keep $c_i$ bounded. For example, for the estimator SCH the $c_i$ is bounded by $c$ when $h_i = 0$. Unfortunately, for the last estimator W, the value of $c_i$ increases for small $h_i$. Therefore, for the estimator W the effect of "gross" error in the response $y_i$ is not bounded when the point $(x_i, y_i)$ corresponds to very small $h_i$. Logically, a reasonable procedure, optimum for the outliers in the independent variables (large $h_i$), should not fare too badly for observations with small $h_i$, which might be influential.

Mallows found Huber's results[Hub83] sensible and his suggestion was to use the effects of the two types of tuning constants $c$ simultaneously. Consequently, we found it reasonable to modify the problem (2.10) using as $c_i$ the minimum value of the tuning constants of Huber and bounded influence-type estimators. Thus the new proposed estimator QPBI is defined from the solution of the problem (2.10), by replacing the $c_i$ by

$$c_i = c \cdot \min[1, (1 - h_i)/h_i^{1/2}] \tag{2.11}$$

or equivalently by minimizing the following loss function:

$$\sum_{i=1}^{n} (u_i^{*^2} + 2c_{1i} \cdot \sigma \cdot \varepsilon_{1i} + 2c_{2i} \cdot \sigma \cdot \varepsilon_{2i}) \tag{2.12}$$

where $for \quad i = 1, 2, ...n$

$$c_{1i} = c$$
$$c_{2i} = \frac{1 - h_i}{h_i^{1/2}} \tag{2.13}$$

The advantage of our estimator over the previous approaches is that it downweights both:

- bad leverage points (outliers in the independent variables)
- low leverage points which influence badly due to gross errors in the response (outliers in the dependent variable)

The solution of this minimization problem is obtained by developing a new quadratic programming formulation which is presented in section 3. Now, we transfer these ideas for robust estimation in an autoregressive model. The robustness concepts just described are directly applicable to time series for parameter estimation of autoregressive models. Efficiency, robustness, minimax robustness and influence curve are concepts directly applicable .

Consider now the autoregressive AR(p) process model (1.1) under both types of outliers. We wish to extend the above results to obtain estimators for the parameters involved. Given the data $y_1, y_2, ..., y_{n+p}$ of the time series, let $z_t^T = (y_{t-1}, y_{t-2}, ..., y_{t-p})$ and $Z$ a $n \times p$ matrix, $Z = (z_t^T)$ for $t = p+1, ..., n+p$. Obviously, the $z_t$ and $Z$ correspond to independent variable $x_i$ and design matrix $X$ respectively in the linear regression model (2.1).

The models of innovation and additive outliers in the autoregressive process have a similarity to the usual regression model with outliers . For example the innovation outliers have a similarity to the regression outliers in the response. In this case the M-estimators are expected to be robust and efficient.

On the other hand, the additive outliers have a similarity to the regression outliers in the response $y$ and the independent variable $x$, a case in which bounded influence estimators of type (2.10) perform well. Similar generalizations of the M-estimates or bounded influence estimates have been considered by Denby and Martin[DM79].

Thus, by replacing $h_i$ with $h_t = z_t^T [Z^T Z]^{-1} z_t$ we obtain the QPBI estimator of an autoregressive AR(p) model from the solution of the problem

$$\sum_{t=p+1}^{n+p} (u_t^{*2} + 2c_t \varepsilon_t \sigma) \tag{2.14}$$

where

$$u_t^* = \left| y_t - z_t^T \phi \right| - \varepsilon_t. \tag{2.15}$$

Certainly, the tuning constant $c_t$ is

$$c_t = c \cdot \min[1, (1 - h_t)/h_t^{1/2}] \tag{2.16}$$

The basic idea behind the QPBI estimator is the combination of M and GM estimates. The total influence function of the estimator is limited, by bounding the "position" influence similarly with Welsch's weight function and the "residual" influence with a weight function of Huber's type. Using as criterion the minimization of "loss" function (2.14), these weight functions are combined efficiently.

In the next section a quadratic programming (QP) formulation is developed for the solution of the minimization problem (2.14).

# 3  The Optimization Technique

One of the most common problems in scientific computing is that of fitting a curve to a set of empirical data. The weighted least squares curve fitting problem can be stated as follows. Given n points $(y_i, x_{1i}, ..., x_{pi})$ in $p+1$ Euclidean space, we wish to find $\beta^T = (\beta_1, \beta_2, ..., \beta_p)$ to minimize

$$\sum_{i=1}^{n} w_i \cdot (y_i - x_i^T \beta)^2. \tag{3.1}$$

This problem has been expressed as a Mathematical Programming problem and some numerical and theoretical results were presented by Barrodale and Roberts [BR70],[BR73]. They formulated the problem by introducing nonnegative variables $u_i, v_i, \beta_1^T = (\beta_{11}, \beta_{12}, ..., \beta_{1p}), \beta_2^T = (\beta_{21}, \beta_{22}, ..., \beta_{2p})$ and constructing a concave (convex) quadratic function subject to linear constraints:

$$\sum_{i=1}^{n} w_i(u_i^2 + v_i^2) \tag{3.2}$$

subject to:

$$y_i + x_i^T \beta_1 - x_i^T \beta_2 + u_i - v_i = 0$$
$$\beta_1, \beta_2, u_i, v_i \geq 0 \quad for \quad i = 1, ..., n.$$

An estimation of the vector parameter $\beta$ is obtained by solving (3.2) and setting $\beta = \beta_1 - \beta_2$.

The last authors have demonstrated the existence of exactly one best $l_2$-approximation, since the objective function is continuous and the constraints form a closed convex set.

We use a modification of the Barrodale and Roberts[BR70],[BR73] formulation for an autoregressive AR(p) process. Thus, the minimization problem (2.14) for defining a robust estimate of the parameter $\phi$ is expressed as the quadratic programming problem:

$$\sum_{t=p+1}^{n+p} (u_t^{*2} + 2c_{1t}\sigma\varepsilon_{1t} + 2c_{2t}\sigma\varepsilon_{2t}) \tag{3.3}$$

subject to:

$$z_t^T \varphi_1 - z_t^T \varphi_2 + u_t^* + \varepsilon_{1t} + \varepsilon_{2t} \geq y_t$$
$$z_t^T \varphi_1 - z_t^T \varphi_2 - u_t^* - \varepsilon_{1t} - \varepsilon_{2t} \leq y_t$$
$$\varphi_1, \varphi_2, u_t^*, \varepsilon_{1t}, \varepsilon_{2t} \geq 0$$
$$for \quad t = p+1, ..., n+p$$

where

$$\varepsilon_{1t} = \begin{cases} u_t - c_{1t}\sigma & for \quad u_t > c_{1t}\sigma \\ 0 & for \quad u_t \leq c_{1t}\sigma \end{cases}$$

$$\varepsilon_{1t} = \begin{cases} u_t - c_{2t}\sigma & for \quad u_t > c_{2t}\sigma \\ 0 & for \quad u_t \leq c_{2t}\sigma \end{cases}$$

and $\varphi_1^T = (\varphi_{11}, ...., \varphi_{1p}), \varphi_2^T = (\varphi_{21}, ..., \varphi_{2p})$ are nonnegative vector variables. Obviously, an estimation of the parameter $\phi$ is obtained by solving (3.3) and setting $\phi = \varphi_1 - \varphi_2$.

The winsorized residuals $u_t^*$ are obtained by pulling $y_t$ towards the fitted value $z_t^T \cdot \phi$ by a distance $\varepsilon_{1t} + \varepsilon_{2t}$. Clearly, at an optimum solution of (3.3) for each t only one of the $\varepsilon_{1t}$ and $\varepsilon_{2t}$ could be greater than zero. This means that an influential observation at time t is being downweighted either with $\varepsilon_{1t} > 0$ when $c_{1t} < c_{2t}$ or with $\varepsilon_{2t} > 0$ when $c_{1t} > c_{2t}$.

The tuning constants $c_{1t}, c_{2t}$ are used for bounding the "position" and the "residual" influence respectively. As given in the previous section a preferable choice is:

$c_{1t} = c\frac{1-h_t}{h_t^{1/2}}$ to bound the self influence sensitivity

$c_{2t} = c$ to bound the residual influence

The quadratic programming problem (3.3) has a continuous and concave objective function and the constraints form a closed convex set. Therefore, the simplex method stops at the extreme point $(\varphi_1, \varphi_2, u^*, \varepsilon_1, \varepsilon_2)$ which is the unique global minimum, since the objective function is continuous and concave.

The robust properties of the autoregressive estimator QPBI are illustrated numerically with Monte Carlo simulation which is discussed in the following paragraph.

# 4    Monte Carlo Design-Results

In order to study the performance of the new approach we make use of Monte Carlo methods. We describe the design of it which follows the lines of that in Denby and Martin[DM79].

We consider, finally, four different types of robust estimators:
- the Huber M-estimator (H),
- the bounded influence estimator with Hampel-Krasker weights (H-K),
- the bounded influence estimator of Welsch (W),
- the proposed estimator with bounded the "self influence" (QPBI), and
- the ordinary least square estimator (OLS).

We consider simultaneously the two types of outliers. We choose a sample size $n + p = 28$, and an AR(3) process with mean zero, $\mu = 0$, and vector parameter $\phi^T = (1.61, -0.79, 0.12)$.

The errors $u$ were generated from the Normal distribution $N(0, 16)$. We explored the case of one additive and one innovation outlier drawn from the distribution $N(80, 10)$. In every sample an additive outlier has been inserted in a row which corresponds to a low leverage point.

The number of replications for this Monte Carlo study should be $N = 500$, in order to obtain a relative error smaller than 10% with a reasonable confidence level of at least 90% for all the simulation estimates. In the following Table the results are presented only of 100 replications. Hopefully, the results of more replications will be available by the authors soon.

A small problem arises when comparing different approaches. How should the cutoffs be chosen? All the constants of the robust estimates are tuned so that all the yielded average weights are the same. Thus for all methods we downweight approximately the same fraction of the data.

For each value of the parameter $c$ we choose a value for the scale parameter from Huber's estimation. According to the robust estimation literature any other choice for the scale parameter does not change the performance of our proposed robust estimator.

We ran all of the computer programs on a PC IBM machine. Uniform random numbers were generated by using a standard linear congruential method and normal random numbers were obtained from these by the Box-Muller transformation. All computations of the four estimators H, H-K, W and OLS were carried out on the ROBETH-system [Mar80a]. The simplex iterations for the quadratic programming solution were carried out on the same machine using the Mathematical Programming Code (MPCODE) software.

The results are consistent with the fact that the estimators studied downweight the outliers in various degrees. All of the following conclusions were supported by careful examination of the individual estimates:

- The M-estimator of Huber type improves little over least squares as is shown in the Table. This is due to the large additive outlier.

- The Welsch type W and the QPBI estimators yielded similar results.

- Quadratic Programming has improved the performance of the bounded influence estimator W.

- The additive outlier, when it corresponds to "gross" error in the response $y_t$ is downweighted due to variable $\varepsilon_{2t}$ in agreement with the argument of section 3.

- The additive outlier, when it corresponds to outliers in the independed variables, is downweighted due to variable $\varepsilon_{1t}$ in agreement with the formulation (3.3).

- For the bounded approach W , the "gross" error in the response $y_t$ is not downweighted strongly when $h_t$ is small.

- Finally, it is pointed out that in every replication the variables $\varepsilon_{1t}$ and $\varepsilon_{2t}$ enter into the optimum solution of the quadratic programming problem (3.2), in order to bound the "position" and the "residual" influence respectively.

The performance of the robust estimators is measured by the following Monte Carlo estimates of:

1) Mean Square Error of the estimator (MSE): $MSE(\hat{\phi}) = E\left\|\hat{\phi} - \phi\right\|^2$

2) Norm of the bias of $\hat{\phi}$ : $E\left\|\hat{\phi} - \phi\right\|$

3) Trace of the covariance matrix: of $\hat{\phi}$ : $TCV(\hat{\phi}) = E \left\| \hat{\phi} - E(\hat{\phi}) \right\|^2$

4) Mean Square Error of Prediction.

We restrict first our attention to the robust estimator with the best performance in the Table. Huber 's estimator is the best only with the trace of covariance criterion, as was expected from asymptotic results.

The QPBI estimator has decreased the bias compared to the estimators W, K-W, H and OLS and at the same time the variance seems to be reduced. Consequently, as indicated in the Table, the estimator defined by quadratic programming has better MSE. Since from MSE is the sum of variance and square bias, the agreement between Monte Carlo and asymptotic results is rather good.

Finally, there can be no doubt that QPBI and W estimators are the best . The three estimators OLS, H, and H-K are very poor compared with QPBI and W estimators. Clearly our estimator has improved the performance criteria compared to the best competitor W.

TABLE

Summary of Monte Carlo results for AR(3) process
with $\hat{\phi}_1 = 1.61$, $\hat{\phi}_2 = -0.79$, $\hat{\phi}_2 = 0.12$.

Time series has been contaminated by AO and IO outliers

| Estimators | Norm of Bias | MSE of Estimator | MSE of Prediction | Trace of Covariance |
|---|---|---|---|---|
| QPBI | .536 | .406 | 378.88 | .094 |
| W | .547 | .444 | 385.55 | .114 |
| H-K | .624 | .485 | 398.14 | .079 |
| H | .902 | .875 | 483.21 | .052 |
| OLS | 1.121 | 1.446 | 628.47 | .163 |

## 5   Concluding Remarks

The quadratic programming method offers a systematic approach to improve the sample behavior of bounded influence estimators. This technique offers great flexibility in being able to combine two weight functions of M and GM estimators. It is also possible to distinguish the innovation from the additive outliers observing the values of the variables $\varepsilon_1$ and $\varepsilon_2$ in the optimum solution.

Basically, in this work we have motivated the bounded influence estimates, using a bounded "position" weight function for points with small $h_t$, a desirability which has been discussed by Huber and Mallows in 1983.

Since the additive outliers are the most dangerous, we recommend the use of the QPBI estimator which provides the best protection against bias. This estimator is obtained by combining Welsch type "position" and Huber type "residual" weight functions, using an optimum crterion.

Based on the above optimum criterion and results, we conclude that the QPBI estimator works well in all circumstances and is reasonable for additive outliers including heavy tails in the response $y_t$ where $(y_t, z_t)$ is a low leverage point (small $h_t$).

From this study we conclude, finally, that one can gain something by using quadratic programming for the bounded influence autoregressive estimators over the ordinary bounded influence. Further research is needed to determine possible better choices of the combined weight functions using as optimization technique mathematical programming. Also, constraints involving the parameter vector are possible. Another interesting point of research in this context is the use of post-optimality analysis to determine the effect of changing the value of the scale parameter and to identify the most catastrophic outliers in a sample of observations.

# 6 References

[BJ76]       G. E. P. Box and G. M. Jenkins. In *Time Series Analysis Forecasting and Control*. Holden-Day, San Francisco, 1976.

[BR70]       I. Barrodale and D.K. Roberts. Applications of mathematical programming to lp-approximation. In *Nonlinear Programming, J. B. Rosen, O. L.* Mangasarian and K. Ritter, eds., Academic Press, New York, 1970.

[BR73]       I. Barrodale and D. K. Roberts. An improved algorithm for discrete l1 linear approximation. In *SIAM J. Numer. Anal., 10, 839-848*, 1973.

[BY86]       O. H. Bustos and V. J. Yohai. Robust estimates for arma models. In *Journal of the American Statistical Association, 81, 155-168*, 1986.

[dJdW85]   P. J. de Jongh and T. de Wet. Trimmed mean and bounded influence estimators for the parameters of the ar(1) process. In *Commun.Statist.-Theor. Meth., 14, 1361-1375*, 1985.

[DM79]       L. Denby and R. D. Martin. Robust estimation of the first-order autoregressive parameter. In *Journal of the American Statistical Association, 74, 140-146*, 1979.

[DMGR86] R.J. Carroll D. M. Giltinan and D. Ruppert. Some new estimation methods for weighted regression when there are possible outliers. In *Technometrics, 28, 219-230*, 1986.

[EHF75]     F. Schweppe E. Handschin, I. Kohlas and A. Fiechter. Bad data analysis for power system state estimation. In *IEEE Transactions on Power Apparatus and Systems, 2, 329-337*, 1975.

[Ham78]    F. R. Hampel. Optimally bounding the gross-error-sensitivity and the influence of position in factor space. In *1978 Proceedings of the ASA, Statistical Computing Section, 59-64*, 1978.

[Hub73]    P.J. Huber. Robust regression : Asymptotics, conjectures, and monte carlo. In *Annals of Statistics, 1, 799-821*, 1973.

[Hub81]    P. J. Huber. In *Robust Statistics.* John Wiley, New York, 1981.

[Hub83]    P. J. Huber. Minimax aspects of bounded-influence regression. In *Journal of the American Statistical Association, 78, 66-72*, 1983.

[KW82]    W. S. Krasker and R. E. Welsch. Efficient bounded-influence regression estimation. In *Journal of the American Statistical Association, 77, 595-604*, 1982.

[LASR86]    R. J. Carroll L. A. Stefanski and D. Ruppert. Optimally bounded score functions for generalized linear models with applications to logistic regression. In *Biometrica, 73, 2, 413-24*, 1986.

[LM82]    C. H. Lee and R. Martin. M-estimates for arma process. In *Technical Report 23, University of Washington, Dept. of Statistics*, 1982.

[Mal73]    C. L. Mallows. Influence functions. In *National Bureau of Economic Reasearch Conference on Robust Regression in Cambridge, Massachusetts*, 1973.

[Mal75]    C. L. Mallows. On some topics in robustness. In *unpublished memorandum, Bell Telephone Laboratories, Murray Hill, New Jersey*, 1975.

[Mar80a]    A. Marazzi. In *Robust Linear Regression Programs in ROBETH, Reasearch Report No. 24*. Eidgenoessische Technishe Hochschule, Zuerich, Fachgruppe fuer Statistik, 1980.

[Mar80b]    R. D. Martin. Robust of autoregressive models. In *Directions in Time Series*. eds. D. R. Brillinger and G. C. Tiao, Haywood, CA:Institute of Mathematical Statistics, 1980.

[Wel80]    R.E. Welsch. Regression sensitivity analysis and bounded-influence estimation. In *Evaluation of Econometric Models, J. Kmenta and J. B. Ramsey, eds.* Academic Press, New York, 1980.

# A Theory of Wavelet Representation and Decomposition for a General Stochastic Process

Bing Cheng and Howell Tong
Institute of Mathematics and Statistics
University of Kent at Canterbury
Kent CT2 7NF, UK

December 19, 1995

## 1 Introduction

Wavelets offer exciting possibilities for statistical problems. The recent paper read by Donoho *et al.* (1994) and the many references therein highlight some of the possibilities in a statistical context. Numerous interesting applications in signal processing, computer vision, numerical analysis, applied mathematics, physics and others have been reported in the mathematical, engineering and physical literature. Some of these also discuss statistically related problems. (See, e.g., Chui, 1992, Mallat, 1989b, 1989c, 1991, Mallat and Zhong, 1992, Segmen and Zeevi, 1993 and others.)

One of the basic techniques in the vast wavelet literature is the wavelet transform. Advantages provided by such a transform have been demonstrated in terms of the ability to handle long-memory stochastic processes, edge detection with noisy data, time-frequency resolution of random signals (See for example, Flandrin, 1992, Cheng and Kay, 1993, and Basseville *et al* 1992). Behind all the statistically related applications is a stochastic process and this suggests that there is a common theoretical basis, namely a common generic wavelet representation of a stochastic process. Needless to say, such a representation theory is fundamental. The situation is not unlike the spectral representation of a stationary time series.

Cambanis and Masry (1994) used wavelets as basis functions for approximations of stochastic processes. However, there are important differences between their results and ours in the following respects. (1) The more fundamental feature behind the wavelet methodology is its ability to construct a uniform multiresolution representation and decomposition (See, e.g., Mallat, 1989b) rather than its ability to approximate any shaped-signals, which can be equally well provided by such techniques as the radial basis function, the spline, the kernel and others. In particular, the approximation provided by the wavelets for deterministic signals is uniformly function-independent (See, e.g., Daubechies, 1992) and the same result should be expected for random signals. However, the approximation obtained by Cambanis and Masry (*op. cit.*) is apparently function-dependent and has thus under-utilised the power of wavelets. (2) The rate of convergence they have obtained is function-dependent in that they assume that the function to be approximated has the same degree of smoothness as the wavelet functions. However, in practice such a rate of convergence is of very limited use since we rarely know the form of the true function and even if we assume that it is known the degree of smoothness could still vary. (3) The assumptions they have made for the approximation are somewhat obscure and not easy to verify.

The plan of this paper is as follows. In section 2, we give a brief introduction of wavelets for deterministic signals. In section 3, we build our representation thoerem for a general stochastic process (Theorem 3.1) to be followed by a decomposition theorem (Theorem 3.2). In section 4, we give a brief conclusion. In addition, we provide an appendix on Hilbert spaces for general stochastic processes.

## 2 Wavelet for deterministic signals

Let $Z$ and $T$ be the set of integers and the set of real numbers respectively and $L^2(T)$ be the space of measurable, square-integrable functions $X(t), t \in T$. Below is the definition of a multiresolution approximation of $L^2(T)$ due to Mallet (1989a):

**Definition 1:** A multiresolution approximation of $L^2(T)$ is a sequence $\{V_k(T)\}_{k \in Z}$ of subspaces of $L^2(T)$ such that the following hold:

(1) $V_k(T) \subset V_{k+1}(T)$,

(2) $\bigcup_{k=-\infty}^{+\infty} V_k(T)$ is dense in $L^2(T)$ and $\bigcap_{k=-\infty}^{+\infty} V_k(T) = \{0\}$

(3) $X(t) \in V_k(T) \Longleftrightarrow X(2t) \in V_{k+1}(T) \ \forall k \in Z$,

(4) $X(t) \in V_k(T) \Longleftrightarrow X(t - 2^k l) \in V_k(T) \ \forall l \in Z$,

(5) $\forall k \in Z, \ \{X(\cdot - l)\}_{l \in Z}$ is a Riesz base of $V_k(T)$, that is, there exist two positive numbers A and B such that

$$A \sum_{l=-\infty}^{+\infty} c_l^2 \leq \| \sum_{l=-\infty}^{+\infty} c_l X(\cdot - l)\|^2 \leq B \sum_{l=-\infty}^{+\infty} c_l^2, \ \forall \ \{c_l\}_{l \in Z} \in l^2(Z),$$

where $l^2(Z) = \{\{c_l\}_{l \in Z} | \sum_{l=-\infty}^{+\infty} c_l^2 < \infty\}$. Then Mallat (1989a) has shown that there exists a unique function $\phi(t) \in L^2(T)$ such that, for any $k \in Z$, $\{\sqrt{2^k}\phi(2^k t - l)\}_{l \in Z}$ is an orthonormal basis of $V_k(T)$ and $\phi$ satisfies the following basic dilation equation which is a two-scale difference equation:

$$\phi(t) = \sum_l c_l \phi(2t - l). \tag{2.1}$$

The function $\phi(t)$ is called the scaling function of the multiresolution approximation.

Let $W_k$ be the subspace such that $V_{k+1}(T) = V_k \oplus W_k$ for $\forall k \in Z$. Then Mallat (1989b, c) has shown that $\{W_k\}$ constitutes a time-frequency decomposition

$$L^2(T) = \sum_{k=-\infty}^{+\infty} \bigoplus W_k \tag{2.2}$$

and that there exists a unique function $w(t) \in L^2(T)$ such that, for any $k \in Z$, $\{\sqrt{2^k}w(2^k t - l)\}_{l \in Z}$ is an orthonormal basis of $W_k(T)$ and $w$ satisfies the following two-scale difference equation:

$$w(t) = \sum_l d_l w(2t - l), \tag{2.3}$$

where the coefficients $\{d_l\}$ satisfy the relation

$$d_l = (-1)^l c_{1-l}, \tag{2.4}$$

with $\{c_l\}$ as the coefficients in the dilation equation. The function $w$ is called a wavelet. The coefficients $\{c_l\}$ and $\{d_l\}$ are the so-called *quadrature mirror filters* (low-pass filter, high-pass filter respectively, see Burt and Adelson 1983 for example). Using these two quadrature mirror filters, Daubechies (1988) has constructed orthogonal compact-supported wavelets.

## 3 Wavelet representation of a general stochastic process

### 3.1 Stochastic processes at different levels of resolution

Let $(\Omega, F, P)$ denote a probability space. Let

$$L^2(\Omega \times T) = \{X(t) | E \int_T X^2(t) dt < \infty\}.$$

From the Appendix, we know that $L^2(\Omega \times T)$ is a Hilbert space. Now, define a sequence of subspaces, $V_k(\Omega \times T)$ say, of $L^2(\Omega \times T)$ for each $k \in Z = \{0, \pm 1, \ldots\}$ by

$$V_k = V_k(\Omega \times T) = \left\{ X \in L^2(\Omega \times T) : X(t) = \sum_l \xi_{kl} \phi_{kl}(t), \sum_l E\xi_{kl}^2 < +\infty \right\}, \tag{3.1}$$

where $\{\xi_{kl}\}_{l \in z}$ is a squence of random variables and $\phi_{kl}(t)$ is defined by

$$\phi_{kl}(t) = \sqrt{2^k} \, \phi \left( 2^k t - l \right). \tag{3.2}$$

In the following we use a continuous scaling function $\phi$.

**Lemma 3.1**: *If $X \in V_k(\Omega \times T)$, then*
*(i) $EX(t) = \sum_l E\xi_{kl} \phi_{kl}(t)$ in $L^2(T)$;*
*(ii) $R(t, s) = \operatorname{cov}(X(t), X(s))$*

$$= \sum_{l, l'} \operatorname{cov}(\xi_{k,l}, \xi_{k,l'}) \phi_{kl}(t) \phi_{kl'}(s) \quad \text{in} \quad L^2(T \times T).$$

**Proof of (i)**: Define $X^{(L)}(t) = \sum_{|l| \leq L} \xi_{kl} \phi_{kl}(t)$. Then

$$\|EX(\cdot) - EX^{(L)}(\cdot)\|_T^2 = \int_T [EX(t) - EX^{(L)}(t)]^2 dt$$

$$= \int_T \{E[X(t) - X^{(L)}(t)]\}^2 dt \leq \int_T E[X(t) - X^{(L)}(t)]^2 dt.$$

By the Fubini theorem

$$\int_T E[X(t) - X^{(L)}(t)]^2 dt = E \int_T [X(t) - X^{(L)}(t)]^2 dt = E \int_T \left[ \sum_{|l| > L} \xi_{kl} \phi_{kl}(t) \right]^2 dt$$

$$= E\left[\sum_{|l|>L} \xi_{kl}^2\right] \text{ (by the orthonormality of } \phi_{kl}) = \left[\sum_{|l|>L} E\xi_{kl}^2\right] \text{ (by the monotone convergence theorem)}$$

Since $\sum_l E\xi_{kl}^2 < \infty$, $\sum_{|l|>L} E\xi_{kl}^2 \to 0$ as $L \to \infty$. This implies that $\lim_{L\to\infty} ||EX(\cdot) - EX^{(L)}(\cdot)||_T^2 = 0$. Since $EX^{(L)}(t) = \sum_{|l|\leq L} E\xi_{kl}\phi_{kl}(t)$, we obtain $EX(t) = \sum_{l=-\infty}^{+\infty} E\xi_{kl}\phi_{kl}(t)$ in $L^2(T)$.

**Proof of (ii):**

Without loss of generality, we assume $EX(t) = 0$ for each $t$. Still let $X^{(L)}(t) = \sum_{|l|\leq L} \xi_{kl}\phi_{kl}(t)$. Then the covariance function of $X^{(L)}(t)$ is

$$R^{(L)}(t,s) = \text{cov}(X^{(L)}(t)), X^{(L)}(s)) = \sum_{\substack{|l|\leq L \\ |l'|\leq L}} \text{cov}(\xi_{kl}, \xi_{kl'})\phi_{kl}(t)\phi_{kl'}(s).$$

Clearly for each $L$, $R^{(L)}(\cdot, \cdot) \in V_k(T) \otimes V_k(T)$. In the following, we first show that $\{R^{(L)}\}_{L\in Z}$ is a Cauchy sequence in $V_k(T) \otimes V_k(T)$. For $M > L$, we have

$$\left|\left|R^{(L)}(\cdot,\cdot) - R^{(M)}(\cdot,\cdot)\right|\right|_{T\times T}^2 = \int_T \int_T [R^{(L)}(t,s) - R^{(M)}(t,s)]^2 dtds$$

$$= \int_T \int_T \sum_{\substack{L<|l|\leq M \\ L<|l'|\leq M}} \text{cov}(\xi_{kl}, \xi_{kl'})\phi_{kl}(t)\phi_{kl'}(s)]^2 dtds.$$

By the orthonormality of $\{\phi_{kl}\}$, the above reduces to $\sum_{\substack{L<|l|\leq M \\ L<|l'|\leq M}} [\text{cov}(\xi_{kl}, \xi_{kl'})]^2$. However,

$$[\text{cov}(\xi_{kl}, \xi_{kl'})]^2 = [E\{\xi_{kl}\xi_{kl'}\}]^2 \leq E\xi_{kl}^2 E\xi_{kl'}^2.$$

[Note that we have assumed $EX(t) = 0$ and so $E\xi_{kl} = 0$ for each $l$.]

We obtain

$$\left|\left|R^{(L)} - R^{(M)}\right|\right|_{T\times T}^2 \leq \left(\sum_{L\leq|l|\leq M} E\xi_{kl}^2\right)^2 \longrightarrow 0 \quad \text{as} \quad M, L \to \infty.$$

Since $L^2(T \times T)$ is complete and $V_k(T) \otimes V_k(T)$ is closed, there exists an $\tilde{R} \in V_k(T) \otimes V_k(T)$ such that

$$\tilde{R}(t,s) = \lim_{L\to\infty} R^{(L)}(t,s)$$

$$= \sum_{L_1 l'} \text{cov}(\xi_{kl}, \xi_{kl'})\phi_{kl}(t)\phi_{kl'}(s) \quad \text{in} \quad L^2(T \times T).$$

Next,

$$\left\| R - \tilde{R} \right\|_{T \times T}^2$$

$$= \int_T \int_T [R(t,s) - \tilde{R}(t,s)]^2 dt ds$$

$$= \int_T \int_T [\text{cov}(X(t), X(s)) - \text{cov}(X^{(L)}(t)), X^{(L)}(s)]^2 dt ds$$

$$\leq 2 \int_T \int_T [\text{cov}(X(t), X(s) - X^{(L)}(s))]^2 dt ds$$

$$+ 2 \int_T \int_2 [\text{cov}(X(t) - X^{(L)}(t), X^{(L)}(s))]^2 dt ds.$$

However,

$$[\text{cov}(X(t), X(s) - X^{(L)}(s))]^2$$

$$= [EX(t)(X(s) - X^{(L)}(s)))]^2 \leq EX^2(t) E[X(s) - X^{(L)}(s)]^2$$

and

$$[\text{cov}(X(t) - X^{(L)}(t), X^{(L)}(s))]^2 \leq E[X(t) - X^{(L)}(t)]^2 E[X^{(L)}(s)]^2.$$

We obtain $\left\| R - \tilde{R} \right\|_{T \times T}^2 \leq 2 \left\{ \int_T E[X(t) - X^{(L)}(t)]^2 dt \left[ \int_T EX^2(t) dt + \int_T E[X^{(L)}(t)]^2 \right] dt \right\}$. Since $X^{(L)} \longrightarrow X$ in $L^2(\Omega \times T)$ and by Fubini theorem $\int_T EX^2(t) dt = E \int_T X^2(t) dt = \|X\|_{\Omega X T}^2 < \infty$, we have

$$\int_T E[X(t) - X^{(L)}(t)]^2 dt = E \int_T [X(t) - X^{(L)}(t)]^2 dt = \left\| X - X^{(L)} \right\|_{\Omega \times T}^2 \longrightarrow 0 \text{ as } L \to \infty, \text{ and}$$

$$\int_T E[X^{(L)}(t)]^2 dt = E \int_T [X^{(L)}(t)]^2 dt \longrightarrow E \int_T X^2(t) dt < \infty \quad \text{as} \quad L \to \infty.$$

Therefore,

$$\|R - \tilde{R}\|_{T \times T} \leq \|R - R^{(L)}\|_{T \times T} + \|\tilde{R} - R^{(L)}\|_{T \times T} \longrightarrow 0 \quad \text{as} \quad L \to \infty.$$

We have in $L^2(T \times T)$,

$$R(t,s) = \tilde{R}(t,s) = \sum_{l,l'} \text{cov}(\xi_{k,l}, \xi_{kl'}) \phi_{kl}(t) \phi_{kl'}(s)$$

**Lemma 3.2**: Let $X \in L^2(\Omega \times T)$ with covariance function $R(\cdot, \cdot)$. Then the following two conditions are equivalent.

(a) $X \in V_k(\Omega \times T)$ has a representation

$$X(t) = \sum_l \xi_{kl} \phi_{kl}(t);$$

(b) $R(\cdot, \cdot) \in V_k(T) \otimes V_k(T)$ has a representation

$$R(t,s) = \sum_{l,l'} C_{l,l'} \phi_{kl}(t) \phi_{kl'}(s)$$

where $\{C_{l,l'}\}$ is a non–negative definite and symmetric sequence. Specifically, $C_{l,l'} = C_{l',l}$ and for any integer $n$, any $n$ points $l_1, \ldots, l_n$, and any set of $n$ real numbers $\{a_1, \ldots, a_n\}$

$$\sum_{i,j=1}^{n} a_i a_j, C_{l_i,l_j} \geq 0.$$

**Proof:**

"(a) $\Rightarrow$ (b)": This follows from Lemma 3.1. Set $C_{l,l'} = \text{cov}(\xi_{k,l}, \xi_{k,l'})$. Clearly, $\{C_{l,l'}\}$ is symmetric and non–negative.

"(b) $\Rightarrow$ (a)": By Mercer's Theorem, there is a series expansion for $R(\cdot, \cdot)$. Specifically, let $\{\psi_n(t), n = 1, 2, \ldots\}$ be a sequence of normalised eigenfunctions of $R$, and $\{\lambda_n, n = 1, 2, \ldots\}$ a sequence of the corresponding (non–negative) eigenvalues such that

$$R(t, s) = \sum_n \lambda_n \psi_n(s) \psi_n(t), \quad t \in T \tag{3.3}$$

and

$$\int_T \psi_m(t)\psi_n(t)dt = \begin{cases} 1 & \text{if } m = n, \\ 0 & \text{if } m \neq n. \end{cases} \tag{3.4}$$

In the representation $R(t, s) = \sum_n \lambda_n \psi_n(t) \psi_n(s)$, we can, without loss of generality, assume that $\lambda_n \neq 0$ for each $n$.

Otherwise, we have the representation $R(t, s) = \sum_{\{n; \lambda_n \neq 0\}} \lambda_n \psi_n(t) \psi_n(s)$. Now, for each $n$, we have $\psi_n(t) = \lambda_n^{-1} \int_T R(t, s)\psi_n(s)ds$ and since $R(t, s) = \sum_{l,l'} C_{l,l'} \phi_{kl}(t) \phi_{kl'}(s)$, we have

$$\begin{aligned} \psi_n(t) &= \lambda_n^{-1} \sum_{l,l'} C_{l,l'} \phi_{kl}(t) \int_T \psi_n(s) \phi_{kl'}(s)ds \\ &= \sum_l \left( \sum_{l'} C_{l,l'} \int_T \psi_n(s) \phi_{kl'}(s)ds \right) \phi_{kl}(t) \\ &= \sum_l d_{kl}^{(n)} \phi_{kl}(t), \text{say.} \end{aligned} \tag{3.5}$$

Therefore, for each $n$, $\psi_n \in V_k(T)$.

Next, define a Hilbert space $H$ by $H = \{g : g(t) = \sum_n \lambda_n a_n \phi_n(t) \text{ and } \sum_n \lambda_n a_n^2 < +\infty\}$, and an inner product $< \cdot, \cdot >$ by $< g, h >= \sum_n \lambda_n a_n b_n$, $\forall g, h \in H$, where

$$g(t) = \sum_n \lambda_n a_n \phi_n(t) \text{ and } h(t) = \sum_n \lambda_n b_n \phi_n(t).$$

Since $R(t, t) = \sum_n \lambda_n \phi_n^2(t) < \infty$, and $R(t, s) = \sum_n \lambda_n \phi_n(t)\phi_n(s)$, we have that, for each $t \in T$, $R(t, \cdot) \in H$. Since $< R(t, \cdot), R(\cdot, s) >= \sum_n \lambda_n \phi_n(t)\phi_n(s) = R(t, s)$, therefore $H$ is a reproducing kernel Hilbert space $(RKHS)$ with kernel $R$, i.e. $H = H(R)$ which is spanned by $\{R(\cdot, t) | t \in T\}$.

Now, let $L^2(X(t), t \in T)$ be the space spanned by the linear combinations of $\{X(t), t \in T\}$ with the following inner product: $\forall \xi, \eta \in L^2(X(t), t \in T)$, $< \xi, \eta >= E\xi\eta$, where $\xi = \sum_n a_n X(t_n)$ and $\eta = \sum_n b_n X(s_n)$. It is easy to see that $L^2(X(t), t \in T)$ is a Hilbert space. Since

$$< R(\cdot, s), \ R(\cdot, t) >= R(s, t) = EX(s)X(t) = < X(s), X(t) >,$$

by the basic convergence theorem in the Appendix, there is a congruence $J$ from $H(R)$ to $L^2(X(t), t \in T)$ such that

$$J(R(\cdot, t)) = X(t). \tag{3.6}$$

Let $g \in H(R)$ with $g(t) = \sum_n \lambda_n a_n \phi_n(t)$. Since $J$ is a congruence, for each $t \in T$,

$$E[J(g)X(t)] = E[J(g)J(R(\cdot, t))] = \ <J(g), J(R(\cdot, t))> \ = \ <g, R(\cdot, t)> \ = \sum_n \lambda_n a_n \phi_n(t) = g(t) \tag{3.7}$$

Choose a series of coefficients $\{a_n^{(m)}\}$ by

$$a_n^{(m)} = \delta_{mn} = \begin{cases} 1 & \text{if } n = m, \\ 0 & \text{if } n \neq m. \end{cases}$$

Let $\eta_m = J(\lambda_m \phi_m) \in L_2(X(t), t \in T)$. We have

$$E[\eta_n \eta_m] = E[J(\lambda_n \phi_n)J(\lambda_m \phi_m)] = \ <\lambda_n \phi_n, \lambda_m \phi_m> \ = \sum_l \lambda_l \delta_{nl} \delta_{ml} = \begin{cases} \lambda_n & \text{if } m = n, \\ 0 & \text{if } m \neq n. \end{cases}$$

Now, define a map $J'$ from $H(R)$ to $L_2(X(t), t \in T)$ as follows.

$$\forall g \in H(R) \quad \text{with} \quad g(t) = \sum_n \lambda_n a_n \phi_n(t), \quad J'(g) = \sum_n a_n y_n.$$

Next, we show that $J(g) = J'(g), \forall g \in H(R)$. Since $L_2(X(t), t \in T)$ is spanned by $\{X(t), t \in T\}$, to show that $J(g) - J'(g) \equiv 0$, it is sufficient to show that $\forall t \in T$ $(J(g) - J'(g)) \perp X(t)$. That is $E[(J(g) - J'(g))X(t)] = 0$, which is equivalent to $E[J(g)X(t)] = E[J'(g)X(t)]$. However, by (3.7) we have that $E[J(g)X(t)] = S(t)$. Therefore,

$$E[J'(g)X(t)] = E[(\sum_n a_n y_n)X(t)] = \sum_n a_n E[y_n X(t)]$$

$$= \sum_n a_n E[J(\lambda_n \phi_n)J(R(\cdot, t))] = \sum_n a_n \ <J(\lambda_n \phi_n), J(R(\cdot, t))> \ = \sum_n a_n \ <\lambda_n \phi_n, R(\cdot, t)> .$$

Since $<\lambda_n \phi_n, R(\cdot, t)> = \sum_m \lambda_m \delta_{mn} \phi_m(t) = \lambda_n \phi_n(t)$, we obtain $E[J'(g)X(t)] = \sum_n \lambda_n a_n \phi_n(t) = g(t)$. Therefore, we have proved that the congruence $J$ has the expansion $J(g) = \sum_n a_n \eta_n$ for each $g(t) = \sum_n \lambda_n \phi_n(t) \in H(R)$; especially by (3.6) we have $X(t) = J(R(\cdot, t)) = \sum_n a_n \eta_n$ (with $a_n = \phi_n(t)) = \sum_n \eta_n \phi_n(t)$. However, by (3.5) we know that $\phi_n \in V_k(T)$ with $\phi_n(t) = \sum_l d_{k,l}^{(n)} \phi_{kl}(t)$. Thus, we have

$$X(t) = \sum_l \xi_{kl} \phi_{kl}(t), \tag{3.8}$$

where

$$\xi_{kl} = \sum_n d_{k,l}^{(n)} y_n. \tag{3.9}$$

This completes the proof of Lemma 3.2.

**Lemma 3.3**: *For each $k \in Z$, $V_k(\Omega \times T)$ is a closed subspace of $L^2(\Omega \times T)$.*

**Proof:** Suppose $X^{(n)} \in V_k(\Omega \times T)$ and $X \in L^2(\Omega \times T)$ such that $X^{(n)} \longrightarrow X$ in $L^2(\Omega \times T)$. We need to show that $X \in V_k(\Omega \times T)$. Define

$$R_X(t,s) = \text{cov}(X(t), X(s)) \text{ and } R_X^{(n)}(t,s) = \text{cov}(X^{(n)}(t), X^{(n)}(s)).$$

Now,

$$\|R_X - R_{X^{(n)}}\|_{TXT}^2 = \int_T \int_T [R_X(t,s) - R_{X^{(n)}}(t,s)]^2 dt ds$$

$$\leq 2 \int_T \int_T [\text{cov}(X(t), X(s) - X^{(n)}(s))]^2 dt ds + 2 \int_T \int_T [\text{cov}(X(t) - X^{(n)}(s), X^{(n)}(s))]^2 dt.$$

However,

$$\{\text{cov}(X(t), X(s) - X^{(n)}(s))\}^2 \leq \text{VAR}(X(t)) \text{var}(X(s) - X^{(n)}(s))$$

$$\leq 2EX^2(t) \; 2E[X(s) - X^{(n)}(s)]^2 = 4EX^2(t)E[X(s) - X^{(n)}(s)]^2$$

and similarly

$$\{\text{cov}(X(t) - X^{(n)}(t), X^{(n)}(s))\}^2 \leq 4E[X^{(n)}(s)]^2 E[X(t) - X^{(n)}(t)]^2.$$

We have

$$\|R_X - R_{X^{(n)}}\|^2 \leq 8 \int_T E[X(t) - X^{(n)}(t)]^2 dt \left\{ \int_T EX^2(t) dt + \int_T E[X^{(n)}(t)]^2 dt \right\}.$$

By the Fubini theorem and the assumption $X^{(n)} \to X$ in $L^2(\Omega \times T)$, we have

$$\int_T E[X(t) - X^{(n)}(t)]^2 dt \to 0 \quad \text{as} \quad n \to \infty$$

and by the monotone convergence theorem and the fact that $X^{(n)} \to X$ in $L^2(\Omega \times T)$, we have

$$\int_T E[X^{(n)}(t)]^2 dt \to \int_T EX^2(t) dt \quad \text{as} \quad n \to \infty.$$

Therefore, we have proved that $R_{X^{(n)}} \to R_X \text{in} L^2(T \times T)$. Since for each $n$, $X^{(n)} \in V_k(\Omega \times T)$, therefore by Lemma 3.2 $R_{X^{(n)}} \in V_k(T) \otimes V_k(T)$. Since $V_k(T) \otimes V_k(T)$ is a closed subspace of $L^2(T \times T)$ and $R_{X^{(n)}} \to R_X$ in $L^2(T \times T)$, we know that $R_X \in V_k(T) \otimes V_k(T)$. Therefore, by Lemma 3.2, $X \in V_k(\Omega \times T)$. In other words, $V_k(\Omega \times T)$ is a closed subspace of $L^2(\Omega \times T)$.

**Theorem 3.1:** *(Representation Theorem)*

$\{V_k(\Omega \times T)\}_{k \in Z}$ *constitute a multiresolution approximation to the space* $L^2(\Omega \times T)$. *Specifically,*

*(M0)* *for each* $k \in Z$, $V_k = V_k(\Omega \times T)$ *is a closed subspace of* $L^2(\Omega \times T)$;

*(M1)* $K_k \subset V_{k+1}$, $\forall k \in Z$;

*(M2)* $\displaystyle\bigcup_{k=-\infty}^{+\infty} V_k$ *is dense in* $L^2(\Omega \times T)$ *and* $\displaystyle\bigcap_{k=-\infty}^{+\infty} V_k = \{0\}$;

*(M3)* $X(t) \in V_k \Leftrightarrow X(2t) \in V_{k+1}, \forall k \in Z$;

*(M4)* $X(t) \in V_k \Leftrightarrow X(t - Z^{-k}l) \in V_k, \forall l \in Z$.

**Proof:** (M0) follows from Lemma 3.3 immediately. For (M1), note that $\forall X \in V_k$, its covariance function $R \in V_k(T) \otimes V_k(T) \subset V_{k+1}(T) \otimes V_{k+1}(T)$. By Lemma 3.2, $X \in V_{k+1}$. Hence $V_k \subset V_{k+1}$. For (M2), note that $\forall X \in \displaystyle\bigcap_{k=-\infty}^{+\infty} V_k$, we have by Lemma 3.2 that $R \in \bigcap_{k=-\infty}^{+\infty} V_k(T) \otimes$

$V_k(T)$. Therefore, $\forall t, s \in T$ $R(t,s) = 0$; in particular, $R(t,t) = 0$. However, $R(t,t) = E[X(t) - EX(t)]^2$, $\forall t \in T$. Thus, $X(t) \equiv$ const a.s, $\forall t \in T$. Now, since $X \in L^2(\Omega \times T)$, we know that these constants must be zero. Therefore, $\bigcap_{k=-\infty}^{+\infty} V_k = \{0\}$ a.s. Next, suppose that $X \perp \bigcup_{k=-\infty}^{+\infty} V_k$. First, we have, for every integer $k$ or $l$, $E \int_{-\infty}^{+\infty} X(t)\phi_{kl}(t)dt = 0$. By the Fubini theorem, we have $\int_{-\infty}^{+\infty} (EX(t))\phi_{kl}(t)dt = 0$. Since $\bigcup_{k=-\infty}^{+\infty} V_k(T)$ is dense in $L^2(T)$, $EX(t) \perp \bigcup_{k=-\infty}^{+\infty} V_k(T)$ implies that $EX(t) = 0, \forall t \in T$. Second, $\forall k, l \in Z$, define $\xi_{kl} = \int_{-\infty}^{+\infty} X(t)\phi_{kl}(t)dt$ . Then by the definition of $V_k(\Omega \times T)$, we know that $\forall l' \in Z$, $\xi_{kl}\phi_{kl'} \in V_k(\Omega \times T)$. Since $X \perp \bigcup_{k=-\infty}^{+\infty} V_k(\Omega \times T)$ we have $E\left[\int_{-\infty}^{+\infty} X(t)\xi_{kl}\phi_{kl'}(t)\right]dt = 0$ . However,

$$E\int_{-\infty}^{+\infty} X(t)\xi_{kl}\phi_{kl'}(t)dt = E\int_{-\infty}^{+\infty}\int_{-\infty}^{+\infty} X(t)X(s)\phi_{kl}(s)\phi_{kl'}(t)dtds$$

$$= \int_{-\infty}^{+\infty}\int_{-\infty}^{+\infty} R(t,s)\phi_{kl}(s)\phi_{kl'}(t)dtds.$$

Again since $\bigcup_{k=-\infty}^{+\infty} V_k(T) \otimes V_k(T)$ is dense in $L^2(\Omega \times T)$, we have that $\forall t, s \in T$ $R(t,s) = 0$. In particular, $R(t,t) = 0$. That is, $EX^2(t) = 0$ and so $X(t) = 0$ a.s. $\forall t \in T$. Therefore, we have proved that $\bigcup_{k=-\infty}^{+\infty} V_k(\Omega \times T)$ is dense in $L^2(\omega \times T)$. For (M3), we note that $X(t) \in V_k$ implies that $X(t) = \sum_l \xi_{kl}\phi_{kl}(t)$. Hence,

$$X(2t) = \sum_l \xi_{kl}\sqrt{2^k}\phi(2^{k+1}t - l) = \sum_l \xi_{kl}\frac{1}{\sqrt{2}}\phi_{k+1,l}(t) \in V_{k+1}(\Omega \times T).$$

For (M4), we note that $X(t) \in V_k$, implies that $X(t) = \sum_l \xi_{kl'}\phi_{kl'}(t)$. Hence,

$$X(t - 2^{-k}l) = \sum_{l'} \xi_{kl'}\sqrt{2^k}\phi(2^k(t - 2^{-k}l) - l')$$

$$= \sum_{l'} \xi_{kl'}\phi_{k,l+l'}(t) = \sum_{l'} \xi_{k,l'-l}\phi_{k,l'}(t) \in V_k(\Omega \times T).$$

Now, define a series of subspaces $\{W_k\}$ in $L^2(\Omega \times T)$ by $W_k = V_{k+1}\backslash V_k$. Since $W_k(t) = V_{k+1}(t)\backslash V_k(T)$, we have $\phi_{k+1,l}(t) = \sum_{l'} d_{k,l,l'}W_{k,l'}(t) + \sum_{l'} c_{k,l,l'}\phi_{k,l'}(t)$. (M4) is proved.

**Lemma 3.4**: $W_k = \left\{ X \in L^2(\Omega \times T) | X(t) = \sum_i \eta_{kl}W_{kl}(t) \right\}.$

**Proof**: $\forall X \in W_k$, we have $X \in V_{k+1}$ and $X \perp V_k$. Therefore,

$$X(t) = \sum_l \xi_{k+1,l}\phi_{k+1,l}(t)$$

$$= \sum_l\left(\sum_l \xi_{k+1,l}d_{k,l,l'}\right)W_{k,l'}(t) + \sum_l\left(\sum_l \xi_{k+1,l}C_{k,l,l'}\right)\phi_{k,l'}(t).$$

Since $X \perp V_k$, so for each $l$, $\sum_l \xi_{k+1,l}C_{k,l,l'} = 0$ a.s. We have $X(t) = \sum_l \eta_{kl}W_{kl}(t)$.

**Theorem 3. 2**: *(Time-Frequency Decomposition Theorem)*

$$L^2(\Omega \times T) = \sum_k \otimes W_k,$$

*that is* $\forall X \in L^2(\Omega \times T)$, *we have* $X(t) = \sum_k \sum_l \eta_{kl} W_{kl}(t)$, *where* $\eta_{kl} = \int_{-\infty}^{+\infty} X(t) W_{kl}(t) dt$.

**Proof**: The theorem follows immediately from Theorem 1 and Lemma 3.4.

**Definition 1**: For $X \in L^2(\Omega \times T)$, and real numbers $a$ ($a > 0$) and $b$, the wavelet transform $\bigcup_{a,b} X$, say, for the stochastic process $X$ is defined by $\bigcup_{a,b} X = \int_{-\infty}^{+\infty} X(t) W\left(\frac{t-b}{a}\right) dt$. Clearly, if we use compactly supported wavelets such as Daubechies wavelet, then the wavelet transform $\bigcup_{a,b} X$ localizes $X$ in both the time-domain and the frequency-domain. Theorem 2 indicates that we can reconstruct $X$ from wavelet transform over a discrete grid by $X(t) = \sum_k \sum_l \frac{1}{\sqrt{2^k}} \bigcup_{2^{-k}, 2^{-k}l} X W_{kl}(t)$.

# 4   Discussion

In this paper, we have constructed a representation and decomposition theory for a general stochastic process under some conditions. The assumption of the finiteness of $E \int_T X(t)^2 dt$ , which excludes the conventional stationary processes over the infinite time interval, can be weakened by introducing a weighting function, say $s$, such that $E \int_T X(t)^2 s(t) dt < \infty$. The remaining task is to rebuild a Hilbert space along the lines described in the Appendix. However, it may be argued that a stationary time series over *infinite* time interval is more for mathematical convenience; in reality such a time series rarely exists.

The theory that we have developed in this paper points to very fertile statistical areas, both in terms of statistical theory and statistical practice. Elsewhere we have introduced the notion of resolution-dependent stationarity. (Cheng and Tong, 1995.) This goes towards realizing his views on non-stationarity expressed to one of us before his untimely death.

# References

[1] Basseville, M., Benveniste, A., Chou, K.C. Colden, S.A., Nikoukhan, R. and Willsky, A.S. (1992). Modeling and estimation of multiresolution stochastic processes. *IEEE Trans. Inf. Th.*, **38**, 766-784.

[2] Burt, P., and Adelson, E. (1983). The Laplacian pyramid as a compact image code. *IEEE Trans. Comm.*, **31**, 482-540.

[3] Cambanis, S. and Masry, E. (1994). Wavelet approximation of deterministic and random signals: convergence properties and rates. *IEEE Trans on Information Theory* **40** pp. 1013-1029.

[4] Cheng, B. and Kay, J. (1993). Multiresolution image reconstruction using the wavelet transform. Submitted to *Statistics and Computing*

[5] Cheng, B. and Tong, H. (1995). On resolution-dependent stationarity. (under preparation.)

[6] Chui, C.K. (Ed.) (1992) *Wavelet Analysis and its Applications*, Vols. 1 and 2, Academic Press, London.

[7] Daubechies, I. (1988). Orthonormal bases of compactly supported wavelets. *Comm. Pure Appl. Math.*, **41** , 909-996.

[8] Flandrin, P. (1992). Wavelet analysis and synthesis of fractional Brownian motion. *IEEE Trans. Inf. Th.*, **38**, 910-917.

[9] Ledoux, M. and Talagrand, M. (1991). *Probability in Banach Spaces*. Springer-Verlag, London.

[10] Mallat, S. (1989a). Multiresolution approximation and wavelet orthonormal bases of $L^2(R)$. *Trans. Amer Math. Soc.*, **315**, 69-87.

[11] Mallat, S. (1989b). A theory for multiresolution signal decomposition: The wavelet representation. *IEEE Trans. Pattern Analysis and Machine Intelligence*, **11** ,674-693.

[12] Mallat, S. (1989c). Multifrequency channel decomposition of images and wavelet models. *IEEE Trans. Acoustic Speech Signal Proc.*, **37**, 2091-2110.

[13] Mallat, S. (1991). Zero-crossing of a Wavelet transform. *IEEE Trans. Inf. Th.*, **37**, 1019-1033.

[14] Mallat, S. and Zhong, M. (1992). Characterization of signals from multiscale edges. *IEEE Trans. Pattern Analysis and Machine Intelligence*, **14**, 710-732.

[15] Segman, J. and Zeevi, Y. Y. (1993). Spherical wavelets and their applications to image representation. *J. Visual Commu. Image Representation*, **4**, 263-270.

## Appendix: A Hilbert Space for Stochastic Processes

**Definition A.1**: A stochastic process $\{X(t), t \in T\}$ is said to be continuous in quadratic mean on $T$ if, for every $t$ in $T$ $\lim_{s \to t} E[X(s) - X(t)]^2 = 0$ .

**Lemma A.1**: *If $\{X(t), t \in T\}$ is continuous in quadratic mean, then*

*(i) the mean function $EX(t)$ is continuous on $T$,*

*(ii) the covariance function $R(t, s)$ is continuous on $T \times T = T^2$.*

**Proof of (i)**: $|EX(t) - EX(s)| = |E[X(t) - X(s)]| \leq \{E[X(t) - X(s)]^2\}^{\frac{1}{2}}$. Therefore $\lim_{s \to t} EX(s) = EX(t)$.

**Proof of (ii)**: Without loss of generality, we can assume $EX(t) \equiv 0$ for each $t$. Now

$$|R(t', s') - R(t, s)| = |E[X(t') - X(t)]X(s')| + |EX(t)[X(s') - X(s)])$$

$$\leq \{E[X(t') - X(t)]^2\}^{\frac{1}{2}}\{EX^2(s')\}^{\frac{1}{2}} + \{EX^2(t)\}^{\frac{1}{2}}\{E[X(s') - X(s)]^2\}^{\frac{1}{2}}.$$

Since

$$\lim_{t' \to t} E[X(t') - X(t)]^2 = 0, \ \lim_{s' \to s} E[X(s') - X(s)]^2 = 0$$

and

$$\lim_{s' \to s} EX^2(s') = EX^2(s) < \infty,$$

we have

$$\lim_{\substack{t' \to t \\ s' \to s}} R(t', s') = R(t, s).$$

**Assumption A.1**: $\{X(t), t \in T\}$ is a stochastic process defined on a probability space $(\Omega, F, P)$ and is continuous in quadratic mean on $T$, where $T = (-\infty, +\infty)$.

For $-\infty < a < b < +\infty$, suppose for each $t \in [a, b]$, $X(t)$ is a sum of an indicator function on $(\Omega, F, P)$. That is, $\forall \omega \in \Omega$,

$$X(t, \omega) = \sum_j a_j(t) I_{A_j(t)}(\omega),$$

where $\{a_j(t)\}$ are real–valued numbers and $\{A_j(t)\} \subset F$. We define an integration on $[a, b] \times \Omega$ by

$$\int_a^b \int_\Omega X(t, \omega) dt\, dP(\omega) = \lim_{n \to \infty} \sum_{i=0}^n \sum_j a_j(t_j) P(A_j(t_i))(t_i - t_{i-1}),$$

where $a = t_0 < t_1 < \cdots < t_{n-1} < t_n = b$ and $t_i - t_{i-1} \to 0$ as $n \to \infty$ for each $t$.

Let $X$ be a non–negative stochastic process on $(\Omega, F, P)$. So for each $t$, there is a sequence of non–decreasing measurable functions $X^{(m)}(t, \cdot)$ such that (a) $X^{(m)}(t, \omega) \le X^{(m+1)}(t, \omega), \forall \omega \in \Omega$ and (b) $\lim_{m \to \infty} X^{(m)}(t, \omega) = X(t, \omega), \forall \omega \in \Omega$. Then we define the integration of $X$ on $[a, b] \times \Omega$ by

$$\int_a^b \int_\Omega X(t, \omega) dt\, dP(\omega) = \lim_{m \to +\infty} \int_a^b \int_\Omega X^{(m)}(t, \omega) dt\, dP(\omega).$$

For a general stochastic process $X$, let

$$X^+(t, \omega) = \begin{cases} X(t, \omega) & \text{if } X(t, \omega) \ge 0, \\ 0 & \text{otherwise,} \end{cases}$$

and

$$X^-(t, \omega) = \begin{cases} -X(t, \omega) & \text{if } X(t, \omega) < 0, \\ 0 & \text{otherwise.} \end{cases}$$

then the integration of $X$ on $[a, b] \times \Omega$ is defined by

$$\int_a^b \int_\Omega X(t, \omega) dt\, dP(\omega) = \int_a^b \int_\Omega X^+(t, \omega) dt\, dP(\omega) - \int_a^b \int_\Omega X^-(t, \omega) dt\, dP(\omega).$$

**Assumption A.2**: For every $-\infty < a < b < +\infty$,

$$\int_a^b \int_\Omega |X(t, \omega)| dt\, dP(\omega) < +\infty.$$

We recall that $X$ is integrable on $[a, b] \times \Omega$.

Then by the Fubini Theorem, we have the iterated integration formula:

$$\int_a^b \left[ \int_\Omega X(t, \omega) dP(\omega) \right] dt = \int_\Omega \left[ \int_a^b X(t, \omega) dt \right] dP(\omega).$$

**Assumption A.3:** $\displaystyle\int_{-\infty}^{+\infty} [EX(t)]^2 dt < \infty.$

**Assumption A.4:** $\displaystyle\int_{-\infty}^{+\infty}\int_{-\infty}^{+\infty} |R(t,s)| dt\, ds < \infty.$

**Lemma A.2:** *Under the assumption A1 – A4, there is a random variable $\xi$ in $L^2(\Omega, F, P)$ such that*

$$\int_a^b X(t)dt \xrightarrow{L_2} \xi \quad as \quad b \to +\infty \quad and \quad a \to -\infty.$$

*We denote $\xi$ by $\displaystyle\int_{-\infty}^{+\infty} X(t)dt.$*

**Proof:** Without loss of generality, by assumption A3, we can assume $EX(t) \equiv 0$ for each $t$. Now, let

$$S_m = \int_0^m X(t)dt.$$

Then, for $n < m$

$$E[S_m - S_n]^2 = E\left[\int_n^m X(t)dt\right]^2$$

$$= E\left[\int_n^m \int_n^m X(t)X(s)dt\,ds\right] = \int_n^m \int_n^m R(t,s)dtds.$$

So, by assumption A4,

$$\lim_{\substack{m\to\infty \\ n\to\infty}} E[S_m - S_n]^2 = 0 \text{ and } \lim_{\substack{m\to-\infty \\ n\to-\infty}} E[S_m - S_n]^2 = 0.$$

Since $L_2(\Omega, F, P)$ is a complete space, there is a random variable $\xi \in L_2(\Omega, F, P)$ such that

$$\int_n^m X(t)dt \xrightarrow{L^2} \xi \quad as \quad m \to +\infty \quad and \quad n \to -\infty.$$

**Lemma A.3:** *Under the assumption A1 – A4, we have*

$$\int_{-\infty}^{+\infty} EX(t)dt = E\int_{-\infty}^{+\infty} X(t)dt.$$

**Proof:** First, for every $-\infty < a < b < +\infty$, by the Fubini theorem,

$$\int_a^b EX(t)dt = E\int_a^b X(t)dt.$$

For LHS, by assumption A3,

$$\lim_{\substack{b\to+\infty \\ a\to-\infty}} \int_a^b EX(t)dt = \int_{-\infty}^{+\infty} EX(t)dt.$$

For the RHS, by Lemma A.2,

$$\int_a^b X(t)dt \xrightarrow{L^2} \int_{-\infty}^{+\infty} X(t)dt \quad as \quad b \to +\infty \quad and \quad a \to -\infty.$$

So

$$\lim_{\substack{b\to+\infty \\ a\to-\infty}} E\int_a^b X(t)dt = E\int_{-\infty}^{+\infty} X(t)dt.$$

**Lemma A.4:** *Under the assumption A1 – A4,*

$$E\left[\int_{-\infty}^{+\infty} X^2(t)dt\right] < +\infty.$$

**Proof:**

$$\int_{-\infty}^{+\infty} X^2(t)dt \leqq 2\int_{-\infty}^{+\infty} (X(t) - EX(t))^2 + 2\int_{-\infty}^{+\infty} [EX(t)]^2 dt.$$

By assumption A3,

$$\int_{-\infty}^{+\infty} [EX(t)]^2 dt < \infty,$$

and since $(X(t) - EX(t))^2$ is non–negative,

$$E\int_{-\infty}^{+\infty} (X(t) - EX(t))^2 dt = \int_{-\infty}^{+\infty} E[X(t) - E(X(t)]^2 dt = \int_{-\infty}^{+\infty} R(t,t)dt < \infty.$$

The last part above is by virtue of assumption A4. So $E\left[\int_{-\infty}^{+\infty} X^2(t)dt\right] < \infty$. It is well known that $L^2(T)$ with the inner product $< f,g >_{L^2(T)} = \int_{-\infty}^{+\infty} f(t)g(t)dt$ is a Hilbert space. The norm in $L^2(T)$ is defined by

$$\|f\|_{L^2(T)} = \left\{\int_{-\infty}^{+\infty} f^2(t(dt)\right\}^{\frac{1}{2}}.$$

So for a stochastic process $X$ along its path, we can write $\int_{-\infty}^{+\infty} X^2(t)dt$ as $\|X\|_{L^2(T)}^2$.

Define a space of mean-square continuous stochastic processes on $(\Omega, F, P)$ by

$$L^2(\Omega \times T) = \left\{X : E\|X\|_{L^2(T)}^2 < +\infty\right\}$$

and an inner product in $L^2(\Omega \times T)$ by

$$< X,Y > = E < X,Y >_{L^2(T)} = E\int_{-\infty}^{+\infty} X(t)Y(t)dt.$$

It is straightforward to show that $L^2(\Omega \times T)$ is a pre–Hilbert space. In order to show the completeness of $L^2(\Omega \times T)$, we need the probability on a Banach space: $L^2(\Omega \times T)$ is a space of the Banach space $L^2(T)$– valued random variables defined on the probability space $(\Omega, F, P)$. Furthermore $L^2(T)$ is a separable Banach space since $L^2(T)$ has countable basis functions.

**Definition A.2:** Let $(\Omega, F, P)$ be a probability space and $(B, \| \cdot \|_B)$ be a Banach space. $\{X\}$ and $X$ are measurable functions on $(\Omega, F, P)$ and take values in $B$.

(i) $X_n$ converges almost surely to $X$, or symbolically $X_n \overset{a.s.}{\to} X$, if

$$P\{\omega : \lim_{n \to \infty} \|X_n(\omega) - X(\omega)\|_B \neq 0\} = 0.$$

(ii) $X_n$ converges in probability to $X$, or symbolically $X_n \overset{P}{\to} X$, if $\forall \epsilon > 0$,

$$\lim_{n \to \infty} P(\omega : \|X_n(\omega) - X(\omega)\|_B > \epsilon) = 0.$$

(iii) $X_n$ converges in $r$th mean to $X$ if $\lim_{n \to \infty} E\|X_n - X\|_B^r = 0$.

Correspondingly, if $X_n - X_m \overset{a.s.}{\to} 0$ a.s. as $n, m \to \infty$, we say $\{X_n\}$ is an almost surely a Cauchy sequence . If $X_n - X_m \overset{P}{\to} 0$ as $n, m \to \infty$, we say that $\{X_n\}$ is a Cauchy sequence in probability.

**Lemma A.5:**

*(i)* $X_n \overset{a.s.}{\to} X$ *if and only if* $\forall \epsilon > 0$, $P\left(\bigcap_{n=1}^{\infty} \bigcup_{\nu=0}^{\infty} \{\omega : ||X_{n+\nu}(\omega) - X(\omega)||_B \geq \epsilon\}\right) = 0$ *or*

$$\lim_{n \to \infty} P\left(\bigcup_{\nu=0}^{\infty} \{\omega : ||X_{n+\nu}(\omega) - X(\omega)||_B \geq \epsilon\}\right) = 0.$$

*(ii)* $\{X_n\}$ *is an almost surely Cauchy sequence if and only if* $\forall \epsilon > 0$,

$$\lim_{n \to \infty} P\left(\bigcup_{\nu=0}^{+\infty} \{\omega : ||X_{n+\nu}(\omega) - X(\omega)|| \geq \epsilon\}\right) = 0.$$

**Proof**: See Ledoux and Talagrand (1991).

**Lemma A.6**: $\{X_n\}$ *is an almost surely Cauchy sequence if and onlt if theres exist an* $X$ *taking values in* $B$ *such that* $X_n \overset{a.s.}{\to} X$.

**Proof**: See Ledoux and Talagrand (1991).

**Lemma A.7**: *(Riesz Theorem). If* $X_n \overset{P}{\to} X$, *then there exists a subsequence* $\{X_{n_k}, k \geq 1\}$ *of* $\{X_n\}$ *such that* $X_{n_k} \overset{a.s.}{\to} X$ *as* $k \to \infty$.
**Proof**: See Ledoux and Talagrand (1991).

**Lemma A.8**: $\{X_n\}$ *is a Cauchy sequence in probability if and only if there is a* $X$ *taking values in* $B$ *such that* $X_n \overset{P}{\to} X$.

**Proof**: See Ledoux and Talagrand (1991).

**Lemma A.9**: $L^2(\Omega \times T)$ *is a Hilbert space.*

**Proof**: We only need to show the completeness of $L^2(\Omega \times T)$ . Suppose $\{X_n\} \in L^2(\Omega \times T)$ such that $E||X_n - X_m||^2_{L^2(T)} \to 0$ as $n, m \to \infty$. So $X_n - X_m \overset{P}{\to} 0$ (by Chebyshev type inequality). By Lemma A.8, there is a $X \in L_2(\Omega \times T)$ such that $X_n \overset{P}{\to} X$. By Lemma A.7, there is a subsequence $\{X_{n_k}; k \geq 1\}$ of $\{X_n\}$ such that $X_{n_k} \overset{a.s.}{\to} X$. Now,

$$E||X_m - X||^2_{L^2(T)} = \int_\Omega ||X_m - X||^2_{L^2(T)} dP = \int_\Omega \liminf_{k \to \infty} ||X_m - X_{n_k}||^2_{L^2(T)} dP,$$

which by the Fatou Lemma, is $\leq \lim_{k \to \infty} \int_\Omega ||X_m - X_{n_k}||^2_{L^2(T)} dP$. Since

$$\lim_{\substack{m \to \infty \\ k \to \infty}} \int_\Omega ||X_m - X_{n_k}||^2_{L^2(T)} dP = 0,$$

we obtain $\lim_{m \to \infty} E||X_m - X_{n_k}||^2_{L^2(T)} = 0$. So, $L^2(\Omega \times T)$ is a complete space. Since $L^2(\Omega \times T)$ is already a pre–Hilbert space, $L^2(\Omega \times T)$ is a Hilbert space.

# Modeling the Distribution of Highly Volatile Exchange-rate Time Series

George Chobanov
Plamen Mateev
Stefan Mittnik
Svetlozar Rachev

ABSTRACT  Modeling the unconditional distribution of returns on exchange rates is nontrivial, because they are typically fat-tailed and highly peaked around the center. Especially, the exchange rates of many Eastern European countries exhibit a highly volatile and seemingly erratic behavior in the initial phase of moving from controlled to freely floating exchange rates. In this paper we examine to what extend the distributional properties of such time series can be captured by certain types of stable distributions. We consider the standard stable Paretian distribution and special, thin-tailed members of the family of geometric summation stable distributions. An application to the Bulgarian lev/US dollar exchange-rate series reveals that among the candidates considered only the fat-tailed stable Paretian distribution is able to capture the characteristics of this highly volatile time series.

## 1   Introduction

Several empirical studies have examined the distributional properties of exchange rate time series in recent years. As with stock returns, relative to the normal distribution, empirical distributions of exchange-rate returns are typically fat-tailed and more peaked around the origin. Various statistical distributions allowing for such features have been suggested in the literature. The stable Paretian (or $\alpha$-stable) distribution, whose application in financial modeling was originally proposed in [Man63], the Student $t$ distribution and the mixture-of-normals distribution have been used as alternatives to the prevailing normal assumption in finance.

Stable Paretian distributions have several theoretical properties that are attractive in financial modeling. They are stable with respect to addition and scaling, implying that distributions of returns over periods of different lengths have the same *shape*; they have domains of attractions; and, being characterized by four parameters, they are fairly flexible in fitting empirical distributions. Despite these desirable properties, stable Paretian distributions have not been overly successful in empirical modeling of financial asset returns. It should be noted, however, that almost all studies consider only *symmetric* stable Paretian distributions and/or use crude and unreliable estimation techniques.

In this paper, we first examine the goodness of fit of the above candidate distributions for the Bulgarian lev/US dollar exchange rate by comparing the maximum likelihood values of estimated distributions. To estimate both the *symmetric* and *asymmetric* stable Paretian distribution, we use the maximum-likelihood estimation algorithm in [Che91]. Secondly, we also consider certain members of the *random summation stable distributions*—specifically, we fit Laplace and (double) Weibull distributions.

The use of random summation stable distributions for modeling financial asset returns has been proposed in [MR89, MR91, MR93], where the concept of stability is broadened by considering models that are stable with respect to different underlying probabilistic scheme. Letting $X_i$, $i = 1, \ldots, n$, denote identically and independently distributed (iid) random variables, the schemes can be written as

$$X_1 \stackrel{d}{=} a_n(X_1 \circ X_2 \circ \cdots \circ X_n) + b_n, \tag{1.1}$$

where $\stackrel{d}{=}$ denotes equality in distribution; operator $\circ$ stands for *summation, min, max,* or *multiplication* operations; $n$ is a *deterministic* or *random* integer independent of the $X_i$'s; $a_n > 0$; and $b_n \in \mathbf{R}$.

Letting $r$ denote the (random) return on an asset, we say $r$ is *stable under scheme (1.1)*, if, for some $\alpha_{pos}, \alpha_{neg} > 0$,

$$X_1 = \begin{cases} r^{\alpha_{pos}}, & r > 0, \\ -|r|^{\alpha_{neg}}, & r \leq 0, \end{cases}$$

satisfies relationship (1.1). The standard (deterministic) summation scheme, i.e., $\circ$ stands for $+$ and $n$ is a deterministic integer, produces the well-known stable Paretian distribution; the maximum and minimum schemes lead to extreme-value distributions; and the multiplication scheme yields the multiplication stable distribution. If $n$ is random and independent of the $X_i$'s, we obtain *random* summation, maximum, minimum, and multiplication stable distributions, respectively. In the case of the random summation scheme, it can be shown analytically that the only possible choice for $n$ is the class of geometric random variables with support on a general lattice. For this reason [MR89] refer to this scheme as the *geometric summation scheme*.

Here, we focus on the geometric summation scheme, because, when fitting the distributions associated with (1.1) to S&P500 stock-index data, [MR93] find that the Weibull return distribution, a member of the geometric summation and the minimum schemes, clearly dominates all of the other candidates—including the fat-tailed, asymmetric stable Paretian distribution. Given that the S&P500 stock-index is "well-behaved" compared to the volatile return series under consideration here, the appropriateness of the thin-tailed Weibull distribution is questionable.

The paper is organized as follows. The next section presents the geometric summation model and discusses some of its properties. Section 3 reports estimation results for Bulgarian lev/US dollar exchange-rate returns. The final section contains concluding remarks.

## 2   The Geometric Summation Model

To motivate the geometric summation model, let us first consider the standard summation scheme that gives rise to the stable Paretian distribution. Suppose $X_1, X_2, \ldots$ are iid random variables with common distribution function $\mathcal{H}$. $\mathcal{H}$ is said to be stable Paretian (see, for example, [Zol86]), if there exist constants $a_n > 0$ and $b_n \in \mathbf{R}$, such that for any $n$

$$a_n(X_1 + \ldots + X_n) + b_n \overset{d}{=} X_1. \tag{1.2}$$

Except for some special cases, there is no closed-form expression for the distribution function of the stable Paretian distribution. However, the characteristic function has the explicit representation

$$\int e^{itx} dH(x) = \begin{cases} \exp\{-c^\alpha |t|^\alpha [1 - i\beta \operatorname{sign}(t)\tan \frac{\pi\alpha}{2}] + i\delta t\}, & \text{if } \alpha \neq 1, \\ \exp\{-c|t|[1 + i\beta \frac{2}{\pi}\operatorname{sign}(t)\ln |t|] + i\delta t\}, & \text{if } \alpha = 1, \end{cases} \tag{1.3}$$

where $\operatorname{sign}(t)$ equals $1$ if $t > 0$, $0$ if $t = 0$, and $-1$ if $t < 0$. The characteristic exponent $\alpha$ $(0 < \alpha \leq 2)$ is the *index of stability* and can also be interpreted as a *shape* parameter; $\beta$ $(-1 \leq \beta \leq 1)$ is the *skewness* parameter; $\delta$ $(\delta \in \mathbf{R})$ is a *location* parameter; and $c$ $(c > 0)$ is the *scale* parameter.

The idea underlying the *geometric* summation scheme is to allow for the fact that financial markets may with some—typically small—probability substantially change their characteristics in any given period. In the context of exchange-rate modeling we can, for example, think of major political or economic events, which could occur in any period and drastically affect the nature of the exchange-rate process.

To state the geometric summation stable model formally, let $X_i$ denote the change of an exchange rate during period $t = t_0 + i$ and let the $X_i$'s have common distribution function $\mathcal{G}(x) = \Pr(X_i < x)$. In each period we expect with probability $p \in (0, 1)$ the occurrence of an event significantly altering the characteristics of the return process. Letting $T(p)$ denote the number of periods after which such an event is expected to occur, we assume that $T(p)$ is independent of $\{X_i\}$ and geometrically distributed; i.e.,

$$\Pr\{T(p) = k\} = (1 - p)^{k-1} p, \quad k = 1, 2, \ldots. \tag{1.4}$$

Replacing in (1.2) the deterministic variable $n$ by the geometric random variable $T(p)$, we obtain the *geometric sum*

$$G_p = \sum_{i=1}^{T(p)} X_i, \tag{1.5}$$

which represents the accumulation of the $X_i$'s up to the event at time $t_0 + T(p)$, i.e., the total change of the exchange rate over that period. Given that $T(p)$ represents the moment at which the probabilistic structure governing the exchange-rate process breaks down, $T(p)$ can be viewed as an the appropriate investment horizon. The nature of the underlying process after this period is uncertain, so that returns beyond $T(p)$ cannot meaningfully enter current investment decisions.

Distribution function $\mathcal{G}$ is *strictly geometric summation stable*, if, for any $p \in (0, 1)$, there exist constants $a = a(p) > 0$, such that $aG_p \overset{d}{=} X_1$. As with the stable Paretian distribution, there is an explicit representation of characteristic function of the strictly geometric stable distribution, but in general there is no such expression for the distribution function. The characteristic function of the strictly geometric summation stable distribution has the form

$$f(t) = \frac{1}{1 + \lambda |t|^\alpha \exp\left\{-i\frac{\pi}{2}\theta\alpha \operatorname{sign}(t)\right\}}, \tag{1.6}$$

where $0 < \alpha \le 2$, $|\theta| = \min(1, \frac{2}{\alpha} - 1)$, and $\lambda > 0$. If $\alpha = 2$ and $\theta = 0$, then $\mathcal{G}$ is the symmetric Laplace distribution

$$\mathcal{G}(x) = \frac{\lambda}{2} \int_{-\infty}^{x} e^{-\lambda|u|} du,$$

which plays within the class of geometric summation stable distributions a role that is analogous to that of the normal in the class of stable Paretian distributions.

The standard one-sided Weibull distribution,

$$F(x; \alpha, \lambda) = \begin{cases} 1 - \exp\left\{-\lambda x^\alpha\right\}, & \text{if } x \ge 0, \\ 0, & \text{if } x < 0, \end{cases} \tag{1.7}$$

with $\alpha > 0$ and $\lambda > 0$, and its (symmetric, double-sided) version can be viewed as both a member of the geometric summation stable and minimum stable distribution families. The Laplace distribution is the special case of the (symmetric) Weibull distribution where $\alpha$ is unity.

In the infinite variance case, i.e., $\alpha < 2$, the geometric stable law (1.6) belongs to the domain of attraction of the stable Paretian law (1.3) and, thus, the tail behavior of the two laws is identical. Initial evidence indicates that the geometric stable law can provide a better fit around the center of asset return distributions, while still matching stable Paretian tails. Overall, however, their statistical properties are quite similar due to the central limit theorem for stable laws. For this reason we consider the fat-tailed stable Paretian distribution ($0 < \alpha < 2$) as the only infinite-variance candidate distribution. From the class of geometric summation distributions we consider only the thin-tailed members, namely the Laplace and the (double) Weibull distribution. They can be viewed as analogues to the normal distribution (i.e., $\alpha = 2$), the only thin-tailed member of the stable Paretian law.

## 3  The Lev/Dollar Exchange Rate

We examine daily Bulgarian lev/US dollar exchange spot rates covering the period from February 1991 to February 1995, altogether 1021 observations. The returns, $r_t$, are computed by $r_t = 100 \times \log(R_t/R_{t-1})$, where $R_t$ is the level of the exchange rate at time $t$. Level and return data, $R_t$ and $r_t$, are shown in Figures 1 and 2, respectively, at the end of the paper.

We compare the goodness of fit of the stable Paretian, Laplace and Weibull distributions with that of alternatives commonly used in financial modeling. We consider both symmetric and asymmetric versions of these three distributions. The symmetric stable Paretian distribution is specified by setting the skewness parameter $\beta$ in (1.3) equal to zero. The symmetric (double) Weibull distribution is given by

$$\Pr(|r| < x) = F(x; \alpha, \lambda) = 1 - \exp\left\{-\lambda x^{\alpha}\right\}, \quad x > 0, \ \lambda > 0, \ \alpha > 0, \ (1.8)$$

i.e., negative and positive returns are assumed to have the same distributional properties, and the asymmetric (double) Weibull distribution by modeling positive returns with

$$\Pr(r_{pos} < x) = F(x; \alpha_{pos}, \lambda_{pos}) = 1 - \exp\left\{-\lambda_{pos} x^{\alpha_{pos}}\right\}, \quad (1.9)$$
$$x > 0, \ \lambda_{pos} > 0, \ \alpha_{pos} > 0,$$

and the absolute values of negative returns with

$$\Pr(|r_{neg}| < x) = F(x; \alpha_{neg}, \lambda_{neg}) = 1 - \exp\left\{-\lambda_{neg} x^{\alpha_{neg}}\right\}, \quad (1.10)$$
$$x > 0, \ \lambda_{neg} > 0, \ \alpha_{neg} > 0.$$

For the symmetric and asymmetric Laplace distribution we have $\alpha = 1$ and $\alpha_{neg} = \alpha_{pos} = 1$ in (1.8), (1.9) and (1.10), respectively.

As alternative candidates we consider the simple normal distribution with density function

$$f(x; \mu, \sigma^2) = \frac{1}{\sqrt{2\pi}\sigma} \exp\left\{-\frac{(x-\mu)^2}{2\sigma^2}\right\}; \quad (1.11)$$

the mixture of two normals with common mean, i.e.,

$$f(x; \mu, \sigma_1^2, \sigma_2^2, \rho) = \rho f(x; \mu, \sigma_1^2) + (1-\rho)f(x; \mu, \sigma_2^2), \quad (1.12)$$

Table 1
Results of Maximum-Likelihood Estimation

| Distribution | Parameter Estimates | | | | Log-lik |
|---|---|---|---|---|---|
| **Normal:** | $\mu$ | $\sigma_1^2$ | $\sigma_2^2$ | $\rho$ | |
| simple | 0.0839 | 1.5733 | — | — | −1910.91 |
| mixture | 0.0485 | 3.3388 | 0.3415 | 0.2137 | −1187.28 |
| **$\alpha$-stable:** | $\alpha$ | $\beta$ | $c$ | $\delta$ | |
| symmetric | 0.9780 | 0 | 0.2413 | 0.0301 | −1117.40 |
| asymmetric | 0.9535 | 0.1758 | 0.2385 | −0.5426 | −1110.96 |
| **Geo-sum stable:** | $\alpha_{neg}$ | $\lambda_{neg}$ | $\alpha_{pos}$ | $\lambda_{pos}$ | |
| sym. Laplace | 1 | 1.4463 | 1 | 1.4463 | −1350.69 |
| asym. Laplace | 1 | 1.4832 | 1 | 1.4174 | −1350.43 |
| sym. Weibull | 0.6851 | 1.6028 | 0.6851 | 1.6028 | −1171.85 |
| asym. Weibull | 0.6384 | 1.7035 | 0.7344 | 1.5266 | −1165.58 |

with mixing proportion $\rho$, $0 \leq \rho \leq 1$, and $\sigma_1^2 > \sigma_2^2$;[1] the asymmetric stable Paretian distribution with characteristic function (1.3); and the Student $t$ distribution with density

$$f(x; \mu, H, n) = \frac{n^{n/2} H^{1/2}}{B(\frac{1}{2}, \frac{n}{2})} [n + H(x - \mu)^2]^{-(n+1)/2}, \qquad (1.13)$$

where $B(\cdot, \cdot)$ denotes the beta function $B(a, b) = \Gamma(a)\Gamma(b)/\Gamma(a + b)$; and $\mu$, $H$, and $n$ are the location, scale and degrees-of-freedom parameters, respectively.

We employ the log-likelihood values obtained from maximum-likelihood estimation to compare the goodness of fit of the distributions considered.[2] The maximum-likelihood value may be viewed as an overall measure of goodness of fit, allowing us to judge which of the candidate distributions is more likely to have generated the data. From a Bayesian viewpoint, given large samples and assuming equal prior probabilities for two candidates, the ratio of maximum likelihood values of two competing models represents the asymptotic posterior odds ratio of one candidate relative to the other (see [Zel71, BG74]).

Parameter estimates for successfully estimated distributions are shown in Table 1. It turned out that we could not obtain maximum-likelihood

---

[1] Using this convention, mixing proportion $\rho$ reflects the portion of observations that is associated with the normal that has the higher variance.

[2] Note that this comparison does not take into account how many parameters we require to specify a distribution.

estimates for the Student $t$ distribution. The estimation did not converge, because the degree-of-freedom parameter attempted to assume values below two, causing the maximum-likelihood estimation algorithm to fail. This failure is no surprise in view of the estimates of the stability index $\alpha$ for the stable Paretian distribution. Values below 2 imply that no second moments and values below 1 that no first moments exist. Hence, with $\alpha$-estimates below one the degree-of-freedom parameter of the $t$ distribution should be less than unity.

The maximum log-likelihood values of the estimated distributions are shown in the last column of Table 1. The asymmetric and the symmetric stable Paretian distributions achieve the best fit with maximum log-likelihood values of $-1110.96$ and $-1117.40$, respectively. They are followed by the asymmetric ($-1165.58$) and symmetric ($-1171.85$) Weibull distributions, the mixture of normals ($-1187.28$), and the asymmetric ($-1350.43$) and symmetric ($-1350.69$) Laplace distributions. The normal distribution has by far the poorest fit with a maximum log-likelihood value of $-1910.91$.

A visual comparison of empirical and estimated densities confirms the goodness-of-fit results—especially, the dismal fit of the normal distribution. Figure 3 shows substantial deviations between the empirical density (dotted curve) and the estimated normal density (solid curve). The mixture of normals fits considerably better, but has still problems in capturing the peakedness of the empirical distribution (see Figure 4). The symmetric and asymmetric stable Paretian distributions fit very well in the center and the tails (see Figures 5 for the asymmetric case). The Laplace distributions yield less reasonable fits with respect to both the peak and the tails. However, considering the symmetric Laplace distribution (see Figure 6), which involves only one parameter, the goodness of the fit, relative to the normal, is somewhat surprising. The Weibull distributions fit reasonably well (see Figure 7 for the asymmetric case). An exception is the exaggeration of the peak, which occurs, however, only in the narrow return range $\pm\ 0.05\%$—a range that is irrelevant for practical investment decisions.

# 4  Conclusion

Several candidate distributions have been applied to daily Bulgarian lev/US dollar exchange-rate returns. It is a highly volatile time series and typical for currencies in the initial phase of moving from controlled to freely floating exchange rates. We find that the asymmetric stable Paretian distribution dominates all other candidates. The symmetric stable Paretian distribution has the second best fit, but a likelihood-ratio test will reject the symmetry restriction $\beta = 0$.

All other distributions under consideration are thin-tailed and not quite as capable in capturing the distributional characteristics as the fat-tailed stable Paretian distribution. The goodness of fit, measured in terms of the maximum log-likelihood, and a visual comparison of empirical and estimated densities reveal their shortcomings. However, the Weibull, mixture of normals and, to some extent, the Laplace distributions are still reasonable, whereas the normal distribution—the model most widely adopted in financial modeling—appears to be very inappropriate for modeling highly volatile returns on exchange rates like that of the Bulgarian lev against the US dollar.

Given that Laplace and Weibull distributions are only special, thin-tailed members of the geometric summation stable family, future work will examine the fit of fat-tailed members of that family.

*Acknowledgments:* The research of S. Mittnik was supported by the *Deutsche Forschungsgemeinschaft*. Part of the research was conducted while S. Rachev was visiting the University of Kiel with support from the *Alexander-von-Humboldt Foundation*.

# 5 References

[BG74] R.C. Blattberg and N.J. Gonedes, Stable and Student distributions for stock prices. *Journal of Business* 47, 244-280, 1974.

[Che91] Y. Chen, Distributions for Asset Returns. *Unpublished Ph.D. Dissertation, SUNY–Stony Brook, Department of Economics*, 1991.

[Man63] B. Mandelbrot, The variation of certain speculative prices. *Journal of Business* 36, 394-419, 1963.

[MR89] S. Mittnik and S.T. Rachev Stable distributions for asset returns. *Applied Mathematics Letters* 2, 301-304, 1989.

[MR91] S. Mittnik and S.T. Rachev, Alternative multivariate stable distributions and their applications to financial modeling. In: S. Cambanis et al., eds., Stable Processes and Related Topics, *Birkhäuser*, Boston, 107-119, 1991.

[MR93] S. Mittnik and S.T. Rachev, Modeling asset returns with alternative stable distributions. *Econometric Reviews* 12, 261-330, 1993.

[Zel71] A. Zellner, An Introduction to Bayesian Inference in Econometrics. *John Wiley & Sons*, New York, 1971.

[Zol86] V.M. Zolotarev, One-dimensional Stable Distributions. *Vol. 65 of Translations of Mathematical Monographs, American Mathematical Society*, 1986.

Ohridski University (Bulgaria)

Ohridski University (Bulgaria)

University of Kiel

University of California, Santa Barbara

Figure 1

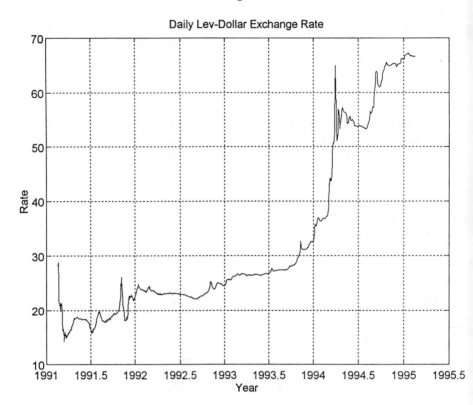

Daily Lev-Dollar Exchange Rate

Figure 2

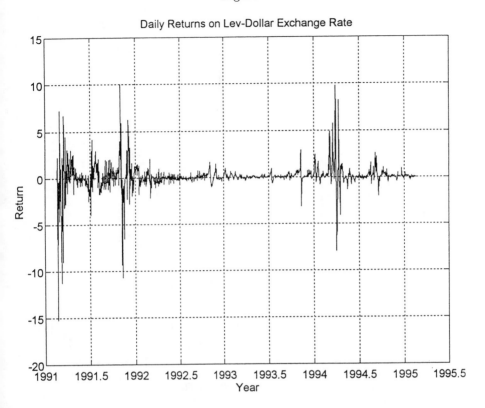

Figure 3

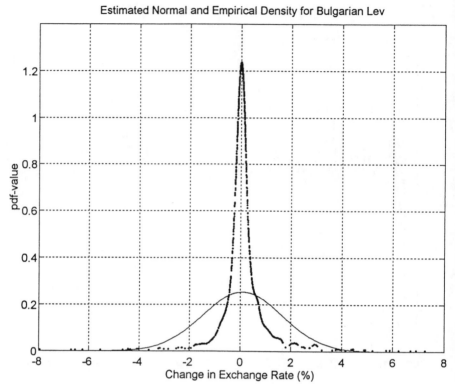

Estimated Normal and Empirical Density for Bulgarian Lev

Figure 4

Estimated Mixed Normal and Empirical Density for Bulgarian Lev

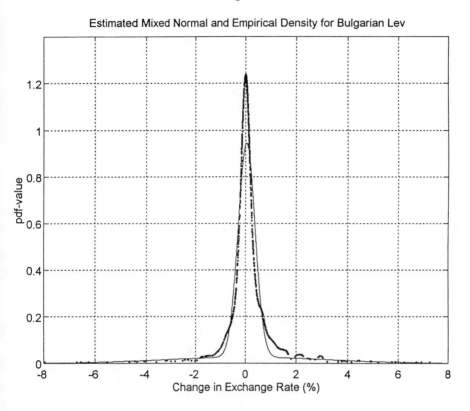

Figure 5

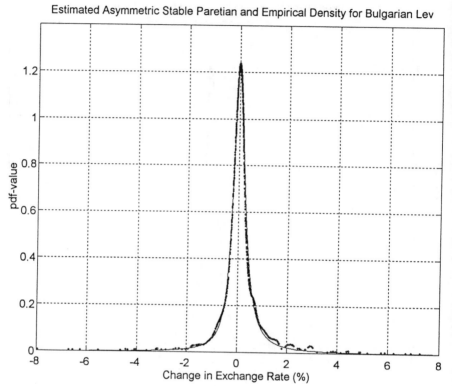

Estimated Asymmetric Stable Paretian and Empirical Density for Bulgarian Lev

Figure 6

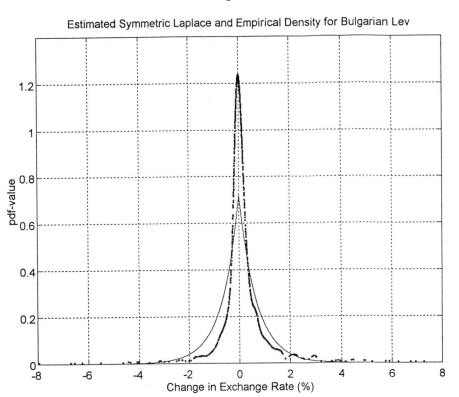

Estimated Symmetric Laplace and Empirical Density for Bulgarian Lev

Figure 7

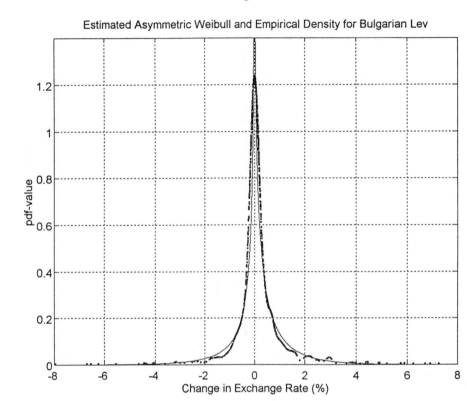

Estimated Asymmetric Weibull and Empirical Density for Bulgarian Lev

# Asymptotic statistical inference for nonstationary processes with evolutionary spectra

R. Dahlhaus[1]

ABSTRACT An asymptotic approach for nonstationary processes which locally show a stationary behaviour is presented. Kernel methods for local estimates of the covariance structure and the time varying spectral density are discussed. In particular, we calculate the bias of the estimates due to nonstationarity and determine the optimal bandwidth. Furthermore, the relation to Priestley's theory of processes with evolutionary spectra is discussed.

## 1 Introduction

Stationarity has always played a major role in time series analysis. For stationary processes there are a large variety of models, such as ARMA-models, and powerful techniques, such as bootstrap methods or methods based on the spectral density. Furthermore, the theory for stationary process is well developed and most of the statistical methods are satisfactorily explained by asymptotic theoretical results.

For nonstationary time series this is different, mainly for two reasons. First, there exists no natural generalization from stationary processes to nonstationary processes, and second, it is often not clear how to set down a reasonable asymptotics for nonstationary processes. Asymptotic considerations are very often contradictory to the idea of nonstationarity since future observations of a general nonstationary process may not contain any information at all on the probabilistic structure of the process at present.

One exception are cointegrated processes where the integrating equations are assumed to be fixed as time tends to infinity and the remainders are assumed to be stationary. Such a setting still allows for meaningful asymptotic considerations and a number of important asymptotic results for cointegrated processes have been derived in the last decade.

[1] This work has been supported by the Deutsche Forschungsgemeinschaft (DA 187/9-1) and by a European Union Capital and Mobility Programme (ERB CHRX-CT 940693).

The situation becomes different when one thinks of a process which is close to a stationary one at a certain time but whose charateristics (covariances, parameters) are gradually changing in an unspecific way as time evolves. Suppose for example a discrete time process stays stationary over a segment of length $L$ and changes afterwards. Then only the covariances up to lag $L - 1$ are identifiable from that segment which means that the process cannot be uniquely determined. Furthermore, classical asymptotic considerations for e.g. the covariance estimator of lag one are meaningless since the true covariance of lag one may change as time passes on.

Nevertheless, the idea of fitting locally a stationary model to time series data may be meaningful and in fact it has been used quite often. An example is in speech coding where the LPC-method (linear predictive coding) consists of fitting stationary AR-models to the speech signal on small data segments (cf. [TS85]).

The idea of locally having an approximately stationary process was made rigorous in Priestley's theory of processes with evolutionary spectra ([Pri65]; overviews may be found in [Pri81], [Pri88]). Priestley considered processes having a time varying spectral representation.

$$X_t = \int_{-\pi}^{\pi} \exp(i\lambda t) A_t(\lambda) d\xi(\lambda), \quad t \in \mathbb{Z} \tag{1.1}$$

with an orthogonal increament process $\xi(\lambda)$ and a time varying transfer function $A_t(\lambda)$. (Priestley mainly looks at continuous time processes, but the theory is the same for processes in discrete time).

Within the approach of Priestley asymptotic considerations (e.g. for judging the efficiency of a local covariance estimator) are not possible (or they are meaningless). As explained above this is due to the nature of nonstationarity. As a consequence such valuable tools as consistency, asymptotic normality, efficiency, LAN-expansions, neglecting higher order terms in Taylor expansions, etc. cannot be used in the theoretical treatment of statistical procedures for such procceses. This is particularly bad since the complicated structure of nonstationary time series requires technical simplifications which could be provided by an asymptotic approach.

To overcome this problem we have introduced in [Dah96a] an asymptotic approach similar to nonparametric regression for processes with evolutionary spectra. The idea can be explained most easily by the example of a time varying $AR(1)$-process

$$X_t = g(t)X_{t-1} + \varepsilon_t \qquad \text{with } \varepsilon_t \text{ iid } \mathcal{N}(0, \sigma^2)$$

which we assume to be observed for $t = 1, ..., T$. Suppose we fit the model $g_\theta(t) = a + bt + ct^2$ to the data. It is easy to construct different estimates for the parameters (e.g. a least squares estimate, a maximum likelihood estimate or a fit to locally estimated $AR(1)$-parameters). Classical asymptotic considerations for comparing these estimates make no sense since with $T$

tending to infinity $g_\theta(t) \to \infty$ while e.g. $|g(t)|$ may be less than one within the observation segment. Analogously to nonparametric regression it therefore seems to be natural to set down the asymptotic theory by assuming that $g(t)$ is "observed" on a finer and finer grid (but on the same interval), i.e. that we observe

$$X_{t,T} = g\left(\frac{t}{T}\right) X_{t-1,T} + \varepsilon_t \tag{1.2}$$

for $t = 1, ..., T$ (where $g$ is now rescaled to the interval $[0, 1]$).

Letting $T$ tend to infinity now means that we have in the sample $X_{1,T}, ..., X_{T,T}$ more and more "observations" for the local structure of $g$ at each time point (if e.g. $g(u) = \chi_{[1/2,1]}(u)$ we have $T/2$ observations, both for the white noise situation ($g = 0$) and for the random walk case ($g = 1$)). If $g$ is constant $X_{t,T}$ is independent of $T$, and the above asymptotic approach becomes identical to the classical asymptotics for stationary processes.

To set down a similar approach for arbitrary processes with evolutionary spectra we may try to rescale the transfer function in (1.1) in exactly the same way and look at processes of the form

$$X_{t,T} = \mu\left(\frac{t}{T}\right) + \int_{-\pi}^{\pi} \exp(i\lambda t) A\left(\frac{t}{T}, \lambda\right) d\xi(\lambda).$$

However, it turns out that the equations (1.2) have not exactly but only approximately a solution of this form which leads to the following more general definition which was first given in [Dah96a].

**Definition 1.1** *A sequence of stochastic processes $X_{t,T}(t = 1, ..., T)$ is called locally stationary with transfer function $A^o$ and trend $\mu$ if there exists a representation*

$$X_{t,T} = \mu\left(\frac{t}{T}\right) + \int_{-\pi}^{\pi} exp(i\lambda t) A_{t,T}^o(\lambda) d\xi(\lambda) \tag{1.3}$$

*where*
*(i) $\xi(\lambda)$ is a stochastic process on $[-\pi, \pi]$ with $\overline{\xi(\lambda)} = \xi(-\lambda)$ and*

$$cum\{d\xi(\lambda_1), ..., d\xi(\lambda_k)\} = \eta\left(\sum_{j=1}^{k} \lambda_j\right) g_k(\lambda_1, ..., \lambda_{k-1}) d\lambda_1 ... d\lambda_k$$

*where cum $\{...\}$ denotes the cumulant of $k - th$ order, $g_1 = 0$, $g_2(\lambda) = 1$, $|g_k(\lambda_1, ..., \lambda_{k-1})| \le const_k$ for all $k$ and $\eta(\lambda) = \sum_{j=-\infty}^{\infty} \delta(\lambda + 2\pi j)$ is the period $2\pi$ extension of the Dirac delta function.*
*(ii) There exists a constant $K$ and a $2\pi$-periodic function $A : [0, 1] \times \mathbb{R} \to \mathbb{C}$ with $A(u, -\lambda) = \overline{A(u, \lambda)}$ and*

$$\sup_{t,\lambda} |A_{t,T}^o(\lambda) - A\left(\frac{t}{T}, \lambda\right)| \leq KT^{-1} \tag{1.4}$$

*for all T. $A(u, \lambda)$ and $\mu(u)$ are assumed to be continuous in u.*

The smoothness of $A$ in $u$ guarantees that the process has (assymptotically) locally a stationary behaviour. $f(u, \lambda) = |A(u, \lambda)|^2$ is called the time-varying spectral density of the process. In [Dah96b] we have proved that $f(u, \lambda)$ is uniquely determined by the triangular array.

**Example 1.1** *(i)  Suppose $Y_t$ is a stationary process and $\mu, \sigma : [0, 1] \to \mathbb{R}$ are continuous. Then*

$$X_{t,T} = \mu\left(\frac{t}{T}\right) + \sigma\left(\frac{t}{T}\right) Y_t$$

*is locally stationary with $A_{t,T}^o(\lambda) = A(\frac{t}{T}, \lambda)$. If $Y_t$ is an AR(2)-process with (complex) roots close to the unit circle then $Y_t$ shows a periodic behaviour and $\sigma$ may be regarded as a time varying amplitude function of the process $X_{t,T}$. If $T$ tends to infinity more and more cycles of the process with $u = t/T \in [u_o - \varepsilon, u_o + \varepsilon]$, i.e. with amplitude close to $\sigma(u_o)$ are observed.*

*(ii)  Suppose $X_{t,T}$ is an autoregressive moving average process with time varying coefficients, i.e. a solution of*

$$\sum_{j=0}^{p} a_j\left(\frac{t}{T}\right) X_{t-j,T} = \sum_{j=0}^{q} b_j\left(\frac{t}{T}\right) \sigma\left(\frac{t-j}{T}\right) \varepsilon_{t-j}$$

*where $\varepsilon_t$ are iid with mean zero and variance 1 and $a_0(u) \equiv b_0(u) \equiv 1$. Then $X_{t,T}$ is locally stationary with time varying spectral density*

$$f(u, \lambda) = \frac{\sigma^2(u)}{2\pi} \frac{|\Sigma_{j=0}^{q} b_j(u) \exp(i\lambda j)|^2}{|\Sigma_{j=0}^{p} a_j(u) \exp(i\lambda j)|^2}.$$

*This was proved in [Dah96b] (Chapter 3). In this case we only have (1.4) instead of $A_{t,T}^o(\lambda) = A(\frac{t}{T}, \lambda)$.*

An *ARMA*-process with time varying coefficients is an important example for a locally stationary process which is purely defined in the time domain. If its coefficient functions are parametric functions (e.g. polynomials in time) we obtain an example for a parametric locally stationary process. Inference for such processes has been investigated in [Dah96a] and [Dah96c] where maximum likelihood estimates, minimum distance estimates and least squares estimates have been discussed. Furthermore, the aspects of model selection and model misspecification for locally stationary time series models have been investigated. Typically, parametric models are models in the time domain and the time varying spectral density only serves as a tool (e.g. in the proofs).

We denote by

$$c(u, k) := \int_{-\pi}^{\pi} f(u, \lambda) \exp(i\lambda k) d\lambda \tag{1.5}$$

the local covariance of lag $k$ at time $u$. We have

$$
\begin{aligned}
cov(X_{[uT],T}, X_{[uT]+k,T}) &= \int_{-\pi}^{\pi} \exp(i\lambda k) A^{o}_{[uT]+k,T}(\lambda) A^{o}_{[uT],T}(-\lambda) d\lambda \\
&= c(u, k) + O(T^{-1})
\end{aligned}
$$

uniformly in $u$ and $k$ (if e.g. $A(u, \lambda)$ is uniformly differentiable in $u$).

The rescaling property inherent in the definition and the loss of $T$ as time ($T \to \infty$ no longer means looking into the future) may look strange from a first view. Roughly speaking the approach is an abstract setting for processes with evolutionary spectra allowing for a meaningful asymptotic theory. A deeper justification and a comparison with the approach of Priestley can be found in Section 3. In particular we point out why this approach leads to a natural generalization of the asymptotic theory for stationary time series which is contained as a special case (for $\mu$ and $A$ being constant over time).

In Section 2 we show how the approach can be used for the investigation of local covariance and spectral estimates. In particular, we show how the bias due to nonstationarity can be quantified.

## 2   Covariance- and spectral-estimates

In the first example we consider an estimate for the covariance of lag $k$ at time $u$ as defined in (1.5). For simplicity we assume $\mu(u) \equiv 0$. It is natural to consider as an estimate all products of lag $k$ in a neighbourhood of $u$, i.e.

$$\hat{c}_T(u, k) := \frac{1}{b_T T} \sum_t K\left(\frac{u - (t + k/2)/T}{b_T}\right) X_{t,T} X_{t+k,T}$$

where $K : \mathbb{R} \to [0, \infty)$ is a kernel with $K(x) = K(-x)$, $\int K(x) dx = 1$ and $K(x) = 0$ for $x \notin [-1/2, 1/2]$ and $b_T$ is the bandwidth. The following theorem gives a bias expansion for $\hat{c}_T$.

**Theorem 2.1** *Suppose $X_{t,T}$ is locally stationary with mean 0 where $A(u, \lambda)$ is twice differentiable in $u$ with uniformly bounded derivative. Then*

$$E\hat{c}_T(u, k) = c(u, k) + \frac{1}{2} b_T^2 \int x^2 K(x) dx \left[\frac{\partial^2}{\partial^2 u} c(u, k)\right] + o(b_T^2) + O\left(\frac{1}{b_T T}\right).$$

PROOF. From the representation (1.3) we get

$$
\begin{aligned}
E\hat{c}_T(u,k) &= \frac{1}{b_T T}\sum_t K\left(\frac{u-(t+k/2)/T}{b_T}\right)\\
&\quad \times \int_{-\pi}^{\pi}\exp(i\lambda k)A_{t+k,T}^{o}(\lambda)A_{t,k}^{o}(-\lambda)d\lambda\\
&= \frac{1}{b_T T}\sum_t K\left(\frac{u-(t+k/2)/T}{b_T}\right)\\
&\quad \times \int_{-\pi}^{\pi}\exp(i\lambda k)A\left(\frac{t+k}{T},\lambda\right)A\left(\frac{t}{T},-\lambda\right)d\lambda+O(T^{-1})\\
&= \frac{1}{b_T T}\sum_t K\left(\frac{u-(t+k/2)/T}{b_T}\right)\\
&\quad \times \int_{-\pi}^{\pi}\exp(i\lambda k)f\left(\frac{t+k/2}{T},\lambda\right)d\lambda+O(T^{-1}).
\end{aligned}
$$

A second order expansion of $f$ around $u$ shows with the symmetry of the kernel that this is equal to

$$
=\frac{1}{2}b_T^2\int_{-1/2}^{1/2}x^2K(x)dx\int_{-\pi}^{\pi}\exp(i\lambda k)\frac{\partial^2}{\partial u^2}f(u,\lambda)d\lambda+o(b_T^2)+O\left(\frac{1}{b_T T}\right)
$$

which gives the result.    □

Note, that the above bias of order $b_T^2$ is solely due to nonstationarity which is measured by $\frac{\partial^2}{\partial u^2}c(u,k)$. If the process is stationary this second derivative is zero and the bias disappears. The bias of order $(b_T T)^{-1}$ is due to the procedure and can be calculated as well (it varies if the mean is unknown). Note, that $b_T T$ may be viewed as the 'effective sample length' which is used for the caluclation of $\hat{c}_T(u,k)$. The bandwidth $b_T$ may now be chosen to balance these two bias terms. Without proof we remark that

$$
\mathrm{var}(\hat{c}_T(u,k))=\frac{1}{b_T T}\int_{-1/2}^{1/2}K(x)^2dx\sum_{\ell=-\infty}^{\infty}c(u,\ell)[c(u,\ell)+c(u,\ell+2k)].
$$

We now may select $b_T$ such that it minimizes the mean squared error.

Note, that the above estimate of $c(u,k)$ is obtained by applying a stationary estimation procedure to the data $X_{t,T}$ on the time segment $t\in\{[uT-b_T T/2]+1,...,[uT+b_T T/2]\}$. An improvement may be obtained by modelling $c(v,k)$ on the segment $v\in[u-b_T/2,u+b_T/2]$ by a curve, e.g. by a polynomial leading to a local polynomial fit and $\bar{c}_T(u,k):=\hat{c}_o$ where

$$
\hat{c}=\operatorname*{argmin}_c\frac{1}{b_T T}\sum_t K\left(\frac{u-(t+k/2)/T}{b_T}\right)
$$

$$\times \left( X_{t,T} X_{t+k,T} - \sum_{j=0}^{q} c_j \left( u - \frac{t+k/2}{T} \right)^j \right)^2.$$

Such estimates have been considered in the context of covariances depending on an arbitrary covariable by Hyndman and Wand [HW96].

The ordinary (tapered) covariance on a segment of length $N$

$$\tilde{c}_T(u, k) = H_N^{-1} \sum_{1 \leq t, t+k \leq N} h\left(\frac{t}{N}\right) h\left(\frac{t+k}{N}\right) X_{[uT]-N/2+t} X_{[uT]-N/2+t+k}$$

with $H_N = \sum_{t=1}^{N} h(t/N)^2$ has the same bias expansion with $K(x) = h(x)^2$ and $b_t = N/T$. This gives also the bias for the classical covariance estimate where $h(x) = I_{(0,1]}(x)$.

Suppose now we have a time varying $AR(1)$-model

$$X_{t,T} = a\left(\frac{t}{T}\right) X_{t-1,T} + \varepsilon_t$$

with $|a(u)| \leq 1 - c$ for all $u$ where $c > 0$ and $a(u)$ is continuous in $u$. Then it may be shown that there exists a causal solution. Multiplication with $X_{t-1,T}$ and taking expectations gives

$$\text{cov}(X_{t,T}, X_{t-1,T}) = a\left(\frac{t}{T}\right) \text{cov}(X_{t-1,T}, X_{t-1,T})$$

which suggests as an estimate of $a(u)$

$$\hat{a}(u) = \frac{c_T(u, 1)}{c_T(u, 0)}$$

with one of the above covariance estimates.

By using the above bias expansion it can be shown that

$$\begin{aligned}
E\hat{a}(u) &= a(u) - \frac{1}{2} b_T^2 \int x^2 K(x) dx \, c(u, 0)^{-1} \\
&\quad \times \left[ \left( \frac{\partial^2}{\partial u^2} c(u, 0) \right) a(u) - \frac{\partial^2}{\partial u^2} c(u, 1) \right] + \frac{1}{b_T T} c^* + o\left( b_T^2 + \frac{1}{b_T T} \right)
\end{aligned}$$

where the constant $c^*$ depends on the particular covariance estimate. Furthermore,

$$\text{var}(\hat{a}(u)) = \frac{1}{b_T T} \int K(x)^2 dx (1 - a(u)^2) + o\left( \frac{1}{b_T T} \right).$$

Again, the bandwidth $b_T$ and the kernel $K$ may now be selected to balance the (nonstationarity-) bias and the variance. For $c_T(u, k) = \tilde{c}_T(u, k)$ and time-varying autoregressive processes (of arbitrary order) this was done in [DG95] (Theorem 1 + 2).

In the second example we study the behaviour of a kernel estimate for $f(u, \lambda)$ over the segment $\{[uT] - N/2 + 1, [uT] + N/2\}$. Let

$$I_N(u, \lambda) = \frac{1}{2\pi H_N} \left| \sum_{s=1}^{N} h\left(\frac{s}{N}\right) X_{[uT]-N/2+s,T} \exp(-i\lambda s) \right|^2$$

where $h : [0, 1] \to \mathbb{R}$ is a data window. We assume that $h(x) = h(1 - x)$. As Theorem 2.2 below shows

$$K_t(x) := \left\{ \int_0^1 h(x)^2 dx \right\}^{-1} h(x + 1/2)^2, \qquad x \in [-1/2, 1/2] \qquad (1.6)$$

plays the role of a kernel in the time domain and $b_t := N/T$ may be regarded as the bandwidth. Even in the stationary case $I_N(u, \lambda)$ is not a consistent estimate of the spectrum and we have to smooth it over neighbouring frequencies. Let

$$\hat{f}(u, \lambda) = \frac{1}{b_f} \int K_f\left(\frac{\lambda - \mu}{b_f}\right) I_N(u, \mu) d\mu$$

where $K_f : \mathbb{R} \to [0, \infty)$ is a kernel with $K_f(x) = 0$ for $x \notin [-1/2, 1/2]$, $K_f(x) = K_f(-x)$ and $\int K_f(x)dx = 1$, and $b_f$ is the bandwidth in frequency direction.

**Theorem 2.2** *Suppose $X_{t,T}$ is locally stationary with $\mu \equiv 0$ and transfer function $A$ whose derivatives $\frac{\partial^2}{\partial u^2}A$, $\frac{\partial^2}{\partial \lambda^2}A$, $\frac{\partial^2}{\partial u \partial \lambda}A$ are continuous. Then*

*(i)* $EI_N(u, \lambda) = f(u, \lambda) + \frac{1}{2}b_t^2 \int_{-1/2}^{1/2} x^2 K_t(x)dx \frac{\partial^2}{\partial u^2} f(u, \lambda) + o(b_t^2)$

$$+ O\left(\frac{\log(b_t T)}{b_t T}\right),$$

*(ii)* $E\hat{f}(u, \lambda) = f(u, \lambda) + \frac{1}{2}b_t^2 \int_{-1/2}^{1/2} x^2 K_t(x)dx \frac{\partial^2}{\partial u^2} f(u, \lambda)$

$$+ \frac{1}{2}b_f^2 \int_{-1/2}^{1/2} x^2 K_f(x)dx \frac{\partial^2}{\partial \lambda^2} f(u, \lambda) + o\left(b_t^2 + \frac{\log(b_t T)}{b_t T} + b_f^2\right),$$

*(iii)* $var(\hat{f}(u, \lambda)) = (b_t b_f T)^{-1} f(u, \lambda)^2 \int_{-1/2}^{1/2} K_t(x)^2 dx \int_{-1/2}^{1/2} K_f(x)^2 dx \cdot$
$\cdot (2\pi + 2\pi\{\lambda \equiv 0 \, mod \, \pi\}).$

PROOF. We only give a brief sketch. (1) We have

$$EI_N(u, \lambda) = \frac{1}{2\pi H_N} \sum_{r,s=1}^{N} h\left(\frac{r}{N}\right) h\left(\frac{s}{N}\right) \int_{-\pi}^{\pi} A^o_{[uT]-N/2+r,T}(\mu)$$

$$\times A^o_{[uT]-N/2+s,T}(-\mu)\exp(-i(\lambda - \mu)(r - s))d\mu.$$

We now replace $A^o_{[uT]-N/2+r,T}(\mu)$ by $A(u + \frac{r-N/2}{T}, \lambda)$. Let $L_N : \mathbb{R} \to \mathbb{R}$ be the periodic extension (with period $2\pi$) of

$$L_N(\alpha) := \begin{cases} N, & |\alpha| \le 1/N \\ 1/|\alpha|, & 1/N \le |\alpha| \le \pi. \end{cases}$$

As in the proof of Lemma A.5 of [Dah96a] we obtain with integration by parts that

$$\left| \sum_{r=1}^{N} h\left(\frac{r}{N}\right) \left[ A^o_{[uT]-N/2+r,T}(\mu) - A(u, \lambda) \right] \exp(-i(\lambda - \mu)r) \right|$$

$$\le K\left(\frac{N}{T} + |\lambda - \mu| L_N(\lambda - \mu)\right) \le K$$

with some constant $K$ while

$$\left| \sum_{s=1}^{N} h\left(\frac{s}{N}\right) A^o_{[uT]-N/2+s,T}(-\mu) \exp(i(\lambda - \mu)s) \right| \le K L_N(\lambda - \mu).$$

The replacement error therefore is of order $O(\frac{\log N}{N})$. In the same way we then replace $A^o_{[uT]-N/2+s,T}(-\mu)$ by $A(u + \frac{s-N/2}{T}, -\lambda)$ and integrate over $\mu$ leading to

$$EI_N(u, \lambda) = \frac{1}{H_N} \sum_{r=1}^{N} h\left(\frac{r}{N}\right)^2 f\left(u + \frac{r - N/2}{T}, \lambda\right) + O\left(\frac{\log N}{N}\right).$$

A Taylor expansion of $f$ around $u$ now gives part (i). (ii) follows from (i), also with a Taylor expansion. The proof of (iii) is omitted. It is similar to the proof of Lemma A.9 in [Dah96a]. □

Note, that the first bias term of $\hat{f}$ is due to nonstationarity while the second is due to the variation of the spectrum in frequency direction.

We now may minimize the relative mean squared error $\text{RMSE}(\hat{f}) := E(\hat{f}(u, \lambda)/f(u, \lambda) - 1)^2$ with respect to $b_f$, $b_t$ (i.e. $N$), $K_f$ and $K_t$ (i.e. $h$). Similarly, we can treat the integrated mean squared error (integrated over time and frequency). Let

$$\Delta_u := \frac{\partial^2}{\partial u^2} f(u, \lambda)/f(u, \lambda), \qquad \Delta_\lambda := \frac{\partial^2}{\partial \lambda^2} f(u, \lambda)/f(u, \lambda),$$

$$v_t = \int x^2 K_t(x) dx, \quad k_t = \int K_t(x)^2 dx, \quad v_f = \int x^2 K_f(x), \quad k_f = \int K_f(x)^2 dx.$$

**Theorem 2.3** *The mean squared error RMSE($\hat{f}$) is minimal for*

$$K_t^{opt}(x) = K_f^{opt}(x) = 6(1/4 - x^2), \qquad -1/2 \le x \le 1/2,$$

$$b_t^{opt} = T^{-1/6}(576\pi)^{1/6}\left(\frac{\Delta_\lambda}{\Delta_u^5}\right)^{1/12},$$

*and*

$$b_f^{opt} = T^{-1/6}(576\pi)^{1/6}\left(\frac{\Delta_u}{\Delta_\lambda^5}\right)^{1/12},$$

*leading to*

$$RMSE(\hat{f}^{opt}) = \frac{3^{7/3}}{25}\pi^2(\Delta_u\Delta_\lambda)^{1/3}T^{-2/3}.$$

PROOF. Again we only give a brief sketch. The relative mean squared error is of the form $(c_1 b_t^2 + c_2 b_f^2)^2 + c_3/(b_t b_f)$. Minimisation with respect to $b_t$ and $b_f$ leads to

$$b_t^{opt} = \left(\frac{c_3 c_2^{1/2}}{8c_1^{5/2}}\right)^{1/6}, \qquad b_f^{opt} = \left(\frac{c_3 c_1^{1/2}}{8c_2^{5/2}}\right)^{1/6}$$

with an optimal RMSE of

$$3(c_1 c_2 c_3^2)^{1/3} = \text{const.}(v_t v_f k_t^2 k_f^2)^{1/3}.$$

In order to minimize this with respect to the kernels we can minimize $v_t k_t^2$ and $v_f k_f^2$ separately (under the restriction $\int K_t(x)dx = \int K_f(x)dx = 1$) leading to the optimal kernel $K_t(x) = K_f(x) = 6(1/4 - x^2)$ (cf. [Pri81], Chapter 7.5). For this kernel we get $v_t = v_f = 1/20$ and $k_t = k_f = 6/5$ leading to the result.    □

The relations $b_t = N/T$ and (1.6) immediately lead to the optimal segment length and the optimal data window $h$. The result of Theorem 2.3 is quite reasonable: If the degree of nonstationarity is small, then $\Delta_u$ is small and $b_t^{opt}$ gets large. If the variation of $f$ is small in frequency direction, then $\Delta_\lambda$ is small and $b_t^{opt}$ gets smaller (more smoothing is put in frequency direction than in time direction). This is another example, how the bias due to nonstationarity can be quantified with the approach of local stationarity and balanced with another bias term and a variance term.

As shown by Neumann and von Sachs [NvS96] the above rate for the spectral density estimate is optimal. In the same paper they study the behaviour of a wavelet estimate for the time varying spectral density (see also von Sachs and Schneider [vSS96] and Priestley [Pri96]). Riedel [Rie93] has discussed smoothing of the log-periodogram in the framework of Priestley.

# 3   Local stationarity and Priestley's theory of evolutionary spectra

We now discuss the relation between our approach and Priestley's theory of nonstationary processes with evolutionary spectra in some depth. Furthermore, we give a stronger justification why our approach leads to a reasonable asymptotic theory for nonstationary processes.

In his theory Priestley investigates processes having a time varying spectral representation.

$$X_t = \int_{-\pi}^{\pi} \exp(i\lambda t) A_t(\lambda) d\xi(\lambda), \qquad t \in \mathbb{Z}.$$

Typically, such a representation is not unique with respect to $A_t(\lambda)$.

If one takes the view of $|A_t(\lambda)|^2$ being some local spectrum of time $t$ there is a <u>heuristic</u> explanation for this lack of uniqueness. Even if $A_t(\lambda)$ were constant over a time period $t \in \{T_1, ..., T_2\}$ we could only identify the corresponding covariances up to lag $T_2 - T_1 + 1$ from the process, i.e. the spectral density being the Fourier transform of all covariances, must remain unidentifiable. Taking the expectation of the periodogram over this segment shows however, that a convolution of $|A_t(\lambda)|^2$ with the Fejer kernel is uniquely determined from the process.

Although the exact properties are a bit more difficult to derive than in the above heuristics the result is roughly the same: Not the spectrum itsself but only an 'average' within same band of frequencies around $\lambda_0$ and an interval of time around $t_0$ is uniquely determined (cf. [Pri81], Section 11.2, "Determination of evolutionary spectra"). This average again is obtained as the expectation of a windowed periodogram over some time segment. Roughly we can say that this segment has to be smaller as what Priestley calls (and defines) the 'bandwidth of stationarity' (the exact setting in Priestley is a bit more difficult). If this bandwidth of stationarity were allowed to tend to infinity (as a characteristics of the process it is not, however!!) the above mentioned average would converge with an increasing time segment to the local spectrum at frequency $\lambda_0$ and at time $t_0$ leading in particular to the uniqueness of $|A_t(\lambda)|^2$.

For the same reason it is not possible to set down a meaningful asymptotic theory of statistical inference in the classical sense leading to such valuable properties as asymptotic unbiasedness, consistnecy or efficiency. As pointed out in the introduction we have only a finite number of observations with (approximately) the same local covariance structure and letting the sample size tend to infinity of e.g. the classical covariance estimator does not make sense since the covariance structure of the forthcoming observations will be different. The only way out of this limitation again would be an increase of the bandwidth of stationarity (with the sample size) which is not possible within the framework of Priestley's theory.

For stationary processes these problems do not occur: the spectrum is well defined and letting the sample size tend to infinity leads to a meaningful asymptotic theory.

However, if we look as a statistician at a finite number of observations from a stationary process, the situation becomes very similar to the situation for nonstationary processes as described above. Again we only can identify finitely many covariances and the spectrum therefore is not uniquely determined. As above we may take the expectation of the periodogram over that sample to see that a convolution of the spectrum is uniquely determined. And again we have only a finite number of observations to estimate a covariance.

The asymptotic theory for stationary processes which describes what happens when the process physically moves on can also be looked at in a different way. We may regard it as the description of what happens under the abstraction 'Suppose we had more data (a longer segment) of the same structure available' which, in the stationary case, is not that much different. However, we believe that in nearly all cases this is the correct point of view (since we usually only investigate a data set of fixed size). This abstraction now leads to an (asymptotically) uniquely defined spectral density and a meaningful asymptotic theory for statistical inference.

For nonstationary processes with evolutionary spectra the corresponding abstraction would be: 'Suppose we had more data of each local structure available' which heuristically would lead to an increasing bandwidth of stationarity as the sample size increases. This abstraction also would lead to an (asymptotically) uniquely defined spectrum (at each time point) and to a meaningful asymptotic theory for nonstationary processes.

In our approach of local stationarity presented in this paper we take exactly this viewpoint. As already discussed in the introduction with increasing $T$ more and more data of each local structure become available allowing for meaningful asymptotic investigations of statistical procedures. Theorem 2.2 of [Dah96b] makes the above considerations rigorous by proving that the spectrum at each time point is asymptotically uniquely defined.

Therefore, the asymptotic approach presented in this paper is a natural generalisation of the classical asymptotic approach for stationary time series. In fact, for $A(u, \lambda)$ and $\mu(u)$ being constant in $u$, the triangular array $X_{t,T}$ reduces to a stationary sequence and the asymptotic theory for stationary time series is obtained as a special case. An example is the behaviour of the MLE for stationary time series which can be obtained as a corrolary from Theorem 2.4 of [Dah96c] where the locally stationary case is treated. This is an important property of the approach.

The big difference to stationary time series is that letting $T$ tend to infinity no longer has a physical interpretation. The approach purely is an abstraction for judging statistical inference. As a consequence it makes for example no sense to ask for a real data example that fulfills the rescaling property of the triangular array $X_{t,T}$ inherent in Definition 1.1.

It should be remarked that our rescaling approach is standard in asymptotic statistical theory: The statistical experiment is rescaled in such a way that the parameters of interest remain asymptotically the same (in our situation the curves $\mu(u)$ and $A(u, \lambda)$ or e.g. the curves $a_1(u)$, $\sigma(u)$ in the time-varying $AR$-case). Examples are nonparametric regression or the consideration of data coming from a local alternative $X_{t,T} \sim P_{\theta_o + h/\sqrt{T}}$ (iid) if one uses asymptotics to evaluate the power of a test ($h$ now becomes the local parameter which remains asymptotically the same).

One may raise the question whether it makes sense to fit statistical models to data that cannot even be identified uniquely from the probabilistic structure of the data set (which we do when using a nonstationary model with an unspecified evolutionary spectrum). However, in a lot of cases this seems to be very reasonable to us. In particular when we have physical reasons for assuming a time varying spectral representation (such as in signal processing) or reasons for assuming that the process should be locally close to a stationary one it seems to be reasonable to fit such a model even if we have not enough data locally to identify the fitted structure completely. It then also makes sense to speak of 'the spectrum' (which would be well defined if we had more local data) and to investigate the question how good our estimate is.

In other situations we might prefer a model whose parameters are uniquely defined by the probabilistic structure of the data (e.g. if we are only looking for a nonstationary parsimonious model). An example for such a model is the time varying $AR$-model whose parameters are uniquely identifiable by the time varying Yule-Walker equations. We still have a model with an evolutionary spectrum - however, now our spectrum is a parametric one of lower dimension.

While our concern is to find a meaningful asymptotic theory for processes with evolutionary spectra and to use it for the investigation of statistical methods, Priestley's concern was mainly to find physically meaningful representations for the process itsself. In order to have a representation where $|A_t(\lambda)|^2$ can be interpreted as the power of frequency $\lambda$ at time $t$, Priestley requires that $\exp(i\lambda t)A_t(\lambda)$ is an 'oscillatory function', i.e. that the Fourier transform of $A_t(\lambda)$ in $t$ has an absolute maximum at zero. A process with such a spectral representation is called 'oscillatory process'. Nevertheless, there usually still exist several evolutionary spectral representations for a specific oscillatory process and Priestley suggests to take that representation where $\{A_t(\lambda)\}$ is 'most slowly varying' in $t$ which is measured by the degree of concentration of the Fourier transform of $A_t(\lambda)$ (in $t$) around zero. It is heuristically clear that the rescaling of $A(u, \lambda)$ in the definition of local stationarity with a smooth $A(u, \lambda)$ asymptotically leads to the most slowly varying representation. This means that any estimate of the time varying spectrum should be regarded as an estimate for the most slowly varying $|A_t(\lambda)|^2$ in the finite sample situation.

So far we have not defined in our approach a bandwidth of stationarity. A definition in the spirit of Priestley (although not completely analogous) could be $B_x^{(T)} := T(\sup_\lambda \int_0^1 |\frac{\partial}{\partial u} A(u, \lambda)| du)^{-1}$. Analogously to the proof of Theorem 2.2 (i) we get with a differentiable $A(u, \lambda)$ and

$$H_N(\lambda) := \left\{ 2\pi \sum_{s=1}^N h\left(\frac{s}{N}\right)^2 \right\}^{-1} \left| \sum_{s=1}^N h\left(\frac{s}{N}\right) \exp(-i\lambda s) \right|^2,$$

$$
\begin{aligned}
EI_N(u, \lambda_0) &= \int_{-\pi}^{\pi} H_N(\lambda) f(u, \lambda + \lambda_0) d\lambda + O\left(\frac{N}{B_x^{(T)}}\right) \\
&= f(u, \lambda_0) + O\left(\frac{N}{B_x^{(T)}}\right) + O(N^{-1}).
\end{aligned}
$$

The first equation is completely analogous to Priestley's Theorem 11.2.3 ([Pri81]) where $N$ is proportional to his width $B_g$ (it is the bandwidth of the kernel in ordinary time). In order to keep the $O(\frac{N}{B_x^{(T)}})$-term small Priestley requires that $N \leq \varepsilon B_x$ with some fixed $\varepsilon > 0$. Since the bandwidth of stationarity $B_x$ stays fixed in Pristley's setting he could still write down the second equation - but it is not possible to make both remainders arbitrarily small at the same time.

Summarizing, Priestley's theory of evolutionary spectra and our approach for locally stationary time series are much more related than it might look from a first view. While the former is mainly concerned with physically meaningful representations of the process (and of course the statistical analysis) the latter provides the framework for a meaningful asymptotic analysis of statistical procedures allowing to retain such powerful tools as consistency, asymptotic normality, efficiency, LAN-expansions, etc. for processes with evolutionary spectra. The benefits of this asymptotic approach go even further. For example, with this approach

$$f(u, \lambda) := \frac{\sigma^2(u)}{2\pi} \left| \sum_{j=0}^p a_j(u) \exp(i\lambda j) \right|^{-2}$$

becomes (asymptotically) the spectrum of a time varying $AR$-process at time $u$ while this is not true for a finite sample size (cf. [Dah96b] and [MS89]). The classical asymptotics for stationary processes arises as a special case. Furthermore, the proofs of some results are similar to the analogous results in the stationary case (although technically more difficult).

*Acknowledgments:* This paper was written while the author was visiting the Royal Melbourne Institute of Technology. He is very grateful to the institute and in particular Professor Peter Brockwell for providing an ideal working environment and for some interesting discussions.

# 4   References

[Dah96a]  R. Dahlhaus. Fitting time series models to nonstationary processes. *Ann. Statist.*, 1996. To be published.

[Dah96b]  R. Dahlhaus. Maximum likelihood estimation and model selection for locally stationary processes. *J. Nonpar. Statist.*, 1996. To be published.

[Dah96c]  R. Dahlhaus. On the Kullback-Leibler information divergence of locally stationary processes. *Stochastic Processes and their Applications*, 1996. To be published.

[DG95]  R. Dahlhaus and L. Giraitis. The bias and the mean squared error in semiparametric models for locally stationary time-series. Preprint, Universität Heidelberg, 1995.

[HW96]  R.J. Hyndman and M.P. Wand. Nonparametric autocovariance function estimation, 1996. Unpublished manuscript.

[MS89]  G. Mélard and A. Herteleer-de Schutter. Contributions to evolutionary spectral theory. *J. Time Ser. Anal.*, 10(1):41–63, 1989.

[NvS96]  M.H. Neumann and R. von Sachs. Wavelet thresholding in anisotropic function classes and application to adaptive estimation of evolutionary spectra. *Annals of Statistics*, 1996. To be published.

[Pri65]  M.B. Priestley. Evolutionary spectra and non-stationary processes. *J. Roy. Statist. Soc. Ser. B*, 27:204–237, 1965.

[Pri81]  M.B. Priestley. *Spectral Analysis and Time Series*. Academic Press, London, 1981.

[Pri88]  M.B. Priestley. *Non-linear and Non-stationary Time Series Analysis*. Academic Press, London, 1988.

[Pri96]  M.B. Priestley. Wavelets and time-dependent spectral analysis. *J. Time Series Anal.*, 17:85–103, 1996.

[Rie93]  K.S. Riedel. Optimal data-based kernel estimation of evolutionary spectra. *IEEE Trans. Signal Proc.*, 41:2439–2447, 1993.

[TS85]  P.J. Thomson and P. Souza. Speech recognition using LPC distance measures. In E.J. Hannan, P.R. Krishnaiah, and M.M. Rao, editors, *Handbook of Statistics*, volume 5, pages 389–412. Elsevier Science Publishers B.V., 1985.

[vSS96]  R. von Sachs and K. Schneider. Wavelet smoothing of evolutionary spectra by non-linear thresholding. *Applied Computational Harmonic Analysis*, 1996. To be published.

University of Heidelberg

# Inference for Seasonal Moving Average Models With a Unit Root

Richard A. Davis, Colorado State University
Meiching Chen, Colorado State University
William T.M. Dunsmuir, U. of New South Wales

ABSTRACT

This paper is concerned with estimation and testing of a unit root which is on or near the unit circle in a seasonal moving average model. Two models are considered. The first is a pure seasonal moving average model of degree one in the seasonal lag $s > 1$. The asymptotic distributions of the maximum likelihood estimator and the likelihood ratio statistic are derived under a sequence of local alternatives converging to the unit circle at rate $1/T$. The power of the likelihood ratio test corresponding to a unit root hypothesis is compared with the locally best invariant unbiased test. It is also observed that as the seasonal lag increases the probability of the maximum likelihood estimate being on the unit circle decreases to a lower bound of .5 while the power of the test of the null hypothesis that there is a root on the unit circle decreases quite substantially with increasing seasonal lag. The second model considered contains a root on or near the unit root at lag $s \geq 1$ in the moving average polynomial of a mixed autoregressive-moving average model. It is shown that the asymptotic distribution of the estimate of the root on or closest to the unit circle obtained by maximizing the likelihood using a process derived by filtering the observed process by the autoregressive moving average model (without the unit root component in the moving average) is the same using consistent estimates of the parameters in the autoregressive moving average operator as it is using their true values. This result provides the basis of a likelihood ratio test in testing for the presence of a unit root in the MA operator of a general ARMA model.

## 1 Introduction

The principal objective of this paper is to extend the limit results of Davis and Dunsmuir (1996) and Davis, Chen and Dunsmuir (1995) for the unit problem in an MA(1) model to the seasonal moving average model specified by

(1.1)
$$Y_t = (1 - \Theta_0 B^s)\epsilon_t = \epsilon_t - \Theta_0 \epsilon_{t-s}.$$

Here, the parameter of interest, $\Theta_0$, is on or near the unit circle, $B$ is the backward shift operator, and $\{\epsilon_t\} \sim \text{IID}(0, \sigma^2)$. Using the parameterization $\Theta_0 = (1 - \gamma/T)$, where $T = ns$ is the sample size, we establish the convergence in distribution of the maximum likelihood estimator of $\Theta_0$. The main idea in this derivation is to show the sequence of processes, $L_T(\beta) - L_T(0) = \ell_T(1 - \beta/T) - \ell_T(1)$ converges in distribution on $C[0, \infty)$ to a limit process $Z_{\gamma,s}$, where $\ell_T(\Theta)$ is the log-likelihood of the data, $Y_1, \ldots, Y_T$ after the variance of $Y_1$ has been concentrated out. As a byproduct of this functional convergence, the maximum likelihood estimator $\hat{\beta}_{MLE}$, under this parameterization, converges in distribution to the maximizer, $\tilde{\beta}_{MLE,\gamma}$, of the limit process $Z_{\gamma,s}$. Returning to the original parameterization, this implies that

$$T(1 - \hat{\Theta}_{MLE}) \overset{d}{\to} \tilde{\beta}_{MLE,\gamma}.$$

As a second application of the functional convergence, the likelihood ratio test statistic for testing $\Theta_0 = 1$ ($\gamma = 0$) versus $\Theta_0 < 1$ is in fact $L_T(\hat{\beta}_{MLE}) - L_T(0)$, which converges in distribution to $Z_{0,s}(\tilde{\beta}_{MLE,0})$.

Section 2 contains the proofs of these results which are simple consequences of our earlier results. One only needs to observe that the data $Y_1, \ldots, Y_T$ can be regarded as $s$ independent realizations of length $n$ from an MA(1) model. The independent realizations are given by $\{Y_{k+sj}, j = 0, \ldots, n-1\}$, $k = 1, \ldots, s$.

In Section 3, we show that the pile-up effect corresponding to the MLE estimate being equal to 1 is less severe as $s$ increases. The limiting probability that $\hat{\Theta}_{MLE} = 1$ in the case $\Theta_0 = 1$ is given by $P[\tilde{\beta}_{MLE,0} = 0]$ which decreases to .5 as $s \to \infty$. An explanation for this is provided in Section 3. We also compare the limiting power for testing $\Theta = 1$ versus $\Theta_0 < 1$ for the tests based on the maximum likelihood estimator, the generalized likelihood ratio (GLR) test statistic and the locally best invariant unbiased (LBIU) test statistics. The latter was derived by Tanaka (1990) and further discussed by Tam and Reinsel (1995). The GLR and LBIU tests dominate the one based on the MLE as least for $\Theta_0$'s near 1. The LBIU test is only marginally better than the GLR test for $\Theta_0$'s close to 1, but falls off quickly for $\Theta_0$'s further away from 1. More interestingly, however, is the sharp decrease in power as the number of seasons $s$ increases. This decrease is larger than one might have expected.

In Section 4, we extend the results of Section 2, to the case of an ARMA model which has a root on or near the unit circle in the moving average polynomial. This case was also considered by Tanaka (1990) and Tam and Reinsel (1995). In this setting, consider observations $W_1, \ldots, W_T$ from the ARMA$(p, q + s)$ model given by

(1.2) $$\phi(B)W_t = (1 - \Theta_0 B^s)\theta(B)\epsilon_t,$$

where the parameter of interest, $\Theta_0$, is on or near the unit circle, $\phi(z) = 1 - \phi_1 z - \cdots - \phi_p z^p$ and $\theta(z) = 1 - \theta_1 z - \cdots - \theta_q z^q$ are the AR and MA polynomials, respectively, with no common zeros and no zeros inside or on the unit circle. Under these assumptions, the process

$$Y_t := \phi(B)\theta^{-1}(B)W_t$$

follows the seasonal MA(1) model as specified in (1.1). If the parameter vector $\boldsymbol{\delta} = (\phi_1, \ldots, \phi_p, \theta_1, \ldots, \theta_q)$, and hence the operator $\phi(B)\theta^{-1}(B)$, can be estimated consistently, then one might expect the statistical properties of the filtered process,

$$Y_t^* := \hat{\phi}(B)\hat{\theta}^{-1}(B)W_t,$$

to roughly mirror those of $\{Y_t\}$. One way to obtain the consistent estimates required is to maximize the likelihood over all parameters in the model (1.2). Using Pötscher (1991) it follows that the estimates which maximize the Gaussian likelihood based on the model (1.2) are consistent even if the moving average parameter has a root on the unit circle. Inferences about $\Theta_0$ can then be made through the likelihood of the estimated filtered data $Y_1^*, \ldots, Y_T^*$ based on a Gaussian seasonal MA(1) model. In Section 4, we show that the inference procedures for $\Theta_0$ based on the likelihood function using $Y_1^*, \ldots, Y_T^*$ as the data are asymptotically the same as those based on knowing the true value of $\boldsymbol{\delta}$, or equivalently, $Y_1, \ldots, Y_T$.

## 2   Asymptotic Theory

Let $Y_1, \ldots, Y_T$ be observations from the seasonal MA(1) model (1.1)

$$(2.1) \qquad\qquad Y_t = \epsilon_t - \Theta_0\epsilon_{t-s},$$

where $T = ns$, $\Theta_0 = 1 - \gamma/T$, $\gamma \geq 0$, and $\{\epsilon_t\} \sim \text{IID}(0, \sigma^2)$ with $E\epsilon_t^4 < \infty$. Since the data $\mathbf{Y} = (Y_1, \ldots, Y_T)'$ can be rearranged as $s$ independent realizations, $\mathbf{V}_k = (Y_k, Y_{k+s}, \cdots, Y_{k+s(n-1)})'$ of length $n$ from the MA(1) model with moving average parameter $\Theta_0$, the logarithm of the Gaussian likelihood based on the observed vector $\mathbf{Y} = (Y_1, \ldots, Y_T)'$ is given by

$$\sum_{k=1}^{s}\left(-\frac{n}{2}\ln(2\pi\gamma(0)) - \frac{1}{2}\ln|G| - \frac{1}{2\gamma(0)}\mathbf{V}_k'G^{-1}\mathbf{V}_k\right)$$

where $\gamma(0) = \text{Var}(Y_1)$ and $G$ is the covariance matrix of $\mathbf{V}_i/\gamma^{1/2}(0)$. After concentrating $\gamma(0)$ out of the likelihood and deleting constant terms, the reduced log likelihood becomes

$$L_T(\beta) = -s\ln|G| - ns\ln(\sum_{k=1}^{s}\mathbf{V}_k'G^{-1}\mathbf{V}_k),$$

under the parameterization $\beta = T(1 - \Theta)$, where we view $L_T$ as a random element of $C[0, \infty)$, the space of continuous functions on $[0, \infty)$ where convergence is defined as uniform convergence on compact sets. The derivative of $L_T$ can be written as

$$L_T'(\beta) = -s\frac{d\ln|G|}{d\beta} - \frac{ns}{\sum_{k=1}^{s}\mathbf{V}_k'G^{-1}\mathbf{V}_k}\sum_{k=1}^{s}\mathbf{V}_k'\frac{dG^{-1}}{d\beta}\mathbf{V}_k.$$

If $L_{n,k}(\beta)$ denotes the corresponding reduced log-likelihood based on $\mathbf{V}_k$ under the parameterization $\beta = T(1 - \Theta)$, then from (2.13) and (2.14) in Davis and Dunsmuir (1996), we have

$$L_T'(\beta) = \sum_{k=1}^{s} L_{n,k}'(\beta) + o_p(1).$$

The independence of the $\mathbf{V}_k$'s implies that

$$(L_{n,k}'(\beta), k = 1, \ldots, s) \xrightarrow{d} \frac{\beta}{2s^2}(Y_{\gamma/s}^{(k)}(\beta/s), k = 1, \ldots, s)$$

in $C^s[0, \infty)$, where

$$Y_\gamma^{(k)}(\beta) = \sum_{t=1}^{\infty} \frac{4(\pi^2 t^2 + \gamma^2)X_{tk}^2}{(\pi^2 t^2 + \beta^2)^2} - \sum_{t=1}^{\infty} \frac{4}{\pi^2 t^2 + \beta^2}$$

with $\{X_{tk}, t \geq 1, k \geq 1\} \sim$ IID N(0, 1) (see Theorem 2.1 of Davis and Dunsmuir (1996)). Thus, by the continuous mapping theorem,

$$L_T'(\beta) \xrightarrow{d} \frac{\beta}{2s^2} \sum_{k=1}^{s} Y_{\gamma/s}^{(k)}(\beta/s)$$

$$=: \frac{\beta}{2s^2} Y_{\gamma/s,s}(\beta/s),$$

where

$$Y_{\gamma,s}(\beta) = -s\sum_{t=1}^{\infty} \frac{4}{\pi^2 t^t + \beta^2} + \sum_{t=1}^{\infty} \frac{4(\pi^2 t^t + \gamma^2)\xi_t}{(\pi^2 t^2 + \beta^2)^2},$$

and the $\xi_t$'s are IID chi-squared random variables with $s$ degrees of freedom. Applying the continuous mapping theorem once again, we obtain

$$L_T(\beta) - L_T(0) = \int_0^\beta L_T'(\tau)\, d\tau$$

$$\xrightarrow{d} \int_0^\beta (\tau/(2s^2))Y_{\gamma/s,s}(\tau/s)\, d\tau$$

$$= s\sum_{t=1}^{\infty} \ln \frac{\pi^2 t^2}{\pi^2 t^2 + (\beta/s)^2} + \sum_{t=1}^{\infty} \frac{(\beta/s)^2(\pi^2 t^2 + (\gamma/s)^2)\xi_t}{(\pi^2 t^2 + (\beta/s)^2)\pi^2 t^2}$$

$$(2.2) \qquad =: Z_{\gamma,s}(\beta).$$

An argument similar to the one given for $L_T'$ can be used to establish convergence of $L_T''(\beta)$ to $Z_{\gamma,s}''(\beta)$. We summarize these results in the following theorem.

**Theorem 2.1** *Suppose $Y_1, \ldots, Y_T$ are observations from model (2.1) with true parameters $\Theta_0 = 1 - \frac{\gamma}{T}$ for some $\gamma \geq 0$. Then, as $T \to \infty$,*

$$(L_T(\beta) - L_T(0), L_T'(\beta), L_T''(\beta)) \xrightarrow{d} (Z_{\gamma,s}(\beta), Z_{\gamma,s}'(\beta), Z_{\gamma,s}''(\beta))$$

on $C^3[0, \infty)$. In addition,
(i) if $\hat{\beta}_{LM} = \inf\{\beta \geq 0: L'_T(\beta) = 0$ and $L''_T(\beta) < 0\}$ (i.e., $\hat{\beta}_{LM}$ is the local maximum of $L_T(\cdot)$ closest to 0), then

$$\hat{\beta}_{LM} \xrightarrow{d} \tilde{\beta}_{LM,\gamma}$$

where

$$\tilde{\beta}_{LM,\gamma} = \inf\{\beta \geq 0 : Z'_{\gamma,s}(\beta) = 0 \quad and \quad Z''_{\gamma,s}(\beta) < 0\},$$

and
(ii) $T(\hat{\Theta}_{MLE} - 1) \xrightarrow{d} -\tilde{\beta}_{MLE,\gamma}$
where $\tilde{\beta}_{MLE,\gamma}$ is the global maximizer of $Z_{\gamma,s}(\beta)$.

**Remark 2.1.** The proofs of (i) and (ii) follow the same arguments given in Davis and Dunsmuir (1996) and Davis, Chen and Dunsmuir (1995).
**Remark 2.2.** Part (i) of the theorem implies that

$$T(\hat{\Theta}_{LM} - 1) \xrightarrow{d} -\tilde{\beta}_{LM,\gamma}$$

where $\hat{\Theta}_{LM}$ is the local maximizer of the likelihood function which is closest to 1. As in the $s = 1$ case, $\hat{\Theta}_{LM}$ and $\hat{\Theta}_{MLE}$ have different limiting distributions. For a discussion on the differences between $\hat{\Theta}_{LM}$ and $\hat{\Theta}_{MLE}$ see Davis and Dunsmuir (1996).

# 3    Comparison of Tests Based on the Local Maximum Estimators and the Generalized Likelihood Ratio

In this section we use simulation to derive type I error probabilities and power functions for testing the null hypothesis that $H_0 : \Theta_0 = 1$ versus $H_A : \Theta_0 < 1$ for the model considered in Section 2. Tests based on $\hat{\beta}_{LM}$, the likelihood ratio statistic and Tanaka's score statistics are considered in this section.

The asymptotic theory of Section 2 also allows us to approximate the nominal power of tests based on $\hat{\Theta}_{LM}$ or $\hat{\Theta}_{MLE}$ against local alternatives of the form:

$$H_A : \Theta = \Theta_A$$

where $\Theta_A = 1 - \gamma/T$. The tests considered here all have asymptotic power equal to 1 against any fixed local alternative.

To describe the test based on the generalized likelihood ratio let $Z_T(\beta) = L_T(\beta) - L_T(0)$ be the -2log of the likelihood ratio. Define the generalized likelihood ratio statistic as $\hat{Z}_T = Z_T(\hat{\beta}_{MLE})$. Also, let $\tilde{Z}_{\gamma,s} = Z_{\gamma,s}(\tilde{\beta}_{MLE,\gamma})$

denote the limit random variable of $\hat{Z}_T$ when $\gamma$ is the true value. The $(1 - \alpha)$th asymptotic quantile $b_{GLR}(\alpha)$ is defined as:

$$P(\tilde{Z}_{0,s} > b_{GLR}(\alpha)) = \alpha.$$

In the results to follow the tests are defined using these asymptotic quantiles to define the critical region.

To describe the test based on the LM point estimate we define the following asymptotic quantile. Let $b_{LM}(\alpha)$ be the $(1 - \alpha)$th quantiles defined as:

$$P(\tilde{\beta}_{LM,0} > b_{LM}(\alpha)) = \alpha$$

In order to find the values of $b_{GLR}(\alpha)$ and $b_{LM}(\alpha)$ using the asymptotic results of Section 2 the following simulation method was used. The infinite sums required in $Z_{\gamma,s}(\beta)$ and $Y_{\gamma,s}(\beta)$ are approximated by truncating them at $k = 1000$. For all results reported below 10,000 replications were used. For each replicate the three statistics were evaluated thereby reducing the between replicate variability as a component in the comparison of the three methods. For finite sample results further details on the methods used to compute the likelihood based estimates are given in Davis and Dunsmuir (1996). Probabilities reported below are accurate to $\pm 0.01$ with 95% confidence. The standard errors displayed in Table 3.2 were based on estimates of the asymptotic variance of the sample $p$th quantile given by $\frac{p(1-p)}{Nf^2(\hat{\xi}_p)}$, where $\hat{\xi}_p$ is the sample $p$th quantile, $\hat{f}(\cdot)$ is the estimated pdf computed using the function DENSITY in SPLUS, and $N$ is the number of replications.

Table 3.1 compares the limit probabilities that the MLE and LM estimates of $\beta$ are equal to zero (or equivalently that the MLE and LM estimates of $\Theta$ are 1) for different $s$ under the true value $\gamma = 0$.

**Table 3.1.** Comparison of probabilities of $\tilde{\beta}_{LM,\gamma}$ and $\tilde{\beta}_{MLE,\gamma}$ being zero under $\gamma = 0$ with Tanaka and Satchell's limiting value of $P_1$.

| $s$ | $P(\tilde{\beta}_{LM,\gamma} = 0)$ | $P(\tilde{\beta}_{MLE,\gamma} = 0)$ | $P_1$ |
|-----|------------------------------------|-------------------------------------|-------|
| 1   | .6570 | .6516 | .6574 |
| 2   | .6163 | .6154 | .6167 |
| 3   | .5960 | .5959 | .5964 |
| 4   | .5844 | .5843 | .5841 |
| 5   | .5773 | .5772 | .5755 |
| 6   | .5712 | .5712 | .5690 |
| 7   | .5636 | .5636 | .5639 |
| 8   | .5611 | .5611 | .5599 |
| 9   | .5577 | .5577 | .5565 |
| 10  | .5532 | .5532 | .5536 |
| 11  | .5523 | .5523 | .5511 |
| 12  | .5471 | .5471 | .5490 |

Also given in Table 3.1 are the corresponding limiting probabilities $P_1$ contained in Table 2 of Tanaka and Satchell (1989). Note that in Table 3.1 the probabilities of the local maximum being on the unit circle are at least as large as the probabilities of the global maximum being on the unit circle but that for $s > 5$ there is no difference to 4 decimal places. An explanation for this is provided in Davis and Dunsmuir (1996) at least for the case where $s = 1$. Although these probabilities are accurate to $\pm.01$ with 95% confidence the values in the first two columns are much closer than this accuracy because they are calculated from the same realizations in the simulations.

Note also that the probabilities of being on the unit circle decrease as $s$ increases. The explanation for this is straightforward. For the case where $s = 1$, as explained in Remark 2.3 of Davis and Dunsmuir(1996), $\tilde{\beta}_\gamma = 0$ if and only if $Y_\gamma(0) < 0$. The same argument holds for the case $s > 1$. In this case, using the notation of Section 2, we can write $Y_{\gamma/s,s}(0) = Y_{\gamma/s}^{(1)} + \cdots + Y_{\gamma/s}^{(s)}(0)$ where each element, $Y_{\gamma/s}^{(j)}(0)$, is independent and identically distributed with $E(Y_0^{(1)}(0)) = 0$ and $P(Y_0^{(1)}(0) < 0) > 0.5$. Consequently $Y_{0,s}(0)$ consists of a sum of $s$ negatively skewed random variables. For large $s$ this sum, $Y_{0,s}(0)$, will be closer to being normally distributed which implies that it has a distribution which is more symmetric. As a result, if $s \to \infty$, $P(Y_{0,s}(0) < 0) \to 0.5$ as is apparent in the above table. It is notable however that the rate at which this probability of the pile up tends to one half appears to be rather slow.

Tam and Reinsel (1995) consider tests for a unit root at a seasonal lag in ARIMA models. They present a series of results for the cases $s = 4$ and $s = 12$. They investigate the locally best invariant unbiased (LBIU) test and point optimal invariant tests. In order to compare the power of the tests based on the LM estimate and the GLR approach with those of Tam and Reinsel we will report the quantiles, achieved type I error rates and power using $s = 4$ and $s = 12$.

Tables 3.2 a&b provide the $b_{LM}(\alpha)$ and $b_{GLR}(\alpha)$ quantiles for commonly used values of $\alpha$ and $s = 4, 12$. Tables 3.3 a&b compare the finite sample achieved significance level using these asymptotic quantiles with the nominal significance levels. Clearly the quantiles derived from the asymptotic distribution provide accurate approximations to the actual finite sample quantiles. Note that the accuracy is not strongly influenced by the number of data but appears to depend on the number of years, $n$, of data.

Tables 3.4 a&b compare the power of the tests based on the LM and the GLR derived above with those for the LBIU test described in Tam and Reinsel (1995). The GLR procedure exhibits slightly greater power than the test based on the local maximum estimate. When $s = 4$ the GLR and LBIU tests have almost identical power curves for values of $\Theta$ close to the unit circle but for alternatives $\Theta < .8$ or so the GLR test is more powerful than the LBIU. Also when $s = 4$ the GLR is about 30% more powerful than the LM test method for close alternatives $\Theta > 0.90$. Otherwise the two procedures have quite similar power levels. When $s = 12$ the LM, GLR and LBIU procedures are quite similar in their power characteristics with, for $\Theta > 0.8$, the LBIU dominating the GLR which in turn dominates the LM test. However the differences are small. Tables 3.5 a&b and 3.6 a&b compare the finite sample powers with those using the asymptotic approximations. In all cases the asymptotic distribution gives slightly higher power values than the actual finite sample values with the degree of difference between the finite sample and the asymptotic values tending to be greatest for the moderate alternatives of $0.7 < \Theta < 0.8$.

Table 3.7 provides the approximations to power obtained from the limiting distributions for the LM and GLR statistics. These power values are plotted in Figure 1 (LM based test) and Figure 2 (GLR test). Note that in these tables and figures an identical sample size, $T = 96$, is used in order to illustrate the considerable reduction in power as $s$ increases. In particular, as is clear in Figures 1 and 2, even with the same amount of data the fact that seasonal process is effectively equivalent to observing $s$ independent moving averages of degree one, each with $T/s$ data points, means that there is a lot less information about the moving average parameter as $s$ increases. One obvious conclusion that can be drawn from this is that inference about the moving average parameter close to the unit circle is more difficult in the seasonal case.

**Table 3.2.a.**  $(1-\alpha)$ quantiles for the distribution of $\tilde{\beta}_{LM,0}$ and $\tilde{Z}_{0,s}$ when $s = 4$.

| Quantiles | $\alpha$ | | | |
|---|---|---|---|---|
| | 0.01 | 0.025 | 0.05 | 0.1 |
| $b_{LM}(\alpha)$ | 21.04 (.348) | 17.48 (.256) | 15.00 (.152) | 12.52 (.124) |
| $b_{GLR}(\alpha)$ | 4.75 (.206) | 3.17 (.096) | 2.18 (.062) | 1.21 (.035) |

**Table 3.2.b.**  $(1-\alpha)$ quantiles for the distribution of $\tilde{\beta}_{LM,0}$ and $\tilde{Z}_{0,s}$ when $s = 12$.

| Quantiles | $\alpha$ | | | |
|---|---|---|---|---|
| | 0.01 | 0.025 | 0.05 | 0.1 |
| $b_{LM}(\alpha)$ | 41.16 (.456) | 36.36 (.336) | 32.28 (.252) | 27.60 (.240) |
| $b_{GLR}(\alpha)$ | 5.12 (.117) | 3.44 (.093) | 2.31 (.056) | 1.36 (.038) |

The next two tables indicate that the use of asymptotic quantiles gives very accurate values of the size of the test for $n = 20$.

**Table 3.3.a**  Achieved significance levels using the asymptotic quantiles of Table 3.2.a for $n = 20$ when $s = 4$.

| Finite sample power | $\alpha$ | | | |
|---|---|---|---|---|
| of LM and GLR tests | 0.01 | 0.025 | 0.05 | 0.1 |
| $P(\hat{\beta}_{LM} > b_{LM}(\alpha)\|\gamma = 0)$ | 0.0110 | 0.0248 | 0.0522 | 0.1036 |
| $P(\hat{Z}_T > b_{GLR}(\alpha)\|\gamma = 0)$ | 0.0108 | 0.0259 | 0.0517 | 0.1015 |

**Table 3.3.b**  Achieved significance levels using the asymptotic quantiles of Table 3.2.b for $n = 20$ when $s = 12$.

| Finite sample power | $\alpha$ | | | |
|---|---|---|---|---|
| of LM and GLR tests | 0.01 | 0.025 | 0.05 | 0.1 |
| $P(\hat{\beta}_{LM} > b_{LM}(\alpha)\|\gamma = 0)$ | 0.0099 | 0.0245 | 0.0508 | 0.1056 |
| $P(\hat{Z}_T > b_{GLR}(\alpha)\|\gamma = 0)$ | 0.0094 | 0.0253 | 0.0512 | 0.1026 |

**Table 3.4.a**  Comparison of the finite sample ($n = 20, s = 4$) power of the LM, GLR and LBIU tests (the latter taken from Table 4 of Tam of Reinsel (1995)) with type I error $\alpha = 0.05$.

| $\Theta$ | $\gamma$ | $P(\hat{\beta}_{LM} > b_{LM}(\alpha)\|\gamma)$ | $P(\hat{Z}_T > b_{GLR}(\alpha)\|\gamma)$ | LBIU |
|---|---|---|---|---|
| 0.95 | 0.25 | 0.0576 | 0.0742 | .073 |
| 0.9 | 0.5 | 0.1239 | 0.1659 | .164 |
| 0.8 | 1.0 | 0.4988 | 0.5606 | .538 |
| 0.75 | 1.25 | 0.6955 | 0.7381 | .705 |
| 0.6 | 2.0 | 0.9614 | 0.9623 | .934 |

**Table 3.4.b**  Comparison of the finite sample ($n = 20, s = 12$) power of the LM, GLR and LBIU tests with $\alpha = 0.05$.

| $\Theta$ | $\gamma$ | $P(\hat{\beta}_{LM} > b_{LM}(\alpha)\|\gamma)$ | $P(\hat{Z}_T > b_{GLR}(\alpha)\|\gamma)$ | LBIU |
|---|---|---|---|---|
| 0.95 | 0.083 | 0.0871 | 0.0890 | .090 |
| 0.9 | 0.167 | 0.2458 | 0.2668 | . 282 |
| 0.8 | 0.333 | 0.8641 | 0.8658 | .875 |
| 0.75 | 0.417 | 0.9715 | 0.9755 | .970 |
| 0.6 | 0.667 | 1.0000 | 1.0000 | 1.000 |

**Table 3.5.a** Comparison of the limiting power of LM and GLR tests when $s = 4$ and $\alpha = 0.05$.

| $\Theta$ | $\gamma$ | $P(\hat{\beta}_{LM,\gamma} > b_{LM}(\alpha)|\gamma)$ | $P(\tilde{Z}_{\gamma,s} > b_{GLR}(\alpha)|\gamma)$ |
|------|------|------|------|
| 0.95 | 0.25 | 0.0573 | 0.0737 |
| 0.9  | 0.5  | 0.1216 | 0.1604 |
| 0.8  | 1.0  | 0.4894 | 0.5305 |
| 0.75 | 1.25 | 0.6743 | 0.7117 |
| 0.6  | 2.0  | 0.9350 | 0.9375 |

**Table 3.5.b** Comparison of the limiting power of LM and GLR tests when $s = 12$ and $\alpha = 0.05$.

| $\Theta$ | $\gamma$ | $P(\hat{\beta}_{LM,\gamma} > b_{LM}(\alpha)|\gamma)$ | $P(\tilde{Z}_{\gamma,s} > b_{GLR}(\alpha)|\gamma)$ |
|------|------|------|------|
| 0.95 | 0.083 | 0.0862 | 0.0878 |
| 0.9  | 0.167 | 0.2483 | 0.2591 |
| 0.8  | 0.333 | 0.8308 | 0.8326 |
| 0.75 | 0.417 | 0.9521 | 0.9526 |
| 0.6  | 0.667 | 1.0000 | 1.0000 |

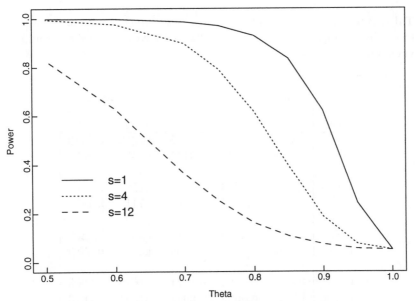

FIGURE 1. *Power curves of LM test for* $s = 1, 4, 12$.

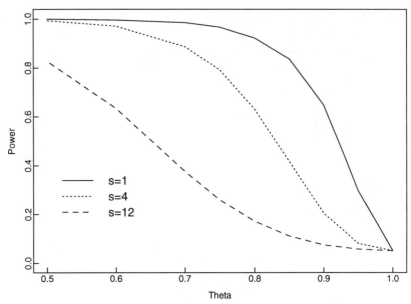

FIGURE 2. *Power curves of GLR test for* $s = 1, 4, 12$.

**Table 3.6.a** Comparison of the limiting power and exact power of the GLR test using quantiles from Table 3.2.a ($s = 4$). Exact values are computed via simulation.

| $\Theta_A$ | | $\alpha$ | | | |
|---|---|---|---|---|---|
| | | 0.01 | 0.025 | 0.05 | 0.10 |
| 0.6 | Exact | 0.9121 | 0.9427 | 0.9623 | 0.9742 |
| | Limit | 0.8812 | 0.9115 | 0.9375 | 0.9507 |
| 0.75 | Exact | 0.5664 | 0.6674 | 0.7381 | 0.8092 |
| | Limit | 0.5376 | 0.6391 | 0.7117 | 0.7859 |
| 0.8 | Exact | 0.3540 | 0.4711 | 0.5606 | 0.6660 |
| | Limit | 0.3298 | 0.4464 | 0.5305 | 0.6389 |
| 0.9 | Exact | 0.0585 | 0.1071 | 0.1659 | 0.2603 |
| | Limit | 0.0559 | 0.0967 | 0.1604 | 0.2613 |
| 0.95 | Exact | 0.0205 | 0.0452 | 0.0742 | 0.1372 |
| | Limit | 0.0202 | 0.0447 | 0.0737 | 0.1365 |

**Table 3.6.b** Comparison of the limiting power and exact power of the GLR test using quantiles from Table 3.2.b $(s = 12)$. Exact values are computed via simulation.

| $\Theta_A$ | | $\alpha$ 0.01 | 0.025 | 0.05 | 0.10 |
|------|-------|--------|--------|--------|--------|
| 0.6  | Exact | 0.9997 | 0.9998 | 1.0000 | 1.0000 |
|      | Limit | 0.9985 | 0.9994 | 0.9995 | 0.9996 |
| 0.75 | Exact | 0.9020 | 0.9612 | 0.9755 | 0.9865 |
|      | Limit | 0.8780 | 0.9238 | 0.9526 | 0.9701 |
| 0.8  | Exact | 0.6943 | 0.7963 | 0.8658 | 0.9272 |
|      | Limit | 0.6619 | 0.7627 | 0.8326 | 0.8878 |
| 0.9  | Exact | 0.1030 | 0.1818 | 0.2668 | 0.3695 |
|      | Limit | 0.0939 | 0.1742 | 0.2591 | 0.3647 |
| 0.95 | Exact | 0.0214 | 0.0488 | 0.0890 | 0.1567 |
|      | Limit | 0.0199 | 0.0474 | 0.0878 | 0.1583 |

**Table 3.7** Comparison of the limiting powers of the tests based on LM and GLR for $\alpha = 0.05$ when $s = 1, 4, 12$ and $T = 96$.

| $\Theta_A$ | $\gamma$ | s=1 LM | GLR | s=4 LM | GLR | s=12 LM | GLR |
|------|------|--------|--------|--------|--------|--------|--------|
| 0.95 | 4.8  | 0.2426 | 0.2986 | 0.0743 | 0.0802 | 0.0546 | 0.0566 |
| 0.9  | 9.6  | 0.6210 | 0.6476 | 0.1866 | 0.2038 | 0.0727 | 0.0743 |
| 0.85 | 14.4 | 0.8352 | 0.8374 | 0.3947 | 0.4167 | 0.1064 | 0.1103 |
| 0.8  | 19.2 | 0.9286 | 0.9227 | 0.6173 | 0.6303 | 0.1602 | 0.1707 |
| 0.75 | 24.0 | 0.9698 | 0.9665 | 0.7898 | 0.7915 | 0.2503 | 0.2582 |
| 0.7  | 28.8 | 0.9855 | 0.9846 | 0.8956 | 0.8874 | 0.3636 | 0.3764 |
| 0.6  | 38.4 | 0.9966 | 0.9963 | 0.9754 | 0.9704 | 0.6262 | 0.6343 |
| 0.5  | 48.0 | 0.9988 | 0.9991 | 0.9946 | 0.9925 | 0.8275 | 0.8262 |

# 4    Inference for a MA Unit Root in ARMA Models

In this section we consider testing for a seasonal unit root in the moving average operator for a general ARMA model. The setting for this problem is as follows. Let $W_1, \ldots, W_T$ be observations from the ARMA$(p, q + s)$ model

$$(4.1) \qquad \phi(B)W_t = (1 - \Theta_0 B^s)\theta(B)\epsilon_t,$$

where the parameter of interest, $\Theta_0$, is on or near the unit circle, $\phi(z) = 1 - \phi_1 z - \cdots - \phi_p z^p$ and $\theta(z) = 1 - \theta_1 z - \cdots - \theta_q z^q$ are the AR and MA polynomials, respectively, and $\{\epsilon_t\} \sim \text{IID}(0, \sigma^2)$ with $E\epsilon_t^4 < \infty$. It is further assumed that $\phi(z)$ and $\theta(z)$ have no common zeros and no zeros inside or on the unit circle. The filtered process,

$$Y_t := \phi(B)\theta^{-1}(B)W_t,$$

follows the seasonal MA(1) model as specified in (2.1). This observation suggests that if the parameter vector $\boldsymbol{\delta} = (\phi_1, \ldots, \phi_p, \theta_1, \ldots, \theta_q)$, and hence the operator $\phi(B)\theta^{-1}(B)$, can be estimated consistently, then one might expect the filtered process,

$$Y_t^* := \hat{\phi}(B)\hat{\theta}^{-1}(B)W_t,$$

to approximate $\{Y_t\}$. Providing this approximation is sufficiently accurate, which will be the case for any consistent estimate of $\boldsymbol{\delta}$, inferences about $\Theta_0$ can then be made through the likelihood of the estimated filtered data $Y_1^*, \ldots, Y_T^*$ based on a Gaussian seasonal MA(1) model. The following theorem shows that the limiting behavior of the likelihood function computed from $Y_1^*, \ldots, Y_T^*$ is identical to the one based on $Y_1, \ldots, Y_T$ described in Theorem 2.1. Hence, inference procedures for $\Theta_0$ based on the likelihood function using $Y_1^*, \ldots, Y_T^*$ as the data are asymptotically the same as those based on knowing the true value of $\boldsymbol{\delta}$. (Consistent estimates of $\boldsymbol{\delta}$ can be found by maximizing the Gaussian likelihood using the ARMA$(p, q + s)$ model in (4.1). If $s = 1$, the estimate of $\boldsymbol{\theta}$ will correspond to the $q$-degree polynomial obtained by selecting the $q$-smallest zeros (in absolute value) from the fitted $q + 1$ degree polynomial.)

**Theorem 4.1** *Let $W_1, \ldots, W_T$ be observations from the ARMA$(p, q + s)$ model (4.1) with $\Theta_0 = 1 - \gamma/T$, $\gamma \geq 0$, and suppose that $\hat{\boldsymbol{\delta}} = (\hat{\phi}, \hat{\theta})$ is a weakly consistent estimate of $\boldsymbol{\delta}$. For $t = 1, \ldots, T$, define*

$$Y_t^* = \sum_{j=0}^{t-1} \pi_j(\hat{\boldsymbol{\delta}}) W_{t-j},$$

*where $\pi_j(\cdot)$ is the coefficient of $z^j$ in the infinite Laurent series expansion of $\phi(z)\theta^{-1}(z) = \sum_{j=0}^{\infty} \pi_j z^j$ with parameter vector $\boldsymbol{\delta} = (\phi_1, \ldots, \phi_p, \theta_1, \ldots, \theta_q)$. If $L_T^*(\beta)$ denotes the concentrated log-likelihood of the data $Y_1^*, \ldots, Y_T^*$ based on the Gaussian seasonal MA(1) model in (1.1) with parameterization $\beta = T(1 - \Theta)$ (see (2.2)), then*

*(i) $L_T^*(\beta) - L_T^*(0) \xrightarrow{d} Z_{\gamma,s}(\beta)$ in $C[0, \infty)$,*

*(ii) $\hat{\beta}_{MLE}^* \xrightarrow{d} \tilde{\beta}_{MLE,\gamma}^*$, where $\hat{\beta}_{MLE}^* = \text{argmax}_{\beta \geq 0} L_T^*(\beta)$ and $\tilde{\beta}_{MLE,\gamma}^* = \text{argmax}_{\beta \geq 0} Z_{\gamma,s}(\beta)$,*
*and*

*(iii) $L_T^*(\hat{\beta}_{MLE}^*) - L_T^*(0) \xrightarrow{d} Z_{\gamma,s}(\tilde{\beta}_{MLE,\gamma}^*)$,*
*where $Z_{\gamma,s}$ is the process given by (2.2).*

**Proof**: We shall only provide a proof in the case $\gamma = 0$ and $s = 1$, the general case being a slight modification of the this argument as in the proof of Theorem 2.1. Define

$$U^*_{t,T} = \sqrt{\frac{2}{T+1}}(1 - d_t)^{-1/2} \sum_{i=1}^{T} Y_i^* \sin(\frac{\pi i t}{T+1}),$$

where $d_t = \cos(\pi t/(T+1))$. From the proof of Theorem 2.1 in Davis and Dunsmuir (1996) (see also Davis, Chen and Dunsmuir (1995)), it suffices to show that

$$(4.2) \qquad \frac{1}{T}\sum_{t=1}^{T}(U^*_{t,T})^2 \xrightarrow{P} 2\sigma^2,$$

$$(4.3) \qquad \sum_{t=1}^{k_T}(U^*_{t,T})^2 = O_p(k_T),$$

$$(4.4) \qquad \sum_{t=1}^{k_T}\frac{(U^*_{t,T})^2}{t^4} = O_p(1),$$

$$(4.5) \qquad \sum_{t=k_T}^{T}(U^*_{t,T})^2 = (T - k_T)O_p(1),$$

and
$$(4.6) \qquad (U^*_{1,T}, \ldots, U^*_{m,T}) \xrightarrow{d} (X_1, \ldots, X_m),$$

where $k_T$ is a sequence of integers satisfying $k_T = o(T)$, $k_T^2/T \to \infty$, and $X_1, \ldots, X_m$ are IID $N(0, 2\sigma^2)$ random variables. As shown in Davis and Dunsmuir (1996), these relations hold for $U_{t,T}$ defined in the same fashion as $U^*_{t,T}$ but with $Y_t^*$ replaced by $Y_t$, so that the strategy in establishing (4.2)–(4.6) will be to show that $U^*_{t,T}$ can be well approximated by $U_{t,T}$.

Writing $\pi(B) = \phi(B)\theta^{-1}(B)$ and $\hat{\pi}(B) = \hat{\phi}(B)\hat{\theta}^{-1}(B)$, we have

$$Y_t^* = \hat{\pi}(B)W_t - \sum_{j=t}^{\infty}\pi_j(\hat{\delta})W_{t-j}$$

$$(4.7) \qquad = Y_t + \sum_{j=0}^{\infty}(\pi_j(\hat{\delta}) - \pi_j(\delta))W_{t-j} - \sum_{j=t}^{\infty}\pi_j(\hat{\delta})W_{t-j}.$$

Expanding $\pi_j(\hat{\delta})$ in a Taylor series about $\delta$, we have

$$(4.8) \qquad |\pi_j(\hat{\delta}) - \pi_j(\delta)| = (\hat{\delta} - \delta)'\frac{\partial\pi_j(\delta_j^\dagger)}{\partial\delta} =: (\hat{\delta} - \delta)'\,\dot{\pi}_j^\dagger$$

where $\|\delta_j^\dagger - \delta\| \leq \|\hat{\delta} - \delta\|$. If $\psi(B) = \sum_{j=0}^{\infty}\psi_j B^j = \pi^{-1}(B) = \theta(B)\phi^{-1}(B)$, then there exist an $\epsilon > 0$ and constants $C > 0$, $r \in (0,1)$ such that the bounds,

$$(4.9) \qquad |\psi_j| \leq Cr^j, \quad |\pi_j(\hat{\boldsymbol{\delta}})| < Cr^j \quad \text{and} \quad \| \overset{.\dagger}{\pi}_j \| < Cr^j, \quad j = 0, 1, \ldots,$$

hold whenever $\|\hat{\boldsymbol{\delta}} - \boldsymbol{\delta}\| < \epsilon$.

¿From (4.7) and (4.8), we can write

$$
\begin{aligned}
Y_i^* - Y_i &= (\hat{\boldsymbol{\delta}} - \boldsymbol{\delta})' \sum_{j=0}^{\infty} \overset{.\dagger}{\pi}_j \, W_{i-j} - \sum_{j=i}^{\infty} \pi_j(\hat{\boldsymbol{\delta}}) W_{i-j} \\
&= (\hat{\boldsymbol{\delta}} - \boldsymbol{\delta})' \sum_{k=0}^{\infty} \sum_{j=0}^{\infty} \overset{.\dagger}{\pi}_j \, \psi_k Y_{i-j-k} - \sum_{k=0}^{\infty} \sum_{j=i}^{\infty} \pi_j(\hat{\boldsymbol{\delta}}) \psi_k Y_{i-j-k}.
\end{aligned}
$$

It follows that
$$(4.10)$$
$$U_{t,T}^* - U_{t,T} = (\hat{\boldsymbol{\delta}} - \boldsymbol{\delta})' \sum_{k=0}^{\infty} \sum_{j=0}^{\infty} \overset{.\dagger}{\pi}_j \, \psi_k A_{j+k,t,T} - \sum_{k=0}^{\infty} \psi_k \sum_{j=1}^{\infty} \pi_j(\hat{\boldsymbol{\delta}}) B_{j,k,t,T},$$

where

$$A_{m,t,T} = \sqrt{\frac{2}{T+1}} (1-d_t)^{-1/2} \sum_{i=1}^{T} Y_{i-m} \sin\left(\frac{\pi t i}{T+1}\right),$$

and

$$B_{j,k,t,T} = \sqrt{\frac{2}{T+1}} (1-d_t)^{-1/2} \sum_{i=1}^{j \wedge T} Y_{i-j-k} \sin\left(\frac{\pi t i}{T+1}\right).$$

Since $E(A_{m,t,T})^2 = EU_{t,T}^2$, we have from (4.9)

$$
E\left( \sum_{k=0}^{\infty} \sum_{j=0}^{\infty} \overset{.\dagger}{\pi}_j \, \psi_k A_{j+k,t,T} 1_{[\|\hat{\boldsymbol{\delta}} - \boldsymbol{\delta}\| < \epsilon]} \right)^2
$$

$$
\leq C^4 \sum_{k=0}^{\infty} \sum_{j=0}^{\infty} \sum_{k'=0}^{\infty} \sum_{j'=0}^{\infty} r^{k+j+k'+j'} E|A_{j+k,t,T} A_{j'+k',t,T}|
$$

$$(4.11) \qquad \leq \frac{C^4}{(1-r)^4} \left( E(U_{t,T})^2 \right)^{1/2} < C_1,$$

(see (A4) of Davis and Dunsmuir (1996)).

Turning to the second term in (4.10), we have using a modification of the decomposition given on p.28 of Davis and Dunsmuir (1996) that,

$$EB_{j,k,t,T}^2 \leq C_2$$

where $C_2$ is a constant independent of $j, k, t$ and $T$. Since $x^2/(1 - \cos x) < C_3$ for $x \in [0, 2\pi]$, we have for $j \leq T^{1/8}$,

$$E(B_{j,k,t,T})^2$$

$$= \frac{2}{(T+1)(1-d_t)} \sum_{i'=1}^{T^{1/8}} \sum_{i=1}^{T^{1/8}} \sin(\frac{\pi t i}{T+1}) \sin(\frac{\pi t i'}{T+1}) E(Y_{i-j-k} Y_{i'-j-k})$$

$$\leq 2(T+1)^{-1} C_3 \left( \sum_{i=1}^{T^{1/8}} i \right)^2 EY_1^2$$

$$\leq C_4(T+1)^{-1}(T^{1/4})^2$$

$$\leq C_4(T+1)^{-1/2},$$

so that

$$E \left( \sum_{k=0}^{\infty} \psi_k \sum_{j=1}^{\infty} \pi_j(\hat{\boldsymbol{\delta}}) B_{j,k,t,T} 1_{[\|\hat{\boldsymbol{\delta}} - \boldsymbol{\delta}\| < \epsilon]} \right)^2$$

$$\leq 2C^4 \sum_{k=0}^{\infty} \sum_{k'=0}^{\infty} r^{k+k'} \sum_{j=1}^{T^{1/8}} \sum_{j'=1}^{T^{1/8}} E|B_{j,k,t,T} B_{j',k',t,T}|$$

$$+ 2C^4 \left( \sum_{k=0}^{\infty} r^k \right)^2 \left( \sum_{j=T^{1/8}}^{\infty} r^j \right)^2 C_2$$

$$\leq C_5 \left( C_4(T+1)^{-1/2} T^{2/8} + C_2 r^{2T^{1/8}}/(1-r)^2 \right)$$

(4.12)  $$\leq C_6 T^{-1/4} \to 0,$$

uniformly in $t$. Combining (4.10)–(4.12), we obtain

$$\sup_{1 \leq t \leq T} E \left( (U_{t,T}^* - U_{t,T})^2 1_{[\|\hat{\boldsymbol{\delta}} - \boldsymbol{\delta}\| < \epsilon]} \right) \leq 2C_1 \epsilon^2 + 2C_6 T^{-1/4},$$

from which the relations (4.2)–(4.6) can now be easily derived.    □

## REFERENCES

Davis, R.A., Chen, M.C. and Dunsmuir, W.T.M. (1995). Inference for MA(1) processes with a root on or near the unit circle. *Probability and Mathematical Statistics* **15** 227–242.

Davis, R.A. and Dunsmuir, W.T.M (1996). Maximum likelihood estimation for MA(1) processes with a root on or near the unit circle. *Econometric Theory* **12** 1–29.

Pötscher, B.M. (1991). Noninvertibility and pseudo maximum likelihood estimation of misspecified ARMA models. *Econometric Theory* **7** 435–449.

Tam, W-K. and Reinsel, G.C. (1995). Tests for seasonal moving average unit root in ARIMA models. Preprint, University of Wisconsin, Madison, Wisconsin, USA.

Tanaka, K. (1990). Testing for a moving average unit root. *Econometric Theory* **6** 433-444.

Tanaka, K. and Satchell, S.E. (1989). Asymptotic properties of the maximum likelihood and nonlinear least-squares estimators for noninvertible moving average models. *Econometric Theory* **5**, 333-353.

# General Kriging for spatial-temporal processes with random ARX–regression parameters

J. Franke and B. Gründer

September 18, 1995

## 1 The data and the model

In the following, we discuss a procedure for interpolating a spatial-temporal stochastic process. We stick to a particular, moderately general model but the approach can be easily transfered to other similar problems. The original data, which motivated this work, are measurements of gas concentrations ($SO_2$, NO, $O_3$) and several meteorological parameters (temperature, sun radiation, precipitation, wind speed etc.). These date have been and are still recorded twice every hour at several irregularly located places in the forests of the state Rheinland-Pfalz as part of a program monitoring the air pollution in forests. Let $\zeta(t, x_j)$ , $j = 1, \ldots, N$ , $t = 0, \ldots T$ , denote the observations of e.g. $SO_2$ concentration which we model as part of a spatial–temporal stochastic process $\zeta(t, x)$, $t \in \mathbb{Z}$, $x \in \mathbb{R}^2$. A particular feature is a large amount of data in the time direction ($T$ very large), but only few locations in the plane where data are available ($N$ small). A more detailed description of the data has been given in Franke and Gründer (1992) and Gründer (1992).

One of the goals which had to be achieved by modelling the data was a procedure for interpolating the gas concentration, i.e. for $x \notin \{x_1, \ldots, x_N\}$, $\zeta(t, x)$ should be estimated from $\zeta(s, x_j)$, $0 \leq s \leq t$, $j = 1, \ldots, N$. The procedure had to be adaptive with respect to new incoming data, and it should allow for the information contained in the meteorological observations. We start from the following model for the gas concentration (or some monotone normalizing transformation of it):

$$\zeta(t, x) = f(x)' \, \beta(t) + \eta(t, x) \ , \quad t \in \mathbb{Z}, \ x \in \mathbb{R}^2 \tag{1}$$

where $f(x)^T$ is a known vector of simple functions of $x$ which allows for systematic differences between the various locations due, e.g., to the topography of the country. The vector $\beta(t)$ of regression coefficients forms a multivariate random time series which is independent of the residual spatial-temporal process $\eta(t, x)$. We assume $\mathcal{E}\eta(t, x) = 0$ and that $\eta(t, x)$ is stationary in $t$ and homogeneous in $x$, i.e. the joint distribution of $\eta(t_1, x_1), \ldots, \eta(t_k, x_k)$ is invariant against a common translation of the arguments. The main assumption of model (1) is the presence of a global time-varying effect represented by $\beta(t)$ which influences the data at all locations $x$. Due to the size of the region and due to the specification of the places where the data come from, this assumption is satisfied for our data. $f(x)'\, \beta(t)$ explains even the major part of variability whereas $\eta(t, x)$ takes care of smaller local fluctuations only.

We assume that $\beta(t)$ is a $m-$variate autoregressive process of order $p$ with an exogenous part of order $r$ or an $\mathrm{ARX}(p, r)-$process:

$$\beta(t) = \sum_{j=1}^{p} A_j\, \beta(t - j) + \delta(t) + \sum_{j=1}^{r} C_j \xi(t - j) \qquad (2)$$

$\delta(t)$ are i.i.d. $m-$dimensional Gaussian random vectors with mean 0 and covariance matrix $\Sigma_\delta$. $\xi(t)$ is the $l-$dimensional vector of exogenous variables representing temperature, precipitation etc. at time $t$. The seasonality of these variables also takes care of the well-known seasonality in gas concentrations. We assume that for

$$A(z) = I_m - \sum_{j=1}^{p} A_j z^j, \quad C(z) = \sum_{j=1}^{p} C_j z^j$$

$\det A(z) \neq 0$ for all $|z| \leq 1$, and that $A(z)$, $C(z)$ have no common left-divisor (up to unimodular matrices).

Due to prevailent wind directions in the area under consideration, we cannot assume that the fluctuation process $\eta(t, x)$ is isotropic with respect to its spatial coordinates, i.e. the covariance $cov(\eta(t, x), \eta(t, y))$ does not only depend on the distance $||y - x||$. Looking at the rather scarce information on spatial dependence we cannot consider general homogeneous processes. We therefore follow Vecchia (1988) and consider so-called $(\alpha, \lambda)-isotropy$, i.e., for fixed $t$, the covariance between $\eta(t, x)$ and $\eta(t, y)$ depends only on the norm of the suitably rotated and dilated difference $y - x$ :

$$cov(\eta(t, x), \ \eta(t, y)) = c_\eta(||S_\lambda R_\alpha(y - x)||)$$

with

$$R_\alpha = \begin{pmatrix} \cos\alpha & -\sin\alpha \\ \sin\alpha & \cos\alpha \end{pmatrix} \text{ and } S_\lambda = \begin{pmatrix} \lambda & 0 \\ 0 & \lambda^{-1} \end{pmatrix} \text{ for } 0 \leq \alpha < \pi, \ \lambda > 0.$$

In practice, $c_\eta$ would be a given function up to few unknown parameters which have to be estimated from the data.

With respect to the temporal dependence structure of $\eta(t, x)$, we assume that it is a *temporal Markovian process*, i.e. for all $(s_1, y_1), (s_2, y_2), \ldots \in \{t, t-1, \ldots\} \times \mathbb{R}^2$, $(s_j, y_j) \neq (t, x)$ for all $j$, we have

$$\mathcal{E}\{\eta(t, x)|\eta(s_j, y_j), \ j \geq 1\} = \mathcal{E}\{\eta(t, x)|\eta(s_j, y_j), \ s_j = t - 1 \text{ or } s_j = t\}.$$

We even assume that $\eta(t, x)$ has an autoregressive structure of order 1 if we consider it only at a finite number of locations $x_1, \ldots, x_N$ :

$$Y(t) = L\, Y(t-1) + \varepsilon(t) \ \text{ with } \ Y(t)' = (\eta(t, x_1), \ldots, \eta(t, x_N)). \quad (3)$$

$\varepsilon(t)$ is $N$-variate, mean-zero Gaussian white noise with covariance $\Sigma_\varepsilon$, and $L$ satisfies the stationarity condition $\det(I_N - Lz) \neq 0$ for all $|z| \leq 1$.

# 2 Estimating the random regression coefficients

To interpolate $\zeta(t, x)$ using the past and present data at $x_1, \ldots, x_N$ we have to estimate $\beta(t)$. For this purpose, we consider a state-space representation of this ARX–process. Let $Z(t)' = (\zeta(t, x_1), \ldots, \zeta(t, x_N))$ be the observations at time $t$, $F = (f(x_1), \ldots, f(x_N))$ the $m \times N$–matrix of regression functions evaluated at $x_1, \ldots x_N$, and $Y(t)$ be, as above, the fluctuation process at time $t$ evaluated at $x_1, \ldots, x_N$. Then we have from (1) the observation equation

$$Z(t) = F'\,\beta(t) + Y(t) \quad (4)$$

The ARX–equation (2) has the state–space representation (compare, e.g., Priestley, 1981)

$$b(t+1) = \mathbf{A}\, b(t) + \mathbf{D}\,\delta(t+1) + \mathbf{C}\,\xi(t) \quad (5)$$

with coefficient matrices $\mathbf{A}, \mathbf{D}, \mathbf{C}$ of dimensions $mq \times mq, \ mq \times m, \ mq \times l$ :

$$\mathbf{A} = \left( \begin{array}{c|ccc} A_1 & I_m & \cdots & 0_m \\ A_2 & 0_m & & 0_m \\ \vdots & \vdots & \ddots & \vdots \\ A_{q-1} & 0_m & \cdots & I_m \\ \hline A_q & 0_m & \cdots & 0_m \end{array} \right), \quad \mathbf{D} = \left( \begin{array}{c} I_m \\ 0_m \\ \vdots \\ 0_m \end{array} \right), \quad \mathbf{C} = \left( \begin{array}{c} C_1 \\ \vdots \\ \vdots \\ C_q \end{array} \right)$$

Here, $q := \max(p, r)$, and we set $A_j = 0_m$ if $p < j \leq q$ and $C_j = 0$ if $r < j \leq q$. The first $m$ coordinates of the state vector $b(t) \in \mathbb{R}^{mq}$ coincide with $\beta(t)$ : $b_j(t) = \beta_j(t)$, $j = 1, \ldots, m$. The remaining part of $b(t)$ can be easily written down recursively from (5), starting with

$$\begin{aligned} (b_{(m-1)q+1}(t), \ldots, b_{mq}(t))' &= A_q\, \beta(t-1) + C_{q-1}\xi(t) \\ (b_{(m-2)q+1}(t), \ldots, b_{(m-1)q}(t))' &= A_{q-1}\beta(t-1) + A_q\,\beta(t-2) \\ &\quad + C_{q-2}\xi(t) + C_{q-1}\xi(t-1) \end{aligned}$$

and so on. Setting $\mathbf{H} = F'(I_m 0_m \ldots 0_m)$ we get from (4) the equation

$$Z(t) = \mathbf{H}\, b(t) + Y(t). \tag{6}$$

For some $s$ let $b(t|s)$ denote the best (in mean-square sense) estimate for $b(t)$ based on the observations $Z(k)$ and the exogenous variables $\xi(k)$, $0 \le k \le s$. $b(t|s)$ may be calculated recursively in time using the Kalman filter, though the linear system given by (5), (6) is non-standard due to the presence of exogenous variables and the serial dependence of the observational noise $Y(t)$. We assume that we observe the system from time 0 onwards, that the initial state $b(0)$ is a Gaussian vector independent of $\rho(t)$, $t \ge 1$, and $Y(t)$, $t \ge 0$. Following Hannan and Deistler (1988, ch. 3.2), we treat the exogenous variables $\xi(t)$ as deterministic, i.e. all calculations are done conditionally on these data.

To formulate the algorithm, we need some notation: For $s < t$, $Z(t|s)$ is the best estimate for $Z(t)$ given $Z(k)$, $\xi(k)$, $k \le s$.

$$d_b(t|s) = b(t) - b(t|s), \quad d_z(t|s) = Z(t) - Z(t|s)$$

are the corresponding estimation errors with covariance and cross covariance matrices

$$P(t|s) = \mathcal{E}\, d_b(t|s)d_b'(t|s)\,, \quad R(t|s) = \mathcal{E}\, d_z(t|s)\, d_z'(t|s)\,, \quad Q(t|s) = \mathcal{E}\, b(t)\, d_z'(t|s).$$

**Theorem 2.1** *Let $b(0|-1) = \mathcal{E}\, b(0)$ and $P(0|-1)$ be the covariance matrix of $b(0)$. Under the conditions stated above for the linear system (5), (6), we have for $t \ge 1$*

$$
\begin{aligned}
b(t|t-1) &= \mathbf{A}\, b(t-1|t-1) + \mathbf{C}\, \xi(t-1) \\
b(t|t) &= b(t|t-1) + K(t)d_z(t|t-1)
\end{aligned}
$$

$$
\begin{aligned}
P(t|t-1) &= \mathbf{A}\, P(t-1|t-1)\mathbf{A}' + \mathbf{D}\, \Sigma_\delta \mathbf{D}' \\
P(t|t) &= P(t|t-1) - K(t)R(t|t-1)K(t)'
\end{aligned}
$$

$$
d_z(t|t-1) = Z(t) - (\mathbf{H}\,\mathbf{A} - L\,\mathbf{H})\, b(t-1|t-1) - L\, Z(t-1) - \mathbf{H}\,\mathbf{C}\,\xi(t-1)
$$

$$
\begin{aligned}
K(t) &= Q(t|t-1)\, R(t|t-1)^{-1} \\
R(t|t-1) &= (\mathbf{H}\,\mathbf{A} - L\,\mathbf{H})\, P(t-1|t-1)(\mathbf{H}\,\mathbf{A} - L\,\mathbf{H})' \\
&\quad + \mathbf{H}\,\mathbf{D}\, \Sigma_\delta \mathbf{D}'\mathbf{H}' + \Sigma_\varepsilon \\
Q(t|t-1) &= \mathbf{A}\, P(t-1|t-1)(\mathbf{H}\,\mathbf{A} - L\,\mathbf{H})' + \mathbf{D}\Sigma_\delta \mathbf{D}'\,\mathbf{H}'
\end{aligned}
$$

*For starting the algorithm the relations*

$$d_z(0| - 1) = Z(0) - \mathbf{H}\, b(0| - 1) \, , \quad R(0| - 1) = \mathbf{H}\, P(0| - 1)'\, \mathbf{H}' + \Sigma_0 \, ,$$

$$Q(0| - 1) = P(0| - 1)\, \mathbf{H}'$$

*may be used, where $\Sigma_0$ denotes the covariance matrix of $Y(0)$.*

**Proof:** Let $\mathcal{P}_s$ denote the orthogonal projection onto the span of $Z(0), \ldots, Z(s)$, $\xi(0), \ldots, \xi(s)$. Using (3) and (6), we have

$$
\begin{aligned}
d_y(t|t - 1) \quad &:= \quad Y(t) - \mathcal{P}_{t-1} Y(t) \\
&= \quad Y(t) - \mathcal{P}_{t-1}(L\, Z(t - 1) - L\, \mathbf{H}b(t - 1) + \varepsilon(t)) \\
&= \quad Y(t) - L\, Z(t - 1) + L\, \mathbf{H}b(t - 1|t - 1) \\
&= \quad -L\, \mathbf{H}d_b(t - 1|t - 1) + \varepsilon(t)
\end{aligned}
$$

Using (5) we immediately have

$$b(t|t - 1) = \mathbf{A}b(t - 1|t - 1) + \mathbf{C}\xi(t - 1)$$

and therefore the following three relations, using also (6),

$$d_b(t|t - 1) \quad = \quad \mathbf{A}d_b(t - 1|t - 1) + \mathbf{D}\delta(t) \tag{7}$$

$$
\begin{aligned}
Z(t|t - 1) \quad &= \quad \mathbf{H}b(t|t - 1) + \mathbf{Y}(t|t - 1) \\
&= \quad \mathbf{H}(\mathbf{A}b(t - 1|t - 1) + \mathbf{C}\xi(t - 1)) + L\, Z(t - 1) \\
&\quad\ - L\, \mathbf{H}b(t - 1|t - 1) \\
&= \quad \mathbf{\Delta}b(t - 1|t - 1) + \mathbf{HC}\xi(t - 1) + L\, Z(t - 1) \\
d_z(t|t - 1) \quad &= \quad \mathbf{\Delta}d_b(t - 1|t - 1) + \mathbf{HD}\delta(t) + \varepsilon(t) \, , \tag{8}
\end{aligned}
$$

where $\mathbf{\Delta} = \mathbf{HA} - L\, \mathbf{H}$. (8) and the independence of $d_b(t - 1|t - 1)$, $\delta(t)$ and $\varepsilon(t)$ imply immediately the recursions for $R(t|t - 1)$ and $Q(t|t - 1)$. Now, the rest of the algorithm follows analogously as in Theorem 3.2.3 of Hannan and Deistler (1988) which includes our result for the special case $L = 0$. ∎

182

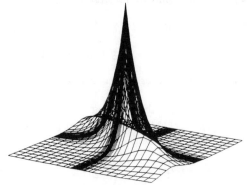

**Figure 1a:** $c_0(u) = \text{cov}(\eta(t,x),\eta(t,x+u))$

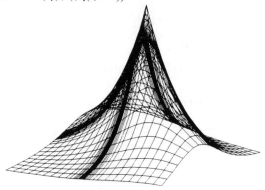

**Figure 1b:** $c_1(u) = \text{cov}(\eta(t,x),\eta(t+1,x+u))$

In Gründer (1993) it is shown that the error covariance matrices $P(t|t)$ converge to some limit $P_\infty$ in the strong sense:

$$\sum_{t=0}^{\infty} ||P(t|t) - P_\infty|| < \infty,$$

and the same type of limit behaviour is shared by the Kalman gain $K(t)$ and by the matrices $R(t|t-1)$, $Q(t|t-1)$. As a consequence of these limit results it follows that the spectral radius, i.e. the maximum of the absolute values of eigenvalues, of the matrices $\mathbf{A} - K(t) (\mathbf{H A} - L \mathbf{H})$ is less than 1 for all $t$ large enough. This implies the asymptotic stability of the algorithm.

The following result allows for fixed interval smoothing, i.e. for calculating $b(t|s)$ for all $t \leq s$, $s$ fixed. It is analogous to Theorem 3.2.2 of Hannan and Deistler (1988) with some slight modification necessary due to the serial dependence of $Y(t)$. As there, $P(t+1|t+1)^{-1}$ is a generalized inverse of the error covariance matrix.

**Theorem 2.2** *Let $\boldsymbol{\Delta} = \mathbf{H A} - L\, \mathbf{H}$, and let the assumption of Theorem 2.1 be satisfied. Then, for $s > t$*

$$
\begin{aligned}
b(t|s) &= b(t|t+1) + F(t)\{b(t+1|s) - b(t+1|t+1)\} \\
F(t) &= P(t|t)\{\mathbf{A} - K(t+1)\,\boldsymbol{\Delta}\}'\, P(t+1|t+1)^{-1} \\
b(t|t+1) &= b(t|t) + P(t|t)\,\boldsymbol{\Delta}'\, R(t+1|t)^{-1} dz(t+1|t).
\end{aligned}
$$

**Proof:** We give the proof only for the case $C \equiv 0$. Then, the general case follows exactly as in the proof of Theorem 3.2.3 of Hannan and Deistler (1988). As abbreviations, we use $e(t) = d_z(t|t-1)$ with the corresponding covariance matrix $R_t = R(t|t-1)$. From Theorem 2.1, (7) and (8) we have

$$
\begin{aligned}
d_b(t+1|t+1) &= d_b(t+1|t) - K(t+1)\, e(t+1) \\
&= \{\mathbf{A} - K(t+1)\,\boldsymbol{\Delta}\}\, d_b(t|t) + \{I - K(t+1)\mathbf{H}\}\, \mathbf{D}\delta(t+1) \\
&\quad - K(t+1)\,\varepsilon(t+1)
\end{aligned}
$$

Using this relation and the orthogonality of $\delta(t+1)$, $\varepsilon(t+1)$ and $b(t)$ we get

$$
\mathcal{E}b(t)\, d_b'(t+1|t+1) = \mathcal{E}b(t)\, d_b'(t|t)\, \{\mathbf{A} - K(t+1)\,\boldsymbol{\Delta}\}' = P(t|t)\, \{\mathbf{A} - K(t+1)\,\boldsymbol{\Delta}\}'
$$

As, by definition, $\mathcal{E}b(t+1)\, d_b'(t+1|t+1) = P(t+1|t+1)$ we have

$$
\mathcal{E}\{b(t) - F(t)b(t+1)\}\, d_b'(t+1|t+1) = 0
$$

By iteration, using the above equation for $d_b(t+1|t+1)$ repeatedly,

$$
\mathcal{E}\{b(t) - F(t)\, b(t+1)\}\, e'(j) = 0 \quad \text{for all } j > t+1. \tag{9}
$$

As $e(t+j)$ spans the linear space, representing the new information at time $t+j$ and orthogonal to the past, we have for $t < s$

$$
b(t|s) - b(t|t+1) = \sum_{j=t+2}^{s} \mathcal{E}\{b(t)\, e'(j)\}\, R_j^{-1} e(j)
$$

and, using (9),

$$
\begin{aligned}
F(t)\{b(t+1|s) - b(t+1|t+1)\} &= F(t) \sum_{j=t+2}^{s} \mathcal{E}\{b(t+1)\, e'(j)\}\, R_j^{-1} e(j) \\
&= b(t|s) - b(t|t+1).
\end{aligned}
$$

Finally, again using (8),

$$\begin{aligned} b(t|t+1) &= b(t|t) + \mathcal{E}\left\{b(t)\, e'(t+1)\right\} R_{t+1}^{-1} e(t+1) \\ &= b(t|t) + P(t|t)\, \Delta' R_{t+1}^{-1} e(t+1) \end{aligned}$$

∎

In practice, we do not know the parameters of the linear system given by (3), (5) and (6), but we have to work with estimates. Let us assume that we use the algorithm of Theorem 2.1 with approximations $\overline{A}, \overline{L}, \overline{C}, \overline{\Sigma}_\delta, \overline{\Sigma}_\varepsilon$ replacing the true system matrices. Then, the additional error in estimating the state $b(t)$ is of the same size as the approximation errors $||\overline{A} - A||$, $||\overline{L} - L||$ etc. To make this statement precise let us assume that we have sequences $\overline{A}_n, \overline{L}_n, \overline{C}_n, \overline{\Sigma}_{\delta,n}$ and $\overline{\Sigma}_{\varepsilon,n}$ converging for $n \to \infty$ to $A, L, C, \Sigma_\delta$ and $\Sigma_\varepsilon$ at least with the rate $O(\nu_n)$ for some sequence $\nu_n \to 0$. Let $\overline{P}_n(t|t)$ be the error covariances matrices resulting in the use of $\overline{A}_n$ etc. instead of $A$ etc. in the Kalman filter. Then:

**Theorem 2.3** $\displaystyle\lim_{t\to\infty} ||P(t|t) - \overline{P}_n(t|t)|| = O(\nu_n)$

# 3  Spatial-temporal Kriging

Knowing how to estimate the random regression coefficients $\beta(t)$ of (1), we return to the original interpolation problem. We want to estimate $\zeta(t, x)$ from $\xi(s)$, $\zeta(s, x_j)$, $j = 1, \ldots, N$, $s = 0, \ldots, T$. Let $\mathcal{A}_T$ denote the $\sigma-$algebra generated by the latter random variables. Furthermore, we use the following notation:

$$\mathbf{Z}(t) = \begin{pmatrix} Z(0) \\ \vdots \\ Z(t) \end{pmatrix} = (\zeta(0, x_1), \ldots, \zeta(0, x_N), \ldots, \zeta(t, x_N))'$$

$$\mathbf{Y}(t) = \begin{pmatrix} Y(0) \\ \vdots \\ Y(t) \end{pmatrix} = (\eta(0, x_1), \ldots, \eta(0, x_N), \ldots, \eta(t, x_N))'$$

$$\boldsymbol{\beta}(t) = \begin{pmatrix} \beta(0) \\ \vdots \\ \beta(t) \end{pmatrix} = (\beta_1(0), \ldots, \beta_m(0), \ldots, \beta_m(t))'$$

$$\mathbf{F}_t = \begin{pmatrix} F & 0 & \cdots & 0 \\ 0 & F & \cdots & 0 \\ \vdots & & \ddots & \vdots \\ 0 & 0 & \cdots & F \end{pmatrix} \quad \text{as a } m(t+1) \times N(t+1) - \text{matrix}$$

In particular, we have from (4) the relation

$$\mathbf{Z}(t) = \mathbf{F}'_t \boldsymbol{\beta}(t) + \mathbf{Y}(t) \ .$$

Under the normality assumptions which we have imposed in paragraphs 1 and 2, the best estimate of $\zeta(t, x)$ is by (1)

$$\hat{\zeta}_T(t, x) = \mathcal{E}\{\zeta(t, x)|\mathcal{A}_T\} = f(x)' \, \mathcal{E}\{\boldsymbol{\beta}(t)|\mathcal{A}_T\} + \mathcal{E}\{\eta(t, x)|\mathcal{A}_T\}$$

where with $\mathbf{S}_t$ denoting the covariance matrix of $\mathbf{Z}(t)$

$$
\begin{aligned}
\hat{\eta}_T(t, x) = \mathcal{E}\{\eta(t, x)|\mathcal{A}_T\} &= \mathcal{E}(\eta(t, x)\, \mathbf{Z}'(T))\, \mathbf{S}_T^{-1}\mathbf{Z}(T) \\
&= \mathcal{E}(\eta(t, x)\, \mathbf{Y}'(T))\, \mathbf{S}_T^{-1}\mathbf{Z}(T) \\
&=: \ g'_T(t, x)\, \mathbf{S}_T^{-1}\mathbf{Z}(T)
\end{aligned}
$$

Analogously, we have for the vector of sample residuals

$$
\begin{aligned}
\hat{\mathbf{Y}}(T) &= (\hat{\eta}_T(0, x_1), \ldots, \hat{\eta}_T(T, x_N))' = \mathcal{E}\{\mathbf{Y}(t)|\mathcal{A}_T\} \\
&= \mathbf{G}_T\, \mathbf{S}_T^{-1}\, \mathbf{Z}(T)
\end{aligned}
$$

where $\mathbf{G}_t$ denotes the covariance matrix of $\mathbf{Y}(t)$. Therefore, $\hat{\eta}_T(t, x)$ can also be written as a linear combination of the sample residuals

$$\hat{\eta}_T(t, x) = g'_T(t, x)\, \mathbf{G}_T^{-1}\hat{\mathbf{Y}}(T)$$

Otherwise, as

$$\hat{\eta}_T(t, x) = \mathcal{E}\{\zeta(t, x) - f'(x)\beta(t)|\mathcal{A}_T\}$$

we have

$$
\begin{aligned}
\hat{\mathbf{Y}}(T) &= \mathcal{E}\{\mathbf{Z}(T) - \mathbf{F}'_T \boldsymbol{\beta}(T)|\mathcal{A}_T\} \\
&= \mathbf{Z}(T) - \mathbf{F}'_T\, \hat{\boldsymbol{\beta}}(T)
\end{aligned}
$$

where $\hat{\boldsymbol{\beta}}(T)$ denotes the best estimate of $\boldsymbol{\beta}(T)$ given the information in $\mathcal{A}_T$. Finally, we get as the desired interpolator for $\zeta(t, x)$

$$
\begin{aligned}
\hat{\zeta}_T(t, x) &= f(x)'\, \hat{\beta}(t|T) + \hat{\eta}_T(t, x) \\
&= f(x)'\, \hat{\beta}(t|T) + g'_T(t, x)\, \mathbf{G}_T^{-1}\hat{\mathbf{Y}}(T) \\
&= f(x)'\, \hat{\beta}(t|T) + g'_T(t, x)\, \mathbf{G}_T^{-1}(\mathbf{Z}(T) - \mathbf{F}'_T\, \hat{\beta}(T)) \quad (10)
\end{aligned}
$$

(10) generalizes the interpolation procedure known in geostatistics as "Kriging" (compare, e.g., Ripley, 1982). If our model (1) is reduced to

$$\zeta(x) = f(x)'\, \beta + \eta(x)$$

without any time dependence and with deterministic regression parameter $\beta$ then (10) is just the well-known "Universal Kriging". If $f(x)'\, \beta \equiv \mu$ even

186

reduces to a constant, then (10) is the original "Kriging".

For the original practical problem we are mostly interested in estimating $\zeta(t, x)$ using all data up to time $t$, i.e. in calculating $\hat{\zeta}_t(t, x)$. For this purpose, the first summand $f(x)'\ \hat{\beta}(t|t)$ on the right-hand side of (10) can be calculated recursively in $t$ using the Kalman filter of Theorem 2.1. In the second summand $\hat{\beta}(t)$ appears which consists of the subvectors $\hat{\beta}(0|t), \ldots, \hat{\beta}(t|t)$. They can be calculated efficiently using Theorem 2.2. The resulting procedure works quite well, and it is fast enough to allow real-time calculations of $\hat{\zeta}(t, x)$, $t = 0, 1, 2, \ldots$ for all knots $x$ of an equispaced lattice in the plane such that the spatial-temporal evolution of the process can be studied as an animated graphic in detail.

For sake of illustration we consider one particular example with simulated data. Figure 1a and 1b show $c_0(y - x) = \mathcal{E}\, \eta(t, x)\, \eta(t, y)$ and $c_1(y - x) = \mathcal{E}\, \eta(t, x)\, \eta(t + 1, y)$. This corresponds to a nonisotropic dependency where the peak of $c_1(y - x)$ is shifted in a direction corresponding to the prevailing wind direction. Observations were taken at 16 irregularly spaced locations, and as a function representing general topographic trends, only linear functions were allowed, i.e. $f(x) = (1, x_1, x_2)'$. Figure 2a and 2b show two snapshots of the spatial-temporal interpolations $\hat{\zeta}(t, x)$ for $t = 4$ and $t = 18$.

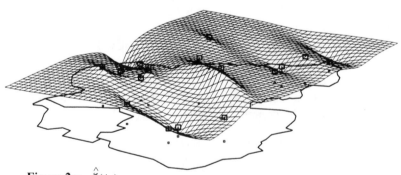

**Figure 2a:** $\hat{\zeta}(4, \cdot)$

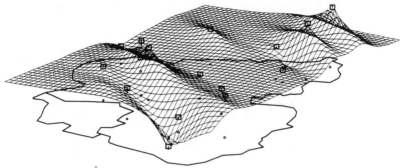

**Figure 2b:** $\hat{\zeta}(18,\cdot)$

# 4 Some remarks on estimation of model parameters

Up to now we have pretended to know the model parameters like the ARX–coefficient matrices $A_1, \ldots, A_p$, $C_0, \ldots, C_r$ of (2), the AR(1)–matrix $L$ of (3) or the noise covariance matrices $\Sigma_\delta$ and $\Sigma_\varepsilon$. As in the known procedures for Kriging of purely spatial data, these parameters have to be estimated (compare, e.g., Ripley, 1982). We do not want to go into details but point out some special features which follow from our modelling approach.

Hannan and his coworkers have given over the years an extensive theoretical treatment of maximum likelihood estimates for the parameters of multivariate Gaussian ARMA– and ARMAX–systems. A compilation of the relevant results is provided by Hannan and Deistler (1988). In our case, we cannot use this estimation theory directly as we do not observe the ARX–process $\beta(t)$. However, let $F^-$ denote a pseudo inverse of $F'$, e.g. $F^- = (FF')^{-1}F$ for $m \leq N$ and $FF'$ invertible, and let $U^{-1}$ be the one time unit backshift operator: $U^{-1}\beta(t) = \beta(t-1)$. Then, we have from (2) and (4):

$$
\begin{aligned}
A(U^{-1})\,F^-Z(t) &= A(U^{-1})\,F^-(F'\,\beta(t) + Y(t)) \\
&= A(U^{-1})(\beta(t) + F^-Y(t)) \\
&= A(U^{-1})\,F^-Y(t) + \delta(t) + C(U^{-1})\xi(t)
\end{aligned}
$$

If $L^s$ decreases fast enough for $s \to \infty$ such that $Y(t) \approx \sum_{s=0}^q L^s\varepsilon(t-s)$, then we have for the observable time series $Z^*(t) = F^-Z(t)$

$$
A(U^{-1})\,Z^*(t) \approx A(U^{-1})\,F^- \sum_{s=0}^q L^s\varepsilon(t-s) + \delta(t) + C(U^{-1})\,\xi(t)
$$

As the first two summands of the right-hand side form a $(p+q)$–dependent time series we have for suitable white noise $\psi(t)$ and matrix coefficients $B_0, \ldots, B_{p+q}$

$$A(U^{-1}) \, Z^*(t) \approx B(U^{-1}) \, \psi(t) + C(U^{-1}) \, \xi(t), \quad B(z) = \sum_{j=0}^{p+q} B_j \, z^j,$$

i.e. $Z^*(t)$ is approximately an ARMAX$(p, p+q, r)$–process with the same autoregressive and exogenous part $A(U^{-1})$, $C(U^{-1})$ as the ARX–process $\beta(t)$. Therefore, fitting an ARMA$(p, p+q, r)$–model to $Z^*(0), \ldots, Z^*(T)$ as described in Hannan and Deistler (1988) provides estimates of $A_1, \ldots, A_p$, $C_1, \ldots, C_r$.

In a similar manner, the parameters of the AR(1)–model (3) may be estimated. If we could observe the time series $Y(t)$ directly, we could estimate the autocovariances

$$\Sigma_0 = \mathcal{E} \, Y(t) \, Y(t)' \text{ and } \Sigma_1 = \mathcal{E} \, Y(t) \, Y(t+1)'$$

directly by the corresponding sample autocovariances, and then, using

$$L = \Sigma_1 \Sigma_0^{-1} \text{ and } \Sigma_\epsilon = \Sigma_0 - \Sigma_1 \Sigma_0^{-1} \Sigma_1$$

we would have estimates of $L$ and $\Sigma_\epsilon$, too. Now, let

$$
\begin{aligned}
\tilde{Z}(t) &= (I_N - F'F^-) \, Z(t) = (I_N - F'F^-)(F' \, \beta(t) + Y(t)) \\
&= (I_N - F'F^-) \, Y(t)
\end{aligned}
$$

as $F' \, F^- \, F' = F'$. Therefore, we may consider the sample autocovariances of the observable time series $\tilde{Z}(t)$ and calculate from them estimates of $\Sigma_0$, $\Sigma_1$. Here, the properties of $I_N - F' \, F^-$ have to be exploited, in particular the fact that it has eigenvalues 1 and 0 only and that its rank is $N - m$. The details of the somewhat involved procedure have been given by Gründer (1993).

**Acknowledgement:** This work has been funded by the Volkswagen-stiftung as part of the Center for Practical Mathematics. We thank A. Roeder of the Forest Research Center of Rheinland-Pfalz for providing the problem, the data and continuous advice.

# References

[1] Franke, J., and B. Gründer (1992). *Stochastic modelling for analyzing immission data in forests.* In: Tagungsberichte der Arbeitsgruppe Bio-

metrie in der Ökologie der Deutschen Region der Internationalen Biometrischen Gesellschaft, Wuppertal.

[2] Gründer, B. (1992). *Gasdistribution as a space-time model with regard to meteorological data.* In: Mathematical Modelling of Forest Ecosystems, ed. J. Franke and A. Roeder, J.D. Sauerländer's Verlag, Frankfurt a.M. .

[3] Gründer, B. (1993). *Stochastische Modelle für die Luftschadstoffverteilung in Waldgebieten.* Ph.D. Thesis, University of Kaiserslautern.

[4] Hannan, E.J., and M. Deistler (1988). *The Statistical Theory of Linear Systems*, Wiley, New York.

[5] Kalman, R.E. (1960). A new approach to linear filtering and prediction problems. Trans. ASME J. Basic Engr., Series D, 82, 35-45.

[6] Priestley, M.B. (1981). *Spectral Analysis and Time Series*, vol. 2, Academic Press, London, New York.

[7] Ripley, R.D. (1981). Spatial Statistics. Wiley, New York.

[8] Vecchia, A.V. (1988). *Estimation and model identification for continuous spatial processes.* J.R. Statist. Soc. B, 50, 297-312.

Corresponding author's address:

Prof. Dr. J. Franke
Department of Mathematics
University of Kaiserslautern
P.O. Box 3049
D-67653 Kaiserslautern
F.R. of Germany

Dr. B. Gründer
TecMath GmbH
D-67661 Kaiserslautern

# FRACTIONAL STOCHASTIC UNIT ROOT PROCESSES

Clive W.J. Granger, Namwon Hyung, and Yongil Jeon
Department of Economics
University of California, San Diego
9500 Gilman Drive
La Jolla, CA 92093-0508

August 1995

## 1. Introduction

One way in which the field of time series analysis develops is by considering processes of increasing complexity, hopefully producing models which can still be analyzed whilst remaining interpretable and useful in applied work. This development is often of a stepping-stone form, with one model naturally evolving into the next or two familiar models merging into something less familiar. The simple autoregressive model of order one

$$x_t = a x_{t-1} + \varepsilon_t \qquad , \quad |a| < 1 \tag{1.1}$$

where, throughout, $\varepsilon_t$ will be taken to be i.i.d. and zero mean, leads immediately to the random walk, by taking $a = 1$ but this change produces a series with dramatically different properties

$$(1-B)x_t = \varepsilon_t . \tag{1.2}$$

An interesting class of processes which generalize (1.1) are the doubly stochastic processes, such as

$$x_t = a_t x_{t-1} + \varepsilon_t \tag{1.3}$$

where

$$a_t = m + \rho a_{t-1} + \eta_t \quad , \text{ for example }, \tag{1.4}$$

with $\varepsilon_t, \eta_t$ independent. As there are now two stochastic inputs, $\varepsilon_t$ and $\eta_t$, these processes are not invertible. If one knows the model in (1.1) and the values of the output $x_t$, then the values of the inputs $\varepsilon_t$ can be deduced. This is not possible for a doubly stochastic series, in general.

Doubly stochastic processes have been studied by Tjøstheim (1986), Andel (1976), Brandt (1986), Pourahmadi (1988) and others.

A mixture of the unit root (1.2) and a doubly stochastic process produces a stochastic unit root process (or STUR) which takes the form (1.3), i.e. here $a_t$ is stochastic but is such that $x_t$ has some of the properties of a unit root; for example, if $E[a_t] = 1$. These have been discussed by Granger and Swanson (1994) and the results obtained are outlined in Section 2.

A further form of generalization of the unit root process (1.2) is the fractional integrated process

$$(1-B)^d x_t = \varepsilon_t \tag{1.5}$$

where $d$ can be a non-integer. These were introduced in Granger and Joyeaux (1980) and Hosking (1981), but with $\varepsilon_t$ possibly a stationary ARMA process, and are based on earlier work by Mandelbrot (1969) and others. There has been a lot of interest in these processes both by theorists and applied workers. Much of this work is summarized in Beran (1994). The main properties of these processes will be outlined in Section 3. They provide useful intermediate steps between stationary AR and unit root processes and when $d$ is positive are called "long-memory" because they have an infinite peak at zero frequency.

In this paper the stochastic unit root idea is applied to fractionally integrated processes so that $x_t$ is generated by

$$(1-a_t B)^d x_t = \varepsilon_t$$

where $a_t$ is stochastic and will be given some appropriate property, such as $E[a_t] = 1$ and, for ease of analysis, only the case where $\varepsilon_t$ is zero mean i.i.d. is considered. The properties of this process is considered in Section 4. Finally, the effect of making $d$ stochastic is briefly considered in Section 5.

Although Ted Hannan did not publish anything directly dealing with long-memory, a number of his papers on topics such as regressions involving stochastic processes and processes with infinite variance are of immediate relevance to the area, as seen by the citations in Beran (1994). We would like to think that he would approve of the increasing richness of models that are available for consideration and analysis.

## 2. Properties of STUR

It is convenient to write the model as

$$x_t = a_t x_{t-1} + \varepsilon_t \tag{2.1}$$

where

$$a_t = \exp \alpha_t \qquad (2.2)$$

where it is assumed that $\alpha_t$ is a stationary process, normally distributed with mean $\mu$, variance $\sigma_\alpha^2$, autocorrelations $\rho_{\alpha,k}$ and spectrum $f_\alpha(w)$; so that $f_\alpha(0) = \sigma_\alpha^2(1 + 2\sum_{j=1}^{\infty} \rho_{\alpha,j})$. In the following it will be assumed that $f_\alpha(0)$ is positive, although it would be interesting to relax this constraint.

The corresponding solution to (2.1) is

$$x_t = \varepsilon_t + \sum_{j=1}^{k-1} \pi_{t,j} \varepsilon_{t-j} + \pi_{t,k} x_{t-k} \qquad (2.3)$$

for any integer $k$, $0 \le k \le t$, where

$$\pi_{t,j} = \exp\left(\sum_{i=0}^{j-1} \alpha_{t-i}\right). \qquad (2.4)$$

(2.3) is sometimes called the "impulse response" of the variable, measuring the response of $x_t$ to an old shock $\varepsilon_{t-j}$. For an ordinary unit root, such as the random walk (1.2), these response terms $\pi_{t-j}$ will all be constant and equal to one but for STUR they are random variables.

Granger and Swanson (1994) discuss two alternative form of STUR;

STUR-A: For which $E[a_t] = 1$, so that

$$\mu + \tfrac{1}{2}\sigma_\alpha^2 = 0 \qquad (2.5)$$

and STUR-B: such that $E[\pi_{t,j}] = 1$ for $j$ large, so that

$$\mu + \tfrac{1}{2} f_\alpha(0) = 0. \qquad (2.6)$$

Note that the two properties (2.5), (2.6), occur together only if $\sum_{j=1}^{\infty} \rho_{\alpha,j} = 0$. If $f_\alpha(0) > \sigma_\alpha^2$ it is found that under STUR-A, the variance of the process increases exponentially with time, rather than the linear increase observed with a random walk. Further, $E[\pi_{t,j}]$ increases as $j$ increases.

For STUR-B, the variance still increases exponentially but the expected value of the impulses is now constrained to be one. However, other aspects of the distribution of $\pi_{t,j}$ is found to be unusual, as $j$ increases the variance goes to infinity, the median to zero and all $100(1-\alpha)\%$

confidence intervals for positive $\alpha$ also go to zero. Thus, the distribution of $\pi_{t,j}$ becomes a high peak very near zero plus a long tail on the positive side to ensure the unit mean and giving the increasing variance. Going from a pure to a stochastic unit root is seen to produce substantially different properties for the process.

Granger and Swanson (1994) also find that standard unit root tests, such as the well known one proposed by Dickey and Fuller, do not reject a null of a unit root when the data is generated by a stochastic unit root. Leybourne, McCabe and Tremayne (1993) have proposed a test with power to distinguish between pure and stochastic unit roots.

### 3. Fractionally Integrated Processes

It is documented in Beran (1994) and elsewhere that the process generated by (1.5)has finite variance if $| d | < \frac{1}{2}$ and autocorrelations

$$\rho_k = \frac{\Gamma(1-d)}{\Gamma(d)} \cdot \frac{\Gamma(k+d)}{\Gamma(k+1-d)} \tag{3.1}$$

which for $k$ large can be approximated by

$$\rho_k \approx A_1(d)k^{2d-1} . \tag{3.2}$$

These produce a spectrum which is proportional to $w^{-2d}$ at low frequency $w$.

The equivalent in AR($\infty$) representation to (1.5) is

$$\sum_{j=0}^{\infty} a_j x_{t-j} = \varepsilon_t , \quad a_o = 1 , \tag{3.3}$$

with

$$a_j = \frac{\Gamma(j-d)}{\Gamma(-d)\Gamma(j+1)} \quad j \geq 1 , \tag{3.4}$$

so that for $j$ large

$$a_j \approx A_2(d)j^{-(1+d)} . \tag{3.5}$$

Similarly, the MA($\infty$) representation is

$$x_t = \sum_{j=0}^{\infty} b_j \varepsilon_{t-j} , \quad b_o = 1 , \tag{3.6}$$

and

$$b_j = \frac{\Gamma(j+d)}{\Gamma(d)\Gamma(j+1)} \quad j \geq 1 \tag{3.7}$$

$$\approx A_3(d)j^{d-1} \quad \text{for } j \text{ large} \tag{3.8}$$

where here $A_1, A_2, A_3$ are specific functions of $d$, in particular $A_3 = [\Gamma(d)]^{-1}$. Squaring both sides of (3.5), taking expectations, and assuming that the process starts at time $t=0$, gives

$$var\ (x_t) = \sigma_\varepsilon^2 [1 + \sum_{j=1}^{t-1} b_j^2]. \tag{3.9}$$

Using the approximation for $b_j$ (3.8), approximating the sum in (3.9) by an integral and then performing the integration gives

$$var\ (x_t) = O(t^{2d-1}) \text{ for } d > \tfrac{1}{2}. \tag{3.10}$$

It can be shown that this variance is $O(\log t)$ for $d = \tfrac{1}{2}$.

## 4. Fractional Stochastically Integrated Processes

The process considered here is

$$(1-a_t B)^d x_t = \varepsilon_t \tag{4.1}$$

where $a_t = \exp \alpha_t$ and $\alpha_t$ is a stationary, stochastic process where $\varepsilon_t$ is i.i.d., zero mean independent of $\alpha_t$. Two such cases will be analyzed, with $\alpha_t$ constrained to obey the STUR-A condition (2.5) or the STUR-B condition (2.6), as these were found to be particularly interesting for the stochastic unit root process.

Writing (4.1) in its moving average form one gets

$$x_t = \varepsilon_t + \sum_{j=1}^{t} \pi_{t,j} b_j \varepsilon_{t-j} \tag{4.2}$$

where $\pi_{t,j}$ is given by (2.4) and $b_j$ by (3.7). As this expression is rather difficult to analyze in general, just the case where $\alpha_t$ is generated by

$$\alpha_t = m + \rho \alpha_{t-1} + \eta_t \tag{4.3}$$

will be considered. Some complicated algebra gives

$$cov\ (x_t, x_{t+k}) \approx \sigma_\varepsilon^2 \sum_{j=0}^{t-1} b_j b_{j+k} \exp\ [(k+2j)m + \tfrac{1}{2} f_\alpha(o)\{k+4j \tag{4.4}$$

$$+ \frac{4\rho}{1-\rho^2}(1-\rho^j)(1-\rho^k)\}] \ .$$

Using the approximation (3.8) and the STUR-A constant, $m = -\frac{1}{2}\sigma^2/(1-\rho^2)$ so that $E[\exp\alpha_t] = 1$, gives

$$var_A(x_t) \approx \frac{\sigma_\varepsilon^2}{(\Gamma(d))^2} \left[ \sum_{j=1}^{t-1} j^{2(d-1)} \exp(\theta_1 j) + 1 \right] \qquad (4.5)$$

where $\theta_1 = \sigma_\eta^2[2/(1-\rho)^2 - 1/(1-\rho^2)]$ and with the STUR-B constraint, so that $E[\pi_{tj}]$ tends to 1 for large $j$ has $m = -\frac{1}{2}\sigma_\eta^2/(1-\rho)^2$ and gives

$$var_B(x_t) = \frac{\sigma_\varepsilon^2}{(\Gamma(d))^2} \left[ \sum_{j=1}^{t-1} j^{2(d-1)} \exp(\theta_2 j) + 1 \right] \qquad (4.6)$$

where $\theta_2 = \sigma_\eta^2/(1-\rho)^2$, so $var_B(x_t) \leq var_A(x_t)$ if $\rho > 0$. Expressing these variances as approximate integrals gives

$$v(t) \approx c \int_0^t s^{2(d-1)} \exp(\lambda s) ds$$

then approximation results for these integrals suggests

$$v(t) \approx ct^{2(d-1)} \exp(\lambda t) \qquad (4.7)$$

where $\lambda = \theta_1$ for STUR-A and $\lambda = \theta_2$ for STUR-B. However, as it is clear that $v(t)$ is non-decreasing with $t$, so that $dv(t)/dt$ is positive, this approximation cannot be true unless

$$t > t_o = \frac{2(1-d)}{\theta} \ .$$

To explore how the variance of these processes increase with time, values were estimated from (4.5) and (4.6) with $d_t$ generated by (4.3) and with the following:

| Coefficient | Case 1 | 2 | 3 | 4 |
|---|---|---|---|---|
| $\rho$ | 0.3 | 0.7 | 0.3 | 0.7 |
| $\sigma_\eta^2 \times 1000$ | 0.1 | 0.1 | 1.6 | 1.6 |
| 95% interval | (0.978 to | (0.972 to | (0.919 to | (0.843 to |
| for $d_t$ | 1.021) | 1.028) | 1.086) | 1.117) |
| $t_0(d = 0.25)$ | 6024 | 741 | 377 | 46 |
| $t_0(d = 0.75)$ | 2008 | 247 | 125.5 | 15.4 |

Each case is repeated for $d = 0.25$ and $d = 0.75$ and under the STUR-A and STUR-B constraints.

The table also shows the theoretical 95% confidence interval for $a_t = \exp\alpha_t$, assuming $\alpha_t$ is

Gaussian, and also the value for $t_o$ for the STUR-A cases. The values of log $v(t)$ for STUR-A are illustrated in Figure 1, plotted against $t$ up to time 1000. For $d = 0.25$ and cases 1 and 2, which have $\sigma_\eta^2$ small, log $v(t)$ is actually or apparently converging, which is what would occur if $d$ is less than 0.5 and the unit root was not stochastic. In all other cases, $v(t)$ is seen to be an exponential function of $t$, as predicted by (4.7). For the non-stochastic I(d) process $v(t)$ is $O(t^{2d-1})$ for $d > \frac{1}{2}$. Thus, one consequence of introducing a stochastic unit root is to move to exponentially increasing trends, as was found for the STUR models. The figures for STUR-B are not shown as they are similar but not identical. Figures 2 and 3 illustrate the plots of $\rho_{k,t} = \dfrac{cov\ (x_t, x_{t+k})}{var\ (x_t)}$ and log $\rho_{k,t}$, where the quantities in the nominator and denominator are given by (4.4), for various values of $k$ and $t$, under the STUR-A constraint, for $d = 0.25$ and for cases (1) and (2). In Figure 2 the plots against $k$ are shown for $t = 500$, 1000, 1500, and 2000. It is seen that for Case (1), with $\rho = 0.3$, convergence is not a big problem, but has still not happened with a sample of 2000. However, for case 2, with $\rho = 0.7$, convergence is much slower, with apparently longer memory as the length of sample increases. Figure 3 shows for cases 1 and 2, with $t = 1000$, the log autocorrelations and superimposed are the same values for the deterministic case, with $\sigma_\eta^2 = 0$. For the case with $\rho = 0.3$ but particularly when $\rho = 0.7$, the stochastic model has longer memory, with more slowly declining autocorrelations, compared to the deterministic model with the same $d$-value.

## 5. Stochastic Fractionally Integrated Process

The deterministic I(d) process has apparently been observed in a number of scientific and economic series as discussed by Beran (1994) and Bailey (1994). A particularly clear example of long-memory is given by Ding, Granger, and Engle (1993) using the absolute returns of the daily Standard and Poors Stock Market Index over a period of 64 years, and so contain 17,054 daily observations. For the whole sample period, an estimate of $d$ given by a techniques suggested by Geweke and Porter-Hudak (1983) based on the log of the periodogram, gave $\hat{d} = 0.474$.

However, if the full sample is arbitrarily broken into ten samples of equal length, each containing 1705 observations, the estimated $d$ values were (with associated standard errors) 0.358 (.133), 0.405 (.131), 0.438 (.097), 0.336 (.144), 0.156 (.106), 0.445 (.113), 0.518 (.088), 0.714 (.105), 0.436 (.110), and 0.752 (.070), with an average of 0.4126 and a range of 0.156 to 0.714, as reported in Granger and Ding (1995). Although there may be some bias problems with the estimation technique, the evidence does imply that the value of $d$ changes, through time, suggesting that consideration be given to the model

$$(1-B)^{d_t} x_t = \varepsilon_t \tag{5.1}$$

where $\varepsilon_t$ is iid and $d_t$ is a stochastic process with a mean $\bar{d}$, possibly a white noise, or perhaps with temporal properties. (5.1) can be taken as a short-hand notation for the time-changing parameter moving average

$$x_t = \varepsilon_t + \sum_{j=1}^{t-1} b_{j,t} \varepsilon_{t-j} \tag{5.2}$$

where

$$b_{j,t} = \frac{\Gamma(j+d_t)}{\Gamma(d_t)\Gamma(j+1)} \tag{5.3}$$

which, using Stirling's formula, can be approximated as

$$b_{j,t} \approx j^{d_t-1} / \Gamma(d_t) . \tag{5.4}$$

Analytically, this is difficult to handle exactly, but a useful approximation is suggested by the series expansion provided by Abramowitz and Stegum (1964), page 256, which was originally found by H.T. Davis,

$$1/\Gamma(x) = x + c_2 x^2 + c_3 x^3 + ... \tag{5.5}$$

where $c_2 = 0.577$, $c_3 = -0.656$, $c_4 = -0.042$, $c_5 = 0.166$ and thereafter the constants decline quite rapidly. For $0 \leq x \leq 1$, the approximation $1/\Gamma(x) = x + 0.577 x^2 - 0.656 x^3$ holds with an error of less than 2%. For $0 \leq x \leq 0.65$, the approximation $1/\Gamma(x) = 1.14 x$ holds with under a 1% error, as was found by direct plotting of this function. If $d_t = \bar{d} + e d_t$, when $E[d_t] = \bar{d}$ and noting that the theory being considered is most appropriate for $\bar{d} < \frac{1}{2}$, where the deterministic I(d) process is stationary and where the majority of estimates of $d$ occur, as seen above, it seems that a range of 0 to 0.65 is appropriate for consideration of the variance of the process.

Using the approximation (5.4) and for $1/\Gamma(d)$ in (5.2), squaring and taking expectations gives

$$\text{var } x_t \approx \sigma_\varepsilon^2 [1 + \sum_{j=1}^{t-1} c^2 j^{2(\bar{d}-1)} E[(\bar{d} + ed_t)^2 j^{2ed_t}]] \tag{5.6}$$

where $c = 1.14$.

Writing $j^x = \exp(x \log j)$ and noting that $E[x^k e^{\lambda x}] = \dfrac{d^k m(\lambda)}{d\lambda^k}$, gives $\text{var } x_t$ is the summation of terms such as

$$\sigma_\varepsilon^2 c^2 j^{2(\bar{d}-1)} [\bar{d}^2 m(2 \log j) + 2\bar{d}\dot{m}(2 \log j) + \ddot{m}(2 \log j)] \tag{5.7}$$

where $\dot{m}(2 \log j)$ is the first derivative of $m(\lambda)$ as $\lambda = 2 \log j$ and $\ddot{m}$ is the second derivative. If $ed_t$ is Gaussian $N(0, \sigma^2)$, then $m(\lambda) = \exp(-\frac{1}{2}\lambda^2 \sigma^2)$, and the slowest decreasing component of (5.7) takes the form

$$\sigma_\varepsilon^2 c^2 \sigma^4 (2 \log j)^2 j^{2(\bar{d}-1)} \exp(-\frac{1}{2}\sigma^2 (\log j)^2). \tag{5.8}$$

This suggests that if $\sigma^2 > 0$, the variance of $x_t$ tends to a constant for all values of $\bar{d} < \frac{1}{2}$. A direct estimation of the variance in (5.6), using (5.7) and the Gaussian distribution for $d_t$ with $\bar{d} = 0.25$ and $\sigma = .125$ found that var $x_t$ converged quite quickly, being virtually constant after 200 terms (See Figure 4). An expresion for the autocorrelations was also derived in this case with $d_t$ i.i.d. and the correlogram was again found to converge quickly.

## 6. Conclusion

It has been shown that the standard I(d) model can be generalized by letting the unit root become stochastic or $d_t$ be stochastic and certain theoretical and simulation results achieved. These models are less easy to use than I(d) but are likely to better present real data.

## References

Abramowitz, M. and I.A. Stegum (1964) , Handbook of Mathematical Functions, National Bureau of Standards, Washington, D.C.

Andel, J. (1976), "Autoregressive Series With Random Parameters," *Mathematical Operationsfursch und Statistik* 7, 735-741.

Bailey, R.T (1994), "Long-Memory Processes and Fractional Integration In Economics and Finance," to appear, *Journal of Econometrics.*

Brandt, A. (1986), "The Stochastic Equation $Y_{n+1} = A_n Y_n + B_n$ With Stationary Coefficients," *Advances In Applied Probability* 18, 211-220.

Beran, J. (1994), Statistics for Long-Memory Processes , Chapman-Hall, New York.

Ding, Z, C.W.J. Granger, and R.F. Engle (1993), "A Long-Memory Property of Stock Market Returns and New Model," *Journal of Empirical Finance* 1, 83-106.

Geweke, J. and S. Porter-Hudak (1983), "The Estimation and Application of Long-Memory Time Series Models," *Journal of Time Series Analysis* 4, 221-237.

Granger, C.W.J. and Z. Ding (1995), "Varieties of Long-Memory Models," to appear, *Journal of Econometrics.*

Granger, C.W.J. and R. Joyeux (1980), "An Introduction to Long-Range Time Series Models and Fractional Differencing," *Journal of Time Series Analysis* 1, 15-30.

Granger, C.W.J. and N.R. Swanson (1994), "An Introduction to Stochastic Unit Root Processes." Discussion paper 92-53R, Department of Economics, UCSD.

Hosking, J.R.M. (1981), "Fractional Differencing," *Biometrika* 68, 165-176.

Leybourne, S.J., B.P.M. McCabe, and A.R. Tremayne (1993), "How Many Time Series are Difference Stationary?" Working Paper, University of British Columbia.

Mandelbrot, B.B. (1969), "Long-Run Linearity, Locallyn HGaussian Process, $H$-spectra and Infinite Variance," *International Economic Review* 10, 82-113.

Pourahmadi, M. (1988), "Stationarity of the Soluction of $X_t = A_t X_{t-1} + \varepsilon_t$ and the Analysis of Non-Gaussian Dependent Random Variables," *Journal of Time Series Analysis* 9, 225-230.

Tjøstheim, D. (1986), "Some Doubly Stochastic Time Series Models," *Journal of Time Series Analysis* 7, 51-72.

Figure 1 Variances under STUR-A

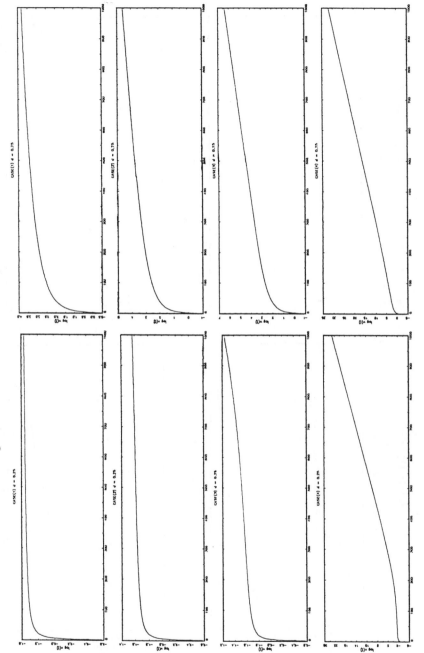

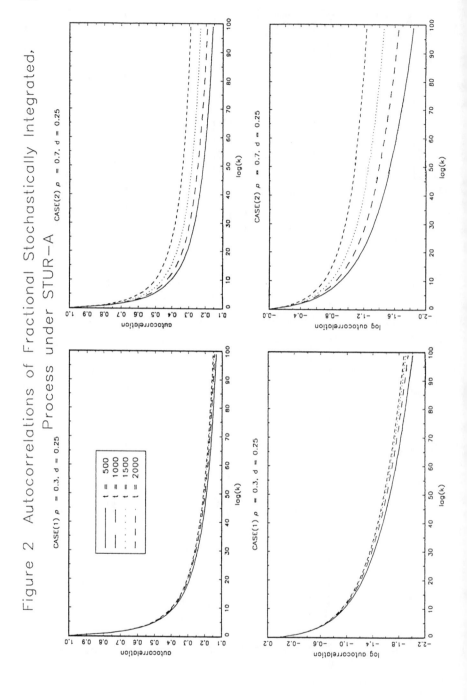

Figure 2 Autocorrelations of Fractional Stochastically Integrated, Process under STUR-A

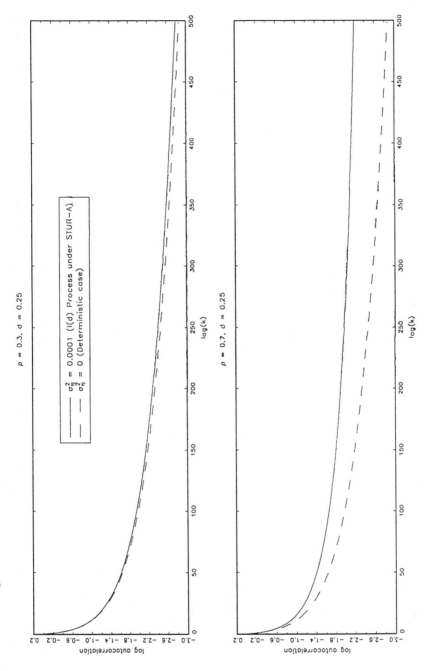

Figure 3 Autocorrelations under STUR–A vs Deterministic case

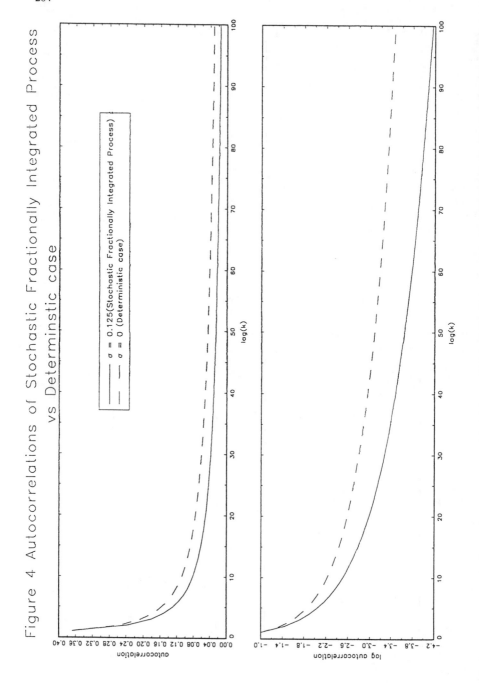

Figure 4 Autocorrelations of Stochastic Fractionally Integrated Process vs Determinstic case

# Design of Moving-Average Trend Filters using Fidelity and Smoothness Criteria

## Alistair Gray
## Peter Thomson

ABSTRACT The development of a flexible family of finite moving-average filters from specified smoothness and fidelity criteria is considered. These filters are based on simple dynamic models operating locally within the span of the filter. They are shown to generalise and extend the standard Macaulay and Henderson filters used in practice. The properties of these filters are determined and evaluated both in theory and in practice.

## 1    Introduction

This paper is concerned with the design of local trend estimation filters for non-seasonal time series using finite moving-averages that are derived from specified fidelity and smoothness criteria. A particular objective is to try and improve current methods of trend estimation within seasonal adjustment procedures where non-seasonal trend filters play a key role, especially at the ends of series. However local trend estimation is also important in its own right and is routinely used, for example, in the analysis of financial and other time series. In this paper we focus on filters for the body of the series.

The design and use of finite moving-averages for local trend estimation has a long history, particularly within the actuarial literature where it is referred to as graduation. Much of this development has been posited on simple structural models of the form $\phi(y_t) = T_t + S_t + \epsilon_t$ where $y_t$ denotes the time series observed, $\phi(x)$ denotes some appropriate transformation, $T_t$ denotes the trend, $S_t$ the seasonal and $\epsilon_t$ the irregular or noise. Typically the unobserved trend and seasonal components are assumed to evolve slowly with more abrupt changes, including calendar effects, handled via simple adjustments. Recently, after somewhat of a lull, more attention has been focussed on both structural models and the need for improved finite moving-average trend filters. Impetus to the design of finite moving-average filters has also come from work in scatterplot smoothing and local regression models. For example smoothers such as loess are, in the time series context, finite moving-average trend filters. The recent seasonal decomposition procedure STL ([CCMT90]), is built around the non-seasonal loess filter.

Structural models have been exploited in a number of global parametric seasonal adjustment procedures. These include BAYSEA ([Aka80]) and other related model-based methods (see [GK83], [Har89] and [Sch81] for example). An explicit and important feature of much of this work is the use of criteria that optimally weight specific measures of fidelity and trend smoothness. The latter is in the spirit of [Hen24] and [Whi23].

In practice the use of finite moving-average trend filters in seasonal adjustment and trend estimation remains the dominant technology. This emphasis needs some justification and explanation. Firstly, the use of moving data windows forms the basis for simple non-parametric models that apply locally and whose assumptions can, in many cases, be justified by simple graphical analysis. As a consequence, they have served official statisticians and many others well over a long period of time. Secondly, they directly control and limit the revisions to historical trend values as additional data is obtained. This is an important and over-riding requirement for most official statisticians. Finally, moving local non-parametric models have the potential to capture both evolutionary change and non-evolutionary structural change in time series in a more direct and transparent way than global parametric models.

However the two technologies should not be seen as competing. The non-parametric local procedures are frequently used as exploratory tools prior to the fitting of a more sophisticated parametric global model. This paper attempts to combine some of the virtues of the global parametric models within the framework of a finite moving-average trend estimation procedure.

# 2 Local dynamic model

We follow the conventional paradigm and consider a moving window of $n = 2r + 1$ consecutive observations within which an estimate of the trend is to be calculated for the central time point. Within the finite window, model the observations as

$$y_t = g_t + \epsilon_t \tag{1.1}$$

where $\epsilon_t$ is white noise with variance $\sigma^2$, the trend $g_t$ is given by

$$g_t = \sum_{j=0}^{p} \beta_j t^j + \xi_t \tag{1.2}$$

and the zero mean stochastic process $\xi_t$ is assumed to be correlated, but uncorrelated with $\epsilon_t$. It is assumed that $\xi_t$, $\sigma^2$ are not both zero for all $t$. In particular we consider the situation where the $\beta_j$ and $\sigma^2$ are parameters local to the window, but $p$, $n$ and the model for $\xi_t/\sigma$ involve global parameters which are constant across windows. Thus, although the parameters involved with the mean and variance of $y_t$ vary across windows, the autocorrelation structure of $y_t$ will be a function of time invariant parameters in addition to time itself.

Loosely speaking, the finite polynomial is intended to capture deterministic low order polynomial trend whereas $\xi_t$ is intended to capture smooth deviations from the polynomial trend. Note that it is the incorporation of $\xi_t$ which distinguishes this local model from the standard situation where it is zero.

Because the window is not likely to be large the model will need to involve as few parameters as possible on the one hand, while allowing for a sufficiently flexible family of forms for $g_t$ on the other. *With these points in mind we choose to model $\xi_t$ as a (possibly integrated) random walk with initial value zero.* Typically $\xi_t$ will be an $I(d)$ process or the sum of $I(d)$ processes. This seems an appropriate and parsimonious model which should account for smooth deviations from the deterministic polynomial trend component. It also provides a dynamic trend model for $g_t$ which is essentially of the same form as that used in the ARIMA structural models that have been successfully applied to economic and official data. (See [Bel93], [Har89] and [KD82] for example.) Note that the conventional model is recovered by eliminating the random walk component $\xi_t$ which is achieved by setting its innovation variance to zero.

## 3  Trend filter design

Given a finite window of width $n = 2r + 1$ points centred at time point $t$ in the body of the time series, we now choose to estimate $g_t$ by the finite moving-average

$$\hat{g}_t = \sum_{s=-r}^{r} w_s y_{t+s}$$

where the $w_s$ are constrained by the requirement that $E\{\hat{g}_t - g_t\} = 0$ and $y_t$ follows the local dynamic model given by (1.1) and (1.2). Thus $\hat{g}_t$ is an unconditionally unbiased predictor of $g_t$. Note that this condition is equivalent to the requirement that the $w_s$ satisfy

$$\sum_{s=-r}^{r} w_s = 1 \qquad \sum_{s=-r}^{r} s^j w_s = 0 \qquad (0 < j \le p) \qquad (1.3)$$

so that the moving average filter passes polynomials of degree $p$.

To assist in the choice of filter weights $w_j$ we consider the criteria

$$F = E\{(\hat{g}_t - g_t)^2\}, \qquad S = E\{(\Delta^{p+1}\hat{g}_t)^2\} \qquad (1.4)$$

where $F$ measures the *fidelity* of $\hat{g}_t$ as an estimator of $g_t$ and $S$ measures its *smoothness*. Here the expectation operator is with respect to the particular local dynamic model adopted and $\Delta$ denotes the (backwards) difference operator.

The smaller $F$ is the better $\hat{g}_t$ is as an estimator of $g_t$ whereas the smaller $S$ is the closer the $\Delta^{p+1}\hat{g}_t$ are to zero and the closer $\hat{g}_t$ is to a smooth polynomial of degree $p$ in $t$. Note that $F$ is the familiar mean squared error criterion whereas $S$, appropriately normalised, is referred to as the $R^2_{p+1}$ criterion in the actuarial graduation literature. (See [Lon85] for example.)

In the spirit of [Hen24] and [Whi23] we further define the compromise criterion

$$Q = \theta F + (1 - \theta)S \qquad (0 \le \theta \le 1)$$

where $\theta$ is some user specified value. Note that $Q$ includes both $F$ and $S$ as special cases. We shall adopt the principle that the smaller these three criteria are the better and determine the filter weights that optimise $Q$. While minimising $F$ is clearly a reasonable criterion to adopt, it is not so clear that this is appropriate for $S$. It could be argued that the aim should be to simply control the level of $S$ rather than minimise it. However this approach again leads to $Q$ as will be shown in Theorem 3.

A primary advantage of defined measures of fidelity and smoothness is that we can now begin to quantify and classify the effects of competing moving-average trend filters and to consider trade-offs between fidelity and smoothness. In particular we can determine the values of $w_j$ that satisfy (1.3) and minimise either $F$, $S$ or $Q$. Before establishing these general results we first consider some special cases.

Consider minimising $F$ in the case where $g_t$ is the conventional model given by (1.2) with $\xi_t$ equal to zero. From the Gauss-Markov theorem, minimising $F$ subject to (1.3) is equivalent to estimating $g_t$ by $\hat{g}_t = \sum_{j=0}^{p} \hat{\beta}_j t^j$ where the estimates $\hat{\beta}_j$ are obtained from fitting the local polynomial model by ordinary least squares. The latter procedure was initially advocated by [Mac31] and forms the basis of many trend estimation filters. (See [Ken73] for example.) Thus minimising $F$ with $\xi_t$ equal to zero yields the Macaulay filters many of which are tabulated in [Ken73]. The case where $\xi_t$ is not equal to zero leads to local polynomial fitting using generalised least squares. An example of the latter is loess (see [CCMT90]) which uses the tricube function for least squares weights. *In general we shall refer to trend filters obtained by minimising $F$ as Macaulay filters irrespective of the choice of local dynamic model.*

Again assume that the model is given by (1.2) with $\xi_t$ equal to zero and consider the case $p = 2$. Then minimising $S$ gives the central Henderson filters that are extensively used in the X-11 seasonal adjustment procedure (for details and tabulated values see [SYC67]). The proof that this is so follows the treatment given in [KD82] with some key differences, the most important of which is that the $w_s$ are not assumed to be symmetric a priori, this property being a consequence of the optimisation process. The fact that the filter also passes a cubic is to be regarded as serendipitous and arising from the fact that this optimal central filter is symmetric. *In general we refer to trend filters obtained by minimising S as Henderson filters irrespective of the choice of local dynamic model.*

We now determine the filter weights $w_s$ that optimise $F$, $S$, and $Q$ in the case where $\xi_t$ is not necessarily zero. Clearly it is sufficient to consider the compromise criterion $Q$ since $F$ and $S$ can be recovered from $Q$ by setting $\theta = 0$ or $\theta = 1$ respectively.

Further notation is needed to establish and present the results. For the local dynamic model specified by (1.1) and (1.2) define covariance matrices $\Omega$ and $\Gamma$ with typical elements

$$\Omega_{jk} = \mathrm{cov}(\xi_{t+j} - \xi_t, \xi_{t+k} - \xi_t), \quad \Gamma_{jk} = \mathrm{cov}(\Delta^{p+1}\xi_{t+j}, \Delta^{p+1}\xi_{t+k}) \quad (1.5)$$

where $-r \leq j, k \leq r$ and $t$ indexes the central time point of the window concerned. Furthermore, define the $n \times (p+1)$ dimensional matrix $\mathbf{C}$ and the $p+1$ dimensional vector $\mathbf{c}$ by

$$\mathbf{C} = \begin{pmatrix} 1 & -r & \cdots & (-r)^p \\ 1 & -r+1 & \cdots & (-r+1)^p \\ \cdot & \cdot & \cdots & \cdot \\ \cdot & \cdot & \cdots & \cdot \\ 1 & r-1 & \cdots & (r-1)^p \\ 1 & r & \cdots & r^p \end{pmatrix} \qquad \mathbf{c} = \begin{pmatrix} 1 \\ 0 \\ \cdot \\ \cdot \\ 0 \\ 0 \end{pmatrix} \qquad (1.6)$$

and let $\sigma^2 \mathbf{B}_k$ denote the covariance matrix of a sequence of $n$ observations from the stationary moving average process $\Delta^k \epsilon_t$. Observe that $\mathbf{B}_k$ does not involve $\sigma^2$ the variance of $\epsilon_t$. Finally, let $\mathbf{I}$ denotes the $n$-dimensional identity matrix. We now have the following result.

**Theorem 1** *Let $y_t$ follow the local dynamic model specified by (1.1) and (1.2). Then the values of $w_s$ that minimise $Q$ subject to (1.3) are given by $\mathbf{w} = (w_{-r}, \ldots, w_r)^T$ where*

$$\mathbf{w} = \mathbf{E}_\theta^{-1} \mathbf{C} (\mathbf{C}^T \mathbf{E}_\theta^{-1} \mathbf{C})^{-1} \mathbf{c}$$

*and*

$$\mathbf{E}_\theta = \theta(\sigma^2 \mathbf{I} + \Omega) + (1 - \theta)(\sigma^2 \mathbf{B}_{p+1} + \Gamma).$$

**Proof**
First note that (1.3) is equivalent to the requirement that $\mathbf{w}$ satisfy

$$\mathbf{C}^T \mathbf{w} = \mathbf{c}. \qquad (1.7)$$

Given this condition,

$$F = \mathbf{w}^T(\sigma^2 \mathbf{I} + \Omega)\mathbf{w}, \quad S = \mathbf{w}^T(\sigma^2 \mathbf{B}_{p+1} + \Gamma)\mathbf{w}, \quad Q = \mathbf{w}^T \mathbf{E}_\theta \mathbf{w},$$

and the minimum of $Q$ subject to (1.7) is now found using Lagrange multipliers.

□

Given a local dynamic model, it is of interest to consider the nature of the variation of the optimal $F$ and $S$ with $\theta$. The following result shows, in particular, that the optimal measures of fidelity $F$ and smoothness $S$ decrease and increase respectively as $\theta$ increases. Thus the filter with the best smoothness measure $S$ has the worst fidelity measure $F$ and vice-versa.

**Theorem 2** Let $y_t$ follow the local dynamic model specified by (1.1) and (1.2). If $\mathbf{w}$ is given by Theorem 1 and $F$, $S$ given by (1.4) then

$$\theta \frac{dF}{d\theta} + (1 - \theta) \frac{dS}{d\theta} = 0, \quad \frac{dF}{d\theta} \leq 0 \quad \frac{dS}{d\theta} \geq 0$$

where $0 \leq \theta \leq 1$. In particular $F$ and $S$ have zero gradients at $\theta = 1$ and $\theta = 0$ respectively.

**Proof**

Differentiating $F$ and $S$ with respect to $\theta$ we obtain

$$\theta \frac{dF}{d\theta} + (1 - \theta) \frac{dS}{d\theta} = 0 \tag{1.8}$$

and

$$\frac{dS}{d\theta} - \frac{dF}{d\theta} \geq 0 \tag{1.9}$$

The latter follows since the left hand side can be written as $\mathbf{u}^T \mathbf{P} \mathbf{u}$, where $\mathbf{P}$ is a projection matrix. From (1.8) we observe that $F$ has gradient zero at $\theta = 1$ and $S$ has gradient zero at $\theta = 0$. Moreover, from (1.8) and (1.9), we obtain $\frac{dS}{d\theta} \geq 0$ and $\frac{dF}{d\theta} \leq 0$, for $0 \leq \theta \leq 1$ as required.

□

As mentioned before, if the aim is to control or impose smoothness on the trend estimate $\hat{g}_t$, then an alternative strategy might be to minimise $F$ subject to (1.3) and a given level of smoothness $S = s$. This is equivalent to minimising $Q$ as we shall see in Theorem 3.

First, however, note that minimising $F$ subject to (1.3) alone determines the best linear unbiased predictor (BLUP) of $g_t$ from the observations $y_t$ in the window. Thus, minimising $F$ subject to (1.3) and $S = s$ gives a constrained BLUP for $g_t$. This estimator will have the desired smoothness property, but at the cost of a higher value of $F$ than before. Now let $s_0$ denote the smoothness of the optimum smoothness filter and $s_1$ denote the smoothness of the optimum fidelity filter so that

$$s_0 = \mathbf{w}_0^T (\sigma^2 \mathbf{B}_{p+1} + \mathbf{\Gamma}) \mathbf{w}_0, \quad s_1 = \mathbf{w}_1^T (\sigma^2 \mathbf{B}_{p+1} + \mathbf{\Gamma}) \mathbf{w}_1$$

where $\mathbf{w}_0$ and $\mathbf{w}_1$ are the values of $\mathbf{w}$ given by Theorem 1 for $\theta = 0$ and $\theta = 1$ respectively. If a trend estimator $\hat{g}_t$ is sought with better smoothness properties then the user will need to constrain the smoothness $S$ to lie in the interval $[s_0, s_1]$.

**Theorem 3** *Let $y_t$ follow the local dynamic model specified by (1.1) and (1.2). Then minimising $F$ subject to both (1.3) and $S = s$ in $[s_0, s_1]$ is equivalent to minimising $Q$ subject to (1.3) alone where $\theta$ is a solution of*

$$\mathbf{c}^T (\mathbf{C}^T \mathbf{E}_\theta^{-1} \mathbf{C})^{-1} \mathbf{C}^T \mathbf{E}_\theta^{-1} (\sigma^2 \mathbf{B}_{p+1} + \mathbf{\Gamma}) \mathbf{E}_\theta^{-1} \mathbf{C} (\mathbf{C}^T \mathbf{E}_\theta^{-1} \mathbf{C})^{-1} \mathbf{c} = s,$$

$\mathbf{E}_\theta$ *is given by Theorem 1 and $\theta$ lies in $[0, 1]$. The minimisations are with respect to the filter weights $w_s$.*

**Proof**

We seek to minimise $\tilde{\mathbf{Q}} = F + \phi(S - s) - 2\mu^T (\mathbf{C}^T \mathbf{w} - \mathbf{c})$ where the scalar $\phi$ and the vector $\mu$ are Lagrange multipliers. Differentiating and solving the resulting equations yields $\mathbf{w} = \tilde{\mathbf{E}}_\phi^{-1} \mathbf{C} (\mathbf{C}^T \tilde{\mathbf{E}}_\phi^{-1} \mathbf{C})^{-1} \mathbf{c}$ where $\phi$ is a solution of

$$\mathbf{c}^T (\mathbf{C}^T \tilde{\mathbf{E}}_\phi^{-1} \mathbf{C})^{-1} \mathbf{C}^T \tilde{\mathbf{E}}_\phi^{-1} (\sigma^2 \mathbf{B}_{p+1} + \mathbf{\Gamma}) \tilde{\mathbf{E}}_\phi^{-1} \mathbf{C} (\mathbf{C}^T \tilde{\mathbf{E}}_\phi^{-1} \mathbf{C})^{-1} \mathbf{c} = s \quad (1.10)$$

and $\tilde{\mathbf{E}}_\phi = (\sigma^2 \mathbf{I} + \mathbf{\Omega}) + \phi(\sigma^2 \mathbf{B}_{p+1} + \mathbf{\Gamma})$. In almost exactly the same way as was done in Theorem 2, it can be shown that the left hand side of (1.10) is a non-increasing function of $\phi$ whose limit, as $\phi \to \infty$, is $s_0$. Moreover $\phi = 0$ yields $s = s_1$. Since $s_0 \le s \le s_1$ we conclude that a solution of (1.10) can be found for $\phi \ge 0$. The result of Theorem 1 is now retrieved by setting $\theta = 1/(1 + \phi)$.

□

In fact, in much the same way as for Theorem 3, it can be shown that minimising smoothness subject to (1.3) and a given level of fidelity $F = f$ is also equivalent to minimising $Q$. However this seems a less relevant criterion to adopt in practice.

In practice $\xi_t$ will typically comprise one or other or possibly the sum of the random walk components $\xi_t^{(1)}$ and $\xi_t^{(2)}$ where

$$\xi_t^{(1)} = \sum_{j=1}^{t} \nu_j \qquad \xi_t^{(2)} = \sum_{j=1}^{t} \sum_{k=1}^{j} \eta_k$$

and $\nu_t, \eta_t$ are mutually uncorrelated white noise processes with $E(\nu_t^2) = \sigma_\nu^2$, $E(\eta_t^2) = \sigma_\eta^2$. If $\xi_t$ is a linear combination of random walk components such as the above whose levels of integration do not exceed $p + 1$, then $\mathbf{\Gamma}$ will be a linear combination of covariance matrices $\mathbf{B}_k$ where $k$ is a nonnegative integer such that $k \le p$.

The matrix $\mathbf{\Omega}$ can be constructed directly from the covariance matrices of the random walk components that make up $\xi_t$. However some simplification is possible. For example, when $\xi_t = \xi_t^{(1)}$ the matrix $\mathbf{\Omega}$ is given by $\mathbf{\Omega}^{(1)}$ which has typical element

$$\Omega_{jk}^{(1)} = \sigma_\nu^2 \min(|j|, |k|) \quad (1.11)$$

for $-r \leq j, k \leq -1, 1 \leq j, k \leq r$ and zero otherwise. When $\xi_t = \xi_t^{(2)}$ the matrix $\Omega$ is given by $\Omega^{(2)}$ which has typical element

$$\Omega_{jk}^{(2)} = \sigma_\eta^2 \left\{ \tfrac{1}{2}|j||k| + \sum_{s=1}^{\min(|j|,|k|)} (|j| - s)(|k| - s) \right\} \qquad (1.12)$$

for $-r \leq j, k \leq -1, 1 \leq j, k \leq r$ and zero otherwise.

**Theorem 4** *The values of $w_s$ given by Theorem 1 correspond to a symmetric filter of length $n = 2r + 1$.*

**Proof**

Let $\mathbf{u} = (w_{-1}, \ldots, w_{-r})^T$, $\mathbf{v} = (w_1, \ldots, w_r)^T$ and $\mathbf{d}_j = (1^j, 2^j, \ldots, r^j)$ for $j$ an integer such that $0 \leq j \leq p$.

Consider first the fidelity measure $F$ and the particular case where $\xi_t = \xi_t^{(2)}$. Then the constraints (1.7) become

$$w_0 + \mathbf{d}_0^T \mathbf{u} + \mathbf{d}_0^T \mathbf{v} = 1, \quad \mathbf{d}_j^T \mathbf{u} = \mathbf{d}_j^T \mathbf{v} \qquad (0 < j \leq p)$$

and so $F$ can be written as

$$\sigma^2(w_0^2 + \mathbf{u}^T\mathbf{u} + \mathbf{v}^T\mathbf{v}) + \sigma_\eta^2(\tfrac{1}{2}(\mathbf{d}_1^T\mathbf{u})^2 + \tfrac{1}{2}(\mathbf{d}_1^T\mathbf{v})^2 + \mathbf{u}^T\Omega^{(2)}\mathbf{u} + \mathbf{v}^T\Omega^{(2)}\mathbf{v}).$$

Note that both the constraints and $F$ are symmetric in the vectors $\mathbf{u}$ and $\mathbf{v}$. A similar argument holds for more general forms of $\xi_t$ involving sums of integrated random walks of any finite order.

Now consider the smoothness measure $S$ which is of the form $S = \mathbf{w}^T \mathbf{A} \mathbf{w}$ where $\mathbf{A}$ is a Toeplitz matrix with spectral density $f(\omega)$. Thus

$$S = \int_{-\pi}^{\pi} |w_0 + \sum_{j=1}^{r} u_j e^{-ij\lambda} + \sum_{j=1}^{r} v_j e^{ij\lambda}|^2 f(\omega) d\omega.$$

is also symmetric in the vectors $\mathbf{u}$ and $\mathbf{v}$. Since $Q = \theta F + (1 - \theta)S$ inherits the same symmetry property we conclude that the value of $\mathbf{w}$ minimising $Q$ satisfies $\mathbf{u} = \mathbf{v}$ and corresponds to a symmetric filter.

$\square$

The properties of these filters are discussed in Section 4.

## 4 Properties of the filters

In this section we focus attention on two particular local dynamic models likely to be used in practice. These are the *local linear model* ($p = 1$) and the *local quadratic model* ($p = 2$) given by (1.1) and (1.2) where, in both cases, $\xi_t$ is a random walk satisfying

$$\xi_t = \xi_{t-1} + \eta_t$$

with $\xi_0 = 0$ and the white noise process $\eta_t$ has variance $\sigma_\eta^2 = \lambda\sigma^2$. Preliminary analysis indicates that these can be regarded as representative of other more general models of the type discussed in Section 2. Unless otherwise stated, we also restrict attention to the case where the window is of length 13 so that $r = 6$.

The impulse response functions of the filters given by Theorem 1 for the local linear and quadratic models are plotted in Figure 1 for selected values of $\theta$ and $\lambda$. Note that the 13 point X-11 Henderson filter corresponds to the case where $\theta = 0$, $\lambda = 0$, $p = 2$ and is shown in the bottom left hand plot of Figure 1.

As $\lambda$ varies the Henderson filters that optimise smoothness $S$ $(\theta = 0)$ differ only slightly whereas the Macaulay filters that optimise fidelity $F$ $(\theta = 1)$ show much greater variation. In particular the Macaulay filters become more adaptive as $\lambda$ increases with the weights at central lags being progressively increased at the expense of those for extreme lags. The mixture of smoothness and fidelity corresponding to $\theta = 0.5$ yields a compromise between the Henderson and Macaulay filters as expected.

Figure 2 shows the gain functions of the filters given by Theorem 1 for the local linear and the quadratic models for selected values of $\theta$ and $\lambda$. The gain function of the 13 point X-11 Henderson is shown in the bottom left hand plot of Figure 2 and corresponds to the case where $\theta = 0$, $\lambda = 0$ and $p = 2$.

Note that the Henderson filters that optimise smoothness $S$ $(\theta = 0)$ show

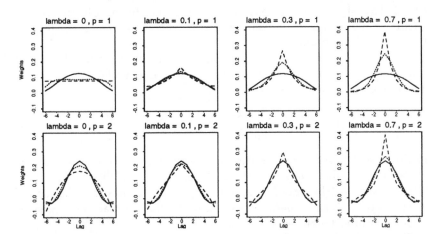

FIGURE 1. Plots of the impulse response functions of the 13 point filters given by Theorem 1 for the specified local linear and quadratic models and selected values of $\theta$ and $\lambda$. The solid lines correspond to $\theta = 0$ (smoothness criterion only), the dashed lines to $\theta = 1$ (fidelity criterion only) and the dotted lines to $\theta = 0.5$ which gives a comprise between smoothness and fidelity. The 13 point X-11 Henderson filter corresponds to the case where $\theta = 0$, $\lambda = 0$, $p = 2$ and is shown in the bottom left hand plot.

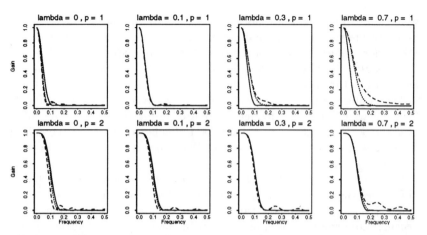

FIGURE 2. Plots of the gain functions of the 13 point filters given by Theorem 1 for the specified local linear and quadratic models and selected values of $\theta$ and $\lambda$. The solid lines correspond to $\theta = 0$ (smoothness criterion only), the dashed lines to $\theta = 1$ (fidelity criterion only) and the dotted lines to $\theta = 0.5$ which gives a comprise between smoothness and fidelity. The 13 point X-11 Henderson filter is shown in the bottom left hand plot and corresponds to the case where $\theta = 0$, $\lambda = 0$ and $p = 2$.

little leakage over the frequency band $[0.1, 0.5]$ when $p = 1$ and $[0.2, 0.5]$ when $p = 2$. By comparison the Macaulay filters that optimise fidelity $F$ ($\theta = 1$) have significant side lobes in these frequency bands. Clearly optimal fidelity comes at a price. Again the mixture corresponding to $\theta = 0.5$ has the expected compromise effect.

The following relative efficiency measures compare gains in fidelity and smoothness across different values of $\theta$ for the same model, indexed by $\lambda$. These measures are defined as

$$F_{rel}(\theta|p, \lambda) = \frac{F(1|p, \lambda)}{F(\theta|p, \lambda)}, \quad S_{rel}(\theta|p, \lambda) = \frac{S(0|p, \lambda)}{S(\theta|p, \lambda)}. \qquad (1.13)$$

In this case $F_{rel}$ measures the efficiency of the trend estimate $\hat{g}_t$ in terms of its fidelity gains relative to those of the optimal fidelity or Macaulay filter for the model indexed by $\lambda$. Similarly $S_{rel}$ measures the efficiency of $\hat{g}_t$ in terms of its smoothness gains relative to the optimal smoothness or Henderson filter for the same model.

The trade-off between fidelity relative efficiency and smoothness relative efficiency for given $\theta$ and selected values of $\lambda$ is illustrated in Figure 3. For both $p = 1$ and $p = 2$ fidelity relative efficiency is less variable than smoothness relative efficiency. In particular, fidelity relative efficiency exceeds 90%, approximately, for $\theta > 0.3$. In general smoothness relative efficiency decreases more rapidly than fidelity relative efficiency as $\theta$ increases with the effect more marked for $p = 1$ than $p = 2$. For $\theta < 0.3$ smoothness

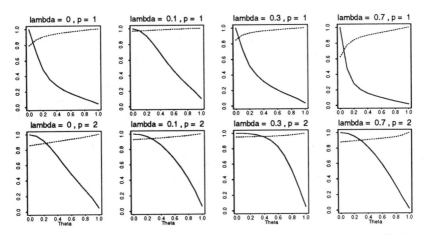

FIGURE 3. Plots of the fidelity relative efficiency $F_{rel}$ and the smoothness relative efficiency $S_{rel}$ of the 13 point filters given by Theorem 1 for the specified local linear and quadratic models and selected values of $\theta$ and $\lambda$. Here $F_{rel}$ (dashed lines) and $S_{rel}$ (solid lines) are defined by (56). The plots illustrate the trade-off between fidelity relative efficiency and smoothness relative efficiency for given $\theta$.

relative efficiency exceeds 80% when $\lambda$ is near 0.1 ($p = 1$) and for all values of $\lambda$ considered ($p = 2$).

Finally, we consider the performance of the X-11 Henderson filter ($\lambda = 0$, $p = 2$) in terms of both fidelity and smoothness. Figure 4 plots the fidelity and smoothness of the X-11 Henderson 13 point filter relative to the optimal fidelity and smoothness filters respectively for the specified local dynamic models considered.

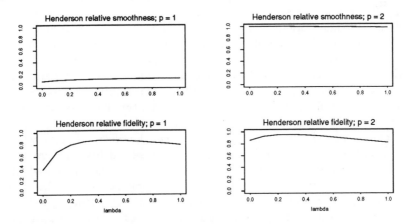

FIGURE 4. Plots of the fidelity and smoothness of the X-11 Henderson 13 point filter relative to the optimal fidelity and smoothness filters respectively for the specified local dynamic models considered. The latter are indexed by $p$ and $\lambda$.

In terms of smoothness, it is clear in the case $p = 2$ that there is little to differentiate the X-11 Henderson ($\lambda = 0$) from its generalisation based on the local dynamic model where $\lambda > 0$. However, in the case $p = 1$, the smoothness of the X-11 Henderson filter is substantially less than that of the Henderson filter tailored for the local linear model concerned. This is not entirely unexpected since the conventional Henderson filter is designed to accommodate quadratic rather than linear trends and so its trends exhibit greater variation in curvature.

In terms of fidelity, the X-11 Henderson filter performs poorly when $\lambda$ is less than 0.1 and $p = 1$, but otherwise its relative fidelity exceeds 80% for $\lambda$ in excess of around 0.2 ($p = 1$) and for all $\lambda$ considered ($p = 2$).

# 5   Practical study

We briefly report on the outcomes of a more extensive study on selected official time series where the filters given by Theorem 1 were compared with conventional Macaulay or Henderson filters. Full details of the study are given in [GT96]. All the time series considered were seasonal and needed to be seasonally adjusted and modified for outliers prior to applying the filters. This was done using the X-11 seasonal adjustment method. The resulting non-seasonal times series presented different types of short and long term trend and their local dynamic models included random walk components with a wide range of variances.

To study the filters given by Theorem 1, the window length $n = 2r + 1$, the order of the polynomial $p$, and a value of $\lambda$ must be determined from the data. These parameters specify the local dynamic model concerned. Typically $n$ and $p$ are identified from a simple graphical analysis of the data although more formal methods could be used. Given $n$ and $p$, $\lambda$ can be determined by a variety of methods. These range from trial and error and simple variational arguments based on quantities like X-11's $I/C$ ratio, through to likelihood analysis based on fitting the local dynamic model within non-overlapping windows or fitting global ARIMA models. In the series we examined, values for $\lambda$ ranged from 0 through to $\infty$ and in about two thirds of the cases $\lambda$ was less than 3.

Once the local dynamic model has been identified, the user will need to decide on a value for $\theta$, the balance between fidelity and smoothness. Although subjective, the choice of $\theta$ will typically be influenced by the overall objectives of the smoothing, as is to some extent, the choice of $n$.

The results obtained are as expected from the discussion given in Section 4. An example is given in Figure 5. Here, as $\lambda$ varies, the Henderson filters which optimise smoothness $S$ ($\theta = 0$) differ only slightly. However, the filters based on a mixture of smoothness and fidelity become more adaptive as $\lambda$ increases with this effect becoming more marked the closer $\theta$

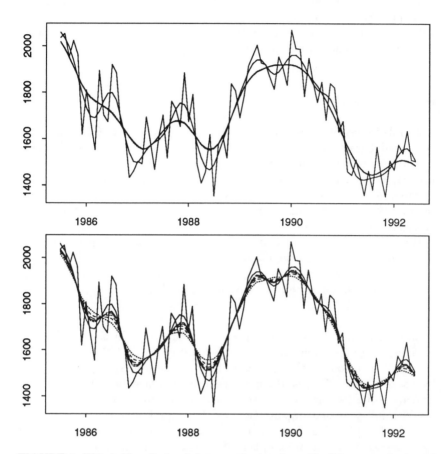

FIGURE 5. This is New Zealand data on the number of building permits issued by Local Authorities for the construction of private houses and flats. The data has been seasonally adjusted by $X - 11$ and been modified to remove any large outliers. The data is characterized by many local turning points, with a general downward drift. In the top graph, the data is filtered by the 13 point Henderson filter used in $X - 11$, and the 13 point filters given by Theorem 1 for the local linear model, for $\theta = 0$. In the bottom graph the data is filtered by the 13 point Henderson filter used in $X - 11$, and the 13 point filters given by Theorem 1 for the local linear model, for $\theta = .5$. In both graphs, the solid line corresponds to the 13 point Henderson, the dotted line to $\lambda = .1$ and the dashed lines to the $\lambda$'s .2, .3, .4, .5. The 'optimal' $\lambda$ was about .3.

is to 1. Notice as well that the effect of varying $\lambda$ is greatest at the turning points. Figure 5 also shows that filters based on a local linear model with large $\lambda$ and $\theta > 0$ approach the X-11 Henderson filter which is based on a local quadratic model with $\theta = 0$. In this sense, the local dynamic model is able to explore trend variation in the window that lies somewhere between the pure linear and pure quadratic case.

# 6  Conclusions

A flexible family of semi-parametric, finite, moving-average filters has been developed based on local dynamic models and specified by fidelity and smoothness criteria. The local dynamic model is determined by the data, whereas $\theta$, the balance between fidelity and smoothness, is user specified. Through $\theta$ the analyst can (subjectively) impose an estimate of trend on the data whose smoothness conforms to other criteria such as the overall objectives of the analysis.

Work is currently being written up on the extension of these filters to the ends of series. There the use of local dynamic models with lower order polynomial trends leads to enhanced performance. Work is also in progress on extending the filters to handle data with outliers.

*Acknowledgments:* This research project was supported by the New Zealand Foundation for Research Science and Technology (Project Number VIC-93-36-039). The second author also gratefully acknowledges support provided by an ASA/NSF/Census Research Fellowship which he held at the US Bureau of the Census. Both authors wish to record their gratitude to Dr David Findley of that organisation for the helpful advice and encouragement he has provided throughout the project. Data was kindly provided by Statistics New Zealand and the US Bureau of the Census.

Finally, we take this opportunity to acknowledge the contributions made by Ted Hannan to this and other related projects both directly (as reviewer of our FRST proposal) and indirectly as mentor and friend.

## 7  REFERENCES

[Aka80]  H. Akaike. Seasonal adjustment by a Bayesian modeling. *Journal of Time Series Analysis*, 1(1):1–13, 1980.

[Bel93]  W. Bell. Empirical comparisons of seasonal ARIMA and ARIMA component (structural) time series models. Research Report CENSUS/SRD/RR-93/10, Statistical Research Division, Bureau of the Census, Washington, D.C. 20233-4200, 1993.

[CCMT90] R. B. Cleveland, W. S. Cleveland, J. E. McRae, and I. J. Terpenning. STL: A seasonal-trend decomposition procedure based on Loess. *Journal of Official Statistics*, 6:3–73, 1990.

[GK83] W. Gersch and G. Kitagawa. The prediction of time series with trends and seasonalities. *Journal of Business and Economic Statistics*, 1:253–264, 1983.

[GT96] A. G. Gray and P. J. Thomson. Design of moving-average trend filters using fidelity and smoothness criteria. Research Report CENSUS/SRD/RR-96/1, Statistical Research Division, Bureau of the Census, Washington, D.C. 20233-4200, 1996.

[Har89] A. C. Harvey. *Forecasting, structural time series models and the Kalman filter*. Cambridge University Press, Cambridge, 1989.

[Hen24] R. Henderson. A new method of graduation. *Transactions of the Actuarial Society of America*, 25:29–40, 1924.

[KD82] P. B. Kenny and J. Durbin. Local trend estimation and seasonal adjustment of economic and social time series. *Journal of the Royal Statistical Society, Series A*, 145:1–41, 1982.

[Ken73] M. G. Kendall. *Time-Series*. Hafner Press, New York, 1973.

[Lon85] D. London. *Graduation: The Revision of Estimates*. ACTEX, Abington, Connecticut, 1985.

[Mac31] F.R. Macaulay. *The Smoothing of Time Series*. National Bureau of Economic Research, New York, 1931.

[Sch81] E. Schlicht. A seasonal adjustment principle and a seasonal adjustment method derived from this principle. *Journal of the American Statistical Association*, 76:374–378, 1981.

[SYC67] J. Shiskin, A. H. Young, and J. C.Musgrave. The X-11 variant of the Census method II seasonal adjustment program. Technical Paper 15, Bureau of the Census, U.S. Department of Commerce, Washington, D.C., 1967.

[Whi23] E. T. Whittaker. On a new method of graduation. *Proceedings of the Edinburgh Mathematical Society*, 41:63–75, 1923.

Victoria University, Wellington

# Bandwidth Choice in Gaussian Semiparametric Estimation of Long Range Dependence

M. Henry and P.M. Robinson
London School of Economics

ABSTRACT  We consider covariance stationary processes with spectral density which behaves according to a power law around zero frequency, where it can be infinite (long range dependence), finite and positive (short range dependence), or zero (antipersistence). This behaviour is governed by a self-similarity parameter which can be estimated semiparametrically by one of several methods, all of which require a choice of bandwidth. We consider a Gaussian estimate which seems likely to have good efficiency, and whose asymptotic distributional properties have already been determined. The minimum mean squared error optimal bandwidth is heuristically derived and feasible approximations to it are proposed, these being assessed in Monte Carlo experiments and applied to financial and Nile river data.

## 1   Introduction

Long-range dependent time series typically exhibit a singularity in the spectrum at frequency zero. One is therefore concerned with the estimation of the slope (as opposed to the value) of the logged spectral density against log frequency around the singularity. A recent literature survey of long range dependence can be found in Robinson [Rob94c].

In semiparametric estimation of the slope of the spectral density in a neighbourhood of frequency zero, the choice of bandwidth, or number of harmonic frequencies used, considerably affects the quality of the estimation. As bias tends to increase, and variance decrease, with bandwidth, a theory of optimal bandwidth usually relies on the minimization of a form of mean-squared error which balances bias and imprecision.

To date, asymptotic properties of three classes of semiparametric estimates of the self-similarity parameter (which determines the slope of the logged spectral density in a neighbourhood of frequency zero) have been established: an averaged periodogram estimate in Robinson [Rob94b]; an estimate based on the least squares estimation of the regression of the logged periodogram against a simple function of the harmonic frequencies in Geweke and Porter-Hudak [GPH83] and Robinson [Rob95b], as well as a more efficient estimate involving pooling of adjacent periodograms prior to logging (Robinson [Rob95b]); and a local Gaussian estimate proposed by Künsch [K87] which was shown to be asymptotically more efficient than the others by Robinson [Rob95a].

Robinson [Rob94a] proposed an optimal bandwidth theory based on an analogue of the mean squared error of smooth spectral density estimates. Delgado and Robinson [DR96] assessed feasible approximations to this optimal bandwidth based on the averaged periodogram estimate of the self-similarity parameter.

The present paper concerns the Gaussian estimate. Heuristic expressions for the mean squared error and the corresponding optimal spectral bandwidth are derived in section 2. Infeasible and feasible approximations of this optimal bandwidth are considered in a series of Monte Carlo studies in section 3 and applied in section 4 to Nile river and financial data.

## 2 Optimal Spectral Bandwidth

Let $x_t$ be a discrete parameter covariance stationary time series with lag-j autocovariances $\gamma_j$ and spectral density $f(\lambda)$ satisfying

$$\gamma_j = \int_{-\pi}^{\pi} f(\lambda) \cos(j\lambda) d\lambda.$$

For a realization of size n, the estimation relies on the periodogram $I(\lambda) = \frac{1}{2\pi n} |\sum_{t=1}^{n} x_t \exp(it\lambda)|^2$ computed at the harmonic frequencies $\lambda_j = \frac{2\pi j}{n}$ for $j$ ranging on a subset of $[1, \frac{n}{2})$.

We require $f(\lambda)$ to satisfy the basic long range dependence specification

$$f(\lambda) \sim G\lambda^{1-2H} \ as \ \lambda \to 0^+ \tag{1.1}$$

where $G$ is a finite positive constant and $H$ is the self-similarity parameter, required here to lie within the interval $(0, 1)$, bounded away from its limit values. For $0 < H < 1/2$, $f(0) = 0$ (antipersistence); for $1/2 < H < 1$, $f(0) = \infty$ (long range dependence); for $H = 1/2$, $f(0)$ is finite and positive (short range dependence).

An estimate of the self-similarity parameter, denoted $\hat{H}$, maximizes a local form of Gaussian log-likelihood. After concentrating $G$ out, it remains to minimize

$$R(m, H) = \log\Big[\frac{1}{m} \sum_{j=1}^{m} \frac{I(\lambda_j)}{\lambda_j^{1-2H}}\Big] - (2H - 1)\frac{1}{m} \sum_{j=1}^{m} \log[\lambda_j]$$

where $m$ is the number of harmonic frequencies used in the estimation and $H \in [\Delta_1, \Delta_2], 0 < \Delta_1 < \Delta_2 < 1$, see Robinson [Rob95a]. For simplicity of notation, we do not distinguish between the true value of $H$ and other admissible values in $[\Delta_1, \Delta_2]$.

The choice $m = [n/2]$ would yield a Gaussian estimate of $H$ in the parametric model $f(\lambda) = G|\lambda|^{1-2H}$ for $\lambda \in (-\pi, \pi]$, the asymptotic properties of which can be found in Fox and Taqqu [FT86] and Giraitis and Surgailis [GS90] (apart from the fact that they consider an integral rather than a sum over the Fourier frequencies). In the more robust semiparametric estimation considered here, the bandwidth $m$ is required to diverge at a strictly smaller rate than $n$. In that case, and under a smoothness condition for the spectral density and a homogeneity condition for the martingale difference innovations in the Wold decomposition of $x_t$, $\hat{H}$ was shown to be consistent in Robinson [Rob95a]. When specification 1.1 is refined to $f(\lambda) \sim G\lambda^{1-2H}[1 + O(\lambda^\beta)]$ as $\lambda \to 0^+$ for some $\beta \in (0, 2]$, the bandwidth satisfies $\lim_{n\to\infty} \frac{m^{1+2\beta}(\log m)^2}{n^{2\beta}} + \frac{1}{m} = 0$, and existence of fourth moments and some other regularity conditions are assumed, asymptotic normality also holds, as shown in Robinson [Rob95a]. This asymptotic normality result, namely $\sqrt{m}(\hat{H} - H) \xrightarrow{D} N(0, 1/4)$ implies for $\hat{H}$ a rate of convergence of $n^{-r}M_n$ where $r = \frac{\beta}{1+2\beta}$ and $(\log m)^{-\frac{1}{1+2\beta}}M_n = o(1)$.

For the sake of optimal bandwidth determination, the long memory local specification 1.1 is further extended (as in Robinson [Rob95a] and Delgado and Robinson [DR96]) to

$$f(\lambda) = G\lambda^{1-2H}\left[1 + E_\beta(H)\lambda^\beta + o(\lambda^\beta)\right] \text{ with } 0 < |E_\beta(H)| < \infty \quad (1.2)$$

as $\lambda \to 0^+$, for some $\beta \in (0, 2]$. Giraitis, Robinson and Samarov [GRS95] showed that there exists a lower bound for the rate of convergence of semiparametric estimates on a class of spectral densities including the above specified. They showed, moreover, that this lower bound is attained by Geweke and Porter-Hudak [GPH83]'s log-periodogram estimate in modified form (cf Robinson [Rob95b]). We suggest that the same property holds for the Gaussian estimate $\hat{H}$.

As is familiar from other uses of smoothed nonparametric estimates, this optimal rate of convergence lies outside the asymptotic normality range for the local Gaussian estimate $\hat{H}$ considered here, as $M_n$ is not free to diverge arbitrarily slowly. Nonetheless, heuristically, the asymptotic variance remains equal to $1/4m$ and the optimal bandwidth is the one that yields that same rate of convergence for the squared bias.

More precisely, the Mean Value Theorem yields:

$$\hat{H} - H = \frac{\frac{\partial R(m,H)}{\partial H}}{\frac{\partial^2 R(m,\tilde{H})}{\partial^2 H}} \text{ where } |\tilde{H} - H| \leq |\hat{H} - H|.$$

Now, as shown in Robinson [Rob95a], $\frac{\partial^2 R(m,\tilde{H})}{\partial H^2} \xrightarrow{p} 4$. The first two moments of $\frac{\partial R(m,H)}{\partial H}$ can be treated heuristically as follows. As $n \to \infty$,

$$\frac{\partial R}{\partial H} \sim_p \frac{2}{m} \sum_{j=1}^{m} \nu_j \left[\frac{I(\lambda_j)}{G\lambda_j^{1-2H}} - 1\right] \text{ where } \nu_j = \log j - \frac{1}{m} \sum_{k=1}^{m} \log j,$$

the notation $\sim_p$ meaning that the ratio of left and right sides tends to 1 in probability. We thus suggest that the expectation of $\frac{\partial R}{\partial H}$ (assuming it exists) can be approximated by

$$\frac{2}{m} \sum_{j=1}^{m} \nu_j \left[\frac{f(\lambda_j)}{G\lambda_j^{1-2H}} - 1\right],$$

so under 1.2 the bias may be approximated by

$$\frac{1}{2m} E_\beta(H) \sum_{j=1}^{m} \nu_j \lambda_j^\beta \sim \frac{1}{2}\left(\frac{2\pi}{n}\right)^\beta E_\beta(H) \frac{\beta}{(\beta+1)^2} m^\beta$$

$$= \frac{2\beta}{(\beta+1)^2} E_\beta(H) \lambda_m^\beta. \tag{1.3}$$

Likewise we suggest that the variance of $\frac{\partial R}{\partial H}$ can be approximated by

$$\frac{4}{m^2} \sum_{j=1}^{m} \nu_j E\left[\frac{I(j)}{G\lambda_j^{1-2H}} - \frac{E[I(j)]}{G\lambda_j^{1-2H}}\right]^2 \sim \frac{4}{m^2} \sum_{j=1}^{m} \nu_j^2 \sim \frac{4}{m} \tag{1.4}$$

Thus we suggest, from 1.3 and 1.4, that the mean squared error $E(\hat{H} - H)^2$ is dominated by

$$\frac{1}{4}\left[\frac{1}{m} + E_\beta(H)^2 \frac{\beta^2}{(\beta+1)^4} \lambda_m^{2\beta}\right], \tag{1.5}$$

from which an "optimal bandwidth" as $n$ tends to infinity can be derived by straightforward calculus:

$$m_{opt} = \left[\frac{(\beta+1)^4}{2\beta^3 E_\beta(H)^2 (2\pi)^{2\beta}}\right]^{\frac{1}{1+2\beta}} n^{\frac{2\beta}{1+2\beta}}. \tag{1.6}$$

This optimal bandwidth may be compared to one relevent to the averaged periodogram estimate of $H$ in Robinson [Rob94b]: see Robinson [Rob94a] and Delgado and Robinson [DR96].

When we consider the smoothest specification, i.e. $\beta = 2$, the heuristic optimal spectral bandwidth becomes: $m_{opt} = \left(\frac{3n}{4\pi}\right)^{\frac{4}{5}} |E_2(H)|^{-\frac{2}{5}}$. In a fractional differencing representation of the form $f(\lambda) = |1-\exp(i\lambda)|^{1-2H} h(\lambda)$, where $h$ is twice continuously differentiable and $h(0)$ is positive and finite, $E_2(H)$ can be approximated by $\frac{h''(0)}{2h(0)} - \frac{1}{24}(1-2H)$, as shown by Delgado and Robinson [DR96]. Figure 1 plots $m_{opt}$ for $0 \le n \le 4000$ and fractionally integrated $ARIMA(p, H - 1/2, q)$ processes with $p = q = 0$; $p = 1, q = 0$ and autoregressive coefficient 0.5; $p = 0, q = 1$ and moving average coefficient 0.5; $p = 2, q = 0$ and autoregressive coefficients 0.5 and 0.1. The variability of the optimal bandwidth in case of an $ARIMA(0, H - 1/2, 0)$ is very small for most values of $H$ and most of the $n$ values considered except in a neighbourhood of $H = 0.5$ where the divergence is very fast. The optimal bandwidth is even less variable in $H$ and exhibits no singularity in the case of an $ARIMA(p, H - 1/2, q)$ for $max(p, q) > 0$. As might be expected for series exhibiting autoregressive moving average features, the optimal bandwidth is far smaller than in the previous case, fewer harmonic frequencies being used to avoid flawing the estimates. Note that in the $ARIMA(1, H - 1/2, 0)$ case, $\frac{h''(0)}{2h(0)} = -\frac{a}{(1-a)^2}$, see Delgado and Robinson [DR96], so that the optimal bandwidth tends to zero as the autoregressive coefficient $a$ tends to one.

# 3 Approximations to the Optimal Bandwidth

Again considering the smoothest specification in the class above, namely $\beta = 2$, an approximation to the optimal bandwidth $m_{opt} = (\frac{3n}{4\pi})^{\frac{4}{5}} |E_2(H)|^{-\frac{2}{5}}$ relies on a preliminary approximation of the unknown $E_2$. The latter's dependence on the self-similarity parameter naturally points to an iterative procedure whereby

$$\hat{H}^{(k)} = argmin_{H \in [\Delta_1, \Delta_2]} R(\hat{m}^{(k)}, H),$$
$$\hat{m}^{(k+1)} = (\frac{3n}{4\pi})^{\frac{4}{5}} |E_2(\hat{H}^{(k)})|^{-\frac{2}{5}},$$

starting from an ad hoc value $\hat{m}^{(0)} = n^{\frac{4}{5}}$ and with $\Delta_1 = 0.001$ and $\Delta_2 = 0.999$.

First we consider an infeasible procedure in which

$$E_2(\hat{H}^{(k)}) = \frac{h''(0)}{2h(0)} + \frac{1}{24}(2\hat{H}^{(k)} - 1), \qquad (1.7)$$

where $\frac{h''(0)}{2h(0)} = -\frac{a}{(1-a)^2}$ is taken as known. Table 1.1 presents Monte Carlo results for this infeasible procedure on $ARIMA(1, H - 1/2, 0)$ series with autoregressive coefficient $a = 0.5$, simulated using the Davies and Harte algorithm [DH87] The optimal bandwidth converges immediately to its optimal value, in this case 61.

One feasible approximation to the optimal bandwidth proposed in Delgado and Robinson [DR96] is based on an expansion of the semiparametric spectral density $f(\lambda) = |1 - \exp(i\lambda)|^{1-2H} h(\lambda)$. $h(0)$ and $h''(0)$ are taken to be respectively the first and last coefficient in the least squares regression of the periodogram $I(\lambda_j)$ against $|1 - \exp(i\lambda_j)|^{1-2\hat{H}^{(0)}} ( 1, \lambda_j, \frac{\lambda_j^2}{2})$

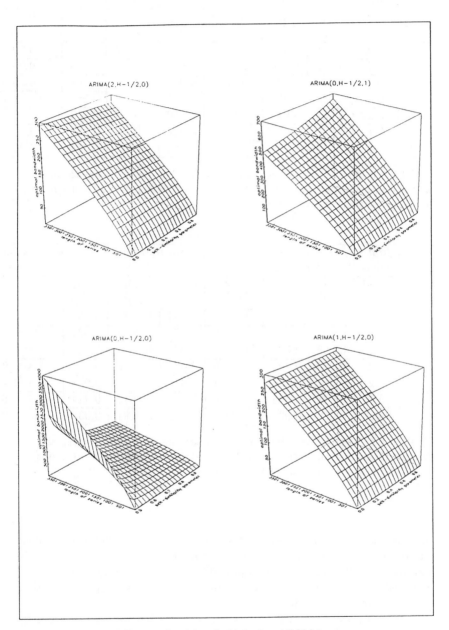

FIGURE 1. optimal bandwidths for $ARIMA$ processes

| H | first iteration | | second iteration | | third iteration | | convergence value | |
|---|---|---|---|---|---|---|---|---|
| | bias | mse | bias | mse | bias | mse | bias | mse |
| 0.1 | 0.11 | 0.013 | 0.009 | 0.0044 | 0.008 | 0.0045 | 0.008 | 0.0044 |
| 0.2 | 0.18 | 0.033 | 0.032 | 0.0069 | 0.029 | 0.0069 | 0.028 | 0.0069 |
| 0.3 | 0.22 | 0.048 | 0.041 | 0.0074 | 0.038 | 0.0073 | 0.037 | 0.0073 |
| 0.4 | 0.24 | 0.058 | 0.049 | 0.0081 | 0.046 | 0.0080 | 0.045 | 0.0079 |
| 0.5 | 0.25 | 0.065 | 0.049 | 0 0076 | 0.046 | 0.0077 | 0.048 | 0.0077 |
| 0.6 | 0.23 | 0.053 | 0.039 | 0.0062 | 0.051 | 0.0076 | 0.051 | 0.0076 |
| 0.7 | 0.22 | 0.051 | 0.047 | 0.0077 | 0.060 | 0.0085 | 0.058 | 0.0081 |
| 0.8 | 0.19 | 0.037 | 0.064 | 0.0091 | 0.064 | 0.0088 | 0.064 | 0.0088 |
| 0.9 | 0.099 | 0.010 | 0.056 | 0.0057 | 0.055 | 0.0057 | 0.054 | 0.0057 |

TABLE 1.1. Monte Carlo biases and mean squared errors of $\hat{H}$ using the infeasible approximation to the optimal bandwidth on $ARIMA(1, H - 1/2, 0)$ series with 1000 observations and 1000 replications.

| H | first iteration | | second iteration | | third iteration | | convergence value | |
|---|---|---|---|---|---|---|---|---|
| | bias | mse | bias | mse | bias | mse | bias | mse |
| 0.1 | 0.11 | 0.013 | 0.059 | 0.0079 | 0.058 | 0.0075 | .057 | 0.0074 |
| 0.2 | 0.18 | 0.033 | 0.090 | 0.013 | 0.090 | 0.013 | 0.090 | 0.013 |
| 0.3 | 0.22 | 0.048 | 0.055 | 0.0080 | 0.054 | 0.0080 | 0.054 | 0.0083 |
| 0.4 | 0.24 | 0.058 | 0.035 | 0.0070 | 0.035 | 0.0071 | 0.034 | 0.0076 |
| 0.5 | 0.25 | 0.065 | 0.026 | 0.0090 | 0.026 | 0.0092 | 0.025 | 0.0091 |
| 0.6 | 0.23 | 0.053 | 0.025 | 0.010 | 0.025 | 0.010 | 0.024 | 0.010 |
| 0.7 | 0.22 | 0.051 | 0.039 | 0.014 | 0.038 | 0.014 | 0.038 | 0.014 |
| 0.8 | 0.19 | 0.037 | 0.046 | 0.018 | 0.046 | 0.018 | 0.044 | 0.016 |
| 0.9 | 0.099 | 0.010 | 0.036 | 0.0090 | 0.036 | 0.0091 | 0.036 | 0.0090 |

TABLE 1.2. Monte Carlo biases and mean squared errors of $\hat{H}$ using the first feasible approximation to the optimal bandwidth on $ARIMA(1, H-1/2, 0)$ series with 1000 observations and 1000 replications.

for $j = 1$ to $\hat{m}^{(0)}$. The Monte Carlo results for this feasible procedure are summarized in Table 1.2. The results happen to be worse when the approximations of $h(0)$ and $h''(0)$ are updated at each iteration and they are not reported here. Table 1.3 gives the values of the successive approximations for the optimal bandwidth.

Table 1.4 presents results from a "direct" feasible approximation of the optimal bandwidth, "direct" in the sense that it relies only on the local spectral specification 1.2 . Again we rely on a simple periodogram regression to estimate $E_2(H)$. We take the value of $H$ in the current iteration, but we also need an estimate for the scale factor $G$. From Robinson [Rob95a]

| $H$ | $\hat{m}^{(0)}$ | $\hat{m}^{(1)}$ | $\hat{m}^{(2)}$ | $\hat{m}^{(\infty)}$ | $m_{opt}$ |
|---|---|---|---|---|---|
| 0.1 | 256 | 159 | 157 | 156 | 61 |
| 0.2 | 256 | 142 | 139 | 140 | 61 |
| 0.3 | 256 | 89 | 88 | 88 | 61 |
| 0.4 | 256 | 60 | 60 | 60 | 61 |
| 0.5 | 256 | 44 | 44 | 44 | 61 |
| 0.6 | 256 | 40 | 40 | 41 | 61 |
| 0.7 | 256 | 38 | 38 | 38 | 61 |
| 0.8 | 256 | 42 | 43 | 43 | 61 |
| 0.9 | 256 | 55 | 56 | 56 | 61 |

TABLE 1.3. approximations to the optimal bandwidth using the first feasible procedure on $ARIMA(1, H-1/2, 0)$ series with 1000 observations and 1000 replications.

| $H$ | first iteration | | second iteration | | third iteration | | convergence value | |
|---|---|---|---|---|---|---|---|---|
| | bias | mse | bias | mse | bias | mse | bias | mse |
| 0.1 | 0.11 | 0.013 | 0.068 | 0.073 | 0.064 | 0.0082 | 0.054 | 0.0070 |
| 0.2 | 0.18 | 0.033 | 0.11 | 0.016 | 0.097 | 0.015 | 0.086 | 0.0088 |
| 0.3 | 0.22 | 0.048 | 0.13 | 0.020 | 0.10 | 0.017 | 0.090 | 0.016 |
| 0.4 | 0.24 | 0.058 | 0.14 | 0.021 | 0.11 | 0.016 | 0.091 | 0.016 |
| 0.5 | 0.25 | 0.065 | 0.14 | 0.024 | 0.10 | 0.017 | 0.093 | 0.015 |
| 0.6 | 0.23 | 0.053 | 0.15 | 0.025 | 0.11 | 0.018 | 0.096 | 0.016 |
| 0.7 | 0.22 | 0.051 | 0.14 | 0.025 | 0.10 | 0.017 | 0.095 | 0.016 |
| 0.8 | 0.19 | 0.037 | 0.11 | 0.016 | 0.10 | 0.016 | 0.084 | 0.013 |
| 0.9 | 0.099 | 0.10 | 0.077 | 0.0070 | 0.066 | 0.0068 | 0.058 | 0.0071 |

TABLE 1.4. Monte Carlo biases and mean squared errors of $\hat{H}$ using the "direct" approximation of the optimal bandwidth on $ARIMA(1, H - 1/2, 0)$ series with 1000 observations and 1000 replications.

| $H$ | $\hat{m}^{(0)}$ | $\hat{m}^{(1)}$ | $\hat{m}^{(2)}$ | $\hat{m}^{(\infty)}$ | $m_{opt}$ |
|---|---|---|---|---|---|
| 0.1 | 256 | 201 | 200 | 193 | 61 |
| 0.2 | 256 | 183 | 170 | 166 | 61 |
| 0.3 | 256 | 170 | 156 | 148 | 61 |
| 0.4 | 256 | 164 | 140 | 138 | 61 |
| 0.5 | 256 | 162 | 139 | 131 | 61 |
| 0.6 | 256 | 160 | 136 | 134 | 61 |
| 0.7 | 256 | 156 | 133 | 128 | 61 |
| 0.8 | 256 | 132 | 130 | 123 | 61 |
| 0.9 | 256 | 110 | 118 | 109 | 61 |

TABLE 1.5. approximations to the optimal bandwidth with the "direct" procedure on $ARIMA(1, H - 1/2, 0)$ series with 1000 observations and 1000 replications.

the Gaussian estimate of $G$ is $\hat{G} = G(m, \hat{H})$, where

$$\hat{G}(m, H) = \frac{1}{m} \sum_{j=1}^{m} \frac{I(\lambda_j)}{\lambda_j^{1-2H}}.$$

Therefore, with the current values $\hat{m}^{(k)}$, $\hat{H}^{(k)}$ and $\hat{G}^{(k)} = \hat{G}(\hat{m}^{(k)}, \hat{H}^{(k)})$, the following least squares regression is performed:

$$\frac{I(\lambda_j)}{\hat{G}^{(k)}\lambda_j^{1-2\hat{H}^{(k)}}} = \hat{a}^{(k)} + \hat{b}^{(k)} \lambda_j^2 \; for \; j = 1 \; to \; \hat{m}^{(k)},$$

and $\hat{b}^{(k)}$ is taken as the current value for $E_2$ in the next iteration's optimal bandwidth determination. The approximations to 1.7 are reported in Table 1.5. A slightly modified version of the "direct" method, where the first value for $G$ is kept fixed, results in faster convergence of the values of the bandwidth but considerably higher convergence mean squared errors (the results are not reported here).

As was expected, the Monte Carlo mean squared errors of $\hat{H}$ are lowest where the infeasible procedure is used, but biases are smaller for long memory values of $H$ when the first feasible procedure is used, as convergence values of the bandwidth are smaller. It is to be noted that the direct procedure performs better only for extreme values of $H$. For $H = 0.9$ and $H = 0.8$, the convergence values for the bandwidth, being on either side of the optimal value, are not directly comparable; but it is slightly surprising for the low values of $H$ where mean squared errors are smaller with the "direct" procedure notwithstanding convergence values of the bandwidth closer to the optimal ones with the first feasible procedure.

A final study summarized in Table 1.6 is a comparison of the performances of the infeasible and the "direct" procedures on a pure fractional Gaussian noise series with autocovariances

$$\gamma_j = \frac{1}{2}(|j+1|^{2H+1} - 2|j|^{2H+1} + |j-1|^{2H+1}). \tag{1.8}$$

The data were simulated using the algorithm of Davies and Harte [DH87] applied to the autocovariances 1.8. One has to bear in mind the fact that the infeasible and the first feasible procedures were based on a fractional differencing specification. This partly explains why the first feasible procedure performs less well than the "direct" procedure (results for the first feasible procedure are not reported here). Nevertheless, the infeasible procedure always performs slightly better except for the two extreme values of the self-similarity parameter $H = 0.1$ and $H = 0.9$, where the variance is curbed by the censoring of estimates at both bounds $\Delta_1$ and $\Delta_2$.

| $H$ | bias | | mse | | bandwidth | |
| --- | --- | --- | --- | --- | --- | --- |
| | *infeasible* | *direct* | *infeasible* | *direct* | *infeasible* | *direct* |
| 0.1 | -0.086 | -0.067 | 0.0077 | 0.0066 | 317 | 190 |
| 0.2 | -0.047 | -0.039 | 0.0031 | 0.0045 | 356 | 314 |
| 0.3 | -0.025 | -0.014 | 0.0013 | 0.0020 | 418 | 436 |
| 0.4 | -0.012 | -0.0081 | 0.0008 | 0.0016 | 511 | 485 |
| 0.5 | -0.00062 | -0.00035 | 0.0003 | 0.00093 | 511 | 464 |
| 0.6 | 0.0088 | 0.00030 | 0.0006 | 0.0012 | 511 | 473 |
| 0.7 | 0.0066 | 0.0010 | 0.0007 | 0.0023 | 418 | 449 |
| 0.8 | 0.0096 | 0.0048 | 0.0008 | 0.0031 | 356 | 408 |
| 0.9 | 0.0072 | 0.0063 | 0.0009 | 0.0044 | 317 | 386 |

TABLE 1.6. Comparison of the infeasible and the "direct" procedures on pure fractional Gaussian noise series with 1000 observations and 1000 replications.

# 4 Empirical Illustrations

Both feasible procedures are first applied to the Nile River annual minimum levels recorded between 622 and 1284 A.D. in Toussoun [Tou25]. Using the averaged periodogram estimator, Robinson [Rob94b] found estimates of the self-similarity parameter for this series between 0.832 and 0.859 for choices of bandwidth between 20 and 180. Using the same range of bandwidths, the present estimator yields values of the self-similarity parameter between 0.758 and 0.857. These are graphed against bandwidth in Figure 2. The convergence value for the optimal bandwidth using the direct procedure is $\hat{m}^{(\infty)} = 71$ and the corresponding estimate is $\hat{H}^{(\infty)} = 0.838$. The first feasible procedure yields a smaller approximation of the optimal bandwidth, $\hat{m}^{(\infty)} = 53$, and a larger value for $\hat{H}$, $\hat{H}^{(\infty)} = 0.854$. Robinson [Rob95a] reported rather larger estimates of $H$ using both $\hat{H}$ and the log periodogram estimate for $m$ values of 41 and 164.

Two types of financial series are investigated in this framework: a series of daily stock prices for Guinness Ltd, with 1000 observations between 1988 and 1990 on the London Stock Exchange, and two intra-day IBM equities transaction prices series for two consecutive trading days on the New York Stock Exchange, of 4000 observations each.

The loss function $R(m, H)$ against both arguments, and against $H$ at the estimated optimal bandwidth, as well as $\frac{\partial R}{\partial H}$ against $H$ are reported for the Guinness data on Figure 3. The corresponding values of $\hat{H}$ and the optimal bandwidth are $\hat{H}^{(\infty)} = 0.785$ and $\hat{m}^{(\infty)} = 75$.

The IBM transaction prices series for the two consecutive trading days (6 and 7 January 1993) yield results very similar to one another (they are summarized in Table 1.7). There is evidence of a unit root in the series which is confirmed by the estimation on the first-differenced series.

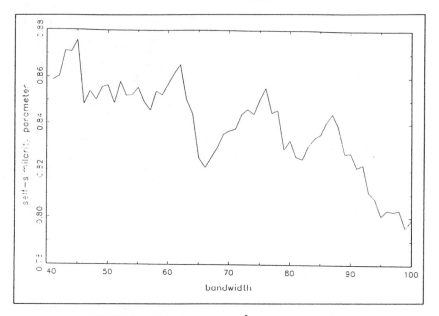

FIGURE 2. Nile River Data $\hat{H}$ against bandwidth

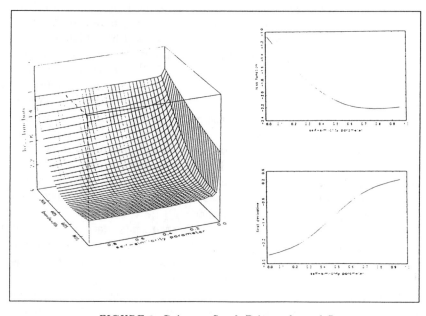

FIGURE 3. Guinness Stock Prices: plots of $R$

| | 6/1/93 | | 7/1/93 | |
|---|---|---|---|---|
| | $\hat{H}^{(\infty)}$ | $\hat{m}^{(\infty)}$ | $\hat{H}^{(\infty)}$ | $\hat{m}^{(\infty)}$ |
| *original series* | | | | |
| first feasible procedure | 0.999 | 30 | 0.963 | 147 |
| "direct" procedure | 0.999 | 87 | 0.999 | 63 |
| *first differenced* | | | | |
| first feasible procedure | 0.500 | 1042 | 0.496 | 164 |
| "direct" procedure | 0.500 | 2047 | 0.469 | 361 |

TABLE 1.7. IBM Transactions Prices.

# 5  Acknowledgements

This research was carried out in the Department of Economics of the London School of Economics and the Financial Markets Group. Financial support from ESRC grant R000235892 and from the Ecole polytechnique in the form of an A.M.X. grant is gratefully acknowledged.

# 6  References

[DH87]  R.B. Davies and D.S. Harte. Tests for the Hurst Effect. *Biometrika*, 74:95–101, 1987.

[DR96]  M.A. Delgado and P.M. Robinson. Optimal spectral bandwidth for long memory. *Statistica Sinica*, 1996. forthcoming.

[FT86]  R. Fox and M.S. Taqqu. Large sample properties of parameter estimates for strongly dependent stationary Gaussian time series. *Annals of Statistics*, 14:517–132, 1986.

[GPH83]  J. Geweke and S. Porter-Hudak. The estimation and application of long memory time series models. *Journal of Time Series Analysis*, 4:221–238, 1983.

[GRS95]  L. Giraitis, P.M. Robinson, and A. Samarov. Rate optimal semiparametric estimation of the memory parameter of the Gaussian time series with long range dependence. Preprint, 1995.

[GS90]  L. Giraitis and D. Surgailis. A central limit theorem for quadratic forms in strongly dependent linear variables and its application to asymptotic normality of Whittle's estimate. *Probability Theory and Related Fields*, 86:87–104, 1990.

[K87]  H.R. Künsch. Statistical aspects of self-similar processes. In *Proceedings of the First World Congress of the Bernouilli Society*, pages 67–74. VNU Science Press, 1987.

[Rob94a]  P.M. Robinson. Rates of convergence and optimal bandwidth in spectral analysis of processes with long range dependence. *Probability Theory and Related Fields*, 99:443–473, 1994.

[Rob94b]  P.M. Robinson. Semiparametric analysis of long memory time series. *Annals of Statistics*, 22:515–539, 1994.

[Rob94c]  P.M. Robinson. Time series with strong dependence. In C.A. Sims, editor, *Advances in Econometrics*, volume 1. Cambridge University Press, 1994.

[Rob95a]  P.M. Robinson. Gaussian semiparametric estimation of long range dependence. *Annals of Statistics*, 1995. forthcoming.

[Rob95b]  P.M. Robinson. Log periodogram regression of time series with long range dependence. *Annals of Statistics*, 23:1048–1072, 1995.

[Tou25]  O. Toussoun. *Mémoire sur l'Histoire du Nil*, volume 9 of *Mémoires de l'Institut d'Egypte*. Institut d'Egypte, 1925. Le Caire.

# Some limit theorems on stationary processes with long-range dependence

Yuzo Hosoya

ABSTRACT  The paper provides central limit theorems on multivariate stationary processes with long-range dependence as a natural extension of the corresponding theory on short-range dependent processes. In order to establish those theorems, the paper imposes weak assumptions on conditional moments of innovation processes, dispensing with the usual assumptions of exact Martingale difference or the contemporaneously transformed Gaussianity.

## 1  Introduction

In recent years there has been the development of probability theory to deal with long-range dependent stationary processes, and now as far as Gaussian (or contemporaneously transformed Gaussian) long-range stationary processes are concerned, the limit theory for basic statistics and the large-sample estimation theory seem to have been well framed [see Breuer and Major (1983), Fox and Taqqu (1986, 87) and Dahlhaus (1989) among others]. With the aim to extend those results to non-Gaussian multivariate long-range processes, this paper establishes the limit theorems for sample sum, serial covariances, based on weak restrictions on the innovation process involved, presenting explicit evaluation of the contribution of fourth-order innovation spectra.

Sections 2 and 3 give central limit thorems for sample sum [Theorem 2.1] and sample serial covariances [Theorems 3.1 and 3.2], respectively. Those theorems are proved based on the assumption that the innovation processes concerned satisfy certain mixing conditions for their conditional second or fourth-order moments, and the assumptions of exact Martingale difference [Hannan (1976) for example] or the transformed Gaussianity for the innovation processes [Breuer and Major (1983)] are dispensed with. Theorem 1.1 gives a central limit theorem for sample sum in such a form that spectral densities of individual process $\{z_\alpha(t)\}, \alpha = 1, \cdots, q$ may possibly have an unbounded peak at the origin. Theorems 3.1 and 3.2 deal with the serial covariances for the case the individual spectral density $f_{\alpha\alpha}$ may possibly not be square-integrable. The latter theorem shows that serial covariances have different asymptotic performance according as peaks of spectral density occur at the origin or at some other frequencies. Section 4 gives the proofs of the lemmas and theorems.

Throughout the paper, $J$ denotes the set of all integers; $\delta(\cdot, \cdot)$ indicates a function such that $\delta(x, y) = 1$ if $x = y$ and $\delta(x, y) = 0$ otherwise. $I(C)$ is the indicator function of a set $C$. The conjugate transpose of a matrix $A$ is denoted $A^*$ (the notation is retained for the transpose of real $A$ also). In the proofs of Section 4, $c_1, c_2, \cdots$ indicate positive constants relevant to respective context.

## 2  A central limit theorem for sample sum

Let $\{z(t); t \in J\}$ be a real vector-valued linear process generated by

$$z(t) = \sum_{j=0}^{\infty} G(j)e(t - j), \quad t \in J, \tag{1.1}$$

where the $z(t)$'s are $q$-vectors and the $e(t)$'s are $p$-vectors such that $E\{e(m)e(n)^*\} = \delta(m, n)K$ for $K$ a nonsingular $p$ by $p$ matrix; the matrices $G(j)$ are $q$ by $p$ and the components of $z, e$ and $G$ are all real. Assume throughout the paper that

$$\sum_{j=0}^{\infty} tr G(j)K G(j)^* < \infty$$

so that the process $\{z(t)\}$ is a second order stationary process and has a spectral density matrix $f(\omega)$ representable as

$$f(\omega) = \frac{1}{2\pi}k(\omega)Kk(\omega)^*, \quad -\pi \leq \omega \leq \pi,$$

where $k(\omega) = \sum_{j=0}^{\infty} G(j)e^{i\omega j}$. Denote the $(\alpha, \beta)$ component of $G(j)$ by $G_{\alpha\beta}(j)$ and the $\alpha$-th component of $z(t)$ and $e(t)$ by $z_\alpha(t)$ and $e_\alpha(t)$. Let $K_n(\omega)$ be the Fejér kernel defined by $K_n(\omega) = [sin(n\omega/2)/sin(\omega/2)]^2/n$. For an integrable possibly matrix-valued function $g$ define $L_n(g)$ by

$$L_n(g) = \frac{n}{2\pi} \int_{-\pi}^{\pi} K_n(\omega)g(\omega)d\omega$$

and define $\phi_\alpha(n)$ by $L_n(f_{\alpha\alpha}) = n\phi_\alpha(n)$ where $f_{\alpha\alpha}$ is the $(\alpha, \alpha)$ component of the spectral density matrix $f$. Let $D_n$ be the $q$ by $q$ diagonal matrix with $L_n(f_{\alpha\alpha})^{1/2}$ as the $(\alpha, \alpha)$th element. Let $\{m(n)\}$ be a sequence of positive integers. Set $k^{(2)}(\omega) = \sum_{j=m_2(n)+1}^{\infty} G(j)e^{i\omega j}$.

**Assumption 2.1.** There exists a choice of $m_1(n)$ and $m_2(n)$ such that

(i)    $\lim\limits_{n\to\infty} m_1(n) = \lim\limits_{n\to\infty} m_2(n) = \lim\limits_{n\to\infty} \dfrac{n}{m_1(n) + m_2(n)} = \infty,$

   $\lim\limits_{n\to\infty} m_1(n)/m_2(n) = \infty,$

(ii)   $\lim\limits_{n\to\infty} D_n^{-1} L_n(k^{(2)} K k^{(2)*}) D_n^{-1} = 0,$

(iii)  $\lim\limits_{n\to\infty} \phi_\alpha(m_1(n))/\phi_\alpha(n) = 1,$

(iv)   $\phi_\alpha(n)/n$  is bounded away from  0.

**Assumption 2.2.**    $D_n^{-1} L_n(k K k^*) D_n^{-1}$ tends to a positive-definite matrix $\Omega$ as $n$ tends to infinity.

Let $\mathcal{F}(t)$ be the $\sigma$-field generated by the random variables $\{e_\alpha(s);\ 1 \le \alpha \le p, s \le t\}$.

**Assumption 2.3.**    For any $t_1, t_2 > t$, there is $\delta > 0$ such that uniformly in $t$ and $\alpha, \beta$,

(i) $E\,|\,E\{e_\alpha(t_1)\,|\,\mathcal{F}(t)\} E\{e_\beta(t_2)\,|\,\mathcal{F}(t)\}\,| = O(|\,(t_1 - t)(t_2 - t)\,|^{-(1+\delta)}),$

(ii) $E\,|\,E\{e_\alpha(t_1)e_\beta(t_2)|\mathcal{F}(t)\} - E(e_\alpha(t_1)e_\beta(t_2))|$

$$= O(|\,(t_1 - t)(t_2 - t)\,|^{-(1+\delta)}).$$

Let $S_{n\alpha}(r) = \sum_{t=1}^{n} z_\alpha(t+r)/L_n(f_{\alpha\alpha})^{1/2}$ and let $S_n(r)$ be the $q$-vector with components $S_{n\alpha}(r)$.

**Theorem 2.1.**    Suppose that Assumptions 2.1 through 2.3 hold and suppose that for any $\varepsilon > 0$, there exists $M_\varepsilon > 0$ such that for all sufficiently large $n$,

$$E\{S_n(r)^* S_n(r) I(S_n(r)^* S_n(r) > M_\varepsilon)\} < \varepsilon \qquad (1.2)$$

uniformly in $r$. Then $S_n(0)$ has the asymptotically normal distribution with mean vector $0$ and covariance matrix $\Omega$.

The proof is given in Section 4.

**Example 2.1.**    Let $\{z(t);\ t \in J\}$ be a scalar-valued linear process which has the representation

$$z(t) = \sum_{j=1}^{\infty} j^{-1}\, e\,(t - j),$$

where $\{e(t); t \in J\}$ is a white noise scalar process such that $E\{e(t)\} = 0$ and $Var\{e(t)\} = 1$. Set $S_n(r) = \sum_{t=1+r}^{n+r} z(t)$ and $z''(t) = \sum_{j=m(n)+1}^{\infty} \frac{1}{j} e(t - j)$ where $\{m(n)\}$ is a sequence of positive integers such that $m(n) \le n$. Set $L_n = Var\{S_n(0)\}$. Then it holds that

$$L_n = (\log n)^2 n(1 + o(1)) \qquad (1.3)$$

and

$$Var\{\sum_{t=1}^{n} z''(t)\}/L_n = O(\{\log(n/m(n))/\log n\}^2). \tag{1.4}$$

The left-hand member of (1.4) tends to 0 as $n \to \infty$ if $m_2(n) = n(\log n)^{-1}$ for example and then Assumption 2.1 (ii) is satisfied. Also if $m_1(n) = n(\log n)^{-1+\varepsilon}(0 < \varepsilon < 1)$, then Assumption 2.1 (i) and (iii) are satisfied.

## 3   Limit theorems for serial covariances

This section considers the process $\{z(t)\}$ given in (1.1) under the additional assumptions that $\{e(t)\}$ is fourth-order stationary and that

$$\sum_{t_1,t_2,t_3=-\infty}^{\infty} | \tilde{Q}_{\beta_1,\cdots,\beta_4}^e (t_1, t_2, t_3) | < \infty, \tag{1.5}$$

where $\tilde{Q}_{\beta_1,\cdots,\beta_4}^e$ is the joint fourth cumulant of $e_{\beta_1}(t), e_{\beta_2}(t + t_1), e_{\beta_3}(t + t_2), e_{\beta_4}(t+t_3)$, so that the process $\{e(t)\}$ has a fourth-order spectral density $Q_{\beta_1,\cdots,\beta_4}^e(\omega_1, \omega_2, \omega_3)$ such that

$$Q_{\beta_1,\cdots,\beta_4}^e(\omega_1, \omega_2, \omega_3)$$

$$= \frac{1}{(2\pi)^3} \sum_{t_1,t_2,t_3=-\infty}^{\infty} exp\{-i(\omega_1 t_1 + \omega_2 t_2 + \omega_3 t_3)\} \tilde{Q}_{\beta_1,\cdots,\beta_4}^e (t_1, t_2, t_3).$$

**Assumption 3.1.**   (i) For any $t_1, t_2, t_3, t_4 > t$ and for some $\delta_2 > 0$,
$$E|E\{e_{\alpha_1}(t_1)e_{\alpha_2}(t_2)|\mathcal{F}(t)\}E\{e_{\alpha_3}(t_3)e_{\alpha_4}(t_4)|\mathcal{F}(t)\}$$
$$-E\{e_{\alpha_1}(t_1)e_{\alpha_2}(t_2)\}E\{e_{\alpha_3}(3)e_{\alpha_4}(t_4)\}|$$
$$= O(\Pi_{j=1}^4(t_j - t)^{-(1+\delta_2)}), \text{ uniformly in } t.$$
(ii) $E|E\{e_{\alpha_1}(t_1)e_{\alpha_2}(t_2)e_{\alpha_3}(t_3)e_{\alpha_4}(t_4)|\mathcal{F}(t)\}$
$$-E\{e_{\alpha_1}(t_1)e_{\alpha_2}(t_2)e_{\alpha_3}(t_3)e_{\alpha_4}(t_4)\}|$$
$$= O(\Pi_{j=1}^4(t_j - t)^{-(1+\delta_2)}), \text{ uniformly in } t.$$
(iii) For any $\varepsilon > 0$ and for any integer $l \geq 0$, there exists $B_\varepsilon > 0$ such that

$$E[U(n, s)^2 I\{U(n, s) > B_\varepsilon\}] < \varepsilon$$

uniformly in $n, s$, where

$$U(n, s) = [ \sum_{\alpha,\beta=1}^{p} \sum_{r=0}^{l} \{\sum_{t=1}^{n} (e_\alpha(t + s)e_\beta(t + s + r) - K_{\alpha\beta}\delta(0, r))/n^{1/2}\}^2]^{1/2}.$$

Denote by $C_{\alpha\beta}^z(r)(1 \leq \alpha, \beta \leq q)$ the serial covariance constructed form the partial realization $\{z(1), \cdots, z(n)\}$; namely, $C_{\alpha\beta}^z(r) = \frac{1}{n} \sum_{t=1}^{n-r} z_\alpha(t)z_\beta(t +$

$r$) for $0 \leq r \leq n-1$ and $C_{\alpha\beta}^z(r) = C_{\alpha\beta}^z(-r)$ for $-n+r \leq r \leq 0$. Set also $\gamma_{\alpha\beta}^z(r) = E(z_\alpha(t)z_\beta(t+r))$. Define $\varphi_{\alpha_1\alpha_2\alpha_3\alpha_4}^{r_1r_2}$, $1 \leq \alpha_i \leq q$, for any integers $r_1, r_2$ by

$$\varphi_{\alpha_1\alpha_2\alpha_3\alpha_4}^{r_1r_2}(\omega) = 2\pi \int_{-\pi}^{\pi} [f_{\alpha_1\alpha_3}(\omega_1 - \omega)\overline{f_{\alpha_2\alpha_4}(\omega_1)}exp\{-i(r_2 - r_1)\omega_1\}$$

$$+f_{\alpha_1\alpha_4}(\omega_1 - \omega)\overline{f_{\alpha_2\alpha_3}(\omega_1)}exp\{i(r_1 + r_2)(\omega_1 - \omega) + ir_1\omega_1\}]d\omega_1$$

$$+2\pi \sum_{\beta_1\cdots\beta_4=1}^{p} \int_{-\pi}^{\pi}\int_{-\pi}^{\pi} exp(ir_1\omega_1 + ir_2\omega_2)k_{\alpha_1\beta_1}(\omega_1 + \omega)k_{\alpha_2\beta_2}(-\omega_1)$$

$$\cdot k_{\alpha_3\beta_3}(-\omega + \omega_2)k_{\alpha_4\beta_4}(-\omega_2)Q_{\beta_1\cdots\beta_4}^e(\omega_1 + \omega, \omega - \omega_2, \omega_2)d\omega_1 d\omega_2.$$

$$(1.6)$$

Set $T_{\alpha_1\alpha_2}^r(n, s) = \sum_{t=1+s}^{n+s-|r|}\{z_{\alpha_1}(t)z_{\alpha_2}(t+r) - E(z_{\alpha_1}(t)z_{\alpha_2}(t+r))\}$. Then in view of Zygmund (1959, p.36), Hannan (1976) and Hosoya and Taniguchi (1982, p.145), there is the relationship:

$$Cov\{T_{\alpha_1\alpha_2}^{r_1}(n, s), T_{\alpha_3\alpha_4}^{r_2}(n, s)\} = L_n(\varphi_{\alpha_1\alpha_2\alpha_3\alpha_4}^{r_1r_2}).$$

Set $N_\alpha(n) = L_n(\varphi_{\alpha\alpha\alpha\alpha}^{00})$, and set $z_\alpha''(t) = \sum_{j=m_2(n)+1}^{\infty}\sum_\beta G_{\alpha\beta}(j)e_\beta(t-j)$.
**Assumption 3.2.** There exists a choice of sequences $\{m_1(n)\}$ and $\{m_2(n)\}$ such that, for $\alpha = 1, \cdots, q$,

(i) $\quad \lim_{n\to\infty} E\{(\sum_{t=1}^{n} z_\alpha''(t)^2)^2\}/N_\alpha(n) = 0,$

(ii) $\quad \lim_{n\to\infty} (n_1 m_1)^{1/2}/m_2^{1+2\delta_2} = \lim_{n\to\infty} m_1/m_2^{1+2\delta} = 0,$

(iii) $\quad g_\alpha(n) = N_\alpha(n)/n^{1/2} > \lambda > 0$ and $\lim_{n\to\infty} g_\alpha(m_1(n))/g_\alpha(n) = 1.$

where $n_1 = [n/(m_1 + m_2)]$.
**Assumption 3.3.** $L_n(\varphi_{\alpha_1\alpha_2\alpha_3\alpha_4}^{r_1r_2})/\{\Pi_{j=1}^4 N_{\alpha_j}(n)\}^{1/4}$ tends to $\Phi_{\alpha_1\alpha_2\alpha_3\alpha_4}^{r_1r_2}$ as $n \to \infty$ for any $r_1, r_2$ and $\alpha_1, \alpha_2, \alpha_3, \alpha_4(1 \leq \alpha_j \leq q)$.

For fixed $l \geq 0$, let $T(n,0)$ be the vector whose components consist of ordered $T_{\alpha_1\alpha_2}^r(n,0)/\{N_{\alpha_1}(n)N_{\alpha_2}(n)\}^{1/4}$ in any fixed order where $\alpha_1, \alpha_2$ and $r$ range over on $1 \leq \alpha_1 \leq \alpha_2 \leq q$ and $0 \leq r \leq l$.
**Theorem 3.1.** *Suppose that Assumptions 3.1 and through 3.3 are satisfied. Then $T(n,0)$ has the limit normal distribution with mean 0 and covariances $\Phi_{\alpha_1\alpha_2\alpha_3\alpha_4}^{r_1r_2}$ $(1 \leq \alpha_i \leq q$, and $0 \leq r_1 \leq r_2 \leq l)$.*
**Example 3.1.** If $f_{\alpha_i\alpha_i}(\omega), i = 1, \cdots, 4$, are square-integrable, Assumption 3.3 is automatically satisfied and it follows that

$$\Phi_{\alpha_1\alpha_2\alpha_3\alpha_4}^{r_1r_2} = \varphi_{\alpha_1\alpha_2\alpha_3\alpha_4}^{r_1r_2}(0)/\{\Pi_{j=1}^4\varphi_{\alpha_j}(0)\}^{1/4},$$

where $\varphi_{\alpha_j}(0) = \varphi^{00}_{\alpha_j,\alpha_j,\alpha_j,\alpha_j}(0)$. See also Hosoya (1993) for an example where teh spectral density is not square-integrable but Assumption 3.2 is satisfied.

The next theorem characterizes the limit normal distribution for specific types of singularity of $f$.

**Condition A.**     A spectral density $f_{\alpha\alpha}(\omega)$ satisfies

$$\int_{(-\varepsilon,\varepsilon)} \mid f_{\alpha\alpha}(\omega) \mid^2 d\omega = \infty \quad \text{and} \quad \int_{(-\pi,-\varepsilon)\cup(\varepsilon,\pi)} \mid f_{\alpha\alpha}(\omega) \mid^2 d\omega < \infty$$

for any $\varepsilon > 0$.

**Condition B.**     A spectral density $f_{\alpha\alpha}(\omega)$ has the unique unbounded peak at the origin and is bounded a.e. for $\omega \in [-\pi, -\varepsilon) \cup (\varepsilon, \pi)$ for $\varepsilon > 0$ and $\int_{-\pi}^{\pi} |f_{\alpha\alpha}(\omega)|^2 d\omega = \infty$.

The next theorem generalizes the corresponding result of Breuer and Major (1983).

**Theorem 3.2.**     (i) If $f_{\alpha\alpha}(\omega)$ satisfies Condition A and if $f_{\beta\beta}(\omega)$ satisfies Condition B, or (ii) if $f_{\alpha\alpha}(\omega)$ is square-integrable and if $f_{\beta\beta}(\omega)$ satisfies Condition A, $\{T^0_{\alpha\beta}(n,0) - T^r_{\alpha\beta}(n,0)\}/\{N_\alpha(n)N_\beta(n)\}^{1/4}$ tends to 0 in mean square for any r; hence the joint distribution of $\{T^0_{\alpha\beta}(n,0) - E(T^r_{\alpha\beta}(n,0))\}/\{N_\alpha(n)N_\beta(n)\}^{1/4}, r = 0, \cdots, L$, asymptotically degenerates.

# 4   Mathematical proofs

## 4.1   Proofs of Section 2

Henceforth for the sake of simplicity, set $m_1 = m_1(n), m_2 = m_2(n)$ and $L_{n\alpha} = L_n(f_{\alpha\alpha})$. Also set $n_1 = [n/(m_1 + m_2)]$ and assume without much loss of generality that $n_1 = n/(m_1 + m_2)$. Let $I_k$ and $J_k$ be the sets of integers such that

$$\begin{aligned} I_k &= \{i : (k-1)(m_1 + m_2) + 1 \le i \le km_1 + (k-1)m_2\}, \\ J_k &= \{j : km_1 + (k-1)m_2 + 1 \le j \le k(m_1 + m_2)\}, \quad k = 1, \cdots, n_1. \end{aligned}$$

Set $z'(t) = \sum_{j=1}^{m_2} G(j)e(t - j)$ and set $\eta_\alpha(k) = \sum_{l \in I_k} z'(l)/(L_{m_1\alpha})^{1/2}$ and $\eta'_\alpha(k) = \sum_{l \in J_k} z'(l)/(L_{m_1\alpha})^{1/2}$. Define the $\sigma$-fields $\mathcal{B}_k$ and $\mathcal{B}'_k$ by $\mathcal{F}\{km_1 + (k-1)m_2\}$ and $\mathcal{F}\{k(m_1 + m_2)\}$. The next two lemmas imply that the partial sum $S_n$ has the same limit distribution as $\sum_{k=1}^{n_1}\{\eta(k) - E(\eta(k)|\mathcal{B}_{k-1})\}/n_1^{1/2}$.

**Lemma 4.1.**     For all $\alpha$,

$$\lim_{n \to \infty} Var\{\sum_{k=1}^{n_1} E(\eta_\alpha(k)|\mathcal{B}_{k-1})/n_1^{1/2}\} = \lim_{n \to \infty} Var\{\sum_{k=1}^{n_1} E(\eta'_\alpha(k)|\mathcal{B}'_{k-1})/n^{1/2}\}$$

$$= 0.$$

**Proof.**　In the equation

$$Var\{\sum_k E(\eta_\alpha(k)|\mathcal{B}_{k-1})\}/n_1 = \sum_k Var\{E(\eta_\alpha(k)|\mathcal{B}_{k-1})\}/n_1$$

$$+2\sum_{k=1}^{n_1}\sum_{l=k+1}^{n_1} Cov\{E(\eta_\alpha(k)|\mathcal{B}_{k-1})E(\eta_\alpha(l)|\mathcal{B}_{l-1})\}/n_1,$$

each of right-hand side members tends to 0 as $n \to \infty$, as shown below:
(i) Since $\sum_{j=0}^{\infty}\sum_{\alpha=1}^{q}\sum_{\beta=1}^{p} G_{\alpha\beta}(j)^2 < \infty$,

$$\sum_{k=1}^{n_1} Var\{E(\eta_\alpha(k)|\mathcal{B}_{k-1})\}/n_1$$

$$\leq \frac{c_1}{L_{m_1\alpha}n_1}\sum_{k=1}^{n_1}\sum_{j=1}^{m_2} E\{\sum_{t\in I_k} E(e(t+k-j)|\mathcal{B}_{k-1})\}^2 \leq \frac{c_2(m_2+m_1)^{1-\delta}}{L_{m_1\alpha}},$$

whence $\sum_{k=1}^{n_1} Var\{E(\eta_\alpha(k)|\mathcal{B}_{k-1})\}/n_1 \to 0$ as $n \to \infty$ in view of Assumption 2.1.

(ii) 　　　　　$$\sum_{k=1}^{n_1}\sum_{l=k+1}^{n_1} Cov\{E(\eta_\alpha(k)|\mathcal{B}_{k-1})E(\eta_\alpha(l)|\mathcal{B}_{l-1})\}/n_1 \quad (1.7)$$

$$\leq \sum_{k=1}^{n_1}\sum_{l=k+1}^{n_1} [E\{E(\eta_\alpha(k)|\mathcal{B}_{k-1})\}^2]^{1/2}[E\{E(\eta_\alpha(l)|\mathcal{B}_{l-1})\}^2]^{1/2}/n_1,$$

whereas

$$\sum_{l=k+1}^{n_1} [E\{E(\eta_\alpha(l)|\mathcal{B}_{l-1})\}^2]^{1/2} \leq c_2\{\sum_{l=1}^{n_1-k} l^{-(l+\delta)}\}(m_1(n)+m_2(n))^{1/2-\delta}/L_{m_1\alpha}^{1/2}.$$

Therefore it follows that the left-hand side of (1.7) in bounded by $c_3(m_1(n)+m_2(n))^{1-3/2\delta}/L_{m_1\alpha}$ which tends to 0 as $n \to \infty$; whence follows $\lim_{n\to\infty} Var\{\sum_{k=1}^{n_1} E(\eta_\alpha(k)|\mathcal{B}_{k-1})/n_1^{1/2}\} = 0$. A fortiori, the second proposition of the lemma holds. □

**Lemma 4.2.**

$$\lim_{n\to\infty} Var[\sum_k \{\eta_\alpha'(k) - E\{\eta_\alpha'(k)|\mathcal{B}_{k-1}'\}\}]/n_1 = 0.$$

**Proof.**　It holds that

$$Var[\sum_{k=1}^{n_1} \{\eta_\alpha'(k) - E(\eta_\alpha'(k)|\mathcal{B}_{k-1}')\}]^2 = \sum_{k=1}^{n_1} Var\{\eta_\alpha'(k) - E(\eta_\alpha'(k)|\mathcal{B}_{k-1}')\}$$

$$= \sum_{k=1}^{n_1}[Var(\eta'_\alpha(k)) + Var\{E\eta'_\alpha(k)|\mathcal{B}'_{k-1})\}],$$

whereas, since $Var(\eta'_\alpha(k)) = L_{m_2\alpha}/L_{m_1\alpha} = (m_2/m_1)(\varphi_\alpha(m_2)/\varphi_\alpha(m_1))$ and since $Var\{E(\eta'_\alpha(k)|\mathcal{B}'_{k-1})\}$ tends to 0 uniformly in $k$ by lines of argument similar to Lemma 4.1. Therefore the result follows. $\square$

Lemmas 4.1 and 2 imply that $S_n(0)$ has the same limiting distribution as $\sum_{k=1}^{n_1}\{\eta(k) - E(\eta(k)|\mathcal{B}_{k-1})\}/n_1^{1/2}$. Also it is evident that $\{\eta(k) - E(\eta(k)|\mathcal{B}_{k-1}), \mathcal{B}_{k-1}, k = 1,\cdots,n_1\}$ is a Martingale difference process with respect to the filtration $\{\mathcal{B}_{k-1}, k = 1,\cdots,n_1\}$. Let $a$ be an arbitrary non-zero real q-vector. Then the next lemma shows that the condition (1) of Brown (1971, p.69) is satisfied for the sequence $\{a^*\eta(k) - E(a^*\eta(k)|\mathcal{B}_{k-1}), \mathcal{B}_{k-1}\}$.

**Lemma 4.3.**    *Suppose $a \neq 0$; then the ratio*

$$\sum_{k=1}^{n_1} E[\{a^*(\eta(k)-E(\eta(k)|\mathcal{B}_{k-1}))\}^2|\mathcal{B}_{k-1}]/ \sum_{k=1}^{n_1} E[\{a^*(\eta(k)-E(\eta(k)|\mathcal{B}_{k-1}))\}^2]$$

*tends to 1 in probability, and*

$$\lim_{n\to\infty} \frac{1}{n_1} \sum_{k=1}^{n_1} E[\{a^*(\eta(k) - E(\eta(k)|\mathcal{B}_{k-1}))\}^2] = a^*\Omega a.$$

**Proof.**    Since $\sum_{k=1}^{n_1} E\{E(a^*\eta(k)|\mathcal{B}_{k-1})\}^2/n_1$ tends to 0 as $n \to \infty$ in view of Lemma 4.1 and since

$$\lim_{n\to\infty} \frac{1}{n_1} \sum_{k=1}^{n_1} E[(a^*\eta(k))^2] = \lim_{n\to\infty} a^*D_n^{-1}L_n(2\pi f(\omega))D_n^{-1}a = a^*\Omega a$$

by Assumption 2.2, the second proposition follows. For the first, it suffices to show that

$$\lim_{n\to\infty} \sum_{k=1}^{n_1} E|E\{(a^*\eta(k))^2|\mathcal{B}_{k-1}\} - E\{(a^*\eta(k))^2\}|/n_1 = 0.$$

It follows from Assumption 2.3 (ii) that

$$E|E\{(a^*\eta_k)^2|\mathcal{B}_{k-1}\} - E\{(a^*\eta_k)^2\}| \leq \frac{c_1}{\min_\alpha L_{n_1\alpha}}|m_1 + m_2|^{1-\delta},$$

Hence the first statement of the lemma follows. $\square$

**Proof of Theorem 1.1.**    Since the condition (1) of Brown (1971) was seen to hold in the previous lemma it is sufficient to show that for any non-zero q-vector $a$ that $u_n(k) = a^*\eta_k - E(a^*\eta_k|\mathcal{B}_{k-1})$ satisfies the Lindeberg condition under the assumption (1.2). But that can be done quite in parallel to Hosoya and Taniguchi (1992). $\square$

## 4.2  Proofs of Section 3

Set the members on the right-hand side of (1.6) respectively as $\varphi_{\alpha_1\alpha_2\alpha_3\alpha_4}^{r_1r_2(1)}$ and $\varphi_{\alpha_1\alpha_2\alpha_3\alpha_4}^{r_1r_2(2)}$ so that

$$\varphi_{\alpha_1\alpha_2\alpha_3\alpha_4}^{r_1r_2}(\omega) \equiv \varphi_{\alpha_1\alpha_2\alpha_3\alpha_4}^{r_1r_2(1)}(\omega) + \varphi_{\alpha_1\alpha_2\alpha_3\alpha_4}^{r_1r_2(2)}(\omega).$$

**Lemma 4.4.**
$$L_n(\varphi_{\alpha_1\alpha_2\alpha_3\alpha_4}^{r_1r_2(2)}) = O(n).$$

**Proof.**    It follows from Lemma 3.3 of Hosoya and Taniguchi (1982) that

$$\lim_{n\to\infty} n^{-1} L_n(\varphi_{\alpha_1\alpha_2\alpha_3\alpha_4}^{r_1r_2(2)}) = 2\pi \sum_{\beta_1,\cdots,\beta_4=1}^{p} \int \int_{-\pi}^{\pi} exp\{im_1\omega_1 + im_2\omega_2\}$$

$$\cdot k_{\alpha_1\beta_1}(\omega_1)k_{\alpha_2\beta_2}(-\omega_1)k_{\alpha_3\beta_3}(\omega_2)k_{\alpha_4\beta_4}(-\omega_2)Q_{\beta_1\cdots\beta_4}^e(\omega_1,-\omega_2,\omega_2)d\omega_1 d\omega_2$$

which is finite if $Q^e$ is bounded and the $k_{\alpha\beta}$ are square-integrable.  □

Set $z'_\alpha(t) = \sum_{j=1}^{m_2} \sum_\beta G_{\alpha\beta}(j)e_\beta(t-j)$ and set $z''_\alpha(t) = z_\alpha(t) - z'_\alpha(t)$. The next lemma implies that $T_{\alpha_1\alpha_2}^r(n,0)/(N_{\alpha_1}(n)N_{\alpha_2}(n))^{1/4}$ has the same asymptotic distribution as $\sum_{t=1}^{n}\{z'_{\alpha_1}z'_{\alpha_2} - E(z'_{\alpha_1}z'_{\alpha_2})\}/(N_{\alpha_1}(n)N_{\alpha_2}(n))^{1/4}$.

**Lemma 4.5.**    For $\alpha, \beta, 1 \le \alpha, \beta \le q$,

$$\lim_{n\to\infty} Var\{\sum_{t=1}^{n} z_\alpha(t)z_\beta(t+r) - \sum_{t=1}^{n} z'_\alpha(t)z'_\beta(t+r)\}/\{N_\alpha(n)N_\beta(n)\}^{1/2} = 0.$$

**Proof.**    It follows from Assumption 3.2 (i) that

$$\lim_{n\to\infty} Var\{\sum z'_\alpha(t)z''_\beta(t+r)\}/\{N_\alpha(n)N_\beta(n)\}^{1/2} = 0,$$

and the first member on the right-hand side is bounded and the second member tends to 0 in view of that assumption. Similarly it holds that

$$\lim_{n\to\infty} Var\{\sum z''_\alpha(t)z'_\beta(t+r)\}/(N_\alpha(n)N_\beta(n))^{1/2} = 0;$$

a fortiori, $\lim_{n\to\infty} Var\{\sum z''_\alpha(t)z''_\beta(t+r)\}/(N_\alpha(n)N_\beta(n))^{1/2} = 0$.
Thus the result follows. □
Set

$$\xi_{\alpha\beta}^r(k) = \sum_{t\in I_k} z'_\alpha(t)z'_\beta(t+r)/\{N_\alpha(m_1)N_\beta(m_1)\}^{1/4}$$

and    $$\eta_{\alpha\beta}^r(k) = \sum_{t\in J_k} z'_\alpha(t)z'_\beta(t+r)/\{N_\alpha(m_1)N_\beta(m_1)\}^{1/4}.$$

**Lemma 4.6.**    If $\lim_{n\to\infty} n_1^{1/2}m_1/\{m_2^{1+\delta_2}N_\alpha(m_1)N_\beta(m_1)\}^{1/4} = 0$,

then    $$\lim_{n\to\infty} E|\sum_{k=1}^{n_1}[E(\xi_{\alpha\beta}^r(k)|\mathcal{B}_{k-1}) - E(\xi_{\alpha\beta}^r(k))]|/n_1^{1/2} = 0,$$

$$and \qquad \lim_{n \to \infty} E| \sum_{k=1}^{n_1} [E(\eta_{\alpha\beta}^r(k)|\mathcal{B}_{k-1}) - E(\eta_{\alpha\beta}^r(k))]|/n_1^{1/2} = 0.$$

**Proof.**    Assume, without loss of generality, $r \geq 0$. By the Schwartz inequality and by Assumption 3.1 (i),

$$\{N_\alpha(m_1)N_\beta(m_2)\}^{1/2} E|E(\xi_{\alpha\beta}^r(1)|\mathcal{B}_0) - E(\xi_{\alpha\beta}^r(1))|$$

$$\leq c_1 \sum_{t=1}^{m_1} \{\sum_{j=1}^{m_2} (t - j + m_2)^{-(2+2\delta_2)}\}^{1/2} \{\sum_{j=1}^{m_2} (t + r - j + m_2)^{-(2+2\delta_2)}\}^{1/2}$$

$$\leq c_2 m_1 / m_2^{1+2\delta_2}.$$

Hence it follows that

$$E| \sum_{k=1}^{n_1} [E(\xi_{\alpha\beta}^r(k)|\mathcal{B}_{k-1}) - E(\xi_{\alpha\beta}^r(k))]|/n_1^{1/2} \leq c_2 m_1 n_1^{1/2} / \{m_2^{1+2\delta_2} N_{\alpha\beta}(m_1)^{1/2}\}.$$

The second proposition of the lemma holds similarly.    □

**Lemma 4.7.**    *If Assumption 3.1 (i) holds and if*
$m_1{}^2/[m_2^{1+2\delta} \{N_\alpha(m_1)N_\beta(m_1)\}^{1/2}] \to 0$, *then*

$$\lim_{n \to \infty} \sum_{k=1}^{n_1} E\{E(\xi_{\alpha\beta}^r(k)|\mathcal{B}_{k-1}) - E(\xi_{\alpha\beta}^r(k))\}^2/n_1 = 0.$$

**Proof.**    It follows from Assumption 3.1(i), by repeated use of the Schwartz inequality that

$$E[E\{\xi_{\alpha\beta}^r(1)|\mathcal{B}_0\} - E(\xi_{\alpha\beta}^r(1))]^2 \leq \frac{c_1}{N_{\alpha\beta}(m_1)} [\sum_{t=1}^{m_1} \sum_{j=0}^{m_2} (t - j + m_2)^{-2(1+\delta_2)}]^2.$$

$$(1.8)$$

Then the line of argument given in the previous lemma shows that the right-hand side term in (1.8) is less then $c_2 m_1^2/(m_2^{1+2\delta_2} \{N_\alpha(m_1)N_\beta(m_1)\}^{1/2})$, whence the result follows. □

**Lemma 4.8.**    *Under the same condition as in Lemma 4.7,*

$$\lim_{n \to \infty} E[\sum_{k=1}^{n_1} \{\xi_{\alpha_1\alpha_2}^{r_1}(k) - E(\xi_{\alpha_1\alpha_2}^{r_1}(k)|\mathcal{B}_{k-1})\} \cdot \{\xi_{\alpha_3\alpha_4}^{r_2}(k) - E(\xi_{\alpha_3\alpha_4}^{r_2}(k)|\mathcal{B}_{k-1})\}/n_1]$$

$$= \Phi_{\alpha_1\alpha_2\alpha_3\alpha_4}^{r_1 r_2}.$$

**Proof.**    The left-hand side expection is equal to

$$\sum_{k=1}^{n_1} E[\{\xi_{\alpha_1\alpha_2}^{r_1}(k) - E(\xi_{\alpha_1\alpha_2}^{r_1}(k))\}\{\xi_{\alpha_3\alpha_4}^{r_2}(k) - E(\xi_{\alpha_3\alpha_4}^{r_2}(k))\}]/n_1 +$$

$$\sum_{k=1}^{n_1} E[\{E(\xi^{\tau_1}_{\alpha_1\alpha_2}(k)-E(\xi^{\tau_1}_{\alpha_1\alpha_2}(k)|\mathcal{B}_{k-1})\}\cdot\{E(\xi^{\tau_2}_{\alpha_3\alpha_4}(k))-E(\xi^{\tau_2}_{\alpha_3\alpha_4}(k)|\mathcal{B}_{k-1})\}]/n_1,$$

where the first sum tends to $\Phi^{\tau_1\tau_2}_{\alpha_1\alpha_2\alpha_3\alpha_4}$ and the second sum tends to 0 in view of the Schwartz inequality and by Lemma 4.7.   □

**Lemma 4.9.**   *Suppose that Assumptions 3.1 (i) and (ii) hold and suppose also $m_1^2/[m_1^{1+2\delta_2}\{N_\alpha(m_1)N_\beta(m_1)\}^{1/4}] \to 0$; then the expectation of*

$$\frac{1}{n_1}\sum_{k=1}^{n_1}|E[\{\xi^{\tau_1}_{\alpha_1\alpha_2}(k)-E(\xi^{\tau_1}_{\alpha_1\alpha_2}(k)|\mathcal{B}_{k-1})\}\cdot\{\xi^{\tau_2}_{\alpha_3\alpha_4}(k)-E(\xi^{\tau_2}_{\alpha_3\alpha_4}(k)|\mathcal{B}_{k-1})\}|\mathcal{B}_{k-1}]$$

$$-E[\{\xi^{\tau_1}_{\alpha_1\alpha_2}(k) - E(\xi^{\tau_1}_{\alpha_1\alpha_2}(k)|\mathcal{B}_{k-1})\}\{\xi^{\tau_2}_{\alpha_3\alpha_4}(k) - E(\xi^{\tau_2}_{\alpha_3\alpha_4}(k)|\mathcal{B}_{k-1})\}]|$$

*tends to 0 as $n \to \infty$.*

**Proof.**   The expection is dominated by

$$\sum_{k=1}^{n_1} E|E(\xi^{\tau_1}_{\alpha_1\alpha_2}(k)\xi^{\tau_2}_{\alpha_3\alpha_4}(k)|\mathcal{B}_{k-1}) - E(\xi^{\tau_1}_{\alpha_1\alpha_2}(k))\xi^{\tau_2}_{\alpha_3\alpha_4}(k))|/n_1$$

$$+2\sum_{k=1}^{n_1} E[\{E(\xi^{\tau_1}_{\alpha_1\alpha_2}(k)|\mathcal{B}_{k-1}) - E(\xi^{\tau_1}_{\alpha_1\alpha_2}(k))\}$$

$$\cdot\{E(\xi^{\tau_2}_{\alpha_3\alpha_4}(k)|\mathcal{B}_{k-1}) - E(\xi^{\tau_2}_{\alpha_3\alpha_4}(k))\}]/n_1.$$

In view of Assumption 3.1 (ii), the first sum above is dominated by $c_2 m_1^2/\{m_2^{1+2\delta_2}(\Pi_{i=1}^4 N_{\alpha_i}(m_1))^{1/4}\}$. As for the second sum, it is seen to tend to 0 in view of Lemma 4.6 and by the Schwartz inequality. □

Similar lines of arguments give:

**Lemma 4.10.**   *Under the same conditions as in Lemma 4.9,*

$$\lim_{n\to\infty} E\sum_{k=1}^{n_1}|E[\{\xi^{\tau_1}_{\alpha_1\alpha_2}(k) - E(\xi^{\tau_1}_{\alpha_1\alpha_2}(k))\}\{\xi^{\tau_1}_{\alpha_1\alpha_2}(k) - E(\xi^{\tau_2}_{\alpha_3\alpha_4}(k))\}|\mathcal{B}_{k-1}]$$

$$-E[\{\xi^{\tau_1}_{\alpha_1\alpha_2}(k) - E(\xi^{\tau_1}_{\alpha_1\alpha_2}(k))\}\{\xi^{\tau_2}_{\alpha_3\alpha_4}(k) - E(\xi^{\tau_2}_{\alpha_3\alpha_4}(k))\}]|/n_1 = 0.$$

**Lemma 4.11.**   *Under the same conditions as Lemmas 4.6 and 4.7,*

$$\lim_{n\to\infty} E|\sum_{k=1}^{n_1}\{\eta^\tau_{\alpha\beta}(k) - E(\eta^\tau_{\alpha\beta}(k))\}|/n_1^{1/2} = 0.$$

**Proof of Theorem 3.1.**   Let $\xi(k)$ be the vector whose components are $\xi^\tau_{\alpha\beta}(k)$ ordered in the same way as $T(n,0)$. Then Lemmas 4.5 and 4.6 imply that $T^\tau_{\alpha\beta}(n,0)$ has the same limit distribution as $\sum_{k=1}^{n_1}\{\xi(k) - E(\xi(k)|\mathcal{B}_{k-1})\}/n_1^{1/2}$ since $\lim_{n\to\infty} n_1 N_\alpha(m_1)N_\beta(m_1)/\{N_\alpha(n)N_\beta(n)\} = 1$ by Assumption 3.2 (iii). So what is necessary to prove is that for the Martingale difference process $\{\xi(k) - E(\xi(k)|\mathcal{B}_{k-1}), \mathcal{B}_k, k = 1, \cdots, n_1\}$ and for any non-zero vector $a$ of dimension $s(s+1)/2 + (l+1)$, $u_n(k) = a^*\xi(k) - E(a^*\xi(k)|\mathcal{B}_{k-1})$ satisfies the conditions for the Martingale central limit theorem. Lemmas 4.6 and 4.7 imply that the condition of Brown (1971) is satisfied for $u_n(k) = \xi_k - E(\xi_k|\mathcal{B}_{k-1})$, whereas the application of lines of arguments as Hosoya and Taniguchi (1993) shows that the Lindeberg type condition is satisfied for $u_n(k)$. This proves Theorem 3.1.    □

**Proof of Theorem 3.2.**    A joint spectral density matrix of the bivariate process $\{z_\alpha(t), z_\beta(t) - z_\beta(t+r)\}$ is given by

$$
\begin{bmatrix}
f_{\alpha\alpha}(\omega) & (1 - e^{-ir\omega})f_{\alpha\beta}(\omega) \\
(1 - e^{ir\omega})f_{\beta\alpha}(\omega) & |1 - e^{ir\omega}|^2 f_{\beta\beta}(\omega)
\end{bmatrix}.
$$

In view of this spectral density, we have

$$
E(T_{\alpha\beta}^0(n,0) - T_{\alpha\beta}^r(n,0))^2 = n \int_{-\pi}^{\pi} K_n(\omega)h(\omega)d\omega + Q_n,
$$

where

$$
h(\omega) = 2\pi \int_{-\pi}^{\pi} [f_{\alpha\alpha}(\omega - \omega_1)|1 - e^{i\omega_1 r}|^2
$$

$$
\cdot f_{\beta\beta}(\omega_1) + f_{\alpha\beta}(\omega - \omega_1)\overline{f_{\beta\alpha}(\omega_1)}(1 - e^{-ir(\omega - \omega_1)})(1 - e^{-it\omega_1})]d\omega_1,
$$

and

$$
Q_n = 2\pi n \int_{-\pi}^{\pi} \cdots \int_{\pi}^{\pi} K_n(\omega_2 + \omega_3) \cdot \sum_{\beta_1,\cdots,\beta_4=\alpha,\beta} k_{\alpha\beta_1}(\omega_1 + \omega_2 + \omega_3)
$$

$$
\cdot k_{\alpha\beta_2}(-\omega_1)k_{\beta\beta_3}(-\omega_2)k_{\beta\beta_4}(-\omega_3)\tilde{Q}^e(\omega_1 + \omega_2\omega_3, \omega_2, \omega_3)d\omega_1 d\omega_2 d\omega_3.
$$

We have $Q_n = O(n)$. Also note that if $f_{\alpha\alpha}(\omega)$ has an unbounded peak only at the origin, $|f_{\alpha\alpha}(\omega)| = O(\omega^{-\alpha})$ $0 < \alpha < 1$ for all $\omega$ since $f_{\alpha\alpha}(\omega)$ is integrable; on the other hand, $|1 - e^{ir\omega}| = O(|\omega|)$ in a neighborhood of the origin. Under the Condition A,

$$
\int_{-\pi}^{\pi} |f_{\alpha\alpha}(\omega - \omega_1)||1 - e^{i\omega_1 r}|^2|f_{\beta\beta}(\omega)|d\omega_1 < c_1 \int_{-\pi}^{\pi} |f_{\alpha\alpha}(\omega - \omega_1)|d\omega_1 < c_2.
$$

Hence

$$
\int_{-\pi}^{\pi} |f_{\alpha\beta}(\omega - \omega_1)\overline{f_{\beta\alpha}(\omega_1)}(1 - e^{-ir(\omega - \omega_1)})|d\omega < c_3.
$$

Thus since $h(\omega)$ is a.e. bounded, $n \int_{-\pi}^{\pi} K_n(\omega)h(\omega)d\omega = O(n)$. On the other hand, since $\{N_\alpha(n)N_\beta(n)\}^{1/2}/n \to \infty$, it follows that

$$
\lim_{n \to \infty} \{N_\alpha(n)N_\beta(n)\}^{-1/2}E|T_{\alpha\beta}^0(n,0) - T_{\alpha\beta}^r(n,0)|^2 = 0.    □
$$

# 5 References

[1] P. Breuer and P. Major. Central limit theorems for non-linear functionals of Gaussian fields. In *J. Multivariate Anal.,*, **13**, pp.425-441, 1983.

[2] B.M. Brown. Martingale central limit theorems. In *Ann. Math. Statist.*, **42**, pp.59-66, 1971.

[3] R. Dahlhaus. Efficient parameter estimation for self-similar processes. In *Ann. Stat.*, **17**, pp.1749-1766, 1989.

[4] R. Fox and M.S. Taqqu. Central limit theorems for quadratic forms in random variables having long-range dependence. In *Probab. Th. Rel. Fields*, **74**, pp.213-40, 1987.

[5] E.J. Hannan. The asymptotic theory of linear time-series models. In *J. Appl. Prob.*, **10**, pp.130-145, 1973.

[6] E.J. Hannan. The asymptotic distribution of serial covariance. In *Ann. Statist.*, **4**, pp.396-99, 1976.

[7] Y. Hosoya. Limit theorems for statistical inference on stationary processes with strong dependence. In *Statistical Sciences and Data Analysis; Proceeding of the Third Pacific Area Statistical Conference, Eds. K. Matsusita et al.*, VSP BV, pp.151-163, 1993.

[8] Y. Hosoya and M. Taniguchi. A central limit thorem for stationary processes and the parameter estimation of linear processes. In *Ann. Statist.*, **10**, pp.193-53, 1982; *Correction*, in *Ann. Statist.*, **21**, pp.1115-17, 1993.

[9] A. Zygmund. *Trigonometric Series*, **I**. Cambridge University Press, 1959.

Tohoku University

# Estimation of the Number of Spectral Lines

## L. Kavalieris

ABSTRACT  We consider the estimation of a mixed spectrum. The frequencies of the spectral lines may be related through a function of known form. Aside from the trivial case where the frequencies are unrelated, the simplest functional relationship gives harmonics of a fundamental frequency. Procedures based on Rissanen's minimum description length principle are used to determine both the number of spectral lines and the order of a model for the absolutely continuous component of the spectrum. Illustrative examples from astronomy and acoustics are given.

## 1  Introduction

Consider a stationary process with a mixed spectrum containing a discrete component which we think of as the signal spectrum and an absolutely continuous part that will be called the noise spectrum. This distinction is for convenience only and might not reflect the reality of the physical system under study. We address the problem of estimating a parametric model for the spectrum from observed data $y(t)$, $t = 1, \ldots, T$. The number of parameters in the model is not known and will be determined.

The signal may be thought of as a sum of sinusoids with frequencies $\lambda$ that are related in some way. This leads to the model

$$y(t) = \sum_{\lambda \in \Lambda} [A_\lambda \cos(\lambda t) + B_\lambda \sin(\lambda t)] + u(t) \qquad (1.1)$$

where $u(t)$ is referred to as noise. The noise is assumed to be a stationary process with spectral density $f_u(\omega)$. For practical purposes we assume little more than the differentiability of $f_u(\omega)$. More precisely we always assume that $u(t)$ has an AR($\infty$) representation

$$u(t) + \sum_{j=1}^{\infty} \kappa_j u(t - j) = \epsilon(t), \quad \sum j|\kappa_j| < \infty. \qquad (1.2)$$

Let $\kappa(z) = 1 + \sum \kappa_j z^j$ and assume that $\kappa(z) \neq 0$ for $|z| \leq 1$. The innovations sequence $\epsilon(t)$ is a stationary sequence of martingale differences where

$$E(\epsilon_t | \mathcal{F}_{t-1}) = 0, \ E(\epsilon_t^2 | \mathcal{F}_{t-1}) = \sigma^2, \ E(\epsilon_t^4) < \infty. \qquad (1.3)$$

Here $\mathcal{F}_t$ is the $\sigma$-algebra generated by $\epsilon(s)$, $s \le t$. The conditions on the autoregressive coefficients in (1.2) are slightly stronger than in [KH94] where only $\sum |\kappa_j| < \infty$ is required. The present conditions allow a more concise statement of the analytic results.

Functional relationships between frequencies may be described by partitioning the frequencies $\lambda \in \Lambda$ into subsets $\Lambda_j$, $j = 1, \ldots, J$ so that within each subset there is a relationship

$$\Lambda_j = \{h(k, \theta_j), k \in r_j\} \tag{1.4}$$

for a vector of integers $r_j$. We will assume that $h(\cdot, \cdot)$ has been deduced from an understanding of the physical system generating the data and only $r_j$ and $\theta_j$ need to be estimated from data. In the simplest case the frequencies are unrelated and then $\Lambda_j$ are singleton sets. Then $r_j = 1$, and $h(1, x) = x$ is a convenient choice in (1.4). The parameters $\theta_j$ are just the frequencies of the sinusoids in the discrete spectrum. Another possibility allows the frequencies $\lambda \in \Lambda_j$ to be harmonics. Then $\theta_j$ may be taken to be the fundamental frequency and $h(k, \theta_j) = k\theta_j$. We refer to a partition where (1.4) holds as a model for the discrete spectrum. Though the physical insight that yields the function $h(\cdot, \cdot)$ also gives information concerning the partition, we will usually be required to determine, from data, the total number of frequencies, the number of subsets and the number of frequencies within each subset. To complete the specification of the mixed spectrum a class of models for the noise needs to be specified. We model noise as an autoregression and then only the autoregressive order will need to be determined.

The modelling of data containing periodic components has a long history beginning with the work of Schuster almost 100 years ago. When the number of periodic components is known, the estimation of frequency has been surveyed in [Bri87]. Traditionally the problem of determining the number of spectral lines, and the associated question of distinguishing spectral lines from narrow spectral peaks has been approached through hypothesis testing procedures, see [Pri81] and references therein. Results that allow for structural relationships between frequencies are sparse. Only [QT91] proposes a test to determine the number of harmonics present in data. More recently [Qui89], [Han93], [Wan93], [KH94] have investigated information theoretic criteria related to AIC and BIC to determine the number of unrelated frequencies.

In Section 2 we describe the use of information criteria to determine both a model for the discrete spectrum as well as the order of the autoregressive noise model. By taking into account functional relationships between frequencies we extend the information theoretic criteria approach of [KH94]. These methods complement the hypothesis testing approach in much the same way as AIC and BIC are used to estimate the complexity of a time series model. Once such structural parameters have been determined the remaining parameters in the model are estimated using more or less standard procedures that are discussed in Section 3. Several applications are developed in Section 4.

## 2  Model Selection

Let $\Lambda$ be a set of frequencies. It will be convenient to think of a model $(M, \Lambda)$ as a partition of $\Lambda$ into subsets $\Lambda_j, j = 1, \ldots, J$ for which (1.4) holds. Estimates of frequencies depend on the choice of the model for $\Lambda$. Define

$$\hat{\Lambda}_j = \{h(k, \hat{\theta}_j), k \in r_j\}. \tag{1.5}$$

where $\hat{\theta}_j$ maximizes an approximation of the likelihood in a neighbourhood of $\theta_j$. Let $\hat{\lambda}_M$ denote the estimate of $\lambda$ defined in (1.5) and $\hat{\Lambda}_M = \cup \hat{\Lambda}_j$. Procedures for the estimation of $\theta_j$ will be discussed in Section 3.

Once $\hat{\lambda}_M \in \hat{\Lambda}_M$ have been estimated, the amplitudes $\hat{A}_{\lambda, M}$, $\hat{B}_{\lambda, M}$ are obtained by regressing $y(t) - \bar{y}$ on $\cos(\hat{\lambda}_M t)$, $\sin(\hat{\lambda}_M t)$ for $\hat{\lambda}_M \in \hat{\Lambda}_M$. Let $\hat{u}_{M, \Lambda}(t)$ be the residuals from this regression. The periodogram $\hat{u}_{M, \Lambda}(t), t = 1, \ldots, T$ is denoted by $I_{M, \Lambda}(\omega)$.

A useful model contains only unrelated frequencies. It is denoted by $(U, \Lambda)$ and partitions $\Lambda$ into singleton sets. For brevity $\hat{\lambda}$, $\hat{u}_\Lambda(t)$, $I_\Lambda(\omega)$ are used to denote $\hat{\lambda}_U$, $\hat{u}_{U, \Lambda}(t)$, $I_{U, \Lambda}(\omega)$ respectively.

### 2.1  Procedures Based on MDL

The minimum description length (MDL) principle [Ris89] seeks a parsimonious model for the observed data by minimizing the length of an optimal coding of the data. Here a 'two part' coding scheme is used. The first part is code for the data given a model while the second part is the code for the model. A model with greater complexity (more parameters) leads to more compact code for the data, but such a saving is offset by an increase in the code length needed to describe the model. The model will contain parameters determining the frequencies and amplitudes of the spectral lines. In addition a parametric (autoregression) or non-parametric description of the absolutely continuous component of the spectrum must be included in the model.

As a working principle the (asymptotic) code length for each parameter is $-2 \times$ log of its standard deviation. The code length for the data, given the model, may be shown to be approximately $T \log \hat{\sigma}^2$ where $\hat{\sigma}^2$ is the variance of the one-step ahead predictor from this model.

Spectral lines can be coded through their frequencies and amplitudes. The maximum likelihood estimate of a frequency has standard deviation $O(T^{-3/2})$ while an amplitude has a standard deviation $O(T^{-1/2})$. Thus when there are $r$ spectral lines at unrelated frequencies the cost of encoding the frequencies and amplitudes is $5r \log T$. If the frequencies are harmonics of a fundamental frequency only one frequency and all amplitudes need to be coded and then the cost is $(3 + 2r) \log T$. When they are related through some parametric function $h(k, \theta_j)$ in (1.4) the parameters in $\theta_j$ and all amplitudes are coded. In general we denote the cost of coding the frequencies in $\Lambda$ through the model $M$ by $n(M, \Lambda) \log T$.

Several alternative models can be proposed for the absolutely continuous part of the spectrum. Following [HR88], [Han93] a non-parametric model approximates $f_u(\omega), \omega \in [0, \pi]$ as a function that is constant in $m$ frequency bands of equal width. In each band $f_u(\omega)$ is estimated as the mean of the periodogram ordinates within the band. This approach is used in [HR88] to derive a criterion for the determination of bandwidth in spectral estimation. We refer to [Han93], [KH94] for its use in the determination of the number of unrelated spectral lines.

An alternative is to model the noise spectrum by an autoregression, [KH94]. As the autoregressive parameters are estimated with standard deviation $O(T^{-1/2})$, we are led to the criterion

$$\phi(M, \Lambda, h) = \log \hat{\sigma}^2_{M,\Lambda}(h) + [n(M, \Lambda) + h] \frac{\log T}{T}. \qquad (1.6)$$

Here $\hat{\sigma}^2_{M,\Lambda}(h)$ is the variance of the one-step predictor from an AR($h$) model for the residuals $\hat{u}_{M,\Lambda}(t)$.

Although the following scheme can be shown to be consistent it should not be thought of as an automatic procedure.

(i) Set $r = 1$. Let $\hat{\lambda}_1$ be the frequency that maximizes the periodogram of $y(t)$. Define

$$\phi_0 = \min_h \{\log \hat{\sigma}^2_y(h) + h \log T / T\}$$

where $\hat{\sigma}^2_y(h)$ is the prediction variance from an AR($h$) model for $y(t)$.

(ii) Put $\Lambda = \{\hat{\lambda}_1, \ldots, \hat{\lambda}_r\}$. For each model $(M, \Lambda)$ estimate $\hat{\lambda}_{M,k}$, $k = 1, \ldots, r$ that satisfies (1.5). Consider a sequence of AR(h) models for $\hat{u}_{M,\Lambda}(t)$, $h = 0, 1, \ldots$.

(iii) Select $\hat{M}$ and $\hat{h}_\Lambda$ by minimizing $\phi(M, \Lambda, h)$. Denote $\phi(\hat{M}, \Lambda, \hat{h}_\Lambda)$ by $\phi_r$. If $\phi_r > \phi_{r-1}$, stop.

(iv) If $\phi_r < \phi_{r-1}$, let $\hat{\lambda}_{r+1}$ be the frequency that maximizes the periodogram of $\hat{u}_\Lambda(t)$. Set $r = r + 1$ and return to step (ii).

The optimal model is the pair $(\hat{M}, \Lambda)$ that minimizes $\phi_r$. The residuals derived from the optimal model are estimates of $u(t)$ and are denoted by $\hat{u}(t)$. Let $\hat{f}_u(\omega)$ denote the AR($\hat{h}_\Lambda$) estimate for the spectrum of $\hat{u}(t)$.

We make several observations that are relevent to the implementation of this procedure.

- In Step (iv) we are usually interested in estimating frequencies of spectral peaks that are large compared to the noise spectrum in their neighbourhood. Asymptotically there is no difficulty as all spectral lines have magnitude $O(T)$ which far exceeds the contribution of the noise to the periodogram. In applications it was found convenient to prewhiten the noise process by filtering $u_\Lambda(t)$ by some crude AR model for the noise process and maximizing the periodogram of the prewhitened series.

- In principle all models $(M, \Lambda)$ should be considered for each $r$, but an understanding of the physical system will reduce the number of such models to a reasonable level. Indeed physical intuition might suggest an ordering of the estimates $\hat{\lambda}_k$ that is different from that determined in Step (iv).

- Methods for estimation of $\hat{\lambda}_{M,k}$ are deliberately left vague at this stage. The analysis in the next section shows that any estimate of frequency that converges at the rate $O(T^{-3/2}(\log T)^{1/2})$ is adequate so that the best possible estimates are not needed from the purely analytic point of view.

## 2.2  Analysis

The conditions on the noise process given in Section 1 are maintained in this section. If, in Step (ii), we ignore functional relationships between frequencies, we revert to the case considered in [KH94]. There it is proved that (1.6) yields a strongly consistent estimate of the number of spectral lines.

Denote the set of frequencies in the true discrete spectrum by $\Lambda_0$. For any frequencies $\Lambda$, let $\Lambda_0 \setminus \Lambda$ contain those frequencies in $\Lambda_0$ but not in $\Lambda$. Recall the definition of $\hat{u}_\Lambda(t)$ at the beginning of Section 2 and define

$$u_\Lambda(t) = \sum_{\lambda \in \Lambda_0 \setminus \Lambda} [A_\lambda \cos(\lambda t) + B_\lambda \sin(\lambda t)] + u(t)$$

[KH94] prove the following lemma for autoregressive modelling of series containing sinusoidal components. A model $M$ for a set of frequencies $\Lambda$ is not relevant as the quantities discussed involve only the true frequencies. Let $\tilde{\sigma}_\Lambda^2(h)$, $\hat{\sigma}^2(h)$ denote variances of one step ahead predictors for AR($h$) models for $u_\Lambda(t)$, $u(t)$, $t = 1, \ldots, T$ respectively. The squared amplitude of a sinusoidal term of frequency $\lambda$ in (1.1) is $\rho_\lambda^2 = A_\lambda^2 + B_\lambda^2$.

**Lemma 1** *Let $H_T = O(T^a)$ for some $a < 1/4$. Uniformly for $h \leq H_T$*

$$\tilde{\sigma}_\Lambda^2(h) = \hat{\sigma}^2(h) + \frac{1}{h} \sum_{\lambda \in \Lambda_0 \setminus \Lambda} \frac{\rho_\lambda^2}{2} + o(h^{-1}), \quad a.s.$$

In the proof of Lemma 1 in [KH94] we derived the expression

$$\tilde{\sigma}_\Lambda^2(h) = \hat{\sigma}^2(h) + R(h) + O(h(\log T/T)^{1/2}), \quad a.s.$$

Here

$$R(h) = \frac{1}{h} \sum_{\lambda \in \Lambda_0 \setminus \Lambda} \frac{\rho_\lambda^2}{2} + o(h^{-1})$$

is a deterministic function. For a fixed set $\Lambda$, AR order is selected by minimizing

$$\phi(h) = \log \tilde{\sigma}_\Lambda^2(h) + h \log T/T$$

for $0 \leq h \leq H_T$, $H_T = O(T^a)$. This is just the relevent part of $\phi(M, \Lambda, h)$ used in Step (iii). If we take $a < 1/6$ it is evident that $\phi(h)$ is a decreasing function of $h$. As a result whenever $\Lambda_0 \setminus \Lambda \neq \emptyset$ and AR($H_T$) model will be selected for $u_\Lambda(t)$. Alternatively, when $\Lambda_0 \setminus \Lambda = \emptyset$, we revert to estimating the order of an AR model for $u(t)$. If the spectrum of $u(t)$ is sufficiently smooth we expect a low order model, indeed if $u(t)$ is ARMA then estimated AR order can be shown to be $O(\log T)$, a.s. This suggests that the order of the AR model for the residuals $\hat{u}_{M,\Lambda}(t)$ will exhibit a marked decrease when all sinusoidal terms have been located.

However additional complications arise when frequencies have been estimated and removed by regression. To overcome such difficulties it is sufficient to show that for any model $(M, \Lambda)$ the autocovariance functions of $\hat{u}_{M,\Lambda}(t)$ and $u_\Lambda(t)$ differ by a term that is $o(\log T/T)$. Then it is easily shown that

$$\hat{\sigma}_{M,\Lambda}^2(h) - \tilde{\sigma}_\Lambda^2(h) = o(h \log T/T)$$

and thus Lemma 1 holds for the quantities $\hat{\sigma}_{M,\Lambda}^2(h)$ that are actually calculated from data.

For a model $(M, \Lambda)$ let $\hat{\gamma}_{M,\Lambda}(h) = T^{-1} \sum \hat{u}_{M,\Lambda}(t) \hat{u}_{M,\Lambda}(t+h)$ and $\gamma_\Lambda(h) = T^{-1} \sum u_\Lambda(t) u_\Lambda(t+h)$. We want to show that, uniformly in $h < H_T$,

$$\hat{\gamma}_{M,\Lambda}(h) - \gamma_\Lambda(h) = o(\log T/T), \quad a.s. \tag{1.7}$$

To simplify notation omit the dependence on the model $M$. The LHS of (1.7) is

$$T^{-1} \sum [\hat{u}_\Lambda(t) - u_\Lambda(t)] u_\Lambda(t+h) + \hat{u}_\Lambda(t)[\hat{u}_\Lambda(t+h) - u_\Lambda(t+h)]$$

Here

$$\hat{u}_\Lambda(t) - u_\Lambda(t) = \sum_{\lambda \in \Lambda} [\hat{A}_\lambda \sin(\hat{\lambda}t) - A_\lambda \sin(\lambda t)] + [\hat{A}_\lambda \cos(\hat{\lambda}t) - A_\lambda \cos(\lambda t)]$$

which is orthogonal (up to terms that are $O(T^{-1})$) to the sinusoidal terms in $u_\Lambda(t + h)$, the latter having frequencies $\lambda \in \Lambda_0 \setminus \Lambda$. Thus

$$\sum [\hat{u}_\Lambda(t) - u_\Lambda(t)] u_\Lambda(t + h) = \sum [\hat{u}_\Lambda(t) - u_\Lambda(t)] u(t + h) + O(1), \text{ a.s.}$$

Consequently we have to bound expressions such as

$$\sum A_\lambda [\cos(\hat{\lambda} t) - \cos(\lambda t)] u(t + h) \tag{1.8}$$

$$\sum [\hat{A}_\lambda - A_\lambda] \cos(\hat{\lambda} t) u(t + h) \tag{1.9}$$

The first of these may be approximated by

$$A_\lambda (\hat{\lambda} - \lambda) \sum t [\cos(\lambda t) + \sin(\lambda t)] u(t + h)].$$

It can be shown [HM86] that

$$\sum t^k \cos(\omega t) u(t) = O(T^{k+1/2} (\log \log T)^{1/2}), \text{ a.s.}$$

Therefore in order to obtain a suitable bound for (1.8), we require that $\hat{\lambda} - \lambda = o(T^{-3/2} g(T))$ where $g(T)(\log \log T)^{1/2} \le \log T$. Thus the only reasonable estimate is one that converges at the rate in the law of the iterated logarithm,

$$\hat{\lambda} - \lambda = O(T^{-3/2} (\log \log T)^{1/2}), \text{ a.s.} \tag{1.10}$$

Then we can also show that $\hat{A}_\lambda - A_\lambda = O(T^{-1/2}(\log \log T)^{1/2})$ and thus (1.9) is also bounded as required. All other terms in (1.7) are bounded in a similar way to complete the proof.

For each $h$ the removal of a spectral line leads to a reduction in (1.6) as long as the frequencies are estimated to the precision required in (1.10). Note that the estimates $\hat{\lambda}_\Lambda$ obtained in Step (iii) satisfy (1.10) whether functional relationships exist between frequencies or not. Therefore (1.6) will consistently estimate $\Lambda_0$. The distinction between different models for $\Lambda_0$ is purely on the basis of parsimony.

The autocovariance functions of $\hat{u}(t)$ and $u(t)$ also differ by a term that is, uniformly $o(\log T/T)$. Thus the extensive theory for autoregressive order estimation can be used to establish consistency of the autoregressive spectral estimate. For example Theorem 2.3 of [HK86] shows that

$$\sup_\omega |\hat{f}_u(\omega) - f_u(\omega)| = o(1), \text{ a.s.}$$

If it assumed that the autocovariance function of $u(t)$ decreases at a geometric rate (without necessarily assuming that $u(t)$ is an ARMA process), the rate of convergence can be significantly improved.

# 3   Estimation methods

Frequencies are estimated using approximate maximum likelihood methods. Suppose frequencies $\lambda_{kj} = h(k, \theta_j)$, $k \in r_j$ are present in the signal. Following [QT91] maximizing the likelihood leads to minimizing, over the parameter vector $\theta$,

$$Q(\theta) = \sum_{k \in r_j} \frac{\left| \sum_{t=1}^{T} y_t e^{it\lambda_{kj}} \right|^2}{f_u(\lambda_{kj})}. \tag{1.11}$$

where $\lambda_k = h(k, \theta)$. Here $f_u(\lambda)$ is the noise spectrum which must be estimated, but even a crude estimate of the spectrum results in very little loss of efficiency. For succesful implementation, the parameter vector $\theta$ must be constrained to lie in a neighbourhood of the true $\theta_j$ which is unknown. However initial frequency estimates $\hat{\lambda}_{kj}$ are available from Step (iv) of our identification scheme. We can use these estimates in the absence of information about $\theta_j$. Maximize $Q(\theta)$ so that $h(k, \theta)$, $k \in r_j$ remains in a neighbourhood of a suitably chosen subset of initial frequency estimates $\hat{\lambda}_{kj}$.

A simple alternative is to fit a model $\lambda_{kj} = h(k, \theta)$ to a subset of the initial estimates $\lambda_{kj}$ using non-linear least squares. As the latter are (asymptotically) independent with variance $O(T^{-3})$ standard regression theory indicates that the estimate of $\theta_j$ will be unbiased with variance that is also $O(T^{-3})$.

Once periodic components are removed by regression the residuals are modelled as a low order AR to obtain an improved estimate of $f_u(\lambda)$ which could be used in (1.11). The frequency estimation and model selection procedures could be repeated.

We have used Burg's method [Bur75] to estimate autoregressions at every step of the model selection procedure. Alternatively ordinary least squares regression could be used to obtain very similar results. These techniques are used in preference to the Levinson recursion to avoid bias inherent in procedures based on a Toeplitz autocovariance matrix.

# 4   Examples

## 4.1   Brightness of a variable star

The data are a series of ten day means from daily observations of the visual brightness of *S. Carinae*, a variable star in the Southern Hemisphere sky. For this illustration a segment of 512 observations was used. The periodogram was calculated after centering and tapering the data with a 12.5% split cosine taper. Inspection of the periodogram (Figure 1) shows

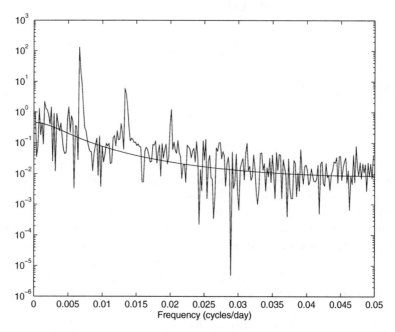

FIGURE 1. Periodogram of 512 observations of the brightness of $S$ $Carinae$. The AR(1) estimate of the noise spectrum is superimposed.

three obvious peaks at frequencies that appear to be harmonics. In [QT91] a hypothesis test was developed and showed that these frequencies may be taken to be harmonics.

Let $\Lambda_r$ denote the 'true' frequencies $\lambda_1, \ldots, \lambda_r$ ordered according to amplitudes. The model $H$ contains only harmonics and $(\hat{H}, \Lambda_r) = \{k\hat{\lambda}, k = 1, \ldots, r\}$. For this model $n(H, \Lambda_r) = 3 + 2r$. Similarly $(\hat{U}, \Lambda_r)$ assumes that the frequencies are unrelated and then they are estimated as $\hat{\lambda}_k$, $k = 1, \ldots, r$ from Step (iv). For these models $n(U, \Lambda_r) = 5r$. The 'Null' model is an AR model for the data $y(t)$.

Table 1.1 compares models $U$ containing only unrelated frequencies with those containing only harmonics. A model containing three harmonics was selected by (1.6) which agrees with the conclusion of [QT91]. This conclusion is hardly suprising as the frequency estimates under the 'unrelated' models $U$ are exceedingly close to those estimated under the harmonic assumption. The noise was modelled as an AR(1) process

$$u(t) - 0.7584u(t - 1) = \epsilon(t), \qquad E(\epsilon^2(t)) = 0.0650$$

The AR(1) spectral estimate for the noise is superimposed on the periodogram of the data in Figure 1. Standard diagnostic checks on the residuals $\hat{u}(t)$ did not indicate any deficiency in this model.

TABLE 1.1. MDL criterion, frequency estimates and AR orders for models $(U, \Lambda_r)$ and $(H, \Lambda_r)$.

| Model | Frequency estimates | | | | $\min \phi$ | $\hat{h}$ |
|---|---|---|---|---|---|---|
| | $\hat{\lambda}_1$ | $\hat{\lambda}_2$ | $\hat{\lambda}_3$ | $\hat{\lambda}_4$ | | |
| Null | | | | | -2.2844 | 17 |
| $(U, \Lambda_1) = (H, \Lambda_1)$ | 0.4204 | | | | -2.4509 | 17 |
| $(U, \Lambda_2)$ | 0.4204 | 0.8409 | | | -2.5244 | 4 |
| $(U, \Lambda_3)$ | 0.4204 | 0.8409 | 1.2614 | | -2.5691 | 1 |
| $(H, \Lambda_2)$ | 0.4204 | 0.8408 | | | -2.5611 | 4 |
| $(H, \Lambda_3)$ | 0.4204 | 0.8408 | 1.2612 | | -2.6423 | 1 |
| $(H, \Lambda_4)$ | 0.4204 | 0.8408 | 1.2612 | 1.6816 | -2.6292 | 1 |

## 4.2   Middle C on a Piano

The sound produced by a piano when middle C (260 Hz) was struck has been sampled at a rate of 7400 Hz. As the data shows large variations in amplitude we take a short series of 512 observations about 1.5 seconds after the note was struck. Visual inspection indicates that the mean and variance are more or less constant in this series. The data is complicated by the fact that the higher notes of a piano are produced by three solid strings, that are tuned to slightly different frequencies, thus the peaks in the spectrum will consist of at least three closely spaced spectral lines. In such a short data set these triads can not be resolved which results in a simplified analysis. Figure 2 shows the periodogram of this data set (the data has been tapered by a 12.5 % split cosine). Periodogram peaks appear at frequencies that are nearly harmonics of 260 Hz. However it has long been recognised they are not exact harmonics and indeed the approximation is poor for the bass strings. Instead the $k^{th}$ partial is given by

$$\lambda_k = k\lambda[(1 + Bk^2)/(1 + B)]^{1/2} \tag{1.12}$$

where $B$ is a constant that can be calculated from the physical properties of the piano wire [Fle64]. For piano strings $B \sim 10^{-4}$. In model (1.4), the function $h(\cdot, \cdot)$ is given by (1.12) where the parameter vector $\theta = \{\lambda, B\}$. Both $\theta$ and the number of partials will be estimated from data.

When many lines are present in the spectrum it is no longer a simple matter to design an automatic procedure for estimating them. As mentioned in Section 2 a useful first step is to find a crude estimate of the noise spectrum after removing the obvious spectral lines by regression. In this example we found the frequencies of peaks of the periodogram in the neighbourhoods of 12 harmonics of the 260 Hz line. Upon removing these lines by regression, the residuals $\hat{u}(t)$ were modelled as an AR(3) process $\tilde{\kappa}(z)$. AR order was selected so that $|\tilde{\kappa}(e^{i\omega})|^{-2}$ captures only the gross features of the periodogram of the residuals. A second or third order model would

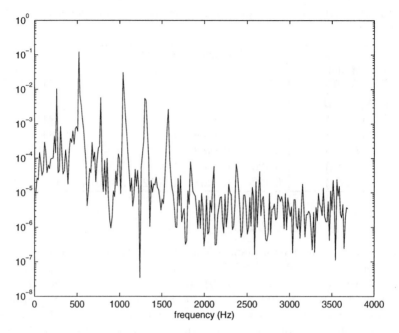

FIGURE 2. Periodogram of 512 observations of a piano note

adequately capture a peak near the origin. Filter $y(t)$ by $\tilde{\kappa}(z)$ to obtain $\tilde{y}(t)$ and proceed to find the frequencies that maximize the periodogram of $\tilde{y}(t)$ as in Step (iv) in Section 2. These frequencies are denoted by $\hat{\lambda}_k$ and correspond to a model that assumes all frequencies are unrelated. Let $\Lambda_r = \{\lambda_k, k = 1, \ldots, r\}$ that are the 'true' frequencies corresponding to the $\hat{\lambda}_k$. In Table 1.2 the models $(\hat{U}, \Lambda_r)$ contain the unrelated frequencies $\hat{\lambda}_k$, $k = 1, \ldots, r$. On the basis of (1.6) seven unrelated frequencies are included in the model and the noise is modelled as an AR(5). Standard diagnostics do not indicate a serious deficiency in this model, however, perusal of the spectrum indicates the presence of peaks at 1841.4 Hz, 2106.6Hz and 2643.9 Hz that may be related to the fundamental frequency of 260.3 Hz through (1.12).

Next we fitted models containing frequencies related by (1.12). Denote such a model by $(\hat{P}, \Lambda_r)$. Parameters $\lambda$ and $B$ in (1.12) were estimated by maximizing (1.11). When $r = 10$ this gave $\hat{\lambda} = 260.4$Hz, $\hat{B} = 3.4560 \times 10^{-4}$. Since two parameters were estimated in model (1.12, $n(P, \Lambda_r) = 6 + 2r$ in (1.6). The optimal model contained the 10 frequencies denote by $\tilde{\lambda}_k$ in Table 1.2. No other frequencies yielded significant peaks. A third order noise model was selected

$$u(t) - 0.8088u(t-1) - 0.0238u(t-2) + 0.2273u(t-3) = \epsilon(t),$$

TABLE 1.2. Frequency estimates, AR order and MDL for various models for a piano note

| $k$ | $\hat{\lambda}_k$ | $\hat{h}$ | $\phi(U, \Lambda_k, \hat{h})$ | $\tilde{\lambda}_k$ | $\phi(P, \Lambda_{10}, \hat{h})$ |
|-----|------|-----|---------|--------|----------|
| 0 | | 29 | -9.8427 | | |
| 1 | 521.4 | 29 | -9.9163 | 521.0 | |
| 2 | 1044.3 | 29 | -10.1315 | 1044.3 | |
| 3 | 1307.9 | 12 | -10.2727 | 1307.3 | |
| 4 | 1571.2 | 13 | -10.3424 | 1571.8 | |
| 5 | 260.2 | 8 | -10.3845 | 260.4 | |
| 6 | 779.3 | 5 | -10.5611 | 782.2 | |
| 7 | 2376.9 | 5 | -10.5634 | 2375.7 | |
| 8 | 2106.6 | 1 | -10.5365 | 2105.7 | |
| 9 | 1841.4 | 3 | -10.5091 | 1837.8 | |
| 10 | 2643.9 | 3 | -10.4751 | 2648.1 | -10.6684 |

in which $E(\epsilon^2(t)) = 1.6346 \times 10^{-5}$. The estimated noise spectrum is superimposed on the periodogram of residuals from the model $(P, \Lambda_{10})$ in Figure 3.

Standard diagnostic checks on the residuals indicate no shortcomings in the model.

# 5  References

[Bri87]   D.R. Brillinger. Fitting cosines: Some procedures and some physical examples. In I.B. MacNeill and G.J. Umphrey, editors, *Applied Probability, Stochastic Processes and Sampling theory*, pages 75–100. Reidel, 1987.

[Bur75]   J.P. Burg. *Maximum entropy spectral analysis*. PhD thesis, Stanford University, 1975.

[Fle64]   H. Fletcher. Normal vibrational modes of a stiff piano string. *Journal of the Accoustical Society of America*, 36:203–209, 1964.

[Han93]   E.J. Hannan. Determining the number of jumps in a spectrum. In T. Subba Rao, editor, *Developments in Time Series Analysis*, pages 127–138. Chapman and Hall, London, 1993.

[HK86]   E.J. Hannan and L. Kavalieris. Regression, autoregression models. *Journal of Time Series Analysis*, 7:27–49, 1986.

[HM86]   E.J. Hannan and M.S. Mackisack. A law of the iterated logarithm for an estimate of frequency. *Stochastic Processes and their Applications*, 22:103–109, 1986.

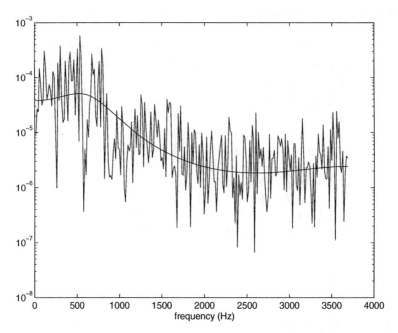

FIGURE 3. Periodogram of the residuals from a model containing 10 related frequencies. The AR(3) estimate of the noise spectrum is superimposed.

[HR88]   E.J. Hannan and J. Rissanen. The width of a spectral window. In J.M. Gani, editor, *A Celebration of Applied Statistics*, pages 301–307. Applied Probability Trust, Sheffield, 1988.

[KH94]   L. Kavalieris and E.J. Hannan. Determining the number of terms in a trigonometric regression. *Journal of Time Series Analysis*, 15:613–626, 1994.

[Pri81]   M.B. Priestley. *Spectral Analysis and Time Series*. Academic Press, London, 1981.

[QT91]   B.G. Quinn and P.J. Thomson. Estimating the frequency of a periodic function. *Biometrika*, 78:65–74, 1991.

[Qui89]   B.G. Quinn. Estimating the number of terms in a sinusoidal regression. *Journal of Time Series Analysis*, 10:71–76, 1989.

[Ris89]   J. Rissanen. *Stochastic complexity in Scientific Inquiry*. World Scientific, Singapore, 1989.

[Wan93]  Xiaobao Wang. An AIC type estimator for the number of cosinusoids. *Journal of Time Series Analysis*, 14:431–440, 1993.

University of Otago, New Zealand

# Self-normalized and randomly centered spectral estimates

C. Klüppelberg[*]
T. Mikosch[†]

ABSTRACT We review some limit theory for the periodogram and for integrated versions of it and explain the use of random normalizing and centering techniques.

## 1 Introduction

Self-normalization is a common technique in time series analysis: let $(X_t)_{t \in \mathbb{Z}}$ be a centered stationary process with finite variance. For instance, natural estimators of the autocorrelations

$$\rho_X(h) = \frac{\gamma_X(h)}{\gamma_X(0)} = \frac{E(X_0 X_h)}{var\,(X_0)} , \quad h \geq 1 ,$$

are the sample autocorrelations

$$\rho_{n,X}(h) = \frac{\frac{1}{n}\sum_{t=1}^{n-h} X_t X_{t+h}}{\frac{1}{n}\sum_{t=1}^{n} X_t^2} = \frac{\sum_{t=1}^{n-h} X_t X_{t+h}}{\sum_{t=1}^{n} X_t^2} , \quad h \geq 1 ,$$

which are nothing but self–normalized (studentized) versions of the sample autocovariances $\widehat{\gamma}_{n,X}(h) = \frac{1}{n}\sum_{t=1}^{n-h} X_t X_{t+h}$. If $EX_t$ is unknown we will not hesitate to replace the $X_t$ by the randomly centered versions $X_t - \overline{X}_n$, where $\overline{X}_n$ denotes the sample mean of the first $n$ observations. Self-normalization has also been proposed in the frequency domain: Bartlett (1954, 1978) uses a sum approximation to the self-normalized integrated periodogram

$$\int_0^\lambda \frac{\widetilde{I}_{n,X}(x)}{f_X(x)} dx , \quad \lambda \in [0, \pi] ,$$

for constructing goodness-of-fit tests of the spectral distribution function. Here

$$I_{n,X}(\lambda) = \frac{1}{n}\left|\sum_{t=1}^{n} X_t e^{-i\lambda t}\right|^2$$

---

[*]Department of Mathematics, University of Mainz, D-55099 Mainz
[†]Department of Mathematics, University of Groningen, P.O. Box 800, NL-9700 AV Groningen

denotes the raw periodogram and

$$\tilde{I}_{n,X}(\lambda) = \frac{I_{n,X}(\lambda)}{\hat{\gamma}_{n,X}(0)} = \frac{\left|\sum_{t=1}^{n} X_t e^{-i\lambda t}\right|^2}{\sum_{t=1}^{n} X_t^2}, \quad \lambda \in [-\pi, \pi], \qquad (1.1)$$

is its self-normalized (studentized) version. Moreover, $f_X$ is the spectral density of the stationary process $(X_t)$. Anderson's paper (1993) is in the same spirit; but he studies the process

$$\int_0^{\lambda} \tilde{I}_{n,X}(x)\,dx, \quad \lambda \in [0, \pi]. \qquad (1.2)$$

Priestley (1981) suggests the self-normalized periodogram $\tilde{I}_{n,X}$ in order to remove the dependence of $I_{n,X}$ on the variance of the noise. It is clear that it is a proportionality factor in the distribution of $I_{n,X}(\lambda)$ and $\hat{\gamma}_{n,X}(0)$.

There are several reasons to use self-normalization and random centering in time series analysis. *The first* is the "plug-in-idea": unknown parameters, centering or normalizing constants are replaced by appropriate estimators. *The second reason* is the general belief that these techniques "robustify" estimators. To give an example: if the variance of $X_0$ is infinite, $\hat{\gamma}_{n,X}(h)$ will behave very irregularly. However, as we will see soon, in fairly general situations the sample autocorrelation $\rho_{n,X}(h)$ is a consistent estimator of the quantity which, in the finite variance case, defines the autocorrelation $\rho_X(h)$ of the process. *The third reason* is that random centering is sometimes necessary for deriving the limit distribution of an estimator. This is, for instance, the case for the integrated periodogram if $var\,(X_0) = \infty$ (see Section 4 below). *The fourth reason*, which is supported by the results formulated below, is that random centering sometimes leads to a "more beautiful" limit distribution or limit process. For example, the limit process of (1.2) is, deterministically centered, a Brownian bridge perturbed by an independent Gaussian term. Using a random centering sequence, the limit process of (1.2) becomes a Brownian bridge. Thus the limit theory for the integrated periodogram parallels very much the empirical process theory for iid observations (see e.g. Shorack and Wellner (1986)). It is our intention to make these parallels more transparent.

## 2 Notation and assumptions

Throughout we consider the linear process

$$X_t = \sum_{j=-\infty}^{\infty} \psi_j Z_{t-j}, \quad t \in \mathbb{Z},$$

under different conditions on the coefficients $(\psi_j)$ and on the iid noise sequence $(Z_t)$. We contrast the limit theory for the periodogram in two situations:

## A: The classical $L^2$-case

$(Z_t)$ is iid white noise with $\sigma^2 = var\ (Z_0) < \infty$ and $\sum_{j=-\infty}^{\infty} |\psi_j| < \infty$.

## B: The infinite variance case

$(Z_t)$ is iid symmetric $\alpha$-stable $(s\alpha s)$ noise for some $\alpha < 2$ and $\sum_{j=-\infty}^{\infty} |j| |\psi_j|^\delta$ $< \infty$ for some $\delta \in (0, \min(1, \alpha))$.

Recall that $Z_0$ is $s\alpha s$ if it has chf $E \exp\{itZ_0\} = \exp\{-c|t|^\alpha\}$ for some $c > 0$, $t \in \mathbb{R}$. A particular consequence is that $P(Z_0 > x) \sim dx^{-\alpha}$ as $x \to \infty$. Hence $var\ (Z_0) = \infty$. The assumption of $\alpha$-stability is for convenience. If one is only interested in weak consistency of the estimators introduced below, then the corresponding results are valid under $E|Z_0|^p < \infty$ for some $p > 0$. In order to derive the asymptotic distribution of the estimators we need that $Z_0$ belongs to the domain of attraction of an $\alpha$-stable distribution which requires unpleasant normalizing sequences depending on the tail of the distribution. To avoid this we restrict ourselves to the $s\alpha s$ case. For the definition of $s\alpha s$ distributions and processes we refer to Samorodnitsky and Taqqu (1994) and for domains of attraction of $\alpha$-stable laws see Feller (1971).

Throughout we write

$$a_n = \left\{ \begin{array}{ll} n^{1/2} & \text{under A}, \\ n^{1/\alpha} & \text{under B}. \end{array} \right.$$

We introduce the properly normalized sample autocovariances and auto-correlations $(h \geq 1)$

$$\gamma_{n,X}(0) = a_n^{-2} \sum_{t=1}^{n} X_t^2\ ,\ \gamma_{n,X}(h) = a_n^{-1} \sum_{t=1}^{n-h} X_t X_{t+h}\ ,\ \rho_{n,X}(h) = \frac{\gamma_{n,X}(h)}{a_n \gamma_{n,X}(0)}\ ,$$

as well as the raw periodogram

$$I_{n,X}(\lambda) = \left| a_n^{-1} \sum_{t=1}^{n} X_t e^{-i\lambda t} \right|^2\ ,\ \lambda \in [-\pi, \pi]\ ,$$

and its self-normalized version (1.1). The normalization $(a_n)$ is motivated by the CLT in the domain of normal attraction of a stable law (see e.g. Feller (1971)). We use the same symbols $\gamma_{n,X}$ and $I_{n,X}$ both under conditions A and B without indicating which $(a_n)$ is used. This will become clear from the context. We also write

$$\psi^2 = \sum_{j=-\infty}^{\infty} \psi_j^2\ .$$

Notice that, under A, $var\ (X_0) = \sigma^2 \psi^2$.

# 3 Some limit theory in the time domain

For completeness, we start with some asymptotic theory for $\gamma_{n,X}(h)$ and $\rho_{n,X}(h)$ which is also the backbone for the proofs of many results in the frequency domain.

Under A, the following classical results can be found e.g. in Brockwell and Davis (1991) and, under B, in Davis and Resnick (1985,1986), see also Brockwell and Davis (1991), Section 12.5.

**Theorem 3.1** *(a) Suppose that A and $\sum_{j=-\infty}^{\infty} \psi_j^2 |j| < \infty$ hold. Then*

$$\gamma_{n,X}(h) \xrightarrow{P} \gamma_X(h) \,, \quad \rho_{n,X}(h) \xrightarrow{P} \rho_X(h) \,, \quad h \geq 1 \,,$$

*and, for every $m \geq 1$,*

$$\sqrt{n}(\rho_{n,X}(h) - \rho_X(h))_{h=1,\ldots,m} \xrightarrow{d} (A_h)_{h=1,\ldots,m} \,,$$

*where*

$$A_h = \sum_{k=1}^{\infty}(\rho_X(k+h) + \rho_X(k-h) - 2\rho_X(h)\rho_X(k))Y_k \,, \qquad (3.3)$$

*and $(Y_k)$ are iid $N(0,1)$ rv's.*
*(b) Suppose that B holds. Set $\rho(h) = \psi^{-2} \sum_{j=-\infty}^{\infty} \psi_j \psi_{j+h}$. Then*

$$\rho_{n,X}(h) \xrightarrow{P} \rho_X(h) \,, \quad h \geq 1 \,,$$

*and*

$$\left(\frac{n}{\ln n}\right)^{1/\alpha} (\rho_{n,X}(h) - \rho(h))_{h=1,\ldots,m} \xrightarrow{d} \frac{1}{Y_0}(A_h)_{h=1,\ldots,m} \,,$$

*where $A_h$ is given by (3.3) with $\rho_X$ replaced by $\rho$ and with iid s$\alpha$s $(Y_k)$. The rv $Y_0$ is positive $\alpha/2$-stable and independent of $(Y_k)$.*

**Remarks** 1) In case B, $\gamma_X$ and $\rho_X$ do not exist. However, $\gamma_{n,X}$, $\rho_{n,X}$ and $\rho$ are well defined. Moreover, in case A, $\rho_X(h) = \rho(h)$ for all $h$. Thus the self–normalized sample autocovariances, i.e. the sample autocorrelations, are consistent estimators of $\rho(h)$ in both cases A and B.

2) In Klüppelberg and Mikosch (1995a) it is shown that $\rho_{n,X}(h)$ does not converge a.s. under B, in contrast to the case A. There is, however, a.s. convergence along a subsequence $(n^{d(\alpha)})$ for some $d(\alpha) > 1$. Moreover, as shown in Klüppelberg and Mikosch (1994), for consistency of $\rho_{n,X}(h)$ the condition $E|Z_0|^p < \infty$ for some $p > 0$ is sufficient.

3) It follows from results by Davis and Resnick (1985,1986) that part B of Theorem 3.1 (with a proper normalization) remains valid in the domain of attraction of an $\alpha$-stable law. However, if $E|Z_0|^\alpha < \infty$, then the $(Y_k)$ in

(3.3) are dependent $\alpha$-stable rv's.

4) Both under A and B, $\gamma_{n,X}(0) = \psi^2 \gamma_{n,Z}(0)(1 + o_P(1))$ . Under A, this implies by the law of large numbers that $\gamma_{n,X}(0) \overset{P}{\to} var\ (X_0) = \sigma^2 \psi^2$. This is in contrast to the case B where $Z_0^2$ is in the domain of attraction of an $\alpha/2$-stable law, hence $\gamma_{n,X}(0) \overset{d}{\to} Y_0 \psi^2$ for a positive $\alpha/2$-stable $Y_0$. Therefore, $\gamma_{n,X}(0) = a_n^{-2} \sum_{t=0}^{\infty} X_t^2$ can be considered as some surrogate of the sample variance. It is appropriate as a self-normalizing sequence in many cases (see below).

Modifications of Theorem 3.1 are necessary if one is interested in joint convergence of $\gamma_{n,X}(0)$ and $\gamma_{n,X}(h)$ for $h \neq 0$. Roughly speaking, three different cases must be considered: (i) $EZ_0^4 < \infty$, (ii) $EZ_0^4 = \infty, \sigma^2 < \infty$ and (iii) $\sigma^2 = \infty$. For a study of these cases see Davis and Resnick (1985,1986). We demonstrate the differences for the noise sequence $(Z_t)$. The results for the noise are fundamental for the understanding of the limit behaviour of the periodogram and its integrated version.

**Theorem 3.2** *(a) Suppose A holds. Then*

$$(\gamma_{n,Z}(h))_{h=1,\ldots,m} \overset{d}{\to} (Y_h)_{h=1,\ldots,m}$$

*and, if $EZ_0^4 < \infty$, then*

$$\left(n^{1/2}(\gamma_{n,Z}(0) - \sigma^2), (\gamma_{n,Z}(h))_{h=1,\ldots,m}\right) \overset{d}{\to} (Y_h)_{h=0,\ldots,m}\ ,$$

*for independent $(Y_h)$, $Y_0$ is $N(0, var\ (Z_0^2))$ and $(Y_h)_{h=1,\ldots,m}$ are iid $N(0, \sigma^4)$. Moreover, if $P(Z_0^2 > x) \sim cx^{-p}$ as $x \to \infty$ for some $p \in (1,2)$, then*

$$\left(n^{1-1/p}(\gamma_{n,Z}(0) - \sigma^2), (\gamma_{n,Z}(h))_{h=1,\ldots,m}\right) \overset{d}{\to} (Y_h)_{h=0,\ldots,m}\ ,$$

*for independent $(Y_h)$, $Y_0$ is p-stable and $(Y_h)_{h=1,\ldots,m}$ are iid $N(0, \sigma^4)$.*
*(b) Suppose B holds. Then*

$$(\gamma_{n,X}(0), (\gamma_{n,Z}(h))_{h=1,\ldots,m}) \overset{d}{\to} (Y_h)_{h=0,\ldots,m}\ ,$$

*for independent $(Y_h)$, $Y_0$ is positive $\alpha/2$-stable and $(Y_h)_{h=1,\ldots,m}$ are iid s$\alpha$s.*

**Remarks** 5) Notice that the limits of $(\gamma_{n,Z}(h))$ determine the limits of $\gamma_{n,X}(h)$ which manifests in the structure of (3.3).

6) Notice furthermore that in the CLT for $\gamma_{n,Z}(0)$ and for $\gamma_{n,Z}(h)$, $h \neq 0$, the same normalization $n^{1/2}$ is needed given $EZ_0^4 < \infty$. This changes if $EZ_0^4 = \infty$. Then $\gamma_{n,Z}(0)$ is significantly larger than the sample autocovariances at the other lags. This is the reason why we will use random centering sequences below: we want to get rid of the dominating term $\gamma_{n,Z}(0)$.

# 4 Some limit theory in the frequency domain

## 4.1 The raw periodogram

### The limit distribution

The proof of a limit result for the periodogram is based on the following decomposition:

$$I_{n,X}(\lambda) = |\psi(\lambda)|^2 I_{n,Z}(\lambda) + R_n(\lambda)$$

where

$$|\psi(\lambda)|^2 = \left| \sum_{j=-\infty}^{\infty} \psi_j e^{-it\lambda} \right|^2 , \quad \lambda \in [-\pi, \pi] ,$$

denotes the power transfer function of $(X_t)$. Notice that, under A, $f_X(\lambda) = \frac{\sigma^2}{2\pi} |\psi(\lambda)|^2$ is the spectral density of $(X_t)$. Under B, such an interpretation is not possible, but $|\psi(\lambda)|^2$ is well defined, and so is $I_{n,X}(\lambda)$. Under A and B, it can be shown that $R_n(\lambda) = o_P(1)$, and so the limit distribution of $I_{n,X}(\lambda)$ is determined by the one of

$$I_{n,Z}(\lambda) = \left( a_n^{-1} \sum_{t=1}^{n} Z_t \cos(\lambda t) \right)^2 + \left( a_n^{-1} \sum_{t=1}^{n} Z_t \sin(\lambda t) \right)^2$$

which means to show that

$$a_n^{-1} \left( \sum_{t=1}^{n} Z_t \cos(\lambda t), \sum_{t=1}^{n} Z_t \sin(\lambda t) \right) \xrightarrow{d} (\alpha(\lambda), \beta(\lambda))$$

for certain rv's $\alpha(\lambda), \beta(\lambda)$. This approach works in both cases A and B. For the case A we refer to Brockwell and Davis (1991), for B to Klüppelberg and Mikosch (1993,1994).

**Theorem 4.1** Let $0 < \lambda_1 < \cdots < \lambda_m < \pi$.
(a) Suppose A holds. Then

$$(\gamma_{n,X}(0), (I_{n,X}(\lambda_j))_{j=1,\ldots,m}) \xrightarrow{d} (var\ (X_0), \sigma^2 |\psi(\lambda)|^2 (\alpha_j^2 + \beta_j^2)_{j=1,\ldots,m}) ,$$

where $\alpha_1, \beta_1, \ldots, \alpha_m, \beta_m, \ldots$ are iid $N(0, 1/\sqrt{2})$ rv's.
(b) Suppose B holds. Then

$$(\gamma_{n,X}(0), (I_{n,X}(\lambda_j))_{j=1,\ldots,m}) \xrightarrow{d} (\psi^2 Y_0, |\psi(\lambda)|^2 (\alpha^2(\lambda_j) + \beta^2(\lambda_j)_{j=1,\ldots,m}) ,$$

where $Y_0$ is positive $\alpha/2$-stable and $\alpha(\lambda_1), \beta(\lambda_1), \ldots, \alpha(\lambda_m), \beta(\lambda_m)$ are jointly $\alpha$-stable in $\mathbb{R}^{2m}$, the distribution of the $\alpha(\lambda_j)$ and $\beta(\lambda_j)$ depends on the frequency $\lambda_j$, and $(Y_0, \alpha(\lambda_j), \beta(\lambda_j), j = 1, \ldots, m)$ does not contain any pair of independent rv's.

A consequence is the following result:

**Corollary 4.2** *(a) Suppose A holds. Then*

$$\left(\widetilde{I}_{n,X}(\lambda_j)\right)_{j=1,\ldots,m} \xrightarrow{d} \psi^{-2}\left(|\psi(\lambda)|^2(\alpha_j^2 + \beta_j^2)\right)_{j=1,\ldots,m} ,$$

*where* $(\alpha_j^2 + \beta_j^2)_{j=1,\ldots,m}$ *are iid standard exponential rv's.*
*(b) Suppose B holds. Then*

$$\left(\widetilde{I}_{n,X}(\lambda_j)\right)_{j=1,\ldots,m} \xrightarrow{d} \psi^{-2}Y_0^{-1}\left(|\psi(\lambda)|^2(\alpha^2(\lambda_j) + \beta^2(\lambda_j))\right)_{j=1,\ldots,m} ,$$

*and the rv's* $T_j = Y_0^{-1}(\alpha^2(\lambda_j) + \beta^2(\lambda_j))$ *are uncorrelated,* $ET_j = 1$ *and* $P(T_j > x) \le e^{-cx}$ *for* $x > 0$ *and a universal constant* $c > 0$.

### Consistent estimates of the power transfer function

From Theorem 4.1 we see that the periodogram is not a consistent estimator of the power transfer function. Nevertheless, the results of Corollary 4.2 ensure that smoothing techniques lead to consistent estimation. For example, consider the discrete average estimator

$$S_n(\lambda) = \sum_{|k|\le m} W_n(k)I_{n,X}\left(\lambda + \frac{2\pi k}{n}\right) , \quad \widetilde{S}_n(\lambda) = \frac{S_n(\lambda)}{\gamma_{n,X}(0)} ,$$

for fixed $\lambda \in (0,\pi)$ and weights $(W_n(k))_{|k|\le m}$, $m = m_n \to \infty, n \to \infty$ and such that

$$W_n(k) = W_n(-k) , \quad \sum_{|k|\le m} W_n(k) = 1 , \quad \sum_{|k|\le m} W_n^2(k) \to 0 .$$

The following is proved e.g. in Brockwell and Davis (1991), under A, and in Klüppelberg and Mikosch (1994), under B.

**Theorem 4.3** *(a) Suppose A holds. Then*

$$S_n(\lambda) \xrightarrow{P} \sigma^2|\psi(\lambda)|^2 , \quad \lambda \in (0,\pi) ,$$
$$\widetilde{S}_n(\lambda) \xrightarrow{P} \psi^{-2}|\psi(\lambda)|^2 , \quad \lambda \in (0,\pi) .$$

*(b) Suppose B holds. Then*

$$\widetilde{S}_n(\lambda) \xrightarrow{P} \psi^{-2}|\psi(\lambda)|^2 , \quad \lambda \in (0,\pi) .$$

**Remark** 1) Under A, discrete average and kernel density estimators have been developed for a long time, uniform convergence results (wrt $\lambda$) were proved and rates of convergence were determined (see e.g. Brockwell and Davis (1991) and Priestley (1981)). Under B, these questions are still open.

## 4.2 The integrated periodogram

For a smooth function $f$ we define the integrated periodogram

$$K_{n,X}(\lambda; f) = \int_{-\pi}^{\lambda} I_{n,X}(x) f(x) dx \ , \quad \widetilde{K}_{n,X}(\lambda; f) = \frac{K_{n,X}(\lambda; f)}{\gamma_{n,X}(0)} \ . \quad (4.4)$$

Notice that in the finite variance case $K_n(\lambda; 1)$ is a natural estimator of the spectral df and therefore it is sometimes called the *empirical spectral df*; other functions $f$ have been used for constructing goodness–of–fit tests and parameter estimators.

The starting point for a limit theory is the decomposition

$$K_{n,X}(\lambda; f) = \int_{-\pi}^{\lambda} I_{n,Z}(x) |\psi(x)|^2 f(x) dx + \int_{-\pi}^{\lambda} R_n(\lambda) f(x) dx \ . \quad (4.5)$$

For a large class of functions $f$ it can be shown that the properly normalized remainder term $\int_{-\pi}^{\lambda} R_n(\lambda) f(x) dx$ converges to zero in probability, uniformly for $\lambda$. Thus it suffices to study the limit of $K_{n,Z}(\cdot; f)$ for the noise $(Z_t)$.

### Whittle's estimate for the parameters of an ARMA model

Whittle's (1953) estimate is based on the integrated periodogram: let $(X_t)$ be an $ARMA(p, q)$ process given by the ARMA equations $\phi(B)X_t = \theta(B)Z_t$, where $B$ denotes the backshift operator and

$$\phi(z) = 1 - \phi_1 z - \cdots - \phi_p z^p \ , \quad \theta(z) = 1 + \theta_1 z + \cdots + \theta_q z^q \ , \quad |z| \le 1 \ .$$

We assume that $(X_t)$ corresponds to the parameter vector $\beta_0$ which belongs to the natural parameter set

$$C = \{\beta = (\phi_1, \ldots, \phi_p, \theta_1, \ldots, \theta_q) \in \mathbb{R}^{p+q} : \theta(\cdot) \text{ and } \phi(\cdot) \text{ do not have}$$
$$\text{common zeros and } \theta(z)\phi(z) \ne 0 \text{ for } |z| \le 1.\}$$

The condition $\beta_0 \in C$ simply means that $(X_t)$ is causal and invertible, in particular, it has representation $X_t = \sum_{j=0}^{\infty} \psi_j Z_{t-j}$, $t \in \mathbb{Z}$. Set

$$\sigma_n^2(\beta) = \int_{-\pi}^{\pi} \frac{I_{n,X}(\lambda)}{g(\lambda; \beta)} d\lambda \ , \quad \widetilde{\sigma}_n^2(\beta) = \frac{\sigma_n^2(\beta)}{\gamma_{n,X}(0)} \ ,$$

where $g(\lambda; \beta) = \left|\sum_{j=0}^{\infty} \psi_j(\beta) e^{-i\lambda j}\right|^2$ denotes the power transfer function corresponding to $\beta \in C$. Formally, replacing the estimator $I_{n,X}(\lambda)$ by the true power transfer function $g(\lambda; \beta_0)$ in $\sigma_n^2(\beta)$, it can be shown (see Brockwell and Davis (1991), Proposition 10.8.1.) that the function $\int_{-\pi}^{\pi} \frac{g(\lambda; \beta_0)}{g(\lambda; \beta)} d\lambda$ has its absolute minimum in $C$ at $\beta = \beta_0$. Thus is makes sense to consider the following estimator of $\beta_0$:

$$\beta_n = argmin_{\beta \in C} \ \sigma_n^2(\beta) = argmin_{\beta \in C} \ \widetilde{\sigma}_n^2(\beta) \ .$$

Under A, the following is proved in Brockwell and Davis (1991), Section 10.8, (there a sum approximation to $\sigma_n^2(\beta)$ is used, which is asymptotically equivalent, but requires slight modifications of the proof) and, under B, in Mikosch et al. (1995):

**Theorem 4.4** *(a) Suppose A holds. Then*

$$\sqrt{n}(\beta_n - \beta_0) \overset{d}{\to} 4\pi W^{-1}(\beta_0) \sum_{j=1}^{\infty} b_j Y_j \ ,$$

*where $W^{-1}(\beta_0)$ is the inverse of the matrix*

$$W(\beta_0) = \int_{-\pi}^{\pi} \left[ \frac{\partial \ln g(\lambda; \beta_0)}{\partial \beta} \right] \left[ \frac{\partial \ln g(\lambda; \beta_0)}{\partial \beta} \right]^T d\lambda \ ,$$

$$b_j = \frac{1}{2\pi} \int_{-\pi}^{\pi} e^{-ij\lambda} g(\lambda; \beta_0) \frac{\partial g^{-1}(\lambda; \beta_0)}{\partial \beta} d\lambda \ ,$$

*and $(Y_j)$ are iid centered Gaussian rv's. Moreover,*

$$\sigma_n^2(\beta_n) \overset{P}{\to} 2\pi\sigma^2 \ , \quad \tilde{\sigma}_n^2(\beta_n) \overset{P}{\to} 2\pi\psi^{-2} \ .$$

*(b) Suppose B holds. Then*

$$\left( \frac{n}{\ln n} \right)^{1/\alpha} (\beta_n - \beta_0) \overset{d}{\to} 4\pi W^{-1}(\beta_0) Y_0^{-1} \sum_{j=1}^{\infty} b_j Y_j \ ,$$

*where $(Y_j)_{j \geq 0}$ are independent, $Y_0$ is positive $\alpha/2$-stable and $(Y_j)_{j \geq 1}$ are iid sas. The vectors $(b_j)$ and the matrix $W(\beta_0)$ are the same as under A. Moreover,*

$$\sigma_n^2(\beta_n) \overset{d}{\to} 2\pi Y_0 \ , \quad \tilde{\sigma}_n^2(\beta_n) \overset{P}{\to} 2\pi\psi^{-2} \ .$$

**Remarks** 2) Self–normalization is not actually needed to derive the asymptotic distribution of $(\beta_n)$. Under B, it is, however, important in order to obtain consistency of $\sigma_n^2(\beta_n)$.

3) Theorem 4.4 has been proved by Kokoszka and Taqqu (1995) for infinite variance fractional ARIMA processes with long–range dependence. In that case, $|\psi(\lambda)|^2$ has a singularity at zero. The limit process is the same as under B.

4) There exist quite a few estimators of the parameters of an ARMA process in the infinite variance case. See Mikosch et al. (1995) for an overview of the recent literature. The general impression on these results is that the rate of convergence compares favourably to the $\sqrt{n}$ in the $L^2$–case, but the limit distribution is unfamiliar.

**Goodness–of–fit tests**

The function $K_n(\lambda; 1)$ can in the $L^2$–case be considered as the empirical spectral distribution function. Its normalized version has many properties in common with the empirical process (see Anderson (1993), Grenander and Rosenblatt (1984) and Priestley (1981) for the corresponding time series literature; Shorack and Wellner (1986) is a general reference to empirical processes). Analogously to empirical process theory, the empirical spectral distribution function is the basis for goodness–of–fit tests. The test statistics are simply continuous functionals of the integrated periodogram. Therefore it is important to determine the limit process of $K(\cdot; f)$.

We concentrate here on the case $f(x) = |\psi(x)|^{-2} > 0$ in (4.3) and write $K_{n,X}(\lambda) = K_{n,X}(\lambda; |\psi(\cdot)|^{-2})$. Then we obtain from (4.5) ($\lambda \in [-\pi, \pi]$)

$$
\begin{aligned}
K_{n,X}(\lambda) &\approx \int_{-\pi}^{\lambda} I_{n,Z}(x)\,dx \\
&= (\lambda + \pi)\gamma_{n,Z}(0) + 2a_n^{-1} \sum_{h=1}^{n-1} \frac{\sin(\lambda h)}{h} \gamma_{n,Z}(h) \ . \quad (4.6)
\end{aligned}
$$

By Theorem 3.2, the properly normalized $(\gamma_{n,Z}(h))$ converge to iid $(Y_h)$ suggesting that

$$
\sum_{h=1}^{n-1} \frac{\sin(\lambda h)}{h}\gamma_{n,Z}(h) \xrightarrow{d} \sum_{h=1}^{\infty} \frac{\sin(\lambda h)}{h} Y_h \equiv \frac{1}{2} S(\lambda) \ . \quad (4.7)
$$

This can indeed be made precise. Moreover, one can replace the unobservable $\gamma_{n,Z}(0)$ in (4.6) by $(2\pi)^{-1} K_{n,X}(\pi)$.

A proof of the following result can be found in Klüppelberg and Mikosch (1995b) under B, and under A it is implicitly stated in Klüppelberg and Mikosch (1995c); in the latter case it can be understood as a version of Bartlett's $T_p$–test.

Denote by $\mathcal{C}[-\pi, \pi]$ the space of the continuous functions on $[-\pi, \pi]$ equipped with the supremum norm and with the corresponding Borel sets.

**Theorem 4.5** *(a) Suppose A holds. Then*

$$
\left(\gamma_{n,X}(0), \frac{K_{n,X}(\pi)}{2\pi}, \sqrt{n}\left(K_{n,X}(\lambda) - \frac{\lambda + \pi}{2\pi} K_{n,X}(\pi)\right)_{\lambda \in [-\pi, \pi]}\right)
$$

$$
\xrightarrow{d} \ (var\,(X_0), var\,(X_0), (S(\lambda))_{\lambda \in [-\pi, \pi]})
$$

*in $\mathcal{C}[-\pi, \pi]$, where $S$ is defined in (4.7) and $(Y_t)$ are iid $N(0, \sigma^4)$ rv's.*
*(b) Suppose B holds and $1 < \alpha < 2$. Then*

$$
\left(\gamma_{n,X}(0), \frac{K_{n,X}(\pi)}{2\pi}, \left(\frac{n}{\ln n}\right)^{1/\alpha}\left(K_{n,X}(\lambda) - \frac{\lambda + \pi}{2\pi} K_{n,X}(\pi)\right)_{\lambda \in [-\pi, \pi]}\right)
$$

$$\xrightarrow{d} \left(Y_0, Y_0, (S(\lambda))_{\lambda \in [-\pi, \pi]}\right) ,$$

in $\mathcal{C}[-\pi, \pi]$, where $S$ is defined in (4.7) and $(Y_h)_{h \geq 0}$ are independent, $Y_0$ is positive $\alpha/2$-stable and $(Y_h)_{h \geq 1}$ are iid s$\alpha$s.

**Remark** 5) The results of Theorem 4.5 allow for self–normalization with $K_{n,X}(\pi)$.

Goodness–of–fit tests are an immediate consequence of Theorem 4.5. For example, for the Kolmogorov–Smirnov test statistic we obtain, under A,

$$\sqrt{n} \sup_{\lambda \in [-\pi, \pi]} \left| K_{n,X}(\lambda) - \frac{\lambda + \pi}{2\pi} K_{n,X}(\pi) \right| \xrightarrow{d} \sup_{\lambda \in [-\pi, \pi]} |S(\lambda)| ,$$

and, under B,

$$\left(\frac{n}{\ln n}\right)^{1/\alpha} \sup_{\lambda \in [-\pi, \pi]} \left| K_{n,X}(\lambda) - \frac{\lambda + \pi}{2\pi} K_{n,X}(\pi) \right| \xrightarrow{d} \sup_{\lambda \in [-\pi, \pi]} |S(\lambda)| .$$

Similar results can be obtained for the Cramér-von Mises test statistic.

**Remarks** 6) Under A, the limit distribution of these test statistics is well known. Indeed, in that case the process $S$ when restricted to $[0, \pi]$ is a Brownian bridge (see e.g. Hida (1980)). In case B, $S$ can be shown to be a harmonisable $\alpha$–stable process with continuous sample paths.

7) Analogous results can be obtained for $K_{n,X}(\cdot; f)$ by partial integration, using the continuous mapping theorem (see Klüppelberg and Mikosch (1995b)).

8) Kokoszka and Mikosch (1995) showed that Theorem 4.5 remains valid for linear processes with long–range dependence both in the finite and in the infinite variance case.

9) Klüppelberg and Mikosch (1995c) prove under $EZ_0^4 < \infty$ a functional central limit theorem for the two–parameter process

$$\frac{1}{n} \int_{-\pi}^{\lambda} \left| \sum_{s=1}^{[nt]} X_s e^{-ixs} \right|^2 dx , \quad \lambda \in [-\pi, \pi], \quad 0 \leq t \leq 1.$$

Using random centering, we obtain a Kiefer process in the limit. This result can be used for change point detection of the spectral distribution function. It is also proved that such results cannot be achieved in the case B.

**Concluding remark.** The theory outlined above shows that many classical estimation and testing procedures in time series analysis can be modified in case there are large values in the noise. It is usually necessary to replace the deterministic normalizing and centering sequences (which contain the variance or an estimator of it) by appropriate random normalizing

270

and centering sequences. This modification leads to consistent estimators in the finite and in the infinite variance case. The limit distributions are different in both cases.

## 5 References

[1] Anderson, T.W. (1993) Goodness of fit tests for spectral distributions. *Ann. Statist.* **21**, 830–847.

[2] Bartlett, M.S. (1954) Problemes de l'analyse spectrale des séries temporelles stationnaires. *Publ. Inst. Statist. Univ. Paris.* **III-3**, 119-134.

[3] Bartlett, M.S. (1978) *An Introduction to Stochastic Processes with Special Reference to Methods and Applications,* 3rd edition. Cambridge University Press, Cambridge (UK).

[4] Billingsley, P. (1968) *Convergence of Probability Measures.* Wiley, New York.

[5] Brockwell, P.J. and Davis, R.A. (1991) *Time Series: Theory and Methods,* 2nd edition. Springer, New York, Berlin.

[6] Davis, R.A. and Resnick, S.I. (1985) Limit theory for moving averages of random variables with regularly varying tail probabilities. *Ann. Probab.* **13**, 179–195.

[7] Davis, R.A. and Resnick, S.I. (1986) Limit theory for the sample covariance and correlation functions of moving averages. *Ann. Statist.* **14**, 533–558.

[8] Feller, W. (1971) *An Introduction to Probability Theory and Its Applications II.* Wiley, New York.

[9] Grenander, U. and Rosenblatt, M. (1984) *Statistical Analysis of Stationary Time Series.* 2nd edition. Chelsea Publishing Co., New York.

[10] Hida, T. (1980) *Brownian Motion.* Springer, New York, Berlin.

[11] Klüppelberg, C. and Mikosch, T. (1993) Spectral estimates and stable processes. *Stoch. Proc. Appl.* **47**, 323-344.

[12] Klüppelberg, C. and Mikosch, T. (1994) Some limit theory for the self-normalised periodogram of stable processes. *Scand. J. Statist.* **21**, 485-491.

[13] Klüppelberg, C. and Mikosch, T. (1995a) On strong consistency of estimators for infinite variance time series. *Probab. Theor. Math. Statist.* To appear.

[14] Klüppelberg, C. and Mikosch, T. (1995b) The integrated periodogram for stable processes. *Ann. Statist.* To appear.

[15] Klüppelberg, C. and Mikosch, T. (1995c) Gaussian limit fields for the integrated periodogram. Preprint.

[16] Kokoszka, P.S. and Mikosch, T. (1995) The integrated periodogram for long memory time series with finite or infinite variance. Preprint.

[17] Kokoszka, P.S. and Taqqu, M.S. (1995) Parameter estimation for infinite variance fractional ARIMA. Preprint.

[18] Mikosch, T., Gadrich, T., Klüppelberg, C. and Adler, R.J. (1995) Parameter estimation for ARMA models with infinite variance innovations. *Ann. Statist.* **23**, 305–326.

[19] Priestley, M.B. (1981) *Spectral Analysis and Time Series.* Academic Press, New York.

[20] Samorodnitsky, G. and Taqqu, M.S. (1994) *Stable non–Gaussian Random Processes. Stochastic Models with Infinite Variance.* Chapman and Hall, London.

[21] Shorack, G.R. and Wellner, J.A. (1986) *Empirical Processes with Applications to Statistics.* Wiley, New York, Chichester.

[22] Whittle, P. (1953) Estimation and information in stationary time series. *Ark. Mat.* **2**, 423-434.

# Asymptotics of M-estimators in non-linear regression with long-range dependent errors

Hira L. Koul[1]

Department of Statistics and Probability
Michigan State University
East Lansing MI 48824

**Running head: Non-linear regression with LRD errors**

**ABSTRACT.** This note first establishes the asymptotic uniform linearity of M-scores in a family of non-linear regression models when the errors are long-range dependent Gaussian or a function of such random variables. This result is then used to obtain the large sample distributions of a class of M-estimators of the underlying parameters for a large class of non-linear regression models. The class of estimators includes analogs of the least square, least absolute deviation and the Huber(c) estimators. The note also gives the asymptotic uniform linearity of the residual empirical processes.

---

[1] Research of this author was also partly supported by the NSF Grant DMS 94-02904.

**AMS (1991) Subject classification.** Primary 62M15, Secondary 62J05.

**Key words.** Huber(c) estimator. Non-normal limiting distributions. Empirical of residuals.

# 1   Introduction

A discrete time stationary stochastic process is said to be long-range dependent (LRD) if its correlations decrease to zero like a power of the lag, as lag tends to infinity, but their sum diverges. This note studies the asymptotic distributions of a class of M-estimators including the least square estimator for a class of non-linear regression models with the errors that are a function of a LRD Gaussian process.

More precisely, let $\eta_1, \eta_2, \ldots$ be a stationary mean zero unit variance Gaussian process with $\rho(j) := E\eta_1\eta_{j+1}, j \geq 0$; $k, n$ and $p$ be positive integers; and $\{\mathbf{x}'_{ni}, 1 \leq i \leq n\}$ be the $i$th row of a rank $p$ known design matrix $\mathbf{X}$ of the order $nxp$. Let $\mathcal{R}^p$ denote the $p-$dimensional Euclidean space, $\mathcal{R}^1 = \mathcal{R}$; $\mathbf{t}'$ denote the transpose of a vector $\mathbf{t} \in \mathcal{R}$; $G$ be a measurable function from $\mathcal{R}$ to $\mathcal{R}$; $\Omega$ be an open subset of $\mathcal{R}^k$ and let $h$ be a known function on $\Omega x \mathcal{R}^p$ to $\mathcal{R}$, such that if $h(\mathbf{t}, \mathbf{x}) = h(\mathbf{s}, \mathbf{x})$, for all $\mathbf{x} \in \mathcal{R}$, then $\mathbf{t} = \mathbf{s}$. Consider the non-linear regression model where the observations $Y_{ni}$ obey the relation

$$Y_{ni} = h(\beta, \mathbf{x}_{ni}) + \epsilon_i, \qquad 1 \leq i \leq n. \qquad (1.1)$$

for some $\beta \in R^p$, with $\epsilon_i = G(\eta_i), i \geq 1$. Note that neither the joint nor the marginal distributions of $\{\epsilon_i\}$ need be Gaussian. We shall further assume that

$$\rho(k) = k^{-\theta}L(k), \text{ for some } 0 < \theta < 1, \text{ for all } k \geq 1, \qquad (1.2)$$

where $L(k)$ is a slowly varying function of $k$ at infinity and positive for large $k$. Because the sum of these covariances diverge, the process $\{\eta_i\}$ is said to be LRD.

The asymptotics of the least square estimator in non-linear regression (NLR) models with the i.i.d. errors have been studied by Jennrich (1969) and Wu (1981), among others. Hannan (1971) studied the asymptotics of this estimator in NLR models when the errors are generated by a stationary weakly dependent time series. The models specified at (1.1) and (1.2) include NLR models with fractional Gaussian and fractional ARIMA errors as defined in Mandelbrot and van Ness (1968), Granger and Joyeux (1980) and Hosking (1981).

The present note obtains the asymptotic distributions of a class of M-estimators for a class of NLR models under (1.1) and (1.2). These estimators include the least square, the least absolute deviation and the Huber(c) estimators. The results are established by first obtaining the asymptotic uniform linearity (AUL) of the M-scores under (1.1) and (1.2) and under minimal assumptions on the function $h$. The note also contains a similar result for the standardized weighted empirical process. The AUL result for non-smooth $\psi$ is obtained with the help of a result from Koul and Mukherjee (1993).

The consistency issue of these estimators is not discussed at any length here. Rather a condition (see (2.12) below) is assumed that ensures that an appropriately standardized M-estimator is bounded in probability and then the above linearity result is used to derive its limiting distribution.

It is observed that if $G(-x) = -G(x)$, $\psi(-x) = \psi(x)$, for all $x \in \mathcal{R}$ and the set $\{x \in \mathcal{R}; G(x) \leq 0\}$ equals either $[0, \infty)$ or $(-\infty, 0]$ then under some additional regularity conditions on $h$, the corresponding M-estimator is asymptotically normally distributed, but not in general. Similar result obtains if $G$ is monotonic and the common error distribution function (d.f.) $F$ has finite Fisher information for location. See Remark 2.2 below for detail. This note thus generalizes the findings of Koul (1992b) and Koul and Mukherjee (1993) in linear regression models with LRD errors to non-linear regression models with similar errors.

The rest of the note is organized as follows. Section 2 contains assumptions, definitions and statements of all the results together with a discussion of their implications. All proofs are deferred to Section 3. The true parameter value is held fixed at $\beta$, $N_b = \{\mathbf{t} \in \Omega; \|\mathbf{t}\| \leq b\}$, for $0 < b < \infty$, and all limits are taken as $n \to \infty$, unless specified otherwise.

# 2 Definitions of estimators and main results

To proceed further we need some more notation. Let $\eta$ be a copy of $\eta_1$, $\epsilon = G(\eta)$, $F$ be the d.f. of $\epsilon$ and $\mathcal{I} = \{x \in \mathcal{R}; 0 < F(x) < 1\} = (c,d)$. Let $\{H_q; q \geq 1\}$ denote the Hermite polynomials; $J_q(x) := E[I(G(\eta) \leq x) - F(x)]H_q(\eta)$, $J_q^+(x) := E[I(G(\eta) \leq x) - F(x)]|H_q(\eta)|$, $x \in \mathcal{I}$, $q \geq 1$; and let $m = \inf\{m(x) : x \in \mathcal{I}\}$, where $m(x) = \min\{q \geq 1; J_q(x) \neq 0\}$. Dehling and Taqqu (1989) call $m$ the Hermite rank of the class of functions $\{I(G(\eta) \leq x) - F(x); x \in \mathcal{I}\}$.

Let $\psi$ be a real valued nondecreasing function on the real line with $E\psi(\epsilon) = 0$. Let $\Gamma_q := E\psi(\epsilon)H_q(\eta) = E\psi(G(\eta))H_q(\eta)$, $q \geq 1$ and let $r$ denote the Hermite rank of the function $\psi(G(\eta))$, i.e., $r = \min\{q \geq 1; \Gamma_q \neq 0\}$. With $\theta$ and $L$ as in (1.2), *assume* $0 < \theta < (1/r) \wedge (1/m)$, and define

$$\nu_n := n^{(1-r\theta)/2} L^{r/2}(n), \qquad \tau_n := n^{(1-m\theta)/2} L^{m/2}(n), \qquad n \geq 1.$$

The following basic assumption is needed to define the M-estimators. For convenience set $h_{ni}(\mathbf{t}) \equiv h(\mathbf{t}, \mathbf{x}_{ni})$. We assume that there exists a k-vector of functions $\dot{\mathbf{h}}$ on $\mathcal{R}^k \mathbf{x} \mathcal{R}^p$ such that, with $\dot{\mathbf{h}}_{ni}(\mathbf{t}) \equiv \dot{\mathbf{h}}_{ni}(\mathbf{t}, \mathbf{x}_{ni})$,

$$\mathbf{H}_n := \sum_{i=1}^{n} \dot{\mathbf{h}}_{ni}(\beta)\dot{\mathbf{h}}_{ni}'(\beta) \text{ is positive definite for all } n \geq k, \qquad (2.1)$$

and for every $\mathbf{s} \in \mathcal{R}^k$, $0 < b < \infty$,

$$\sup_{1 \leq i \leq n, \|\mathbf{A}_n(\mathbf{t}-\mathbf{s})\| \leq b} \{|h_{ni}(\mathbf{t}) - h_{ni}(\mathbf{s}) - (\mathbf{t}-\mathbf{s})'\dot{\mathbf{h}}_{ni}(\mathbf{s})|/\|\mathbf{t}-\mathbf{s}\|\} = o(1), \quad (2.2)$$

where $\mathbf{A}_n = \nu_n^{-1}\mathbf{D}_n$ with $\mathbf{D}_n = \boldsymbol{H}_n^{1/2}$.

Now, let $Y_{nit} \equiv Y_{ni} - h_{ni}(\mathbf{t}), 1 \leq i \leq n$ and define the M-score corresponding to the given $\psi$

$$\mathbf{M(t)} = \sum_{i=1}^{n} \dot{h}_{ni}(\mathbf{t})\psi(Y_{nit}), \qquad \mathbf{t} \in \mathcal{R}.$$

The corresponding M-estimator $\hat{\beta}$ of $\beta$ is defined as a solution of the equation $\mathbf{M(t)} = \mathbf{0}$. Note that the choice of $\psi(x) = x$, $\psi(x) = sign(x)$ and $\psi(x) = |x|I(|x| \leq c) + c\,sign(x)I(|x| > c)$, repsectively, yield the least square, the least absolute deviaiton (LAD) and the Huber(c) type estimators. Computational aspects of these and some other estimators for various $h$ have been addressed in Seber and Wild (1989). Various distributional aspects of analogs of $\hat{\beta}$ for general $\psi$ in the linear regression setup have appeared in Huber (1981), Koul (1992a,b) and Koul and Mukherjee (1993)(**KM**), among others.

The focus here is to obatin the asymptotic distributions of these estimators for the class of models given at $(1.1), (1.2), (2.1)$ and $(2.2)$. To state our results we need additional assumptions. For technical reasons the results for *smooth* $\psi$ and *non-smooth* $\psi$ are stated separately. The smoothness of $\psi$ is clarified in the following condition.

**Condition ($\psi\mathbf{1}$).** $\psi$ is absolutely continuous with $E\psi(\epsilon) = 0$, $E\psi^2(\epsilon) < \infty$. Its a.e. derivative $\psi'$ is such that

$$E\{\psi'(\epsilon)\}^2 < \infty, \qquad E|\psi'(\epsilon - s) - \psi'(\epsilon)| \to 0, \quad \text{as } s \to 0,$$

and for any arrays $\{a_{ni}\}$ of non-negative real numbers with $max_{1 \leq i \leq n}\, a_{ni} \to 0$,

$$max_{1 \leq i \leq n}\, \frac{1}{a_{ni}} \int_{-a_{ni}}^{a_{ni}} |\psi'(\epsilon_i - s)|\, ds = O_p(1).$$

In addition we shall need the following assumptions.

$$\nu_n\|\mathbf{D}_n^{-1}\| \to 0, \quad n^{1/2}max_{1 \leq i \leq n}\|\mathbf{D}_n^{-1}\dot{\boldsymbol{h}}_{ni}(\boldsymbol{\beta})\| = O(1). \qquad (2.3)$$

$$\|\mathbf{D}_n^{-1}\| \sum_{i=1}^{n} \|\mathbf{D}_n^{-1}\dot{h}_{ni}(\beta)\| = O(1). \tag{2.4}$$

$$n \, max_{1 \leq i \leq n}\|\mathbf{D}_n^{-1}\{h_{ni}(\beta + \nu_n\mathbf{D}_n^{-1}\mathbf{t}) - h_{ni}(\beta)\}\|^2 = o(1), \quad \forall \mathbf{t} \in \Omega. \tag{2.5}$$

$$sup_{t \in N_b}\nu_n^{-1}\sum_{i=1}^{n}\|\mathbf{D}_n^{-1}\{h_{ni}(\beta + \nu_n\mathbf{D}_n^{-1}\mathbf{t}) - h_{ni}(\beta)\}\| = O(1), \quad \forall 0 < b < \infty. \tag{2.6}$$

For every $\alpha > 0$, there exists a $\delta > 0$ such that for every $\mathbf{s} \in N_b, 0 < b < \infty$,

$$\limsup_{n} \sup_{t \in N_b, \|t - s\| \leq \delta} \nu_n^{-1}\sum_{i=1}^{n}\|\mathbf{D}_n^{-1}\{\dot{h}_{ni}(\beta + \mathbf{A}_n^{-1}\mathbf{t}) - \dot{h}_{ni}(\beta + \mathbf{A}_n^{-1}\mathbf{s})\}\|$$

$$\leq \alpha. \tag{2.7}$$

For every $\alpha > 0$, there exists a $\delta > 0$ such that for every $\mathbf{s} \in N_b, 0 < b < \infty$,

$$\limsup_{n} P( \sup_{t \in N_b, \|t - s\| \leq \delta} \|\mathbf{Q}_n(\mathbf{t}) - \mathbf{Q}_n(\mathbf{s})\| > \alpha) \leq \alpha, \tag{2.8}$$

where

$$\mathbf{Q}_n(\mathbf{t}) \equiv \nu_n^{-1}\sum_{i=1}^{n}\mathbf{D}_n^{-1}\dot{h}_{ni}(\beta + \mathbf{A}_n^{-1}\mathbf{t})\psi(\epsilon_i).$$

**Remark 2.1.** For the least square score $\psi(x) \equiv x$, the assumption $E\epsilon = 0, E\epsilon^2 < \infty$ imply the condition $(\psi 1)$ trivially while (2.7) and (2.8) are implied by the following: For every $\alpha > 0$, there exists a $\delta > 0$ such that for every $\mathbf{s} \in N_b, 0 < b < \infty$,

$$\limsup_{n} \nu_n^{-1}\sum_{i=1}^{n} \sup_{t \in N_b, \|t - s\| \leq \delta} \|\mathbf{D}_n^{-1}\{\dot{h}_{ni}(\beta + \mathbf{A}_n^{-1}\mathbf{t}) - \dot{h}_{ni}(\beta + \mathbf{A}_n^{-1}\mathbf{s})\}\| \leq \alpha. \tag{2.9}$$

In the case of Huber(c) score, $(\psi 1)$ is satisfied as long as $F$ is continuous at $c$. Also, note that (2.9) implies (2.8) for a bounded $\psi$.

We are now ready to state the asymptotic uniform linearity results.

**Theorem 2.1. (Smooth $\psi$).** In addition to (1.1) and (1.2), assume that $(\psi 1)$, (2.1)-(2.8) hold. Then, for every $0 < b < \infty$,

$$sup_{t \in N_b}\|\nu_n^{-1}\mathbf{D}_n^{-1}[\mathbf{M}(\beta + \nu_n\mathbf{D}_n^{-1}\mathbf{t}) - \mathbf{M}(\beta)] - \mathbf{t}\, E\psi'(\epsilon)\| = o_p(1). \tag{2.10}$$

To state an analogous result for a non-smooth $\psi$, we need the following two assumptions.

**Condition ($F$).** *F has a uniformly continuous a.e. positive density f*

**Condition ($F$1).** *The functions $J_m$, $J_m^+$ are continuously differentiable with respective derivatives $\dot{J}_m$, $\dot{J}_m^+$, satisfying $|\dot{J}_m(x)| \vee \dot{J}_m^+(x) \to 0$ as $|x| \to |c| \vee |d|$.*

Also, let

$$\Psi = \{\psi : \mathcal{R} \, to \, \mathcal{R}, nondecreasing \, right \, continuous, \int \psi dF = 0, \int d\psi \leq 1\}.$$

**Theorem 2.2.** *(Non-smooth $\psi$).* *In addition to (1.1), (1.2), (F) and (F1) assume that (2.1)-(2.7) with $\nu_n$ replaced by $\tau_n$ hold. Then, for every $0 < b < \infty$*

$$sup_{t \in N_b, \psi \in \Psi} \|\tau_n^{-1} \mathbf{D}_n^{-1} [\mathbf{M}(\beta + \tau_n \mathbf{D}_n^{-1} \mathbf{t}) - \mathbf{M}(\beta)] - \mathbf{t} \int f \, d\psi\| = o_p(1). \tag{2.11}$$

We also need the following lemma. Its proof is similar to that of Lemma 1.1 of KM.

**Lemma 2.1.** *Let $\{\epsilon, \epsilon_i, \eta_i; i \geq 1\}$ be as in (1.1) and (1.2). Suppose, additionally that $\psi$ is a measurable function on $\mathcal{R}$ to $\mathcal{R}$ such that $E\psi(\epsilon) = 0$ and $E\psi^2(\epsilon) < \infty$. Then under (2.1) and (2.2),*

$$\nu_n^{-1} \mathbf{D}_n^{-1} \mathbf{M}(\beta) = \nu_n^{-1} \mathbf{D}_n^{-1} \sum_{i=1}^{n} \dot{h}_{ni}(\beta) H_r(\eta_i) \Gamma_r / r! + o_p(1).$$

To use (2.10) to obtain the asymptotic distribution of $\hat{\beta}$, one must first ensure that $\|\nu_n^{-1} \mathbf{D}_n[\hat{\beta} - \beta]\| = O_p(1)$. In view of the fact $\|\nu_n^{-1} \mathbf{D}_n^{-1} \mathbf{M}(\beta)\| = O_p(1)$, which in turn follows from (2.3) and Lemma 2.1, a sufficient condition for this to hold is as follows. For every $\gamma > 0, 0 < \alpha < \infty$, there exists an $N_\gamma$ and a $K \equiv K_{\gamma,\alpha}$ such that

$$P(inf_{\|t\| > K} \|\nu_n^{-1} \mathbf{D}_n^{-1} \mathbf{M}(\beta + \nu_n \mathbf{D}_n^{-1} \mathbf{t})\| \geq \alpha) \geq 1 - \gamma, \quad n > N_\gamma. \tag{2.12}$$

Similarly, under the conditions of Theorem 2.2, one requires an analog of (2.12) with $(\nu_n, r)$ replaced by $(\tau_n, m)$.

**Corollary 2.1.** (A) *In addition to the assumptions of Theorem 2.1, assume that $E\psi'(\epsilon) > 0$, and (2.12) hold. Then*

$$\nu_n^{-1}\mathbf{D}_n(\hat{\beta}-\beta) = \nu_n^{-1}\mathbf{D}_n^{-1}\sum_{i=1}^n \dot{h}_{ni}(\beta)H_r(\eta_i)\,\Gamma_r/(r!\int f\,d\psi)+o_p(1). \quad (2.13)$$

(B) *If, in addition to the assumptions of Theorem 2.2, $\int f\,d\psi > 0$ and (2.12) with $\nu_n$ replaced by $\tau_n$ holds, then (2.13) with $(\nu_n, r)$ replaced by $(\tau_n, m)$ continues to hold for every $\psi \in \Psi$.*

*Sketch of Proof.* Using the fact that $\|\nu_n^{-1}\mathbf{D}_n^{-1}\mathbf{M}(\beta)\| = O_p(1)$ and the linearity result (2.10), one first shows that under $\int \psi'\,dF > 0$ and (2.12), $\|\nu_n^{-1}\mathbf{D}_n[\hat{\beta} - \beta]\| = O_p(1)$. Then (2.13) follows from (2.10) and Lemma 2.1. See Section 5.4 of Koul (1992a) for a general argument. Similarly one obtains (B).

The following result gives the asymptotic distribution of the LAD estimator under weaker conditions than those required by the above corollary.

**Corollary 2.2.** *Assume (1.1), (1.2) and (2.1) to (2.7) hold with $\nu_n$ replaced by $\tau_n$. In addition, if $F$ has a density $f$ in an open neighborhood of 0 that is coninous and positive at 0 and if (2.12) hold for $\psi(x) \equiv sign(x)$, then*

$$\tau_n^{-1}\mathbf{D}_n[\hat{\beta}_{lad} - \beta]\| = \tau_n^{-1}\mathbf{D}_n^{-1}\sum_{i=1}^n \dot{h}_{ni}(\beta)H_{m_0}(\eta_i)\,J_{m_0}/(m_0!\,f(0)) + o_p(1).$$

*where $m_0 = m(0)$.*

**Remark 2.2.** All proofs appear in Section 3. Here we discuss some of the consequences. Arguing as in KM, p540, if $\psi$ and $G$ are skew-symmetric and if $\{x \in \mathcal{R}; G(x) \leq 0\}$ equals either $[0, \infty)$ or $(-\infty, 0]$, then the Hermite ranks $m$, $r$ and $m(0)$ are all equal to 1 and $\nu_n = \tau_n$. It thus follows from the above corollaries that under the additional assumed conditions on $h$, $F$,

$$\Sigma_n^{-1}\nu_n^{-1}\mathbf{D}_n(\hat{\beta} - \beta) \Rightarrow N(\mathbf{0}, \sigma^2\mathbf{I}), \qquad \Sigma_n^{-1}\tau_n^{-1}\mathbf{D}_n(\hat{\beta}_{lad} - \beta) \Rightarrow N(\mathbf{0}, \sigma_0^2\mathbf{I}),$$

where

$$\sigma^2 = \Gamma_1^2/(\int f \, d\psi)^2, \quad \sigma_0^2 = J_1^2(0)/f^2(0), \quad \Sigma_n = \{\nu_n^2 \, \mathbf{D}_n^{-1}\dot{\mathbf{H}}_n' \, \mathbf{R}_n \dot{\mathbf{H}}_n \mathbf{D}_n^{-1}\}^{1/2},$$

and $\mathbf{I}$ is a $k \times k$ identity matrix, with $\mathbf{R}_n$ denoting the correlation matrix of the r.v.'s $(\eta_1, \ldots \eta_n)$ and $\dot{\mathbf{H}}_n$ denoting the $n \times k$ matrix whose $i$th row is $\dot{h}'_{ni}(\beta)$, $1 \leq i \leq n$. Similar conclusions hold if $G$ is strictly monotonic and $F$ has finite Fisher information for location.

In particular if $G(x) \equiv x$, i.e., if the errors are LRD Gaussian, then all of the above estimators are asymptotically equivalent to the least square estimator in probability. It is clear from (2.13) that the limiting distributions of these estimators need not be normal in general. See KM, p540, for some concrete examples.

We end this section by stating a linearity result for the residual empirical process $F_n(., \mathbf{t})$ of the residuals $\{Y_{nit}, 1 \leq i \leq n\}$.

**Proposition 2.1.** *Suppose that in addition to (1.1), (1.2), (2.1)-(2.7), the conditions $(F)$ and $(F1)$ hold. Then, for every $0 < b < \infty$,*

$$sup \, |\nu_n^{-1} n^{1/2} \, [F_n(x, \beta + \nu_n\mathbf{D}_n^{-1}\mathbf{t}) - F_n(x, \beta)] - n^{1/2}\bar{\dot{h}}_n' \mathbf{D}_n^{-1}\mathbf{t} \, f(x)| = o_p(1),$$

*where* $\bar{\dot{h}}_n = n^{-1} \sum_{i=1}^n \dot{h}_{ni}(\beta)$ *and where the supremum is taken over* $x \in \mathcal{R}$, $\mathbf{t} \in N_b$.

The above result is useful in testing the goodness-of-fit hypotheses pertaining to $F$, which can be also construed as testing about $G$. For example testing for $F = N(0, 1)$ is the same as testing if $G(x) \equiv x$. It is also useful in proving the consistency of a kernel type density estimator of $f$ as in KM.

# 3 Proofs

The following facts from Taqqu (1975) and KM will be used often in the proofs. To begin with we have

$$\sum_{i=1}^{n}\sum_{j=1}^{n}|\rho(i-j)|^q = O(n^{2-q\theta}L^q(n)), \qquad (3.1)$$

for any positive integer $q$ with $q < 1/\theta$.

The r.v.'s $\{H_q(\eta_i)\}$ are in $L_2(\mathcal{R}, d\Phi)$ such that $H_0(x) \equiv 1$, $EH_q(\eta) = 0$, and the following holds:

$$\begin{aligned} EH_q(\eta_i)H_s(\eta_j) &= 0, & q \neq s, \text{ for all } i, j, \\ &= q!\rho^q(i-j), & q = r, \text{ for all } i, j. \end{aligned} \qquad (3.2)$$

For any real arrays $\{c_{ni}\}$ and for $q < 1/\theta$,

$$\begin{aligned} Var\{\sum_{i=1}^{n} c_{ni} H_q(\eta_i)\} &= \sum_{i=1}^{n}\sum_{j=1}^{n} c_{ni}c_{nj}\rho^q(i-j) \\ &\leq (n\, max_{1\leq i\leq n}c_{ni}^2)\, O(n^{1-q\theta}L^q(n)). \end{aligned} \qquad (3.3)$$

Now, recall the definition of $\mathbf{A}_n$ from (2.2) and for convenience let

$$d_{ni}(\mathbf{t}) := h_{ni}(\boldsymbol{\beta} + \mathbf{A}_n^{-1}\mathbf{t}) - h_{ni}(\boldsymbol{\beta}), \quad \dot{d}_{ni}(\mathbf{t}) := \dot{h}_{ni}(\boldsymbol{\beta} + \mathbf{A}_n^{-1}\mathbf{t}) - \dot{h}_{ni}(\boldsymbol{\beta}),$$

$$u_{ni} := \|\mathbf{A}_n^{-1}\dot{h}_{ni}(\boldsymbol{\beta})\| + \alpha\,\|\mathbf{A}_n^{-1}\| = \nu_n\{\|\mathbf{D}_n^{-1}\dot{h}_{ni}(\boldsymbol{\beta})\| + \alpha\,\|\mathbf{D}_n^{-1}\|\},$$

for $1 \leq i \leq n$, $\alpha > 0$.

In all the proofs below the indices $i, \mathbf{t}$ in $sup_{i,\mathbf{t}}$ and $max_i$ vary over $1 \leq i \leq n, \mathbf{t} \in N_b$. We shall also write $h_{ni}, \dot{h}_{ni}$ for $h_{ni}(\boldsymbol{\beta}), \dot{h}_{ni}(\boldsymbol{\beta})$, respectively.

**Proof of Theorem 2.1.** Fix an $\alpha > 0$. By (2.2), there exists an $n_1$ such that for all $n > n_1$,

$$sup_{i,\mathbf{t}}|d_{ni}(\mathbf{t}) - (\mathbf{A}_n^{-1}\mathbf{t})'\dot{h}_{ni}| \leq b\alpha\|\mathbf{A}_n^{-1}\|. \qquad (3.4)$$

Therefore,

$$|d_{ni}(\mathbf{t})| \leq bu_{ni}, \quad \text{for all } 1 \leq i \leq n, \mathbf{t} \in N_b, n > n_1. \qquad (3.5)$$

Hence, using (2.3), we obtain

$$max_i\, u_{ni} = n^{-r\theta} L^{r/2}(n)\, n^{1/2} max_i \|\mathbf{D}_n^{-1} \dot{\boldsymbol{h}}_{ni}\| = o(1),$$

$$max_i\, |d_{ni}(\mathbf{t})| = o(1), \qquad \forall\, \mathbf{t} \in N_b. \qquad (3.6)$$

Now, with $\mathbf{B}_n = \nu_n \mathbf{D}_n$, write

$$
\begin{aligned}
\mathbf{Z}_n(\mathbf{t}) &= \mathbf{B}_n^{-1}[\mathbf{M}(\boldsymbol{\beta} + \mathbf{A}_n^{-1}\mathbf{t}) - \mathbf{M}(\boldsymbol{\beta})] \\
&= \mathbf{B}_n^{-1} \sum_{i=1}^{n} \dot{\mathbf{d}}_{ni}(\mathbf{t})\, \psi(\epsilon_i - d_{ni}(\mathbf{t})) \\
&\quad + \mathbf{B}_n^{-1} \sum_{i=1}^{n} \dot{\boldsymbol{h}}_{ni} \{\psi(\epsilon_i - d_{ni}(\mathbf{t})) - \psi(\epsilon_i)\}, \\
&= \mathbf{Z}_{n1}(\mathbf{t}) + \mathbf{Z}_{n2}(\mathbf{t}), \quad \text{say.} \qquad (3.7)
\end{aligned}
$$

Consider the second term first. Rewrite

$$
\begin{aligned}
\mathbf{Z}_{n2}(\mathbf{t}) &= \mathbf{B}_n^{-1} \sum_{i=1}^{n} \dot{\boldsymbol{h}}_{ni} \{\psi(\epsilon_i - d_{ni}(\mathbf{t})) - \psi(\epsilon_i) - d_{ni}(\mathbf{t})\psi'(\epsilon_i)\} \\
&\quad + \mathbf{B}_n^{-1} \sum_{i=1}^{n} \dot{\boldsymbol{h}}_{ni}\{d_{ni}(\mathbf{t}) - \dot{\boldsymbol{h}}_{ni}' \mathbf{A}_n^{-1}\mathbf{t}\}\psi'(\epsilon_i) \\
&\quad + \mathbf{B}_n^{-1} \sum_{i=1}^{n} \dot{\boldsymbol{h}}_{ni}\dot{\boldsymbol{h}}_{ni}' \mathbf{A}_n^{-1}\psi'(\epsilon_i)\, \mathbf{t} \\
&= \mathbf{Z}_{n21}(\mathbf{t}) + \mathbf{Z}_{n22}(\mathbf{t}) + \mathbf{Z}_{n23}\, \mathbf{t}, \qquad \text{say.} \qquad (3.8)
\end{aligned}
$$

Because $\psi'(\epsilon_i)$ is a square integrable function of a Gaussian r.v., its Hermite expansion is given by $\sum_{i=1}^{n} \frac{\alpha_q}{q!} H_q(\eta_i)$, where $\alpha_q = E\psi'(\epsilon)H_q(\eta)$. With $\mathbf{b}_{ni} = \mathbf{D}_n^{-1}\dot{\boldsymbol{h}}_{ni}$, use facts (3.1)-(3.3) to obtain that for every $\lambda \in \mathcal{R}^k$,

$$
\begin{aligned}
E\|\mathbf{Z}_{n23} - \mathbf{I}E\psi'(\epsilon)\|^2 &= \sum_{q=1}^{\infty} \frac{\alpha_q^2}{q!} \sum_{i=1}^{n}\sum_{j=1}^{n} \lambda'\mathbf{b}_{ni}\mathbf{b}_{ni}'\mathbf{b}_{nj}\mathbf{b}_{nj}'\lambda\, \rho^q(i-j) \\
&\leq Var(\psi'(\epsilon))\|\lambda\|^2 max_i \|\mathbf{b}_{ni}\|^4 \sum_{i=1}^{n}\sum_{j=1}^{n} |\rho(i-j)| \\
&= O(n^{-\theta}), \qquad (3.9)
\end{aligned}
$$

by (2.3). Next by using the definitions of $\mathbf{A}_n$ and $\mathbf{B}_n$, for every $n > n_1$,

$$E\,sup_t\, \|\mathbf{Z}_{n22}(\mathbf{t})\| \leq b\alpha \sum_{i=1}^{n} \|\mathbf{D}_n^{-1}\dot{\boldsymbol{h}}_{ni}\|\|\mathbf{D}_n^{-1}\|\, E|\psi'(\epsilon)|,$$

so that, by (2.4),

$$Esup_t \|\mathbf{Z}_{n22}(\mathbf{t})\| = \alpha \, O(1). \qquad (3.10)$$

Use the absolute continuity of $\psi$ and a standard argument to obtain

$$Esup_t \|\mathbf{Z}_{n21}(\mathbf{t})\| \leq b \, \nu_n^{-1} \sum_{i=1}^{n} \|\mathbf{D}_n^{-1}\dot{h}_{ni}\| u_{ni} \cdot$$
$$\cdot max_i \{ v_{ni}^{-1} \int_{-v_{ni}}^{v_{ni}} E|\psi'(\epsilon - s) - \psi'(\epsilon)| \, ds \},$$

for all $n > n_1$, where $v_{ni} = b \, u_{ni}$. But from (2.3) and the definiton of $u_{ni}$ we have

$$\nu_n^{-1} \sum_{i=1}^{n} \|\mathbf{D}_n^{-1}\dot{h}_{ni}\| u_{ni} \leq \sum_{i=1}^{n} \|\mathbf{D}_n^{-1}\dot{h}_{ni}\|^2 + \alpha \|\mathbf{D}_n^{-1}\| \sum_{i=1}^{n} \|\mathbf{D}_n^{-1}\dot{h}_{ni}\|$$
$$= k + \alpha \, O(1),$$

so that from the Condition $(\psi 1)$ we obtain

$$Esup_t \|\mathbf{Z}_{n21}(\mathbf{t})\| = o(1).$$

Upon combining this with (3.10)-(3.8) and the arbitrariness of $\alpha$, we obtain

$$Esup_t \|\mathbf{Z}_{n2}(\mathbf{t}) - E\psi'(\epsilon)\,\mathbf{t}\| = o(1). \qquad (3.11)$$

It remains to show that the first term in the right hand side of (3.7) is negligible. Rewrite

$$\mathbf{Z}_{n1}(\mathbf{t}) = \mathbf{B}_n^{-1} \sum_{i=1}^{n} \dot{d}_{ni}(\mathbf{t}) \{\psi(\epsilon_i - d_{ni}(\mathbf{t})) - \psi(\epsilon_i)\} + \mathbf{B}_n^{-1} \sum_{i=1}^{n} \dot{d}_{ni}(\mathbf{t})\psi(\epsilon_i)$$
$$= \mathbf{Z}_{n11}(\mathbf{t}) + \mathbf{Z}_{n12}(\mathbf{t}), \quad \text{say.} \qquad (3.12)$$

Because $\psi(\epsilon_i)$ is a centered square integrable function of a Gaussian r.v., its Hermite expansion is given by $\sum_{q \geq r}(\Gamma_q/q!)H_q(\eta_i)$. From (3.3), applied with $c_{ni} = \lambda' \mathbf{D}_n^{-1}\mathbf{d}_{ni}$, $q = r$, and arguing as for (3.9), we obtain, in view of (2.5), that $\forall \lambda \in \mathcal{R}^k$, $\forall \mathbf{t} \in N_b$,

$$E\|\lambda'\mathbf{Z}_{n12}(\mathbf{t})\|^2 \leq n \, max_i \|\mathbf{D}_n^{-1}\dot{d}_{ni}(\mathbf{t})\|^2 \, O(1) = o(1). \qquad (3.13)$$

Also, for every $s, t \in N_b$,

$$\|\mathbf{Z}_{n12}(t) - \mathbf{Z}_{n12}(s)\| = \|\mathbf{Q}_n(t) - \mathbf{Q}_n(s)\|.$$

Hence, (3.13), the compactness of $N_b$, (2.8), and a routine argument imply

$$sup_t\|\mathbf{Z}_{n12}(t)\| = o_p(1). \tag{3.14}$$

Now, consider the first term in (3.12). Use the absolute continuity of $\psi$ to obtain that $\forall n > n_1$,

$$\|\mathbf{Z}_{n11}(t)\| \le b \nu_n^{-1} \sum_{i=1}^{n} \|\mathbf{D}_n^{-1}\dot{d}_{ni}(t)\| u_{ni} \cdot max_i \{v_{ni}^{-1} \int_{-v_{ni}}^{v_{ni}} |\psi'(\epsilon_i - s)| ds\}. \tag{3.15}$$

But, with $K_n = sup_t \nu_n^{-1} \sum_{i=1}^{n} \|\mathbf{D}_n^{-1}\dot{d}_{ni}(t)\|$,

$$sup_t \nu_n^{-1} \sum_{i=1}^{n} \|\mathbf{D}_n^{-1}\dot{d}_{ni}(t)\| u_{ni}$$

$$\le sup_t \nu_n^{-1} \sum_{i=1}^{n} \|\mathbf{D}_n^{-1}\dot{h}_{ni}\| \|\mathbf{D}_n^{-1}\dot{d}_{ni}(t)\| + \alpha \|\mathbf{D}_n^{-1}\| K_n$$

$$\le \{n^{1/2} max_i \|\mathbf{D}_n^{-1}\dot{h}_{ni}\| n^{-r\theta} L^{r/2}(n) + \alpha \nu_n \|\mathbf{D}_n^{-1}\|\} K_n$$

$$= o(1), \qquad \text{by (2.3) and (2.6)},$$

which, together with ($\psi$1), (3.14), and (3.15), imply that

$$sup_t\|\mathbf{Z}_{n11}(t)\| = o_p(1).$$

This together with (3.14), (3.12)-(3.10), (3.8) and (3.7) completes the proof of (2.10).

The following Lemma is needed in the proof of Theorem 2.2. Its proof appears in Theorems 1.1 and 2.1 of KM.

**Lemma 3.1.** Let $\eta_i, \epsilon_i; i \ge 1$ be as in (1.2) with $F$ denoting the d.f. of $\epsilon$. Let $\{\gamma_{ni}, \xi_{ni}; 1 \le i \le n\}$ be real arrays satisfying the following conditions.

$$\sum_{i=1}^{n} \gamma_{ni}^2 = 1, \qquad n^{1/2} max_i |\gamma_{ni}| = O(1). \tag{3.16}$$

*Then, with $\bar{\mathcal{I}} = [c, d]$,*

$$sup_{x \in \bar{\mathcal{I}}} \ \tau_n^{-1} |\sum_{i=1}^{n} \gamma_{ni} \{I(\epsilon_i \leq x) - F(x) - (J_m(x)/m!)H_m(\eta_i)\}| = o_p(1).$$

$$(3.17)$$

*If, in addition*

$$max_{1 \leq i \leq n} |\xi_{ni}| \to 0 \qquad (3.18)$$

*then, at every continuity point $x$ of $F$,*

$$\tau_n^{-1} |\sum_{i=1}^{n} \gamma_{ni} \{I(\epsilon_i \leq x + \xi_{ni}) - F(x + \xi_{ni}) - I(\epsilon_i \leq x) + F(x)\}| = o_p(1).$$

$$(3.19)$$

*Suppose, in addition,*

$$\tau_n^{-1} \sum_{i=1}^{n} |\gamma_{ni}\xi_{ni}| = O(1), \qquad (3.20)$$

*then, at every $x \in \mathcal{I}$ at which $F$ has a continuous density $f$,*

$$\tau_n^{-1} |\sum_{i=1}^{n} \gamma_{ni} \{I(\epsilon_i \leq x + \xi_{ni}) - I(\epsilon_i \leq x) - \xi_{ni} f(x)\}| = o_p(1). \quad (3.21)$$

*Finally, if $(F)$ and $(F1)$ hold, then*

$$sup_{x \in \bar{\mathcal{I}}} \ \tau_n^{-1} |\sum_{i=1}^{n} \gamma_{ni} \{I(\epsilon_i \leq x + \xi_{ni}) - F(x + \xi_{ni}) - I(\epsilon_i \leq x) + F(x)\}| = o_p(1).$$

$$(3.22)$$

*and*

$$sup_{x \in \bar{\mathcal{I}}} \ \tau_n^{-1} |\sum_{i=1}^{n} \gamma_{ni} \{I(\epsilon_i \leq x + \xi_{ni}) - I(\epsilon_i \leq x) - \xi_{ni} f(x)| = o_p(1). \quad (3.23)$$

Now, let $\{g_{ni}, q_{ni}\}$ be arrays of real valued functions on $N_b$, and define

$$V(x, t) = \sum_{i=1}^{n} g_{ni}(t)I(\epsilon_i \leq x + q_{ni}(t)),$$

$$v(x, t) = \sum_{i=1}^{n} g_{ni}(t)F(x + q_{ni}(t)),$$

$$U(x, t) = V(x, t) - v(x, t), \qquad x \in \mathcal{R}, \ t \in N_b.$$

and

$$V^*(x, \mathbf{t}) = \sum_{i=1}^{n} g_{ni}(\mathbf{t})I(\epsilon_i \leq x), \qquad v^*(x, \mathbf{t}) = \sum_{i=1}^{n} g_{ni}(\mathbf{t})F(x),$$

$$U^*(x, \mathbf{t}) = V^*(x, \mathbf{t}) - v^*(x, \mathbf{t}), \qquad x \in \mathcal{R}, \mathbf{t} \in N_b.$$

We are now ready to state and prove an analog of Lemma 3.1 where the weights and the noises $\gamma_{ni}$, $\xi_{ni}$ are replaced by functions of $\mathbf{t}$.

**Lemma 3.2.** *In addition to (1.1), (1.2), (F), and (F1), assume that the following hold for every $0 < b < \infty$.*

$$(a) \qquad sup_{1 \leq i \leq n, \mathbf{t} \in N_b} |q_{ni}(\mathbf{t})| = o(1);$$

$$(b) \qquad \tau_n^{-1} \sum_{i=1}^{n} |g_{ni}(\mathbf{t})q_{ni}(\mathbf{t})| = O(1), \quad \forall \mathbf{t} \in N_b. \tag{3.24}$$

$\forall \alpha > 0, \exists \delta > 0, \ni \forall \mathbf{s} \in N_b,$

$$\limsup_{n} \tau_n^{-1} \sum_{i=1}^{n} |g_{ni}(\mathbf{s})| \sup_{\mathbf{t} \in N_b, \|\mathbf{t}-\mathbf{s}\| \leq \delta} |q_{ni}(\mathbf{t}) - q_{ni}(\mathbf{s})| \leq \alpha. \tag{3.25}$$

$$n^{1/2} max_i |g_{ni}(0)| = O(1). \tag{3.26}$$

$$n \, max_{1 \leq i \leq n} |g_{ni}(\mathbf{t}) - g_{ni}(0)|^2 = o(1), \qquad \forall \mathbf{t} \in N_b. \tag{3.27}$$

$\forall \alpha > 0, \exists \delta > 0, \ni \forall \mathbf{s} \in N_b,$

$$\limsup_{n} \sup_{\mathbf{t} \in N_b, \|\mathbf{t}-\mathbf{s}\| \leq \delta} \tau_n^{-1} \sum_{i=1}^{n} |g_{ni}(\mathbf{t}) - g_{ni}(\mathbf{s})| \leq \alpha. \tag{3.28}$$

*Then, for every $0 < b < \infty$,*

$$sup_{x \in \mathcal{I}, \mathbf{t} \in N_b} \tau_n^{-1} |U(x, \mathbf{t}) - U^*(x, \mathbf{t})| = o_p(1), \tag{3.29}$$

$$sup_{x \in \bar{\mathcal{I}}, \mathbf{t} \in N_b} \tau_n^{-1} |U(x, \mathbf{t}) - U(x, 0)| = o_p(1). \tag{3.30}$$

*If additionally, (3.24b) holds uniformly in $\mathbf{t} \in N_b$, then*

$$\sup \tau_n^{-1} |V(x, \mathbf{t}) - V(x, 0) - \sum_{i=1}^{n} g_{ni}(\mathbf{t})q_{ni}(\mathbf{t}) \, f(x)$$

$$- \sum_{i=1}^{n} [g_{ni}(\mathbf{t}) - g_{ni}(0) \, F(x)] = o_p(1). \tag{3.31}$$

*where the supremum is over $x \in \bar{I}, \mathbf{t} \in N_b$.*

**Proof.** Fix a $0 < b < \infty$ and a $\mathbf{t} \in N_b$. In Lemma 3.1 take $\gamma_{ni} \equiv g_{ni}(\mathbf{t})$ and $\xi_{ni} \equiv q_{ni}(\mathbf{t})$. By (3.26) and (3.27),

$$\sum_{i=1}^{n} g_{ni}^2(\mathbf{t}) \leq 2 \sum_{i=1}^{n} \{g_{ni}(\mathbf{t}) - g_{ni}(0)\}^2 + 2n \, max_i g_{ni}^2(0) = O(1),$$

$$n^{1/2} \, max_i |g_{ni}(\mathbf{t})| \leq n^{1/2} \, max_i |g_{ni}(\mathbf{t}) - g_{ni}(0)| + n^{1/2} \, max_i |g_{ni}(0)|$$

$$= O(1).$$

This and assumption (3.24) verify that these $\{\gamma_{ni}\}, \{\xi_{ni}\}$ satisfy (3.16), (3.18) and (3.20) of Lemma 3.1 for every $\mathbf{t} \in N_b$. Hence (3.22) implies that

$$sup_{x \in \bar{I}} \, \tau_n^{-1} |U(x, \mathbf{t}) - U^*(x, \mathbf{t})| = o_p(1), \qquad \forall \, \mathbf{t} \in N_b. \qquad (3.32)$$

To complete the proof of (3.29), because of the compactness of $N_b$ it suffices to show that $\forall \alpha > 0, \exists \delta > 0, n_0 < \infty, \ni \forall \mathbf{s} \in N_b$,

$$P(sup_{x \in \bar{I}, \, \mathbf{t} \in N_b, \, \|\mathbf{t} - \mathbf{s}\| \leq \delta} \, \tau_n^{-1} |D(x, \mathbf{t}) - D(x, \mathbf{s})| > \alpha) \leq \alpha, \qquad \forall n > n_0, \qquad (3.33)$$

where $D(x, \mathbf{t}) \equiv U(x, \mathbf{t}) - U^*(x, \mathbf{t})$.

For convenience, let $\alpha_i(x, \mathbf{t}) \equiv I(\epsilon_i \leq x + q_{ni}(\mathbf{t})) - F(x + q_{ni}(\mathbf{t}))$, $\alpha_i(x) \equiv \alpha_i(x, 0)$. Then

$$D(x, \mathbf{t}) - D(x, \mathbf{s}) = \sum_{i=1}^{n} [g_{ni}(\mathbf{t}) - g_{ni}(\mathbf{s})][\alpha_i(x, \mathbf{t}) - \alpha_i(x)]$$

$$+ \sum_{i=1}^{n} g_{ni}(\mathbf{s}) [\alpha_i(x, \mathbf{t}) - \alpha_i(x, \mathbf{t})]$$

$$= D_1(x, \mathbf{s}, \mathbf{t}) + D_2(x, \mathbf{s}, \mathbf{t}), \qquad \text{say.}$$

It thus suffices to prove analogs of (3.33) for $D_1$ and $D_2$. Because $|\alpha_i(x, \mathbf{t}) - \alpha_i(x)| \leq 1, \forall x, \mathbf{t}$, (3.28) obviously implies this for $D_1$. To prove an analog of (3.33) for $D_2$, write $g_{ni} = g_{ni}^+ - g_{ni}^-$ and $D_2 = D_2^+ - D_2^-$, where $D_2^\pm$ corresponds to $D_2$ with $\{g_{ni}\}$ replaced by $\{g_{ni}^\pm\}$. Thus, by the triangle inequality, it suffices to prove an analog of (3.33) for $D_2^\pm$.

Now, fix an $\alpha > 0$, $\mathbf{s} \in N_b$, and a $\delta > 0$. Let

$$\Delta_{ni} = sup_{\mathbf{t} \in N_b, \|\mathbf{t}-\mathbf{s}\| \leq \delta} |q_{ni}(\mathbf{t}) - q_{ni}(\mathbf{s})|, \qquad 1 \leq i \leq n,$$

and, for $x \in \bar{\mathcal{I}}$, $a \in \mathcal{R}$, let

$$\mathcal{D}^{\pm}(x, \mathbf{s}, a)$$
$$= \sum_{i=1}^{n} g_{ni}^{\pm}(\mathbf{s}) \left[ I(\epsilon_i \leq x + q_{ni}(\mathbf{t}) + a\Delta_{ni}) - F(x + q_{ni}(\mathbf{t}) + a\Delta_{ni}) \right].$$

By (3.24) and (3.28),

$$max_i |q_{ni}(\mathbf{s}) + a\Delta_{ni}| = o(1), \qquad \tau_n^{-1} \sum_{i=1}^{n} g_{ni}^{\pm}(\mathbf{s}) |q_{ni}(\mathbf{s}) + a\Delta_{ni}| = O(1).$$

Thus one more application of (3.22) with $\xi_{ni} \equiv q_{ni}(\mathbf{s}) + a\Delta_{ni}$ and $\gamma_{ni} \equiv g_{ni}^{\pm}(\mathbf{s})$, yields that

$$sup_{x \in \bar{\mathcal{I}}} \tau_n^{-1} |\mathcal{D}^{\pm}(x, \mathbf{s}, a) - \mathcal{D}^{\pm}(x, \mathbf{s}, 0)| = o_p(1), \qquad \forall a \in \mathcal{R}. \qquad (3.34)$$

Use the nonnegativity of $g_{ni}^{\pm}$, the monotonicity of the indicator function and of $F$, and the fact that $D_2^{\pm}(x, \mathbf{s}) \equiv \mathcal{D}_2^{\pm}(x, \mathbf{s}, 0)$, to obtain that $\forall \mathbf{t} \in N_b$, $\|\mathbf{t} - \mathbf{s}\| \leq \delta$,

$$\tau_n^{-1} |D(x, \mathbf{t}) - D(x, \mathbf{s})|$$
$$\leq \tau_n^{-1} |\mathcal{D}^{\pm}(x, \mathbf{s}, 1) - \mathcal{D}^{\pm}(x, \mathbf{s}, 0)| + \tau_n^{-1} |\mathcal{D}^{\pm}(x, \mathbf{s}, -1) - \mathcal{D}^{\pm}(x, \mathbf{s}, 0)|$$
$$+ \tau_n^{-1} \left| \sum_{i=1}^{n} g_{ni}^{\pm}(\mathbf{s}) \left[ F(x + q_{ni}(\mathbf{s}) + \Delta_{ni}) - F(x + q_{ni}(\mathbf{s}) - \Delta_{ni}) \right] \right|.$$

By $(F)$, the last term in this upper bound is no larger than

$$2\|f\| \tau_n^{-1} \sum_{i=1}^{n} |g_{ni}(\mathbf{s})| \Delta_{ni}$$

which, in view of (3.25), can be made smaller than $\alpha$ for sufficiently large $n$ by the choice of $\delta$. This together with (3.34) completes the proof of an analog of (3.33) for $D_2$, and hence of (3.29).

To prove (3.30), rewrite $U(x,t) - U(x,0) = U(x,t) - U^*(x,t) - \mathcal{U}(x,t)$, where

$$\mathcal{U}(x,t) = U^*(x,t) - U(x,0) = \sum_{i=1}^{n} [g_{ni}(t) - g_{ni}(0)]\, \alpha_i(x).$$

Thus, in view of (3.29), it suffices to prove

$$sup_{x,t}\, \tau_n^{-1} |\mathcal{U}(x,t)| = o_p(1). \tag{3.35}$$

From (3.17) applied with $\gamma_{ni} = [g_{ni}(t) - g_{ni}(0)]$, we readily obtain the $\forall\, t \in N_b$,

$$sup_x\, |\tau_n^{-1}\mathcal{U}(x,t) - (J_m(x)/m!)\tau_n^{-1} \sum_{i=1}^{n} [g_{ni}(t) - g_{ni}(0)]\, H_m(\eta_i)| = o_p(1).$$

Moreover, by the fact (3.3) applied with $q = m$ and by the assumption (3.27) it follows that $|\tau_n^{-1} \sum_{i=1}^{n} [g_{ni}(t) - g_{ni}(0)]\, H_m(\eta_i)| = o_p(1)$. Hence, $sup_x\, |\tau_n^{-1}\mathcal{U}(x,t)| = o_p(1)$ for every $t \in N_b$ while the uniformity with respect to $t$ follows from the assumption (3.28). This completes the proof of (3.30), which together with the assumptions $(F)$, (2.3) and the uniform (3.24b) imply (3.31) in an obvious fashion.

**Proof of Theorem 2.2.** Let

$$\boldsymbol{W}(x,t) = \boldsymbol{D}_n^{-1} \sum_{i=1}^{n} \dot{\boldsymbol{h}}_{ni}(\beta + \tau_n \boldsymbol{D}_n^{-1}t)\, I(\epsilon_i \le x + d_{ni})(t), \quad x \in \bar{\mathcal{I}},\ t \in N_b.$$

Then integration by parts shows that a.s., uniformly in $t \in N_b$, $\psi \in \Psi$,

$$\boldsymbol{D}_n^{-1}[\boldsymbol{M}(\beta + \tau_n \boldsymbol{D}_n^{-1}t) - \boldsymbol{M}(\beta)]$$
$$= \int [\boldsymbol{W}(x,t) - \boldsymbol{W}(x,0) - t\, f(x) - \boldsymbol{D}_n^{-1} \sum_{i=1}^{n} \dot{\boldsymbol{d}}_{ni}(t)\, F(x)]\, d\psi(x).$$

The result (2.11) will thus follow if we show that

$$sup_{x,t}\, \tau_n^{-1} \|\boldsymbol{W}(x,t) - \boldsymbol{W}(x,0) - t\, f(x) - \boldsymbol{D}_n^{-1} \sum_{i=1}^{n} \dot{\boldsymbol{d}}_{ni}(t)\, F(x)\| = o_p(1).$$

But this follows from the $k$ applications of the Lemma 3.2, $j$th time applied to the functions $g_{ni}(t) \equiv$ the $j$th coordinate of $\boldsymbol{D}_n^{-1}\dot{\boldsymbol{h}}_{ni}(\beta + \tau_n \boldsymbol{D}_n^{-1}t)$ and

$q_{ni}(t) \equiv d_{ni}(t)$. Verify that the assumed conditions of the theorem imply those of the Lemma 3.2. Details are left out for the sake of brevity.

**Proof of Proposition 2.1.** This again follows trivially from Lemma 3.2 upon now taking $g_{ni}(t) \equiv 1$ and $q_{ni}(t) \equiv d_{ni}(t)$.

## REFERENCES

DEHLING, H. AND TAQQU, M.S. (1989). The empirical process of some long-range dependent sequences with an application to U-statistics. *Ann. Statist.*, **17**, 1767-1783.

GRANGER, C.W.J. AND JOYEUX, R. (1980). An introduction to long-memory time series and fractional differencing. *J. Time Ser.Anal.*, **1**, 15-29.

HANNAN, E.J. (1971). Non-linear regression. *J. Appl.Probab.*, **8**, 767-780.

HOSKING, J.R.M. (1981). Fractional differencing. *Biometrika.*, **68**, 165-176.

HUBER, P.J. (1981). *Robust Statistics.* J. Wiley and Sons, N.Y.

JENNRICH, R.I. (1969). Asymptotic properties of non-linear least square estimators. *Ann. Math. Statist.*, **40**, 633-643.

KOUL, H. L. (1992a). *Weighted Empiricals and Linear Models.* I.M.S. Lecture Notes- Monograph Ser., **21**. Hayward, CA.

KOUL, H. L. (1992b). M-estimators in linear models with long-range dependent errors. *Statist. and Probab. Letters.*, **14**, 153-164.

KOUL, H. L. AND MUKHERJEE, K. (1993). Asymptotics of R-, MD- and LAD-estimators in linear regression models with long-range dependent errors.*Probab. Th. Rel. Fields*, **95**, 535-553.

MANDELBROT, B.B. AND VAN NESS, J.W. (1968). Fractional Brownian motions, fractional noises and applications. *SIAM Rev.*, **10, No. 4**, 422-437.

SEBER, G.A.F. AND WILD, C.J. (1989). *Nonlinear regression .* J. Wiley and Sons. N.Y.

WU, CHIEN-FU. (1981). Asymptotic theory of nonlinear least squares estimation. *Ann. Statist.*, **9**, 501-513.

# ORDER SELECTION, STOCHASTIC COMPLEXITY AND KULLBACK-LEIBLER INFORMATION

Abdelaziz EL MATOUAT
Ecole Normale Supérieure, Fès
Morocco

and

Marc HALLIN [1]
Institut de Statistique
Université Libre de Bruxelles
Belgium

ABSTRACT  The main motivation for Hannan and Quinn's $\varphi$ criterion is the minimality of its $\log \log n$ penalty rate, which is the slowest one compatible with consistency. Unlike Akaike's $AIC$, based on the minimization of some estimated mean Kullback-Leibler distance, or Rissanen and Schwarz's $BIC$, which minimizes an estimate of the expected *complexity*, Hannan and Quinn's $\varphi$ at first sight is not connected with any sound decision-theoretic or statistical principle. The objective of this paper is to provide a stochastic complexity justification and an information-theoretic derivation for $\varphi$. Two generalized $AIC - BIC - \varphi$ criteria also are proposed. Both are derived from Kullback-Leibler distance arguments, and enjoy the same consistency properties as $BIC$ (*strong* consistency) and $\varphi$ (*weak* consistency), respectively.

AMS 1980 subject classification: 62M10.
Key words and phrases. Order selection, time series, information criterion, stochastic complexity.

---

[1]Research supported by the Fonds d'Encouragement à la Recherche de l'Université Libre de Bruxelles and the Human Capital and Mobility contract ERB CT CHRX 940 963.

# 1   Introduction

Denote by $\Theta_1 \subset \Theta_2 \subset \ldots \subset \Theta_K$ a K-tuple of convex subsets of $\mathbf{R}^K$ such that $\Theta_k$ is of dimension $k$. For all $\theta \in \Theta_K$, let $k_0(\theta) = min\{k \mid \theta \in \Theta_k\}$: then

$$\Theta_k = \{\theta = (\theta_1, \ldots, \theta_K)' \in \Theta_K \mid \theta_{k+1} = \ldots = \theta_K = 0\},$$

and $k_0(\theta)$ can be referred to as the *dimension* of $\theta$.

Associated with $\Theta_K$, consider the parametric statistical model

$$(\mathbf{R}^d, \mathcal{B}^d, \mathcal{P} = \{f(.; \theta) \mid \theta \in \Theta_K\}),$$

where $f(.; \theta) = dP_\theta/d\mu$ denotes a probability density defined with respect to the Lebesgue measure $\mu$ on $(\mathbf{R}^d, \mathcal{B}^d)$. Let $\mathbf{X}^{(n)} = (\mathbf{X}_1, \ldots, \mathbf{X}_n)'$ be a sample of independent and identically distributed observations, with common density $f(.; \theta)$ ($\theta$ unspecified): the problem of estimating $k_0(\theta)$ is known as the *order selection* problem.

The same problem also can be formulated for non i.i.d. observations, and has been intensively investigated, in the time-series context, for the selection of the orders of $AR$ and $ARMA$ models.

It is well-known that a crude maximum-likelihood approach, in general, a.s. yields the overestimation $\hat{k} = K$. More subtle estimation principles thus have to be considered, if systematically selecting the highest possible model order is to be avoided. In a seminal paper (Akaike, 1973), which was the starting point of a considerable amount of subsequent research, Akaike proposed choosing $\hat{k}$ as the minimizing value of an estimate of the expected Kullback-Leibler distance

$$E_\theta\left\{K(f(.; \theta), f(.; \theta_k^{(n)}))\right\} = E_\theta\left\{\int \log\left[f(z; \theta)/f(z; \theta_k^{(n)})\right] f(z; \theta)d\mu\right\}$$
$$(1.1)$$

between $f(.; \theta)$ ($\theta$, the "true" unknown parameter value) and $f(.; \theta_k^{(n)})$, where $\theta_k^{(n)}$ denotes the maximum likelihood estimate of $\theta$ under the constraint $\theta \in \Theta_k$ :

$$\prod_{i=1}^n f(\mathbf{X}_i; \theta_k^{(n)}) = \max_{\theta \in \Theta_k}\left\{\prod_{i=1}^n f(\mathbf{X}_i; \theta)\right\}, \quad k = 1, \ldots, K. \quad (1.2)$$

This estimation principle yields the celebrated *Akaike information criterion*

$$AIC(k) = -\sum_{i=1}^n \log f(\mathbf{X}_i; \theta_k^{(n)}) + k \quad (1.3)$$

(the traditional factor 2 is omitted).

Order selection based on $AIC$ thus enjoys the very strong and elegant conceptual justifications inherited from its Kullback-Leibler distance interpretation; unfortunately, it is inconsistent, and asymptotically (in probability) leads to overparametrizing the models under consideration (Shibata 1976; Nishii 1988).

In order to palliate this inconsistency, Akaike (1977), Rissanen (1978) and Schwarz (1978) proposed modifying $AIC$ into

$$BIC(k) = -\sum_{i=1}^{n} \log f(\mathbf{X}_i; \theta_k^{(n)}) + \frac{1}{2}k \log n, \tag{1.4}$$

which is consistent under very general conditions, e.g., in the $AR$ (autoregressive) order selection context (see Nishii (1988) for a more general, *strong consistency* result). The derivation of $BIC$ by Akaike and Schwarz is limited to the $AR$ context. Schwarz's justification relies on Bayesian considerations, whereas Rissanen's derivation is not limited to $AR$ context, and is based on more general (less familiar, though, to most statisticians) *coding theory* arguments.

Let $\log^* k = \log_2 k + \log_2 \log_2 k + \ldots$ and $c = \sum_{k=1}^{\infty} 2^{-\log^* k}$. The *stochastic complexity* associated with the sample $\mathbf{X}_1 \ldots \mathbf{X}_n$ and the submodel corresponding to $\theta \in \Theta_k$ is defined as

$$C_k(\mathbf{X}_1, \ldots, \mathbf{X}_n) = -\sum_{i=2}^{n} \log_2 f(\mathbf{X}_i; \theta_k^{(i-1)}) + \log^* k + \log_2 c . \tag{1.5}$$

Rissanen mainly shows that minimizing an estimate of $E_\theta [C_k(\mathbf{X}_1 \ldots \mathbf{X}_n)]$ yields the $BIC$ criterion.

A third criterion was introduced later by Hannan and Quinn (1979), on the basis of the argument that the rate of increase $\log n$ of the penalty term in the $BIC$ criterion is not the lowest one compatible with strong consistency. Accordingly, they propose selecting $\hat{k}$ by minimizing

$$\varphi(k) = -\sum_{i=1}^{n} \log f(\mathbf{X}_i; \theta_k^{(n)}) + \frac{1}{2}k \log \log n . \tag{1.6}$$

The attractiveness of this criterion lies in the minimality of the penalty rate $\log \log n$, itself related with $LIL$ properties (see also Hannan, 1980). Regrettably, no further justification is provided, and the Hannan and Quinn $\varphi$ criterion at first sight is lacking the strong decisional motivations of its $AIC$ and $BIC$ competitors.

Now, parsimony hardly can be considered as a sound statistical principle. As stated by Findley (1991), *"the utility [of a parsimonious criterion] will be better explained by some deep principle which can be formulated without the assumption that one of the models considered is correct. Both Akaike's entropy maximization principle, motivating AIC, and Rissanen's minimum description length principle* (related with the stochastic complexity concept given above) *have such motivations*". The objective of this paper is to show that Hannan and Quinn's $\varphi$ criterion can be derived from the same decision-theoretic background as *BIC* and *AIC*. A stochastic complexity justification is provided for $\varphi$ in Section 2. In Section 3, two generalized *AIC* criteria are proposed, which both are derived from Kullback-Leibler distance arguments, and are shown to enjoy the same consistency properties as *BIC* (*strong* consistency) and $\varphi$ (*weak* consistency), respectively.

## 2   An information theoretic derivation of Hannan and Quinn's $\varphi$

Denote by $H(\theta)$ the *entropy* associated with $f(.; \theta)$ :

$$H(\theta) = -\int \log f(z; \theta) f(z; \theta) d\mu.$$

The following proposition is an immediate consequence of Rissanen (1986)'s Theorem 1, since $\log n > \log \log n$. All logarithms in this section have basis two.

**Proposition 2.1** . *For all (fixed) $k \in \{1, \ldots, K\}$, and for all $\theta \in \Theta_k$, assume that $\theta_k^{(n)}$ is a strongly consistent, best asymptotically normal estimate satisfying (as $n \to \infty$)*

$$nE_\theta \left[ (\theta_k^{(n)} - \theta)' \mathcal{I}(\theta)(\theta_k^{(n)} - \theta) \right] - k = o(1),$$

*where the Fisher information matrix*

$$\mathcal{I}(\theta) = E_\theta \left[ \left( grad_\theta \log f(\mathbf{X}; \theta) \right) \left( grad_\theta \log f(\mathbf{X}; \theta) \right)' \right]$$

*is assumed to be nonsingular. Then, for any $k \in \{1, \ldots, K\}$ and $\varepsilon > 0$, there exists a sequence $\Theta_k^{(n)}(\varepsilon)$ of subsets of $\Theta_k$ such that the Euclidean volume of $\Theta_k^{(n)}(\varepsilon)$ tends to zero (in $\mathbb{R}^k$), and*

$$\frac{2n}{k \log \log n} \left[ -H(\theta) + \frac{1}{n} E_\theta \left( C_k(\mathbf{X}_1, \ldots, \mathbf{X}_n) \right) \right] > 1 - \varepsilon$$

*for all $\theta \in \Theta_k \setminus \Theta_k^{(n)}(\varepsilon)$.*

Next, we reinforce Rissanen (1986)'s Theorem 3.

**Proposition 2.2.** *Assume that for all $\mathbf{x}$ and all $k \in \{1, \ldots, K\}$, $f(\mathbf{x}; .)$ is of class $C_3$ over the interior $\dot{\Theta}_k$ of $\Theta_k$ (with respect to $\mathbb{R}^k$), and let $\theta_k^{(n)}$*

*satisfy the same assumptions as in Proposition 2.1. Then, for any* $\beta \in (0,1)$ *and* $\varepsilon > 0$, *there exists* $N(\beta; \varepsilon)$ *such that*

$$\frac{2n}{kn^\beta \log \log n} \left[ -H(\theta) + \frac{1}{n} E_\theta \left( C_k(\mathbf{X}_1, \ldots, \mathbf{X}_n) \right) \right] < 1 + \varepsilon$$

*for all* $n \geq N, \theta \in \Theta_k$ *and* $k \in \{1, \ldots, K\}$.

**Proof.** Let $\theta \in \Theta_k$. As in Rissanen (1986)'s proof of Theorem 3, a Taylor expansion of $-\log f(\mathbf{X}_i; \theta_k^{(i-1)})$ yields

$$
\begin{aligned}
-\log f\left(\mathbf{X}_i; \theta_k^{(i-1)}\right) \;+\;& \log f(\mathbf{X}_i; \theta) \\
=\; -\;& \left(\theta_k^{(i-1)} - \theta\right)' grad_\theta \log f\left(\mathbf{X}_i; \theta\right) \\
+\; & \frac{1}{2} \left(\theta_k^{(i-1)} - \theta\right)' \Delta\left(\mathbf{X}_i; \theta\right) \left(\theta_k^{(i-1)} - \theta\right) \\
+\; & R^{(i)}\left(\mathbf{X}^{(i)}\right),
\end{aligned}
$$

where $E_\theta\left(-\log f(\mathbf{X}_i; \theta)\right) = H(\theta)$,

$$\left| E_\theta \left[ R^{(i)}\left(\mathbf{X}_i\right) \right] \right| = O\left( i^{-3/2} \right)$$

as $i \to \infty$ (i.e., there exists $M$ such that $\left| E_\theta \left[ R^{(i)}\left(\mathbf{X}^{(i)}\right) \right] \right| < M i^{-3/2}$, except perhaps for a finite number of values of $i$), and

$$E_\theta \left[ \Delta\left(\mathbf{X}^i; \theta\right) \right] = \mathcal{I}(\theta).$$

Due to the convergence assumptions on $\theta_k^{(i)}$,

$$\left| E_\theta \left[ \left(\theta_k^{(i-1)} - \theta\right)' \mathcal{I}(\theta)(\theta_k^{(i-1)} - \theta) \right] - \frac{k}{i-1} \right| = \frac{\rho^{(i-1)}}{i-1},$$

with $\rho^{(i)} = o(1)$ as $i \to \infty$. It then follows from the independence between $\mathbf{X}_i$ and $\theta_k^{(i-1)}$ that

$$
\begin{aligned}
& \left| \frac{1}{n} E_\theta \left[ C_k(\mathbf{X}_1, \ldots, \mathbf{X}_n) \right] - H(\theta) \right| \\
=\; & \left| \frac{1}{n} \left\{ \sum_{i=2}^{n} E_\theta \left[ -\log f(\mathbf{X}_i; \theta_k^{(i-1)}) + \log f(\mathbf{X}_i; \theta) \right] + \log^* k + \log_2 c \right\} \right| \\
\leq\; & \frac{k}{2n} \sum_{i=2}^{n} \frac{1}{i-1} + \frac{1}{2n} \sum_{i=2}^{n} \left[ \frac{\rho^{(i-1)}}{i-1} + 2M i^{-3/2} \right] + \frac{\log^* k + \log_2 c}{n}.
\end{aligned}
$$

For $n$ sufficiently large,

$$
\frac{1}{n^\beta \log\log n} \sum_{i=2}^{n} \frac{1}{i-1} \leq \frac{1}{\log\log n}\left[1 + \sum_{i=2}^{n-1} \frac{1}{i\log i}\right]
$$

$$
= \frac{1}{\log\log n}\left[\log\log n + o(\log\log n)\right]
$$

$$
= 1 + o(1).
$$

Since $\rho^{(i)} = o(1)$ as $i \to \infty$, it follows that both

$$
\frac{1}{n^\beta \log\log n)} \sum_{i=2}^{n} \frac{\rho^{(i-1)}}{(i-1)}
$$

and

$$
\frac{1}{n^\beta \log\log n)} \sum_{i=2}^{n} i^{-3/2},
$$

are $o(1)$ as $n \to \infty$. The proposition follows.   ∎

Propositions 2.1 and 2.2 imply that, under the assumptions made, for any $\beta \in (0,1)$ and $\varepsilon > 0$, there exists $N(\beta; \varepsilon)$ such that

$$
H(\boldsymbol{\theta}) + (1-\varepsilon)\frac{k\log\log n}{2n} < \frac{1}{n}E_{\boldsymbol{\theta}}\left(C_k(\mathbf{X}_1,\ldots,\mathbf{X}_n)\right)
$$

$$
< H(\boldsymbol{\theta}) + (1+\varepsilon)\frac{k\log\log n}{2n^{1-\beta}}. \quad (1.7)
$$

The right hand side in (2.1) provides an upper bound for the expected complexity $E_{\boldsymbol{\theta}}\left(C_k(\mathbf{X}_1,\ldots,\mathbf{X}_n)\right)$, yielding a one-parameter family of criteria based on the minimization of

$$
\varphi_\beta(k) = -\sum_{i=1}^{n} \log f(\mathbf{X}_i; \boldsymbol{\theta}_k^{(n)}) + \frac{1}{2}kn^\beta \log\log n, \quad \beta \in (0,1/2). \quad (1.8)
$$

The left hand side of (2.1) clearly corresponds to Hannan and Quinn's $\varphi$ criterion. For fixed $\beta$ and $n$, there is no guarantee that these two minimizations lead to the same selection of $k$ and, therefore, none of them can be interpreted as a minimization of the expected complexity. However, for fixed $n$, there exists a $\beta_0$ such that for all $0 < \beta < \beta_0$ the probability that $\varphi_\beta(k)$ and $\varphi(k)$ yield the same order selection (hence, also the value of $k$ minimizing the expected complexity) can be made as close to one as desired since, for all $k$,

$$
\varphi_\beta(k) - \varphi(k) = o_P(1) \quad (1.9)
$$

as $\beta \to 0$. For sufficiently small values of $\beta$, the order selection rules based on $\varphi_\beta(k)$ and $\varphi(k)$ thus enjoy the same stochastic complexity interpretation as the $BIC$ criterion. Our objective of providing a stochastic complexity interpretation for Hannan and Quinn's $\varphi$ criterion is thus fulfilled, since $\varphi(k)$ does not depend on $\beta$. Hannan and Quinn's criterion consequently can be considered as the limit, as $\beta \rightarrow 0$, of a sequence $\varphi_\beta(\,.\,)$ of criteria based on the same information–theoretic argument as $BIC$.

# 3   A Kullback-Leibler approach to $BIC$ and $\varphi$

Still, many a statistician may feel uneasy with the above stochastic complexity justification of Hannan and Quinn's $\varphi$, and, notwithstanding non-consistency, is likely to be more satisfied with the more familiar, and more widely accepted Kullback-Leibler justification of $AIC$. However, the non-consistency problem, with $AIC$, is not to be blamed upon the Kullback-Leibler information approach. The moot point actually lies in considering the Kullback distance (1.1) between $f(.;\theta)$ and $f(.;\theta_k^{(n)})$, where $\theta_k^{(n)}$ denotes a maximum likelihood estimate based on the complete series of observation. As Rissanen (1986, p. 1089) points out, such an estimate, from a predictive point of view, is not a *"honest"* one, and the Kullback distance should rather be taken between $\prod_{i=1}^n f(\mathbf{X}_i; \theta)$ and the *predictive likelihood* $\prod_{i=1}^n f(\mathbf{X}_i; \theta_k^{(i-1)})$. Now,

$$
E_\theta\left[\log\left(\frac{\prod_{i=1}^n f(\mathbf{X}_i; \theta)}{\prod_{i=1}^n f(\mathbf{X}_i; \theta_k^{(i-1)})}\right)\right]
$$
$$
= E_\theta\left[\log\left(\frac{\prod_{i=1}^n f(\mathbf{X}_i; \theta)}{\prod_{i=1}^n f(\mathbf{X}_i; \theta_k^{(n)})}\right)\right] + E_\theta\left[\log\left(\frac{\prod_{i=1}^n f(\mathbf{X}_i; \theta_k^{(n)})}{\prod_{i=1}^n f(\mathbf{X}_i; \theta_k^{(i-1)})}\right)\right].
$$

The first term in the right-hand side of this equality is the Kullback contrast considered by Akaike (1973) in his derivation of $AIC$, whereas the second one is the one appearing in the definition of stochastic complexity. Replacing the Kullback distance $K(f(.;\theta), f(.;\theta_k^{(n)}))$ with the Kullback distance

$$
K(\prod_{i=1}^n f(\,.\,;\theta), \prod_{i=1}^n f(\,.\,;\theta_k^{(i-1)}))
$$

thus leads to the following modifications of $AIC$ and $\varphi_\beta$, $\beta \in (0, 1/2)$, respectively:

$$
AIC^*(k) = -\sum_{i=1}^n \log f(\mathbf{X}_i; \theta_k^{(n)}) + k(1 + \frac{\log n}{2}), \tag{1.10}
$$

and

$$
\varphi_\beta^*(k) = -\sum_{i=1}^n \log f(\mathbf{X}_i; \theta_k^{(n)}) + k(1 + n^\beta \frac{\log\log n}{2}). \tag{1.11}
$$

For $\beta > 0$ sufficiently small, $\varphi_\beta^*$ clearly yields the same order selection as

$$\varphi^*(k) = -\sum_{i=1}^{n} \log f(\mathbf{X}_i; \theta_k^{(n)}) + k(1 + \frac{\log \log n}{2}). \qquad (1.12)$$

$AIC^*$ thus appears as a compromise between $AIC$ and $BIC$, $\varphi^*$ as a compromise between $AIC$ and $\varphi$. Both $AIC^*$ and $\varphi^*$ retain $AIC$'s Kullback-Leibler interpretation. Moreover, they both are consistent. The strong consistency of the estimate $\hat{k}$ associated with $AIC^*$, and the weak consistency of that associated with $\varphi^*$ indeed readily follow from Nishii (1988).

# 4   References

[1] Akaike, H. (1973). Information theory and an extension of the maximum likelihood principle. In *2nd International Symposium of Information Theory*, B.N. Petrov and F.Csaki Eds, Academia Kiado, Budapest, 267-281.

[2] Akaike, H. (1977). An entropy maximisation principle. In *Applications of Statistics*, P.R. Krishnaiah, Ed., North Holland, Amsterdam and New York, 27-41.

[3] El Matouat, A. (1987). Sélection du nombre de paramètres d'un modèle: comparaison avec le critère d'Akaike. *Doctorat d'Université*, Rouen, France.

[4] Findley, D. (1991). Counterexamples to parsimony and BIC. *Ann. Inst. Statist. Math.* **43**, 505-514.

[5] Hannan, E.J. and Quinn, B.G. (1979). The determination of the order of an autoregression. *J.R.S.S.* B **41**, 190-195.

[6] Hannan, E.J. (1980). The determination of the order of an ARMA process. *Ann. Statist.* **8**, 1071-1081.

[7] Nishii, R. (1988). Maximum likelihood principle and model selection when the true model is unspecified. *J. of Mult. Anal.* **27**, 392-403.

[8] Rissanen, J. (1978). Modeling by shortest data description. *Automatica* **14**, 465-471.

[9] Rissanen, J. (1986). Stochastic complexity and modeling. *Ann. Statist.* **14**, 1080-1100.

[10] Shibata, R. (1976). Selection of the order of an autoregressive model by Akaike's Criterion. *Biometrika* **63**, 117-126.

[11] Schwarz, G. (1978). Estimating the dimension of a model. *Ann. Statist.* 6, 461-464.

# EFFICIENCY GAINS FROM
# QUASI-DIFFERENCING UNDER NONSTATIONARITY[*]

Peter C. B. Phillips
*Cowles Foundation for Research in Economics*
*Yale University*

and

Chin Chin Lee
*Department of Economics*
*London School of Economics and Political Science*

16 March 1996

## Abstract

A famous theorem on trend removal by OLS regression (usually attributed to Grenander and Rosenblatt (1957)) gave conditions for the asymptotic equivalence of GLS and OLS in deterministic trend extraction. When a time series has trend components that are stochastically nonstationary, this asymptotic equivalence no longer holds. We consider models with integrated and near-integrated error processes where this asymptotic equivalence breaks down. In such models, the advantages of GLS can be achieved through quasi-differencing and we give an asymptotic theory of the relative gains that occur in deterministic trend extraction in such cases. Some differences between models with and without intercepts are explored.

## 1. Introduction

Grenander and Rosenblatt (1957) analysed asymptotic efficiency conditions in time series regressions with stationary errors. They considered

---

[*]A first draft containing some of the results reported here was written in 1993. The present paper is an abridged version of Phillips and Lee (1996). Our thanks go to the NSF for research support under Grant Numbers SES 9122142 and SBR 94-22922.

univariate regression models with trends such as $y_t = \beta' z_t + u_t$, where $z_t = (1, t, ..., t^k)$ and $u_t$ is stationary with spectral density $f_u(\lambda) > 0$, and demonstrated the asymptotic equivalence of GLS and OLS trend extraction techniques. Hannan (1970, chapter VII) extended the Grenander-Rosenblatt theory to the case of multivariate time series regressions and provided a general treatment of the subject. The Grenander-Rosenblatt result relies on the continuity of the spectrum of $u_t$ at the origin (where the spectral mass of $z_t$ in the above model is concentrated) and it is satisfied in most models that involve stationary time series. But, the condition is violated when there is a unit root in the data generating process of $u_t$. In fact, the condition fails whenever $u_t$ is strongly dependent or integrated of order $d$ with $d > 0$ (denoted as I($d$)). For in that case, the spectral density of $u_t$ behaves like a multiple of $\lambda^{-2d}$ as $\lambda \to 0$ and is unbounded at $\lambda = 0$. In such cases as these, the asymptotic equivalence of GLS and OLS breaks down and we can achieve efficiency gains in estimating the trend coefficients $\beta$ by using GLS methods. When $u_t$ is near-integrated in the sense that it has an autoregressive root that is local to unity, there is again a peak at the origin in its spectrum and we can still expect gains to accrue from the use of GLS estimation.

The present contribution calculates the efficiency gains in GLS trend extraction when $u_t$ is an integrated or near-integrated process. These cases are the most commonly studied in the econometrics literature, they have bearing on the issue of unit root testing, and they lend themselves to simple quasi-differencing formulations that are convenient in practical work. In contrast to the integrated and near-integrated cases, the effects of strong dependence on the efficiency of OLS have received attention in the literature. In particular, Yajima (1988), Beran (1994, ch. 9) and Samarov and Taqqu (1988) study GLS efficiency gains in models with stationary long-memory errors where $0 \leq d < 1/2$. The case where there are nonstationary strongly dependent errors with $1/2 < d < 1$ was analysed in Lee and Phillips (1994).

## 2. Efficiency Gains in Models with Near Integrated Errors

Suppose a time series $y_t$ is generated by

$$y_t = \beta'_k z_{kt} + u_t , \quad t = 1, ..., T , \tag{1}$$
$$u_t = \alpha u_{t-1} + \varepsilon_t , \quad \alpha = 1 + c/T ,$$

where $z_{kt} = (t, ..., t^k)'$, and $c$ is a constant that represents local departures from unity. The parameter setting $\alpha = 1 + c/T$ facilitates efficiency calculations using local-to-unity asymptotics (see Phillips, 1987a, and Chan and

Wei, 1987). The Grenander-Rosenblatt theory applies when $|\alpha| < 1$, and our interest is in the unit root and intermediate cases. Hence, attention here focuses on the domain $c \in (-\infty, 0]$.

Initial conditions for $u_t$ are set at $t = 0$ and $u_0$ may be any random variable with finite variance $\sigma_0^2$. Cases where $\sigma_0^2 \to \infty$ are sometimes of interest and these can correspond to situations where the initial conditions are in the increasingly distant past, although observations on the process $y_t$ are available only from $t = 1$. The effect of such alternative initializations on our results are considered later.

The primary requirement on the shocks $\varepsilon_t$ is that normalized partial sums $S_t = \sum_{s=1}^{t} \varepsilon_s$ of $\varepsilon_t$ satisfy an invariance principle and this will be so under a wide variety of differing conditions on $\varepsilon_t$. The following conditions on $\varepsilon_t$ are sufficient for the limit theory here.

### 2.1 Assumption EC *(Error Conditions)*

(i) $E\varepsilon_t = 0 \ \forall t$; (ii) $\sup_t E|\varepsilon_t|^{b+\delta} < \infty$ *for some* $b > 2$ *and* $\delta > 0$; (iii) $\sigma^2 = \lim E(S_T^2/T)$ *exists, and* $\sigma^2 > 0$; (iv) $\varepsilon_t$ *is strong mixing with coefficients* $\alpha_m$ *that satisfy* $\sum_{m=1}^{\infty} \alpha_m^{1-2/b} < \infty$.

In the following, we use $W(r)$ to denote standard Brownian motion and $J_c(r) = \int_0^r e^{(r-s)c} dW(s)$ to denote a linear diffusion process. Note that $J_c(r)$ satisfies the linear stochastic differential equation $dJ_c(r) = cJ_c(r)dr + dW(r)$. Under Assumption EC we have:

### 2.2 Lemma

(i) $T^{-1/2}S_{[Tr]} \Rightarrow \sigma W(r)$; (ii) $D_{kT}^{-1/2} \sum_{t=2}^{T} z_{kt}\varepsilon_t \Rightarrow \sigma \int_0^1 g_k(r)dW(r)$; (iii) $T^{-1/2}u_{[Tr]} \Rightarrow \sigma J_c(r)$; (iv) $T^{-1}D_{kT}^{-1/2} \sum_{t=1}^{T} z_{kt}u_t \Rightarrow \sigma \int_0^1 g_k(r)J_c(r)dr$;

*where* $D_{kT} = \text{diag}(T^3, T^5, ..., T^{2k+1})$, $g_k(r)' = (r, ..., r^k)$ *and* $\Rightarrow$ *signifies weak convergence.*

Simple least squares regression on (1) leads to the trend coefficient estimator $\widehat{\beta}_{kc} = \left( \Sigma_1^T z_{kt} z_{kt}' \right)^{-1} (\Sigma_1^T z_{kt} y_t)$. GLS regression requires use of the full covariance structure of the error process $u_t$. The Grenander-Rosenblatt theory can be expected to cover contributions to the covariance structure that come from the stationary or weakly dependent components $\varepsilon_t$, but not those that come from the autoregressive root $\alpha = 1 + c/T$ since it is the latter that produces a peak in the spectrum of $u_t$. Hence, as an alternative to OLS, we consider a partial GLS detrending procedure that is based on the quasi-differenced data $\widetilde{z}_{kt} = z_{kt} - \alpha z_{kt-1}$ and $\widetilde{y}_t = y_t - \alpha y_{t-1}$ for $t = 2, ..., T$, combined with the initial observations $\tilde{z}_{k1} = z_{k1}$, $\tilde{y}_1 = y_1$ for $t = 1$. This leads to the estimator $\tilde{\beta}_{kc} = \left(\Sigma_1^T \tilde{z}_{kt} \tilde{z}_{kt}'\right)^{-1} \Sigma_1^T \tilde{z}_{kt} \tilde{y}_t$.

We show that the partial GLS estimator $\tilde{\beta}_{kc}$ of $\beta_k$ in (1) is asymptotically more efficient than $\hat{\beta}_{kc}$ under both a unit root ($c = 0$) and a near unit root ($c < 0$). The following results give the limit distributions of these estimators.

### 2.3 Theorem

$$F_{kT}^{1/2}(\hat{\beta}_{kc} - \beta_k) \Rightarrow \sigma Q_k^{-1} \int_0^1 g_k(r) J_c(r) dr \equiv N(0, V_{kc}^{ols})$$

where $F_{kT}^{1/2} = T^{-1} D_{kT}^{1/2} = \mathrm{diag}(T^{1/2}, T^{3/2}, ..., T^{k-1/2})$, $Q_k = \int_0^1 g_k(r) g_k(r)' dr$ is a $k \times k$ matrix with elements $q_{ij} = 1/(i+j+1)$ and

$$V_{kc}^{ols} = \sigma^2 Q_k^{-1} \int_0^1 \int_0^1 g_k(r) e^{(r+s)c}(1/2c)(1 - e^{-2c(r \wedge s)}) g_k(s)' dr ds Q_k^{-1'}.$$

### 2.4 Theorem

$$F_{kT}^{1/2}(\tilde{\beta}_{kc} - \beta_k) \Rightarrow \sigma \left[\int_0^1 f_{ck}(r) f_{ck}(r)' dr\right]^{-1} \int_0^1 f_{ck}(r) dW(r) \equiv N(0, V_{kc}^{gls}),$$

with

$$V_{kc}^{gls} = \sigma^2 \left[\int_0^1 f_{ck}(r) f_{ck}(r)' dr\right]^{-1} = \sigma^2 \left[\overline{Q}_k + c^2 Q_k - c(\tilde{Q}_k + \tilde{Q}_k')\right]^{-1}.$$

Here, $f_{ck}(r) = g_k^{(1)}(r) - cg_k(r)$, $g_k^{(1)}(r) = (1, 2r, ..., kr^{k-1})'$, and $\overline{Q}_k$ and $\tilde{Q}_k$ are $k \times k$ matrices with elements $\overline{q}_{ij} = ij/(i+j-1)$ and $\tilde{q}_{ij} = i/(i+j)$, respectively.

Define the relative efficiency of $\hat{\beta}_{kc}$ to $\tilde{\beta}_{kc}$ by $R_{kc} \equiv \det(V_{kc}^{ols}) / \det(V_{kc}^{gls})$. To provide some illustrative comparisons, take the case of the linear trend model where $k = 1$ in (1). Then, when $c = 0$, $T^{1/2}(\hat{\beta}_{10} - \beta_1) \Rightarrow N(0, 6\sigma^2/5)$ — a result obtained earlier in Durlauf and Phillips (1988). On the other hand, $T^{1/2}(\tilde{\beta}_{10} - \beta_1) \Rightarrow \sigma W(1) = N(0, \sigma^2)$. Hence, for linear trend extraction there is an asymptotic efficiency gain of 20% from the use of the partial GLS estimator $\tilde{\beta}_{10}$ when $u_t$ is integrated of order 1. When $c \neq 0$, the variances of the limit variates are

$$V_{1c}^{ols} = 9\sigma^2[3e^{2c}(c-1)^2 + 2c^3 + 3c^2 - 3]/6c^5, \quad \text{and} \quad V_{1c}^{gls} = 3\sigma^2/(3 - 3c + c^2).$$

In this case, the relative efficiency $R_{1c} = V_{1c}^{ols}/V_{1c}^{gls}$, is graphed against negative values of $c$ in Figure 1. As $c \to -\infty$, $R_{1c} \to 1$, so there are no gains from the use of GLS–detrending in the limiting case. This is to be expected since $c \to -\infty$ is the limit of the domain of definition of $c$ that corresponds to the stationary case, for which the Grenander–Rosenblatt asymptotic equivalence result holds. Figure 1 also shows that the maximum gains in efficiency from GLS occur for finite $c < 0$, rather than at zero.

In the general case, write the limit variates from theorems 2.3 and 2.4 as $\widehat{Z}_c = \sigma Q_k^{-1} \int_0^1 g_k(r) J_c(r) dr$, and $\widetilde{Z}_c = \sigma \left[ \int_0^1 f_{ck}(r) f_{ck}(r)' dr \right]^{-1} \int_0^1 f_{ck}(r) dW(r)$. Then, as $c \to -\infty$ the asymptotic equivalence of $\widehat{Z}_c$ and $\widetilde{Z}_c$ is given by

**2.5 Theorem** $\sqrt{-c} \left( \widehat{Z}_c - \widetilde{Z}_c \right) \xrightarrow{p} 0$ and $R_{kc} \to 1$ as $c \to -\infty$.

### 3. The Effects of a Fitted Intercept

A constant term is not included in (1) because the intercept is not consistently estimable. Nevertheless, it is usual in empirical work for regression detrending procedures to involve fitted intercepts. So it is of some interest to consider the asymptotic behaviour of the estimators $\widetilde{\beta}_{kc}$ and $\widehat{\beta}_{kc}$ in this case. In related work, Canjels and Watson (1995) studied the case of linear trend extraction with a fitted intercept and near integrated errors. The treatment that follows considers the case of general polynomial trends with fitted intercepts and near integrated errors, and indicates some subtleties in the limit theory that arise as $c \to -\infty$ due to the doubly-infinite triangular array structure of $y_t$.

Consider the following model in place of (1)

$$y_t = \beta_0 + \beta_k' z_{kt} + u_t = \beta' z_t + u_t , \quad t = 1, ..., T . \qquad (2)$$

It turns out that when the localizing parameter $c$ is fixed, the presence of the constant term $\beta_0$ in the regression (2) does not influence the asymptotic distribution of the partial GLS estimator $\widetilde{\beta}_{kc}$. To see this, note that

$$\Sigma_1^T \widetilde{z}_t \widetilde{z}_t' = \begin{bmatrix} 1 + \frac{c^2(T-1)}{T^2} & \widetilde{z}_{k1}' - \frac{c}{T} \Sigma_2^T \widetilde{z}_{kt}' \\ \widetilde{z}_{k1} - \frac{c}{T} \Sigma_2^T \widetilde{z}_{kt} & \Sigma_1^T \widetilde{z}_{kt} \widetilde{z}_{kt}' \end{bmatrix} \qquad (3)$$

where $\tilde{z}_t = z_t - \alpha z_{t-1} = [-c/T, \tilde{z}'_{kt}]' = [-c/T, \Delta z'_{kt} - (c/T)z'_{kt-1}]'$ for $t = 2, ..., T$, and $\tilde{z}_1 = z_1 = [1, \tilde{z}'_{k1}]' = [1, z'_{k1}]'$. Setting $D_T = diag(1, F_{kT})$, we have

$$D_T^{-1/2} \left( \Sigma_1^T \tilde{z}_t \tilde{z}'_t \right) D_T^{-1/2} \rightarrow \begin{bmatrix} 1 & 0 \\ 0 & \int_0^1 f_{ck}(r) f_{ck}(r)' dr \end{bmatrix}. \tag{4}$$

Since this matrix is block diagonal, it follows that the inclusion of a fitted intercept in a partial GLS regression on (2) does not alter the asymptotic distribution of the estimates of the trend coefficient vector $\beta_k$ that is given in theorem 2.4. Thus, the GLS estimates of $\beta_k$ have the same limit distribution whether or not an intercept is included in the regression. This result depends critically on the assumption that the localizing parameter is fixed.

Unlike $\tilde{\beta}_{kc}$, the limit distribution of the OLS trend estimator $\hat{\beta}_{kc}$ is affected by a fitted intercept. In this case, the limit distribution is found to be

$$F_{kT}^{1/2}(\hat{\beta}_{kc} - \beta_k) \Rightarrow \sigma H_k^{-1} \int_0^1 (g_k(r) - h_k)J_c(r)dr = \hat{Z}_{mc}, \text{ say,} \tag{5}$$
$$\equiv N(0, V_{kc}^{fmols}),$$

where $H_k$ is $k \times k$ with elements $h_{ij} = 1/(i+j+1) - 1/(i+1)(j+1)$ and $h_k$ is $k$-vector column with i'th element $1/(i+1)$.

In the linear trend case ($k = 1$), the asymptotic relative efficiency of $\hat{\beta}_{1c}$ to $\tilde{\beta}_{1c}$ is

$$R_{1c}^g = \frac{6c^{-5}(3\exp(2c)(c^2 - 4c + 4) + 12c\exp(c)(c - 2) + 2c^3 + 9c^2 + 12c - 12)}{3/(3 - 3c + c^2)}, \tag{6}$$

which is plotted in Figure 2. Note that the limit of $R_{1c}^g$ as $c \rightarrow -\infty$, does not appear from this figure to be unity, as would be expected from the Grenander-Rosenblatt theory, a point that is unnoticed in the work of Canjels and Watson (1995). In fact, a simple calculation shows that $R_{1c}^g \rightarrow 4$ as $c \rightarrow -\infty$.

We now show why GLS appears to be more efficient than OLS at the boundary of the domain of definition of $c$. First, observe that the asymptotic theory for $\widehat{\beta}_{kc}$ and $\widetilde{\beta}_{kc}$ essentially involves a triangular array limit theory when both $|c|$ and $T$ are large. Limits taken along different diagonals of this array are not necessarily equivalent. To fix ideas, the partial GLS estimator of $\beta_k$ in (2) has the explicit form

$$
\widetilde{\beta}_{kc} = \left[ \Sigma_1^T \widetilde{z}_{kt} \widetilde{z}_{kt}' - \left( z_{k1} - \frac{c}{T} \Sigma_2^T \widetilde{z}_{kt} \right) \left( 1 + \frac{c^2}{T} \right)^{-1} \left( z_{k1} - \frac{c}{T} \Sigma_2^T \widetilde{z}_{kt} \right)' \right]^{-1}
$$
$$
\times \left[ \Sigma_1^T \widetilde{z}_{kt} \widetilde{y}_t - \left( z_{k1} - \frac{c}{T} \Sigma_2^T \widetilde{z}_{kt} \right) \left( 1 + \frac{c^2}{T} \right)^{-1} \left( \widetilde{y}_1 - \frac{c}{T} \Sigma_2^T \widetilde{y}_t \right) \right], \quad (7)
$$

and for $c^2/T$ large this is approximately

$$
\widecheck{\beta}_k = \left[ \Sigma_1^T \widetilde{z}_{kt} \widetilde{z}_{kt}' - T^{-1} \left( \Sigma_1^T \widetilde{z}_{kt} \right) \left( \Sigma_1^T \widetilde{z}_{kt} \right)' \right]^{-1} \left[ \Sigma_1^T \widetilde{z}_{kt} \widetilde{y}_t - T^{-1} \left( \Sigma_1^T \widetilde{z}_{kt} \right) \left( \Sigma_1^T \widetilde{y}_t \right) \right]
$$
$$
= \left[ \Sigma_1^T \left( \widetilde{z}_{kt} - \overline{\widetilde{z}}_k \right) \left( \widetilde{z}_{kt} - \overline{\widetilde{z}}_k \right)' \right]^{-1} \left[ \Sigma_1^T \left( \widetilde{z}_{kt} - \overline{\widetilde{z}}_k \right) \widetilde{y}_t \right], \quad (8)
$$

which is the OLS regression coefficient of $\widetilde{z}_{kt}$ in a regression of $\widetilde{y}_t$ on $\widetilde{z}_{kt}$ with a fitted intercept. Call this estimator the *fitted mean* GLS estimator of $\beta_k$ in (2). Now, in place of theorem 2.4, we get the following asymptotic distribution theory.

**3.1 Theorem** *The limit distribution of the fitted mean GLS estimator $\widecheck{\beta}_k$ in (2) is given by*

$$
F_{kT}^{1/2}(\widecheck{\beta}_k - \beta_k) \Rightarrow \sigma \left[ \int_0^1 \overline{f}_{ck}(r) \overline{f}_{ck}(r)' dr \right]^{-1} \left[ \int_0^1 \overline{f}_{ck}(r) dW(r) \right] \equiv N(0, V_{kc}^{fmgls}),
$$

*where $\overline{f}_{ck}(r) = f_{ck}(r) - \int_0^1 f_{ck}(r) dr$, and $V_{kc}^{fmgls} = \sigma^2 \left[ \int_0^1 \overline{f}_{ck}(r) \overline{f}_{ck}(r)' dr \right]^{-1}$.*

Let $\widecheck{Z}_{mc} = \sigma \left[ \int_0^1 \overline{f}_{ck}(r) \overline{f}_{ck}(r)' dr \right]^{-1} \left[ \int_0^1 \overline{f}_{ck}(r) dW(r) \right]$ be the variate representing the limit distribution of the fitted mean GLS estimator given in theorem 3.1. Define the efficiency ratio of this estimator against OLS as $R_{kc}^m \equiv \det(V_{kc}^{fmols}) / \det(V_{kc}^{fmgls})$. Then, we have accordance with the Grenander-Rosenblatt theory at the limit of the domain of definition of $c$ as follows.

**3.2 Theorem:** $\sqrt{-c} \left( \widehat{Z}_{mc} - \widecheck{Z}_{mc} \right) \xrightarrow{p} 0$ and $R_{kc}^m \to 1$ as $c \to -\infty$.

An intercept in the regression also affects the asymptotics of the partial GLS estimator when $c^2/T \to c_1$, or equivalently, $c/\sqrt{T} \to c_0$, where $c_0$ is some finite negative constant and $c_1 = c_0^2$. In this case when $c/\sqrt{T} \sim c_0$, the partial GLS estimator given in (7) is approximately

$$
\begin{aligned}
\widehat{\beta}_k &= \beta_k + \left[ \Sigma_1^T \widetilde{z}_{kt} \widetilde{z}_{kt}' - \frac{c_1}{1+c_1} \frac{1}{T} \left( \Sigma_2^T \widetilde{z}_{kt} \right) \left( \Sigma_2^T \widetilde{z}_{kt} \right)' \right]^{-1} \\
&\quad \times \left[ \Sigma_1^T \widetilde{z}_{kt} \widetilde{u}_t + \frac{c_0}{1+c_1} \frac{1}{\sqrt{T}} \left( \Sigma_2^T \widetilde{z}_{kt} \right) \left( u_1 - c_0 \frac{1}{\sqrt{T}} \Sigma_2^T \widetilde{u}_t \right) \right].
\end{aligned}
$$

Then, in place of theorem 3.1, we get the following limit theory as $T \to \infty$ for $\widehat{\beta}_k$

$$
F_{kT}^{1/2} (\widehat{\beta}_k - \beta_k) \Rightarrow V^{-1} \left[ \sigma \int_0^1 f_{ck}^{c_1}(r) dW(r) + \frac{c_0}{1+c_1} \left( \int_0^1 f_{ck}(r) dr \right) u_1 \right],
$$

where $V = \int_0^1 f_{ck}(r) f_{ck}(r)' dr - (c_1/(1+c_1)) \left( \int_0^1 f_{ck}(r) dr \right) \left( \int_0^1 f_{ck}(r) dr \right)'$, and $f_{ck}^{c_1}(r) = f_{ck}(r) - (c_1/(1+c_1)) \int_0^1 f_{ck}(s) ds$. This limit distribution is somewhat unusual because the first period error term $u_1$ plays a role in the asymptotics. This is explained by the fact that the normalized second moment matrix (3) is not block diagonal in the limit as $T \to \infty$ when $c/\sqrt{T} \sim c_0$, the intercept in (2) is not consistently estimated, and consequently $u_1$ has an effect on the limit distribution of $\widehat{\beta}_k$.

Setting $\sigma_1^2 = var(u_1)$ and $h_{ck} = \int_0^1 f_{ck}(r) dr$, the variance of the limiting distribution of $\widehat{\beta}_k$ is

$$
V_{kc}^g = V^{-1} \left\{ \sigma^2 \int_0^1 f_{ck}^{c_1}(r) f_{ck}^{c_1}(r)' dr + \sigma_1^2 \frac{c_1}{(1+c_1)^2} h_{ck} h_{ck}' \right\} V^{-1}.
$$

To illustrate, take the case $k = 1$ and suppose $\sigma_1^2 = 0$. The relative asymptotic efficiency $R_{1c}^0 \equiv V_{1c}^{fmols}/V_{1c}^g$ of the partial GLS estimator $\widehat{\beta}_1$ and the OLS estimator is plotted in Figure 3 against $c_0$. As is apparent from the figure, the efficiency curve tends to 4 as $c_0 \to 0$ and tends to 1 as $c_0 \to -\infty$.

Finally, we go back to the direct comparison of OLS and GLS in the model (2). In the general case, define the efficiency ratio of the fitted mean OLS estimator $\widehat{\beta}_{kc}$ and the GLS trend coefficient estimator $\widetilde{\beta}_{kc}$ by $R_{kc}^g \equiv$

$\det(V_{kc}^{fmols})/\det(V_{kc}^{gls})$. The limiting behaviour of this ratio as $c \to -\infty$ is given in the next result.

**3.3 Theorem** $R_{kc}^g \to (k+1)^2$ *as* $c \to -\infty$.

For $k = 1$ this reduces to the earlier result discussed above, where $R_{1c}^g \to 4$. The factor $(k+1)^2$ measures the additional variance in the limit of the OLS procedure that is due to estimating an intercept in the regression.

## 4. Alternative Initializations

The results given above rely on the initial observation $u_0$ having constant variance $\sigma_0^2$, so that $u_0 = O_{a.s.}(1)$ as $T \to \infty$. There is some merit to making assumptions about $u_0$ which give it properties that are analogous to those of $u_t$ itself. This can be done by putting the initial conditions that determine $u_0$ into the increasingly distant past as $T \to \infty$. One way of doing this (*e.g.* Uhlig, 1995, or Canjels and Watson, 1995) is to define $u_0 = u + \sum_{j=0}^{[T\tau]} \alpha^j \varepsilon_{-j}$, for some $\tau \geq 0$, and $u = O_{a.s.}(1)$ and with $\varepsilon_{-j}$ satisfying assumption 2.1. Then $T^{-1/2}u_0 \Rightarrow \sigma J_{c,0}(\tau)$, where $J_{c,0}$ is a diffusion process generated by $dJ_{c,0}(r) = cJ_{c,0}(r)dr + dW_0(r)$, in which $W_0$ is a standard Brownian motion independent of $W$. All of the above theory can be developed for this initialization of $u_0$, with no changes of substance in the limit theory. For example, in place of lemma 2.2 (iii) and (iv) we have:

(iii') $T^{-1/2}u_{[Tr]} \Rightarrow \sigma J_{c\tau}(r)$; (iv') $T^{-1}D_{kT}^{-1/2}\sum_{t=1}^{T} z_{kt}u_t \Rightarrow \sigma \int_0^1 g_k(r)J_{c\tau}(r)dr$; where $J_{c\tau}(r) = J_c(r) + e^{cr}J_{c,0}(\tau)$. For fixed $\tau$, $J_{c,0}(\tau) \equiv N(0, S_c(\tau))$, where $S_c(\tau) = (e^{2\tau c} - 1)/(2c)$ - *e.g.* Phillips (1987a)- and $J_{c,0}(\tau)$ is independent of $J_c(r)$. Then, the limit distribution of the OLS estimator $\widehat{\beta}_{kc}$ is found to be

$$F_{kT}^{1/2}(\widehat{\beta}_{kc} - \beta_k) \Rightarrow \sigma Q_k^{-1} \int_0^1 g_k(r)J_{c\tau}(r)dr$$

$$= \sigma Q_k^{-1}\left\{ \int_0^1 g_k(r)J_c(r)dr + \int_0^1 g_k(r)e^{cr}dr J_{c,0}(\tau) \right\} \equiv N\left(0, V_{\tau kc}^{ols}\right),$$

with

$$V_{\tau kc}^{ols} = V_{kc}^{ols} + \sigma^2 S_c(\tau)Q_k^{-1}\left( \int_0^1 g_k(r)e^{cr}dr \right)\left( \int_0^1 g_k(r)e^{cr}dr \right)' Q_k^{-1},$$

and $Q_k = \int_0^1 g_k(r)g_k(r)'dr$, as before. Similarly, for the partial GLS estimator we get

$$F_{kT}^{1/2}(\widetilde{\beta}_{kc} - \beta_k) \Rightarrow \sigma\left[ \int_0^1 f_{ck}(r)f_{ck}(r)'dr \right]^{-1}\left\{ \left[ \int_0^1 f_{ck}(r)dW(r) \right] + J_{c,0}(\tau)e \right\}$$

$$\equiv N(0, V_{\tau kc}^{gls}),$$

with $V_{\tau k c}^{gls} = V_{kc}^{gls} + \sigma^2 S_c(\tau) \left[\int_0^1 f_{ck}(r) f_{ck}(r)' dr\right]^{-1} ee' \left[\int_0^1 f_{ck}(r) f_{ck}(r)' dr\right]^{-1}$
and $e' = (1, 0, ..., 0)$.

The asymptotic relative efficiency of $\widehat{\beta}_{kc}$ to $\widetilde{\beta}_{kc}$ is now given by the ratio $R_{\tau kc} \equiv \det(V_{kc}^{ols})/\det(V_{\tau kc}^{gls})$, and this depends on the initialization parameter $\tau$. Figure 4 plots the efficiency curves against negative values of $c$ for various $\tau$. Apparently, $\tau$ has little effect on the relative efficiency of $\widehat{\beta}_{kc}$ to $\widetilde{\beta}_{kc}$ for values of $c \leq -4$. However, when $c \in (-4, 0]$, the effect of more distant initial conditions is seen to be substantial. A simple calculation shows that, as $\tau \to \infty$, $S_c(\tau) \to 1/(-2c)$ and $J_{c,0}(\tau) \Rightarrow N(0, 1/(-2c))$. Then, the initial conditions dominate the limit theory of the estimators for $c \sim 0$. In fact, for large $\tau$ and as $c \to 0$ we find that

$$R_{\tau kc} \sim \frac{\det\left[\frac{1}{-2c} Q_k^{-1} \left(\int_0^1 g_k(r) e^{cr} dr\right) \left(\int_0^1 g_k(r) e^{cr} dr\right)' Q_k^{-1}\right]}{\det\left[\frac{1}{-2c} \left[\int_0^1 f_{ck}(r) f_{ck}(r)' dr\right]^{-1} ee' \left[\int_0^1 f_{ck}(r) f_{ck}(r)' dr\right]^{-1}\right]}$$

$$\to \begin{cases} 9/4 & \text{for } k = 1 \\ \infty & \text{for } k > 1 \end{cases}$$

On the other hand, as $c \to -\infty$, $\sqrt{-c}\,[J_{c,\infty}(r) - J_c(r)] \overset{p}{\to} 0$, $S_c(\infty) \to 0$ and $J_{c,0}(\infty) \overset{p}{\to} 0$. Hence, the limit theory for $c \to -\infty$ that is given in theorems 2.5 and 3.2 continues to apply even with the new initialization.

## 5. Conclusion

This paper shows that GLS methods are asymptotically more efficient than OLS in estimating deterministic trend coefficients when the error process is integrated or near-integrated. Maximal gains tend to occur when the localizing parameter $c$ is less than zero in the near integrated case, unless the initial conditions are in the very distant past. If the trend extraction procedures involve a fitted intercept, some interesting subtleties in the limit theory arise as $c \to -\infty$, the lower limit of its domain of definition that corresponds to the case of stationary errors. In this case, $y_t$ is generated by a doubly infinite triangular array, and the limit distribution of the GLS estimator depends on the relative approach to infinity of the two parameters $-c$ and $T$.

The gains in efficiency that accrue from the GLS trend extraction procedures studied here suggest that there are likely to be similar advantages in other models that involve nonstationary processes, such as multiple equation systems with stochastic and deterministic cointegration.

## 6. Proofs

We outline the proofs of the results given in the text. Further details are given in Phillips and Lee (1996).

**6.1 Proof of Lemma 2.2** Parts (i) and (iii) are proved in Phillips (1987a, b). Parts (ii) and (iv) are proved in Phillips and Perron (1988) for $k = 1$. The extension to $k > 1$ is straightforward.

**6.2 Proofs of Theorems 2.3 and 2.4** These follow in a simple way from the form of $F_{kT}^{1/2}(\hat{\beta}_{kc} - \beta_k)$ and $F_{kT}^{1/2}(\tilde{\beta}_{kc} - \beta_k)$ and the results in lemma 2.2.

**6.3 Proof of Theorem 2.5** Since $J_c(r)$ satisfies the differential equation $dJ_c(r) = cJ_c(r)dr + dW(r)$, we have

$$Q_k^{-1} \int_0^1 g_k(r)J_c(r)dr = \frac{1}{-c}Q_k^{-1}\int_0^1 g_k(r)dW(r) + \frac{1}{c}Q_k^{-1}\int_0^1 g_k(r)dJ_c(r) \quad (9)$$

Note that

$$
\begin{aligned}
\int_0^1 g_k(r)dJ_c(r) &= [g_k(r)J_c(r)]_0^1 - \int_0^1 g_k^{(1)}(r)J_c(r)dr \\
&= g_k(1)J_c(1) - \left\{ \frac{1}{-c}\int_0^1 g_k^{(1)}(r)dW(r) + \frac{1}{c}\int_0^1 g_k^{(1)}(r)dJ_c(r) \right\} \\
&= g_k(1)J_c(1) - \frac{1}{c}\left\{ \left[g_k^{(1)}(r)J_c(r)\right]_0^1 - \int_0^1 g_k^{(2)}(r)J_c(r)dr \right\} + O_p\left(\frac{1}{|c|}\right) \\
&= \frac{1}{c}\int_0^1 g_k^{(2)}(r)J_c(r)dr + O_p\left(\frac{1}{|c|}\right) \quad (10)
\end{aligned}
$$

since $J_c(1) \equiv N\left(0, \frac{1}{2c}\left(e^{2c} - 1\right)\right) = O_p\left(1/|c|\right)$ and $(1/c)\int_0^1 g_k^{(1)}(r)dW(r) = O_p\left(1/|c|\right)$ as $c \to -\infty$. Continuing the process leading to (10) until we get to $g_k^{(k+1)}(r) = 0$, we deduce that $\int_0^1 g_k(r)dJ_c(r) = O_p\left(1/|c|\right)$. Hence, from (9) we obtain

$$Q_k^{-1}\int_0^1 g_k(r)J_c(r)dr = \frac{1}{-c}Q_k^{-1}\int_0^1 g_k(r)dW(r) + O_p\left(1/|c|^2\right)$$

Thus,

$$\hat{Z}_c = \sigma Q_k^{-1} \int_0^1 g_k(r) J_c(r) dr = \frac{\sigma}{-c} Q_k^{-1} \int_0^1 g_k(r) dW(r) + O_p\left(1/|c|^2\right). \quad (11)$$

But

$$\begin{aligned}
\breve{Z}_c &= \sigma \left[\int_0^1 (g_k^1(r) - cg_k(r))(g_k^1(r) - cg_k(r))' dr\right]^{-1} \int_0^1 (g_k^1(r) - cg_k(r)) dW(r) \\
&= \frac{\sigma}{-c} \left[\int_0^1 g_k(r) g_k(r)' dr\right]^{-1} \int_0^1 g_k(r) dW(r) + O_p\left(1/|c|^2\right). \quad (12)
\end{aligned}$$

The stated results now follow from (11), (12) and the fact that $Q_k = \int_0^1 g_k(r) g_k(r)' dr$.

**6.4  Proof of Theorem 3.1** The result follows simply from (8) using lemma 2.2 by writing

$$\begin{aligned}
F_{kT}^{1/2}\left(\breve{\beta}_k - \beta_k\right) &= \left[F_{kT}^{-1/2} \Sigma_1^T \left(\bar{z}_{kt} - \bar{\bar{z}}_k\right)\left(\bar{z}_{kt} - \bar{\bar{z}}_k\right)' F_{kT}^{-1/2}\right]^{-1} \left[F_{kT}^{-1/2} \Sigma_1^T \left(\bar{z}_{kt} - \bar{\bar{z}}_k\right) \varepsilon_t\right] \\
&\Rightarrow \sigma \left[\int_0^1 \bar{f}_{ck}(r) \bar{f}_{ck}(r)' dr\right]^{-1} \left[\int_0^1 \bar{f}_{ck}(r) dW(r)\right].
\end{aligned}$$

**6.5  Proof of Theorem 3.2** Observe that as $c \to -\infty$ the function $f_{ck}(r) = g_k^{(1)}(r) - cg_k(r)$ behaves like $-cg_k(r)$. Similarly, $\bar{f}_{ck}(r) = f_{ck}(r) - \int_0^1 f_{ck}(s) ds$ behaves like $-c\left(g_k(r) - \int_0^1 g_k(r) dr\right) = -c(g_k(r) - h_k) = -c\bar{g}_k(r)$, say. It follows that as $c \to -\infty$

$$\begin{aligned}
\breve{Z}_{mc} &= \sigma \left[\int_0^1 \bar{f}_{ck}(r) \bar{f}_{ck}(r)' dr\right]^{-1} \left[\int_0^1 \bar{f}_{ck}(r) dW(r)\right] \\
&\sim \frac{\sigma}{-c} \left[\int_0^1 \bar{g}_k(r) \bar{g}_k(r)' dr\right]^{-1} \left[\int_0^1 \bar{g}_k(r) dW(r)\right]. \quad (13)
\end{aligned}$$

Now the limit distribution of the OLS trend coefficient estimator with a fitted intercept is given in (5), which in the above notation is represented by the variate

$$\hat{Z}_{mc} = \sigma \left[\int_0^1 \bar{g}_k(r) \bar{g}_k(r)' dr\right]^{-1} \left[\int_0^1 \bar{g}_k(r) J_c(r) dr\right].$$

Just as in the proof of theorem 2.5 above, we find that as $c \to -\infty$

$$\hat{Z}_{mc} = \frac{1}{-c} \left[\int_0^1 \bar{g}_k(r) \bar{g}_k(r)' dr\right]^{-1} \int_0^1 \bar{g}_k(r) dW(r) + O_p\left(1/|c|^2\right). \quad (14)$$

Then, $\sqrt{-c}\left(\widehat{Z}_{mc} - \breve{Z}_{mc}\right) \xrightarrow{p} 0$, and $R_{kc}^m \to 1$, as required.

**6.6 Proof of Theorem 3.3** The efficiency ratio in this case is $R_{kc}^g \equiv \det(V_{kc}^{fmols})/\det(V_{kc}^{gls})$. Using (12), the limit variate $\widetilde{Z}_c$ can be written as

$$\widetilde{Z}_c = \frac{\sigma}{-c}\left[\int_0^1 g_k(r)g_k(r)'dr\right]^{-1}\int_0^1 g_k(r)dW(r) + O_p\left(1/|c|^2\right).$$

From this expression and (14) it follows that as $c \to -\infty$ $R_{kc}^g$ has the limit

$$R_{kc}^g = \det\left\{\left(V_{kc}^{gls}\right)^{-1} V_{kc}^{fmols}\right\} \to \det\left[\int_0^1 g_k(r)g_k(r)'dr\right]/\det\left[\int_0^1 \bar{g}_k(r)\bar{g}_k(r)'dr\right].$$

Now

$$\left[\int_0^1 \bar{g}_k(r)\bar{g}_k(r)'dr\right] = \left[\int_0^1 g_k(r)g_k(r)'dr\right] - \left[\left(\int_0^1 g_k(r)dr\right)\left(\int_0^1 g_k(r)dr\right)'\right] = Q_k - h_k h_k'$$

and $\det(Q_k - h_k h_k') = \det(Q_k)\left(1 - h_k'Q_k^{-1}h_k\right) = \left\{\det\left[\int_0^1 g_k(r)g_k(r)'dr\right]\right\}\left(1 - h_k'Q_k^{-1}h_k\right)$. Hence, $R_{kc}^g \to 1/\left(1 - h_k'Q_k^{-1}h_k\right) > 1$. Induction shows that $1 - h_k'Q_k^{-1}h_k = 1/(k+1)^2$, giving the stated result.

# 7. References

Beran, J. (1994). *Statistics for Long-Memory Processes.* Chapman and Hall.

Canjels, E. and M. Watson (1995). "Estimating deterministic trends in the presence of serially correlated errors," forthcoming in *Review of Economics and Statistics.*

Chan, N. H. and C. Z. Wei (1987). "Asymptotic inference for nearly non-stationary AR(1) processes. *Annals of Statistics,* 15, 1050-1063.

Durlauf, S. N. and P. C. B. Phillips (1988) "Trends versus random walks in time series analysis," *Econometrica,* 56, 1333-1354.

Elliott, G., T. J. Rothenberg and J. H. Stock (1995), "Efficient tests of an autoregressive unit root", *Econometrica.*

Grenander, U. and M. Rosenblatt (1957). *Statistical Analysis of Stationary Time Series.* New York: Wiley.

Hannan E. J. (1970). *Multiple Time Series.* New York: Wiley.

Lee C. C. and P. C. B. Phillips (1994). "Efficiency gains using GLS over OLS under nonstationarity," mimeographed, Yale University.

Phillips, P. C. B. (1987a). "Towards a unified asymptotic theory for autoregression," *Biometrika*, 74, 535–547.

Phillips, P. C. B. (1987b). "Time series regression with a unit root," *Econometrica*, 55, 277–301.

Phillips, P. C. B. and C. C. Lee (1996). "Efficiency gains from quasi-differencing under nonstationarity," mimeographed, Yale University.

Phillips, P. C. B. and P. Perron (1988). "Testing for a unit root in time series regression," *Biometrika*, 74, 535–547.

Samarov, A. and M. S. Taqqu (1988). "On the efficiency of the sample mean in long memory noise," *Journal of Time Series Analysis*, 9, 191–200.

Uhlig, H. (1994). "On Jeffreys' prior when using the exact likelihood function," *Econometric Theory*, 10, 633-644.

Yajima, Y. (1988). "On estimation of a regression model with long-memory stationary errors," *Annals of Statistics*, 16, 791–807.

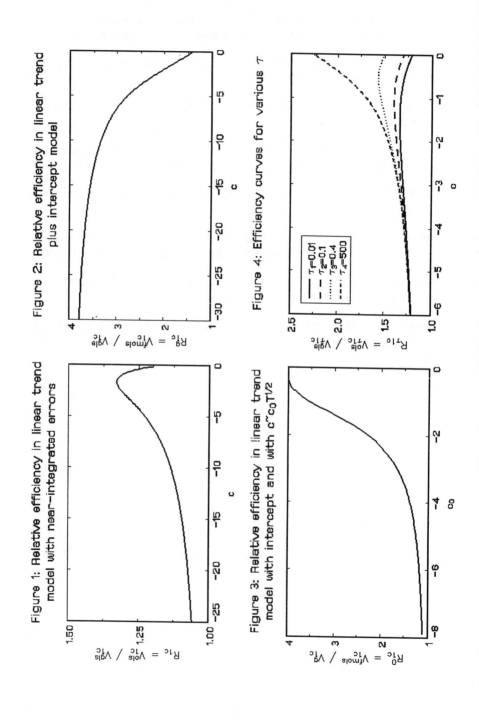

Figure 1: Relative efficiency in linear trend model with near-integrated errors

Figure 2: Relative efficiency in linear trend plus intercept model

Figure 3: Relative efficiency in linear trend model with intercept and with $c \sim c_0 T^{1/2}$

Figure 4: Efficiency curves for various $\tau$

# Estimation of Frequencies

## M. B. Priestley

## 1 Introduction.

Among the many outstanding contributions which Ted Hannan made to Time Series Analysis his interest in the problem of detecting and estimating the frequencies of periodic components was a recurrent theme. He first discussed this topic in his early monograph [Han60], and he continued his studies throughout his research career, his final papers appearing in 1993. In this paper we present a discursive survey of the subject in which we highlight, inter alia, Ted Hannan's seminal contributions. Since the paper is written for a broad readership, not all of whom will be specialists in Time Series Analysis, we first describe briefly the general nature of the problem and recall some early attempts to obtain its solutions.

## 2 The Basic Model.

The basic problem is to search a given data set for the possible presence of periodic components. The simplest model which we can use to describe this situation takes the form,

$$X_t = \sum_{i=1}^{K} (A_i \cos \omega_i t + B_i \sin \omega_i t) + \varepsilon_t, \tag{1.1}$$

where $X_t$ is the observed value at time $t$, $K$, $\{A_i\}$, $\{B_i\}$, $\{\omega_i\}$, $i = 1, \ldots, K$, are unknown constants, and $\{\varepsilon_t\}$ is zero mean random "noise" term. Suppose now that we are given a sample of $N$ observations on $\{X_t\}$ taken at a discrete set of unit spaced time points, i.e. we are given $N$ observations, $X_1, X_2, \ldots, X_N$, say. The problem then is to estimate the unknown parameters in the model (1.1), including, most importantly, the parameter $K$ which determines the number of periodic components in the model. To facilitate the subsequent discussion let us first consider the simpler situation where the frequencies $\{\omega_i\}$ are known, a priori, (so that, by implication, the value of $K$ is known also). Then (1.1) may be regarded as a standard multiple regression model and coefficients $\{A_i\}$, $\{B_i\}$ may be estimated by the method of least-squares, i.e. by minimising

$$Q = \sum_{t=1}^{N} \left\{ X_t - \sum_{i=1}^{K} (A_i \cos \omega_i t + B_i \sin \omega_i t) \right\}^2 \tag{1.2}$$

Differentiating $Q$ w.r.t. $A_i$, $B_i$, and setting the derivatives to zero then gives the usual normal equations. The solution of these equations is considerably simplified if all the $\{\omega_i\}$ are of the form $\omega_i = 2\pi p_i/N$, $i = 1, \ldots, K$, where the $\{p_i\}$ are all integers satisfying $0 \le p_i \le [N/2]$ (i.e. if the data length $N$ is an integral multiple of the period of each of the periodic terms in (1.1)). We can then make use of the orthogonality properties of the sine and cosine functions at these frequencies and, assuming that no $p_i$ is either 0 or $N/2$ ($N$ even), the normal equations have the solutions,

$$\hat{A}_i = \frac{2}{N} \sum_{t=1}^{N} X_t \cos \omega_i t, \quad \hat{B}_i = \frac{2}{N} \sum_{t=1}^{N} X_t \sin \omega_i t. \tag{1.3}$$

# 3   Periodogram Analysis.

In the preceding section we discussed the estimation of the coefficients $\{A_i\}$, $\{B_i\}$, assuming that the frequencies $\{\omega_i\}$ were known. However, in general the $\{\omega_i\}$ would be unknown, (as would be the value of $K$, the number of periodic components), and we would have to treat these as additional unknown parameters. When the $\{\omega_i\}$ are unknown, (1.1) no longer has the form of a "linear model" and we cannot then use the standard least-squares method. It is also apparent from (1.3) that we cannot estimate the $\{A_i\}$ and $\{B_i\}$ until we have first estimated the $\{\omega_i\}$. The usual approach is to locate the $\{\omega_i\}$ by a "search technique" based on a function called the *"periodogram"*. (This function was first introduced by Schuster [Sch98] in connection with what was then termed the model of "hidden periodicities".) The basic idea underlying periodogram analysis may be explained as follows. Suppose we choose a trial value of $\omega_1$, say $\hat{\omega}_1$. We can then compute $\hat{A}_1$, $\hat{B}_1$ from (1.3) using $\hat{\omega}_1$ in place of $\omega_1$. If the trial value $\hat{\omega}_1$ is close to the true value $\omega_1$, then $\hat{A}_1$, $\hat{B}_1$ will be close to $A_1$, $B_1$, and hence the squared amplitude $(\hat{A}_1^2 + \hat{B}_1^2)$ will be non-zero. However, if $\hat{\omega}_1$ is far removed from $\omega_1$ (or any of the other frequencies present in the model), then in effect we are estimating the coefficients of a term which does not exist in the model, and consequently $(\hat{A}_1^2 + \hat{B}_1^2)$ will be close to zero. If we now choose a sufficiently fine grid of trial frequencies, $\hat{w}_1, \hat{w}_2, \hat{w}_3, \ldots$, and plot the estimated square amplitudes $(\hat{A}_p^2 + \hat{B}_p^2)$ against $\hat{\omega}_p$, the ordinates will be non-zero when $\hat{w}_p$ is close to one of the $\omega_i$ but will be close to zero otherwise. We can thus locate the values of the $\{\omega_i\}$ by inspecting the squared amplitudes $\{\hat{A}_p^2 + \hat{B}_p^2\}$ and selecting those whose values that are "appreciably" greater than zero. This is the basis of "periodogram analysis", but in practice we plot the "normalised" amplitudes $\{\frac{N}{2}(\hat{A}_p^2 + \hat{B}_p^2)\}$ rather than the squared amplitudes themselves, the effect of the factor $(N/2)$ being to magnify the difference between "large" and "small" ordinates.

Specifically, given $N$ observations $X_1, X_2, \ldots, X_N$, we define the *periodogram*, $I_N(\omega)$, for all $\omega$ in the range $-\pi \leq \omega \leq \pi$ by

$$I_N(\omega) = \{A(\omega)\}^2 + \{B(\omega)\}^2, \tag{1.4}$$

where $A(\omega) = \sqrt{N/2} \sum_{t=1}^{N} X_t \cos \omega t$, $B(\omega) = \sqrt{N/2} \sum_{t=1}^{N} X_t \sin \omega t$. Equivalently, we may write $I_N(\omega)$ in the form,

$$I_N(\omega) = \frac{2}{N} \left| \sum_{t=1}^{N} X_t e^{-i\omega t} \right|^2 \tag{1.5}$$

Although $I_N(\omega)$ is defined for all $\omega$ in $(-\pi, \pi)$ we cannot evaluate it numerically as a continuous function of $\omega$; we can compute it only at a discrete set of frequencies. We therefore evaluate $I_N(\omega)$ at the set of frequencies 0, $2\pi/N$, $4\pi/N$, $6\pi/N, \ldots$, i.e. we evaluate

$$I_p \equiv I_N(\omega_p), \ \omega_p = 2\pi p/N, \ p = 0, 1, \ldots, [N/2], \tag{1.6}$$

and plot $I_p$ against $p$. We then locate the frequencies of the terms in (1.1) by searching for "large" periodogram ordinates. In fact, if $\{\varepsilon_t\}$ is an uncorrelated (i.e. "white noise") process then it is easily shown that when $\omega_p = \omega_i$,

$$E[I_p] = \frac{N}{2}(A_i^2 + B_i^2) + 2\sigma_\varepsilon^2, \tag{1.7}$$

where $\sigma_\varepsilon^2$ denotes the variance of $\varepsilon_t$. On the other hand, if $\omega_p$ does not fall near one of the $\omega_i$ then setting $A_i^2 + B_i^2 = 0$ in (1.7) we obtain,

$$E[I_p] \doteq 2\sigma_\varepsilon^2 \ \text{if} \ |\omega_p - \omega_i| \gg 0, \ i = 1, \ldots, K.$$

Thus, as $\omega_p$ approaches each of the $\omega_i$ the periodogram ordinate is $O(N)$ and the periodogram exhibits a large peak; otherwise the ordinates are $O(1)$.

More generally, it may be shown (Priestley [Pri81], p399) that for *all* $\omega$ in the range $(-\pi, \pi)$,

$$E[I_N(\omega)] = 2\sigma_\varepsilon^2$$

$$+\frac{1}{2N} \sum_{i=1}^{K} (A_i^2 + B_i^2) \left\{ \frac{\sin^2(\frac{1}{2}N(\omega + \omega_i))}{\sin^2(\frac{1}{2}(\omega + \omega_i))} + \frac{\sin^2(\frac{1}{2}N(\omega - \omega_i))}{\sin^2(\frac{1}{2}(\omega - \omega_i))} \right\} \tag{1.8}$$

Thus, $E[I_N(\omega)]$ takes the form of $2K$ Fejer kernels centered on the points $\omega = \pm\omega_i$, $i = 1, \ldots, K$, together with the constant term $2\sigma_\varepsilon^2$. In examining the ordinates only at $\omega_p = 2\pi p/N$, $p = 0, 1, \ldots, [N/2]$ we in effect place a "grid" with intervals $2\pi/N$ on top of the function (1.8). It is now apparent that if $\omega_i$ is not of the form $2\pi p_i/N$ the height of the observed peak may be substantially reduced. The worst case occurs when $\omega_i$ falls midway between two neighbouring ordinates; in this case the height of the peak is reduced by the factor $4/\pi^2$ (Whittle [Whi52]), and, depending on the magnitude of $(A_i^2 + B_i^2)$, this could have quite a serious effect on our ability to detect the presence of this periodic component.

## 4   Tests for periodogram ordinates.

If we observe several peaks in the periodogram we cannot immediately conclude that each of these corresponds to a genuine periodic component in $X_t$ since even in the null case when $A_i^2 + B_i^2 = 0$, all $i$, it is still possible that peaks may occur in the periodogram ordinates due simply to random fluctuations in the noise term $\varepsilon_t$. We must therefore apply a test to a periodogram peak to see whether it is significantly larger than that which would be likely to arise if no periodic components were present.

If we assume that the $\{\varepsilon_t\}$ are independent zero mean Gaussian random variables then it is easy to show that in the null case, $I_p \equiv I_N(\omega_p)$ are independently distributed and the distribution of each $I_p$ is (for $p \neq 0$) proportional to Chi-squared on two degrees of freedom. This is the basic result which underlies most of the well-known tests for periodogram ordinates. Effectively, we are interested in the distribution of the *maximum* periodogram ordinate, or more specifically, the distribution of the ratio $g = \max I_p / \sum_{p=1}^{[N/2]} I_p$. Approximate large sample tests were constructed by Schuster [Sch98] and Walker [Wal14], but Fisher [Fis29] derived the exact distribution of $g$. Other test procedures were suggested by Hartley [Har49], Grenander and Rosenblatt [GR57], Siegel [Sie80], and Quinn [Qui89]. For a more detailed discussion of these tests, see, e.g., Priestley [Pri81], p398.

## 5   Estimation of frequency.

Suppose that having applied the tests described in the preceding section we conclude that there are $k$ periodic components present in the model. We then have to estimate the corresponding frequencies and amplitudes.

Let us consider the first stage of the analysis; suppose that the test of $\max(I_p)$ gives a significant result. All that we may strictly infer is that the null hypothesis $H_0$ is discredited and hence that there exists a periodic component at *some* frequency $\omega$. However, we would usually wish to conclude more specifically that there exists a periodic component at the frequency corresponding to $\max(I_p)$. Let the maximum periodogram ordinate occur at $p = p'$, corresponding to the frequency $\hat{\omega}_1 = 2\pi p'/N$. Then

Hartley [Har49] has shown that the probability of a misleading result, i.e. the probability that the significance of $\max(I_p)$ is due to a periodic component at some other frequency $\omega$ (where $|\omega - \hat{\omega}_1| \gg 2\pi/N$) is less than $\alpha$, the significance level of the test. Thus, if $\max(I_p)$ is significantly large we may safely estimate the frequency $\omega_1$ (say) of the first detected periodic component by $\hat{w}_1 = 2\pi p'/N$. The corresponding coefficients, $A_1$, $B_1$, may then be estimated by

$$\hat{A}_1 = \sqrt{2/N}A(\hat{\omega}_1), \quad \hat{B}_1 = \sqrt{2/N}B(\hat{\omega}_1), \tag{1.9}$$

where $A(\hat{\omega}_1)$, $B(\hat{\omega}_1)$ are given by (1.4), and the squared amplitude is estimated by $\hat{A}_1^2 + \hat{B}_1^2 = \frac{2}{N}I_{p'}$. If we replace $\hat{\omega}_1$ by its true value $\omega_1$ then, as previously noted, $\hat{A}_1$ and $\hat{B}_1$ become the least-squares estimates of $A_1$, $B_1$, when $\omega_1$ itself is of the form $2\pi p/N$. If $\omega_1$ is not of this form, $\hat{A}_1$ and $\hat{B}_1$ are equivalent to the least squares estimates to $O(1/N)$ provided $\hat{\omega}_1$ and $(\pi - \hat{\omega}_1)$ are each $\gg 2\pi/N$.

In fact, under the assumption that the residuals $\{\varepsilon_t\}$ are Gaussian white noise, the maximum likelihood estimate of $\omega_1$ corresponds to the maximum *over all $\omega$* $(0 \leq \omega \leq \pi)$ of the periodogram function $I_N(\omega)$ given by (1.4). In a pioneering paper Whittle [Whi52] showed that, (on the assumption that the $\{\omega_i\}$ are spaced sufficiently well apart),

$$\text{var}(\hat{\omega}_i) = \frac{24\sigma_\varepsilon^2}{N^3(A_i^2 + B_i^2)} + o(\frac{1}{N^3}) \tag{1.10}$$

This result is remarkable since it shows that, apart from the bias, $\omega_i$ can be estimated to an accuracy of $O(\frac{1}{N^{3/2}})$. If the $\{\omega_i\}$ are estimated from the standard periodogram ordinates $\{I_p, \ p = 0, 1, \ldots, [N/2]\}$ then

$$\text{E}(\hat{\omega}_i) = \omega_i + O(1/N), \tag{1.11}$$

so that the $\{\omega_i\}$ are then estimated to an accuracy of $O(1/N)$. A rigorous derivation of these basic results is given by Walker [Wal71] and Hannan [Han73].

## 5.1   Secondary analysis.

When the true frequencies are not all multiples of $2\pi/N$ it is possible to improve on the estimate $\hat{\omega}_1$ by using a technique called "secondary analysis". This technique is well-known to meteorologists and oceanographers and was developed initially for the purely deterministic case where the noise is absent. The basic idea is to divide the series up into a number of sub series and to perform a separate harmonic analysis for each sub series. The improved estimate of frequency is then derived by noting that the phase of the given frequency component varies (approximately) linearly over the successive sub series, the slope being proportional to the difference between the true frequency and the initial estimate. This technique is a special case of a more general method called *"complex demodulation"*.

To illustrate the method consider the simple model

$$X_t = \cos(\omega_1 t + \phi),$$

(and ignore the residual). Suppose the maximum periodogram ordinate leads to the estimated frequency $\hat{\omega}_1$, and let $\hat{T}_1 = 2\pi/\hat{\omega}_1$ denote the estimated period. Divide the $N$ observations into a number of groups, each group containing $m\hat{T}_1$ observations, say. For the $s^{th}$ group we have,

$$\hat{A}_s = \frac{2}{m\hat{T}_1} \sum_{t=(s-1)m\hat{T}_1}^{sm\hat{T}_1} A\cos(\omega_1 t + \phi)\cos\hat{\omega}_1 t$$

$$\hat{B}_s = \frac{2}{m\hat{T}_1} \sum_{t=(s-1)m\hat{T}_1}^{sm\hat{T}_1} A\cos(\omega_1 t + \phi)\sin\hat{\omega}_1 t$$

It can now be shown (see, e.g., Priestley [Pri81], p413) that if $mT_1$ is sufficiently large and $\omega \gg 0$,

$$\phi_s = \tan^{-1}(\hat{B}_s/\hat{A}_s) \doteq (\hat{\omega}_1 - \omega_1)smT_1 - \phi + \text{const (mod } 2\pi).$$

If we now plot $\phi_s$ against $s$ we obtain a graph which is roughly linear with slope

$$\beta \doteq (\hat{\omega}_1 - \omega_1)mT_1 = 2\pi m(1 - \frac{\omega_1}{\hat{\omega}_1}),$$

so that $\omega_1 \doteq \hat{\omega}_1(1 - \frac{\beta}{2\pi m})$, or $T_1 \doteq \hat{T}_1/(1 - \frac{\beta}{2\pi m})$. Campbell and Walker [CW77] describe an application of Secondary Analysis to the Canadian Lynx data. Here, the maximum periodogram ordinate occurs at frequency $24\pi/112$. Taking $\hat{T}_1 = 28/3$ (years), $m = 3$, and using 4 sub series they obtain $\beta \doteq 33°$, giving a revised estimated period of $T_1 = 9.63$ years.

Hannan and Quinn [HQ89] consider the problem of resolving two closely adjacent frequency components (the problem of "spectral line splitting"). By "closely adjacent" they mean frequencies whose separation is $O(1/N)$, and the model which they consider takes the form

$$X_t = A_1 \cos\omega_1 t + B_1 \sin\omega_1 t$$

$$+A_2 \cos\left(\omega_1 + \frac{a}{N}\right)t + B_2 \sin\left(\omega_1 + \frac{a}{N}\right)t + \varepsilon_t.$$

Hannan and Quinn then propose methods for estimating the parameters $\{A_1, B_1, A_2, B_2, \omega_1, a\}$.

A large number of papers have appeared in the engineering literature on this problem —see, e.g., Pisarenko [Pis73], Chan *et al* [CLP81], Hayes and Clements [HC86], Kay and Marple [KM81], Dragosevic and Stankovic

[DS89]. In particular, Kay and Marple [KM81] and Newton and Pagano [NP83] propose a method of frequency estimation based on the fitting of AR (autoregressive) or ARMA (mixed autoregressive / moving-average) models to the data.

A similar approach was studied by Quinn and Fernandes [QF91] who consider a single frequency model of the form (possibly after some prefiltering of the data),

$$X_t = A\cos(\omega_0 t + \phi) + \varepsilon_t \tag{1.12}$$

Writing $\beta = 2\cos\omega_0$ it follows from (1.12) that

$$(1 - \beta B + B^2)X_t = (1 - \beta B + B^2)\varepsilon_t \tag{1.13}$$

where B denotes the backward shift operator, $BX_t = X_{t-1}$, etc. (Note that the general solution of the difference equation (1.13) is

$$X_t = f(t) + \varepsilon_t$$

where $f(t)$ is the solution of $(1 - \beta B + B^2)X_t = 0$, and hence $f(t)$ has the form of the first term in (1.12).). Quinn and Fernandes estimate $\beta$ by fitting to $\{X_t\}$ a model of the form

$$(1 - \beta B + B^2)X_t = (1 - \alpha B + B^2)\varepsilon_t \tag{1.14}$$

iteratively, with the parameter $\alpha$ ultimately constrained to equal $\beta$. Their fitting technique is based on noting that (1.14) can be regarded as a regression model of the form,

$$X_t + X_{t-2} = \beta X_{t-1} + u_t \tag{1.15}$$

where $u_t = \varepsilon_t - \alpha\varepsilon_{t-1} + \varepsilon_{t-2}$, However, in (1.15) the "residuals" $\{u_t\}$ are correlated, and to correct this effect Quinn and Fernandes introduce a new process $\{\xi_t\}$ defined by

$$X_t = (1 - \alpha B + B^2)\xi_t, \tag{1.16}$$

so that (1.14) can now be written as

$$(1 - \beta B + B^2)\xi_t = \varepsilon_t \tag{1.17}$$

Equation (1.17) may now be written as a regression model,

$$(\xi_t + \xi_{t-2}) = \beta\xi_{t-1} + \varepsilon_t$$

where now the residuals are uncorrelated. The parameter $\beta$ can now be estimated by performing a least-squares regression of $(\xi_t + \xi_{t-2})$ on $\xi_{t-1}$. To implement this technique one needs to start with a preliminary estimate of $\alpha$ (in order to evaluate $\xi_t$ from (1.16)) and then estimate $\beta$ as described above. This value of $\beta$ is then used as the new value of $\alpha$ in a second iteration, and the process continues until the estimated values of $\alpha$ and $\beta$ converge. Huang and Hannan [HH93b] use a generalisation of Quinn and Fernandes' method to track a time varying frequency in a model of the form,

$$X_t = A\cos(t\omega_t + \phi) + \varepsilon_t \qquad (1.18)$$

(Models of the form (1.18) arise, e.g., in the study of phase-locked loop systems). See also Hannan and Huang [HH93a].

# 6 References

[CLP81] Y. T. Chan, J. M. Lavoie, and J. B. Plant. A parametric estimation approach to the estimation of frequencies of sinusoids. *I. E. E. E. Trans. Accoust., Speech and Signal Processing*, ASSP-29:214–229, 1981.

[CW77] M. J. Campbell and A. M. Walker. A survey of statistical work on the MacKenzie river series of annual Canadian lynx trappings for the years 1821-1934, and a new analysis. *J. Roy. Statist. Soc. A*, 140:411–431, 1977.

[DS89] M. Dragosevic and S. S. Stankovic. A general least squares method for frequency estimation. *I. E. E. E. Trans. Accoust., Speech and Signal Processing*, ASSP-37:805–819, 1989.

[Fis29] R. A. Fisher. Tests of significance in harmonic analysis. *Proc. Roy. Soc. A*, 125:54–59, 1929.

[GR57] U. Grenander and M. Rosenblatt. *Statistical Analysis of Stationary Time Series*. Wiley, New-York, 1957.

[Han60] E. J. Hannan. *Time Series Analysis*. Methuen, London, 1960.

[Han73] E. J. Hannan. The estimation of frequency. *J. Appl. Prob.*, 10:510–519, 1973.

[Har49] H. O. Hartley. Tests of significance in harmonic analysis. *Biometrika*, 36:194–201, 1949.

[HC86] M. H. Hayes and M. A. Clements. An efficient algorithm for computing Pisarenko's harmonic decomposition using Levinson's recursion. *I. E. E. E. Trans Accoust., Speech, and Signal Processing*, ASSP-34:489–491, 1986.

[HH93a] E. J. Hannan and D. Huang. On line frequency estimation. *J. Time Series Anal.*, 14:147–162, 1993.

[HH93b] D. Huang and E. J. Hannan. Estimating time varying frequency. Submitted to J. Time Series Anal., 1993.

[HQ89]  E. J. Hannan and B. G. Quinn. The resolution of closely adjacent spectral lines. *J. Time Series Anal.*, 10:13–32, 1989.

[KM81]  S. M. Kay and S. L. Marple. Spectrum analysis – a modern perspective. *Proc. I. E. E. E.*, 69 (11):1380–1419, 1981.

[NP83]  J. H. Newton and M. Pagano. A method for determining periods in time series. *J. Amer. Stat. Assoc.*, 78:152–157, 1983.

[Pis73]  V. F. Pisarenko. The retrieval of harmonics from a covariance function. *Geophics. J. Roy. Astron. Soc.*, 33:347–366, 1973.

[Pri81]  M. B. Priestley. *Spectral Analysis of Time Series (vols I, II)*. Academic Press, London, 1981.

[QF91]  B. G. Quinn and J. M. Fernandes. A fast efficient technique for the estimation of frequency. *Biometrika*, 78:489–497, 1991.

[Qui89]  B. G. Quinn. Estimating the number of terms in a sinusoidal regression. *J. Time Series Anal.*, 10:71–76, 1989.

[Sch98]  A. Schuster. On the investigation of hidden periodicities with application to a supposed 26-day period of meteorological phenomena. *Terr. Mag. Atmos. Elect.*, 3:13–41, 1898.

[Sie80]  A. F. Siegel. Testing for periodicity in a time series. *J. Amer. Statist. Assoc.*, 75:345–348, 1980.

[Wal14]  G. Walker. On the criteria for the reality of relationships or periodicities. *Calcutta Ind. Met. Memo*, 21:9, 1914.

[Wal71]  A. M. Walker. On the estimation of a harmonic component in a time series with stationary independent residuals. *Biometrika*, 58:21–36, 1971.

[Whi52]  P. Whittle. The simultaneous estimation of a time series harmonic components and covariance structure. *Trabafos Estadistica*, 3:43–57, 1952.

University of Manchester Institute of Science and Technology

# Statistical Problems in the Analysis of Underwater Sound

## Barry G. Quinn

ABSTRACT The fundamental problem of estimating the frequency of a sinusoid in the presence of additive noise is discussed. In particular, various techniques for the estimation and tracking of frequency when the signal-to-noise ratio is low but the sample size high are described. These techniques may be used to solve higher-level problems such as the localisation of an acoustic source by single sensors and arrays of sensors.

## 1  Introduction

Many signals (and especially underwater acoustic signals) may be modelled as sums of sinusoids in additive noise. The noise cannot be assumed to be either Gaussian or white. The signal-to-noise ratios (SNRs) may also be very low. However, in these circumstances the instantaneous frequencies of the signals may be fairly constant over many samples. The number of sinusoids may also be quite large, as there will be several periodic signals superimposed, each of which may be approximately decomposed into sinusoids at a fundamental frequency and at harmonics of that frequency. For this reason, fixed length processing needs to be carried out for real time performance and, as many signal processing systems have hardware fixed length Fast Fourier Transforms (FFT's), it is important to use the discrete Fourier coefficients where possible.

This paper looks at the classical problem of estimating the frequency of a single sinusoid; a filtering technique which produces a (Gaussian) efficient estimator; the use of the Fourier coefficients at several frequencies and time blocks to estimate frequency; and the use of the Fourier coefficients to track frequency through time, using a Hidden Markov Model (HMM) approach. It also describes two applications which use frequency and amplitude estimators or tracks to estimate characteristics of a moving acoustic source.

My interest in these problems began when Ted Hannan suggested that I work with him on the problem of the estimation of the frequencies of two added sinusoids and noise, when these frequencies were close together — in fact so close together that they might not be resolved by the periodogram (See [HQ89]). This led me to work on several related problems – testing

for the number of sinusoids, the automatic estimation of the number of sinusoids and other methods for estimating frequency, and, indeed, to leave
academia for nearly five years to pose and solve related problems in mar-
itime Defence. Ted continued to encourage me and stimulated my work by
continuing to solve other problems in the area (See [HH93], [Han93] and
[KH94] , for example). I have adopted the general conditions Ted assumed
so that I might use the results of his important papers on limit theorems in
time series regression (See, for example, [Han79]), which he used so pow-
erfully in his seminal article on the estimation of frequency [Han73]. Ted's
legacy to Time Series Analysis was enormous, but he also touched many
with his wit, generosity and personality. I count myself extremely fortunate
to have known him and been able to work with him.

## 2   Signal model for a single sinusoid

We consider first the case of a single noisy sinusoid. This is represented by
the stochastic process $\{X(t)\}$ which satisfies

$$X(t) = \rho \cos\{\omega t + \phi\} + \varepsilon(t); \ t = 0, 1, \ldots, T - 1$$

where $\{\varepsilon(t)\}$ is a stationary, ergodic, zero mean process with strictly positive spectral density and which satisfies a few other simple conditions which
will not be mentioned here but may be found, for example, in [HQ89]. The
parameter $\rho$ is called the amplitude of the sinusoid, $\omega$ is called the frequency
(and is measured in radians per unit time) and $\phi$ is the initial phase. It
should be mentioned that the introduction of a constant term $\mu$ may be
made with no effect on the development to follow: subtraction of the sample mean results only in small order effects. The $k$th Fourier coefficient,
$k = 0, 1, \ldots, T - 1$, $Y(k)$ of $\{X(0), \ldots, X(T-1)\}$ is given by

$$Y(k) = \sum_{t=0}^{T-1} X(t) \exp\left(-i\frac{2\pi kt}{T}\right)$$

which may be shown to be

$$\frac{\rho}{2} \exp(i\phi) \frac{\exp(iT\omega) - 1}{i(\omega - 2\pi k/T)} + O(1) + U(k)$$

if $\omega - 2\pi k/T$ is $O(T^{-1})$ and where $U_j(k)$ is the $k^{th}$ Fourier coefficient of
$\{\varepsilon(t); t = 0, 1, \ldots, T-1\}$. The least squares estimator of $\omega$ may be shown
to be asymptotically equivalent to the maximiser $\widehat{\omega}$ of the periodogram of
$\{X(0), \ldots, X(T-1)\}$

$$I_X(\lambda) = 2/T |w_X(\lambda)|^2$$

where

$$w_X(\lambda) = \sum_{t=0}^{T-1} X(t) \exp(-i\lambda t)$$

It was shown rigorously by Hannan [Han73], under fairly general conditions, that

$$T^{3/2}(\widehat{\omega} - \omega) \xrightarrow{\mathcal{D}} N(0, 48\pi f(\omega)/\rho^2) \tag{1.1}$$

where $f(\omega)$ is spectral density of $\{\varepsilon(t)\}$. The (Gaussian) maximum likelihood estimator of $\omega$ thus has an asymptotic variance which is inversely proportional to the cube of the sample size. However, the numerical maximisation of $I_X(\lambda)$ is rather difficult. (See, for example, Rice and Rosenblatt [RR88]). In fact, if Newton's method is used to find a zero of $I'_X(\lambda)$, starting at the maximiser of $I_X(2\pi k/T)$, the zero may not equal $\widehat{\omega}$. In fact, Quinn & Fernandes [QF91] show that an initial estimator which is within $O(T^{-1-\varepsilon})$, almost surely, where $\varepsilon > 0$, of $\omega$ is needed to guarantee the convergence of Newton's method. One solution to this problem is to 'zero-pad' the data, effectively increasing $T$. This involves greater computation, however, and is not possible with fixed-length FFT's. Another solution is to use the so-called 'Chirp-Z' transform, which involves the same length FFT's. This requires several passes of the data through a Fourier transform program, unless the frequency is known fairly accurately in advance.

# 3   An iterative filtering technique

In the same way as sinusoids are the solutions of 2nd order constant-coefficient differential equations, they are the solutions of 2nd order constant-coefficient difference equations. In fact,

$$X(t) - 2\cos\omega X(t-1) + X(t-2) = \varepsilon(t) - 2\cos\omega\varepsilon(t-1) + \varepsilon(t-2).$$

It is therefore plausible to fit the following ARMA(2,2) model, even though it is not a model for a stationary and invertible ARMA(2,2) process:

$$X(t) - \alpha X(t-1) + X(t-2) = \varepsilon(t) - \beta\varepsilon(t-1) + \varepsilon(t-2).$$

Such an approach has appeared often in the Engineering literature. Many of these techniques, however, produce inconsistent estimators of $\omega$, while the best produce estimators whose asymptotic variances are only $O(T^{-1})$. The reason for the latter behaviour is that a damping parameter is usually artificially introduced to keep the zeroes of the AR-auxiliary polynomial away from the unit circle. Doing this, however, not only increases the order of the asymptotic variance, but often produces the undesirable property that the estimator is inconsistent unless the noise is white. It is important, therefore, to make use of the constraint that the zeroes are on the unit circle.

As the *true* $\alpha$ and $\beta$ are equal, and $\alpha$ is easily estimated if $\beta$ is known, an obvious technique is to estimate $\alpha$ given an initial estimator of $\beta = 2\cos\omega$, replace $\beta$ with this, and iterate until 'convergence' is reached. We thus formulate the Gaussian likelihood of $\alpha$ for fixed $\beta$, conditional on $X(-2) = X(-1) = 0$, maximize this with respect to $\alpha$, replace $\beta$ with $\alpha$ and iterate:

1. Put $\widehat{\alpha}^{(1)} = 2\cos\widehat{\omega}_c$,

$$\xi_t^{(j)} = X(t) + \widehat{\alpha}^{(j)}\xi_{t-1}^{(j)} - \xi_{t-2}^{(j)}; \quad t = 0, \ldots, T-1$$

where $\xi_t = 0, t < 0, \widehat{\omega}_c = 2\pi\frac{\widehat{k}}{T}$ where $\widehat{k}$ is the maximiser of $|Y(k)|^2$ and

$$\widehat{\alpha}^{(j+1)} = \widehat{\alpha}^{(j)} + 2\frac{\sum_{t=1}^{n} X(t)\xi_{t-1}^{(j)}}{\sum_{t=1}^{n}\left(\xi_{t-1}^{(j)}\right)^2}$$

2. Estimate $\omega$ by $\widehat{\omega} = \arccos(\widehat{\alpha}^{(3)}/2)$.

A careful reader will notice that the multiplicative factor of 2 in the iterative procedure above will not appear in a direct derivation. The factor 2 is the only number which will produce accelerated convergence to the closest zero of $\sum_{t=1}^{n} X(t)\xi_{t-1}(\omega)$, where $\xi_t(\omega) = X(t) + 2\cos\omega\xi_{t-1}(\omega) - \xi_{t-2}(\omega)$. It is proved in Quinn and Fernandes [QF91] that $\widehat{\omega}$ has the same asymptotic distribution as the periodogram maximiser. Thus one obtains, via only two simple iterations, an estimator which has the same asymtptotic standard deviation (which is $O(T^{-3/2})$) as the periodogram maximiser, using an initial estimator which is accurate only to order $O_P(T^{-1})$. The estimator also admits an interpretation as the maximiser of the smoothed version of the periodogram

$$\kappa(\omega) = -\int_{-\pi}^{\pi} \log(2 - 2\cos x) I_X(\omega - x)\, dx$$

The procedure is much more robust than that of maximizing the periodogram. It can be shown, in fact, that if the initial estimator of $\omega$ is accurate to $O(T^{-1/2-\varepsilon})$, almost surely, where $\varepsilon > 0$, then the iterative procedure converges almost surely to the same local maximiser of $\kappa(\omega)$. The 'sidelobe' phenomenon of the periodogram is thus not evident. Figure 1, which shows a time series with $T = 100$ composed of a single sinusoid in white noise, its periodogram, and the functions $\lambda(\omega) = \sum_{t=1}^{n} X(t)\xi_{t-1}(\omega)$ and $\kappa(\omega)$, demonstrates this robustness.

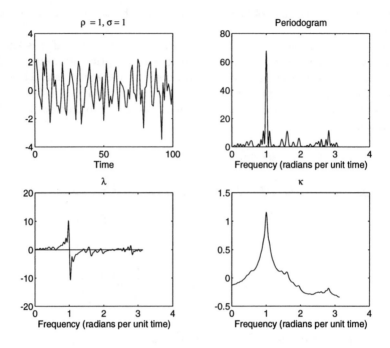

FIGURE 1. Time Series, Periodogram, $\lambda$ and $\kappa$

Truong-Van [TV90] has developed a different iterative procedure which converges to the same estimator of $\omega$. The motivation for his technique is through the amplification of sinusoids: the general solution of the difference equation

$$X_t - 2\cos\omega X_{t-1} + X_{t-2} = \cos(\omega t + \phi)$$

is

$$X_t = a\cos(\omega t + b) + \frac{t\sin\{\omega(t+1) + \phi\}}{2\sin\omega}$$

The search is therefore for a filter $1 - az + z^2$ which 'amplifies' the given data.

## 4  Simple Fourier techniques

If there are many sinusoids, with frequencies typically closer together than $O\left(T^{-1/2-\varepsilon}\right)$, where $\varepsilon > 0$, as is often the case in underwater acoustics, the above procedure has an obvious drawback — stronger signals will smear out weaker ones if the frequencies are relatively close to each other. Periodogram-based procedures will only suffer this problem if the frequencies are much closer to each other. So-called 'high-resolution' procedures may be shown to have nowhere near the resolving power of the periodogram, and, at any rate, when two frequencies are very close together, they must be resolved by a different procedure (See [HQ89]).

Let us suppose that Fourier coefficients are available cheaply. Let $\omega = 2\pi(n + \delta)/T$ where $n$ is an integer, $-\frac{1}{2} < \delta \leq \frac{1}{2}$ and both $n$ and $\delta$ are functions of $T$. Then, as is shown above,

$$Y(n + k) = c_T \frac{\delta}{\delta - k} + U(n + k) + O(1)$$

where

$$c_T = T\rho \exp(i\phi) \frac{\exp(i2\pi\delta) - 1}{4\pi i\delta}$$

Letting $\widehat{n}$ be the maximiser of $|Y(k)|^2$, put

$$\widehat{\alpha}_k = \Re\left\{ \frac{Y(\widehat{n} + k)}{Y(\widehat{n})} \right\}, \ \widehat{\delta}_k = \frac{-k\widehat{\alpha}_k}{1 - \widehat{\alpha}_k} \ ; k = -1, 1$$

$$\widehat{\delta} = \begin{cases} \widehat{\delta}_1 & ; \ \widehat{\delta}_{-1}, \widehat{\delta}_1 > 0 \\ \widehat{\delta}_{-1} & ; \ \text{otherwise} \end{cases}, \widehat{\omega} = 2\pi(\widehat{n} + \widehat{\delta})/T$$

It is shown in [Qui94] that $T^{3/2} \{a(\delta)\}^{-1/2} (\widehat{\omega} - \omega)$ is asymptotically normally distributed with mean zero and variance 1, where $\{a(\delta)\}$ varies between 3.29 and 1.01 times the variance in (1.1). The need to combine two estimators is evident from consideration of the case where $\omega$ is relatively close to a frequency of the form $2\pi(n + 1/2)/T$. An improved method results from combining the $\widehat{\delta}_j$s optimally. The nonlinear combination of $\widehat{\delta}_{-1}$ and $\widehat{\delta}_1$ which has smallest asymptotic variance can be shown to be

$$\widehat{\omega} = 2\pi(\widehat{n} + \widehat{\delta})/T$$

where

$$\widehat{\delta} = \frac{1}{2}\left(\widehat{\delta}_1 + \widehat{\delta}_{-1}\right) + g\left(\widehat{\delta}_1^2\right) - g\left(\widehat{\delta}_{-1}^2\right)$$

and

$$g(x) = \frac{1}{4}\log(3x^2 + 6x + 1) - \frac{\sqrt{6}}{24}\log\left(\frac{x + 1 - \sqrt{\frac{2}{3}}}{x + 1 + \sqrt{\frac{2}{3}}}\right)$$

The asymptotic variance is then between 1.64 and 1.01 the variance in (1.1). It may be shown that this asymptotic variance is the same as that obtained when the asymptotic likelihood function of the parameters given only the Fourier coefficients $\{Y(n - 1), Y(n), Y(n + 1)\}$ is maximised. As this function is highly nonlinear, the closed-form estimator above is to be preferred. Further improvements are gained at the expense of computational complexity through the maximisation of the likelihood of the parameters given more than three of the $Y(k)$s. This is equivalent to minimising with respect to $\delta$

$$\min_{c} \sum_{k=-K}^{K} \left| Y(\hat{n}+k) - c\frac{\delta}{\delta-k} \right|^2$$

## 5  Several time blocks

There will be occasions when a frequency appears to be constant over a lengthy time-period, but only fixed-length FFT processing is available. If the frequency were to be estimated separately over a number $J$ of fixed-length time-blocks, an obvious estimator of the frequency would be the average of these estimators. This would, however, have an asymptotic variance inversely proportional to $JT^3$, while the periodogram maximiser constructed from all $JT$ samples would have one which was inversely proportional to $J^3T^3$. Clearly the higher variance arises because the estimators of the initial phases for each time block have not been used. We shall develop an 'optimal' technique for estimating $\omega$.

If $\omega$ is constant over $JT$ samples, we may compute the Fourier coefficients from each of the $J$ length-$T$ time blocks. Let $Y_j(k)$ be the $k$th Fourier coefficient in the $j$th time-block. Then

$$Y_j(k) = \frac{\rho}{2}\exp\left\{i\left(\phi+Tj\omega\right)\right\}\frac{\exp\left(iT\omega\right)-1}{i(\omega-2\pi k/T)}+O\left(1\right)+U_j(k)$$

Let $\omega = 2\pi\left(n+\delta\right)/T$, where $\delta$ is $O(1)$. We may then estimate $\delta$ by maximising the (asymptotic) likelihood function constructed from

$$\{Y_j(n+k); \; j=0,1,\ldots,J-1; \; k=-K,\ldots,K\}.$$

This is equivalent to minimising

$$\sum_{j=0}^{J-1}\sum_{k=-K}^{K}|Y_j(n+k)-R\exp\left\{i(\phi+2\pi j\delta)\right\}d_k(\delta)|^2$$

where $R = T\rho/2$ and $d_k(\delta) = \frac{e^{i2\pi\delta}-1}{2\pi i(\delta-k)}$. The estimator of $\delta$ is then the maximiser of

$$S_T(\delta) = \frac{\left|\sum_{k=-K}^{K}\left\{\sum_{j=0}^{J-1}Y_j(n+k)\exp(-i2\pi j\delta)\right\}\bar{d}_k(\delta)\right|^2}{J\sum_{k=-K}^{K}|d_k(\delta)|^2}$$

Especially when the computations are done in real time, the $d_k(\delta)$s and denominator of $S_T(\delta)$ may be calculated in advance on a fine grid of $\delta$s and stored, and especially when $J$ is large, $S_T(\delta)$ may be computed efficiently on this grid of $\delta$s by zero padding the $Y_j(n+k)$ and Fourier transforming, separately for each $k$.

# 6 HMM frequency tracking

Suppose that it is known that a signal contains a sinusoidal component with frequency slowly varying in time and that the SNR is so low that it may be impossible to estimate the frequency accurately from a section of the signal for which the frequency could reasonably be assumed to be constant. Although the frequency could be changing in a deterministic fashion, for example, because of relative motion between the source and receiver, there may no obvious mechanism for the variation. A common method of tracking this change is to assume a simple probabilistic mechanism. This idea has been used by Econometricians to describe the slow variation in, say, a regression parameter or autoregressive parameter as well as by Engineers who have applied it in spatial tracking. Inevitably a Kalman filter or extended Kalman filter approach is taken. These techniques, however, have a tendency to wander, and at such low SNRs just do not work. At moderately low SNRs, the use of Fourier coefficients effectively increases the SNR, and looks promising (See La Scala, Bitmead and Quinn [LBQ94]). At very low SNRs, however, there appears to be no alternative other than to discretise the problem and use a Hidden Markov Model (HMM) approach. The following generalizes the approach taken by Streit & Barrett [SB90], and Barrett & Holdsworth [BH93] in tracking frequency.

We shall assume that the frequency of a sinusoid is fixed for $J$ time-blocks, each time-block consisting of $T$ equidistant points in time, and that the sequence of frequencies formed from each cluster of time-blocks constitutes a discrete state Markov Chain, with equidistant frequencies and many states per frequency bin (i.e. there are many frequency states corresponding to the difference between successive Fourier frequencies, namely $\frac{2\pi}{T}$). The model for the signal is then

$$X(t + jT + kJT) = \rho \cos\left\{ \omega_k t + \phi + j\omega_k T + JT \sum_{m=0}^{k-1} \omega_m \right\}$$

$$+\varepsilon(t+jT+kJT);\ t = 0, 1, \ldots, T-1;\ j = 0, 1, \ldots, J-1;\ k = 0, 1, \ldots, K-1$$

where $\{\omega_k\}$ is a Markov Chain with state-space $\left\{ 2\pi \frac{n+u/I}{T}; u = -MI, \ldots, MI \right\}$, say, and $n$ has been chosen a priori. We obtain considerable reduction in the number of computations, if we use the Fourier coefficients near the frequency $2\pi \frac{n}{T}$ : Let

$$Y_{jk}(p) = \sum_{t=0}^{T-1} X(t + jT + kJT)e^{-i2\pi t(n+p)/T};\ p = -P, \ldots, P$$

where $P > M$. Then

$$Y_{jk}(p) = \frac{\rho}{2} \exp\left\{ i\left( \phi + Tj\omega_k + JT \sum_{m=0}^{k-1} \omega_m \right) \right\} \frac{\exp(iT\omega_k) - 1}{i\{\omega_k - 2\pi(n+p)/T\}}$$

$$+ \quad U_{jk}(p) + O(1)$$

almost surely, where

$$U_{jk}(p) = \sum_{t=0}^{T-1} \varepsilon(t + jT + kJT)e^{-i2\pi t(n+p)/T}$$

and the $T^{-1/2}U_{jk}(p)$ are asymptotically complex normal with mean zero and with real and imaginary parts asymptotically independent with variances $\pi f(\omega)$, where $\omega$ is a central frequency. We shall use as 'observations' $Z(k) = \{Y_{jk}(p); p = -P, \ldots, P; j = 0, \ldots, J-1\}$. Note: we are assuming that the possible variation in frequency is $O(T^{-1})$, or that the spectral density is fairly flat near the frequencies of interest.

The model does not fit into the classical HMM framework, for the dependence on past frequencies is through both the frequency and cumulative sum of frequencies. However, if we let $C_k = \left(\omega_k, \sum_{m=0}^{k-1} \omega_m\right)$, and $C_0 = (\omega_0, 0)$, then $\{C_k\}$ is a Markov Chain. At first glance, the dimension of the problem and number of computations seem very large. The dependence on $\sum_{m=0}^{k-1} \omega_m$ is, however, only through the fractional part of $JT\sum_{m=0}^{k-1} \omega_m/(2\pi)$, which take only values from the fractional parts of $\{Ju/I; u = -MI, \ldots, MI\}$. Thus, if we were to chose $I$ to be 20 and $J$ to be 5, there would only be 4 possible values to worry about. Moreover, there will be many zeroes in the transition probability matrix of the augmented Markov Chain $C_k = \left(\omega_k, \mathrm{frac}\left\{JT\sum_{m=0}^{k-1} \omega_m/(2\pi)\right\}\right)$, of which advantage may be taken, since in calculating the transition probabilities there will only be one previous cumulative frequency possible, given the previous and present frequencies and present cumulative frequency. For example, the coding using MATLAB$^{TM}$, which is vector-oriented, is done by rearranging the columns of a matrix which has no zero entries. This is done efficiently by reusing the same matrix at each step, but with a permutation matrix as its argument. The details are too complicated to describe here.

Finally, we assume that the distribution of $\omega_0$ is to be estimated, but that $\{\omega_k\}$ is, in fact, a random walk with independent increments whose common probability function is obtained by discretising a normal random variable with mean zero and variance $\nu^2$. The system parameters to be estimated are then $\rho, \phi, \nu^2$, the initial probability vector $\pi$ and $\sigma^2$, the common asymptotic variance of the real and imaginary parts of $T^{-1/2}U_{jk}(p)$. Initial estimators are provided by estimating the frequencies separately in each block and using simple robust estimates, as these frequency estimators are liable to be inaccurate. The EM algorithm, in conjunction with the Forward-Backward equations, are used to estimate the system parameters. Finally, the Viterbi algorithm is used to obtain the track with highest "probability". For a general treatment of Hidden Markov Models, the reader is referred to [RJ86]. The model is easily modified to take into account changing amplitude between clusters of time-blocks.

A simpler and faster approach, which will not perform as well at low SNRs, is to take ratios of Fourier coefficients within each block of $J$ time-blocks and at each Fourier frequency considered. This reduces the number of system parameters by 2, and the underlying Markov Chain consists only of the frequencies, as the ratios do not depend on the cumulative frequencies. There is quite a large loss of information, however, as there is a 'loss of phase' between the clusters of time-blocks. Details are given in [QBS94].

Figure 2 displays the 'true' and estimated frequencies from a simulation in which the 'true' frequencies are generated according to the random walk model, the block estimation method described above having been used, while Figure 3 displays the 'true' frequencies and the HMM track using the first HMM technique described – it should be emphasised that the HMM technique was fully automated and took approximately fifteen seconds of CPU time using MATLAB on a Pentium PC running at 60Mhz. The figures show that, although the SNR was so low (-20dB in this case) that the estimation technique has gone awry (this is called 'thresholding' and is analogous to the fact that the periodogram will have local maximisers removed from the true frequency if the SNR is low) the HMM technique has held true (See Quinn and Kootsookos [QP94]).

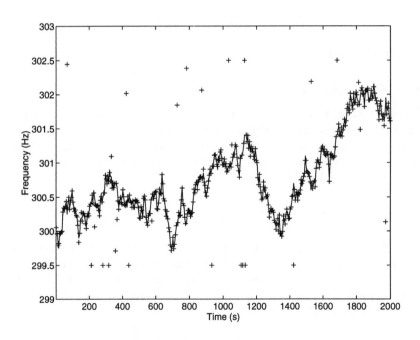

FIGURE 2. Frequency estimates and truth

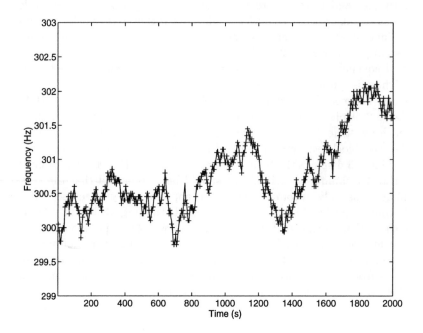

FIGURE 3. HMM track and truth

# 7   Applications

## 7.1   A Doppler example

Suppose that an acoustic source produces a sinusoid at some frequency, due, for example, to rotating machinery. If there is relative motion between the source and a single receiver, the frequency changes through time. If the source and receiver have constant relative velocity, the frequency at time $t$ may be shown to be of the form

$$\omega_t = \alpha + \beta z_t(t_c, s), \; z_t(t_c, s) = \frac{t - t_c}{\sqrt{s^2 + (t - t_c)^2}}$$

where $t_c$ is the time at closest approach if either the source or receiver is at rest. Needless to say, there are many phenomena which will influence the manner in which the received amplitude changes through time, not the least of which is the distance between source and receiver.

In the simplest case, where the source is stationary,

$$s = R/v, \alpha = \omega, \beta = -\alpha v/c,$$

where $R$ is the range at closest approach, $v$ is the speed of the receiver, $\omega$ is the rest frequency of the sinusoid and $c$ is the speed of sound in the medium. When the receiver is stationary,

$$\alpha = \omega/\{1 - (v/c)^2\}, \beta = -\alpha v/c, s = R\sqrt{1 - (v/c)^2}/v$$

If $v \ll c$, as is mostly the case in underwater acoustics, where $c \sim$ 1500ms$^{-1}$, the two cases, and, indeed, all cases, are approximately the same. Added to this sinusoid at the receiver, of course, will be other periodic signals, which are assumed to be at frequencies relatively far from $\lambda$, and additive noise which is assumed to be stationary but not necessarily Gaussian or white. Suppose that estimates of $\hat{\rho}_{t_j}$ and $\hat{\omega}_{t_j}$ of amplitude and frequency are available at times $t_j; j = 1, \ldots, J$ and have been calculated from the same sample sizes. (They may, for example, have been calculated using the estimation or tracking methods described above.) As the frequency and amplitude estimators are asymptotically uncorrelated and normally distributed when they are calculated independently from block to block, and as the asymptotic variance of a frequency estimator is inversely proportional to the square of the amplitude, a reasonable method of estimating $R, v, t_c$ and $\lambda$ is to minimise

$$\sum_{j=1}^{J} \hat{\rho}_{t_j}^2 \left\{ \hat{\omega}_{t_j} - \alpha - \beta z_{t_j}(t_c, s) \right\}^2$$

Since this is quadratic in $\alpha, \beta$, (the reason for parametrising in this way) we can maximise instead a function of $t_c$ and $s$ alone. Initial estimators and a Gauss-Newton method for improving the estimates are given in [Qui95], where the asymptotic statistical properties are also discussed.

### 7.2  Sonobuoy Field

Suppose that a field of acoustic sensors is being used to locate a stealthy quiet acoustic source. It is assumed that the positions of the sensors are known accurately and that the source is moving with constant velocity. We wish to track the source through time. Using thousands (or millions) of samples of the signal from each sensor is usually infeasible as there is no way in which the sensors could communicate with a powerful enough computer for a long enough period of time. Even if a small number of Fourier coefficients were communicated instead, the problem would still be time-consuming. We describe now a technique for tracking the source which uses only the estimates described in the previous example. Other methods of tracking need arrays of sensors or directional sensors, use estimates of bearings from these sensors through time, and require manœuvres from the vessel towing the sensors.

It is easy to describe the mathematical relationship between $\omega, v, t_c, R$ at each sensor and the position at an arbitrary time point and the velocity of the source. Given estimates of these parameters and estimates of the covariance matrix between them, a nonlinear regression model may be formulated. It is possible to produce good (closed-form) initial estimators of position and velocity ( through reparametrisation and linearistation of the problem) and again a Gauss-Newton method may be used to obtain improved estimators.

Figure 4 depicts a simulation in which the SNR is $-23$ dB at 500m (the amplitude of the received sinusoid is 1m when the source is at 500m and the (equivalent white-) noise variance is $100m^2$), the three sensors are 1000m apart and spherical spreading has been assumed. The concentric circles at each sensor represent the estimated range at closest approach, and the range $\pm$ one estimated standard deviation. It is obvious that the geometical solution to the problem which draws tangents to the estimated range circles is prone to large errors. The estimated trajectory using the technique above, however, is so close to the true one that it is difficult to tell it from the true trajectory. Figure 5 shows the estimated and true instantaneous amplitudes and frequencies at the three sonobuoys. Note that although there is quite a large error in estimating the frequency tracks, especially at the furthest sonobuoy, the technique has nevertheless worked well.

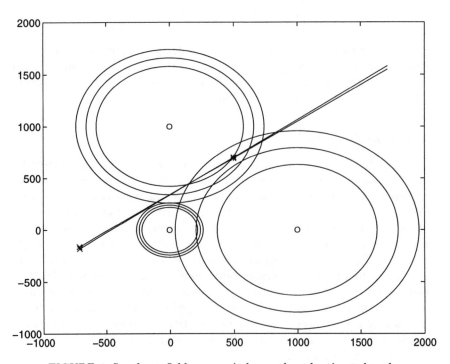

FIGURE 4. Sonobuoy field, range circles, path and estimated path

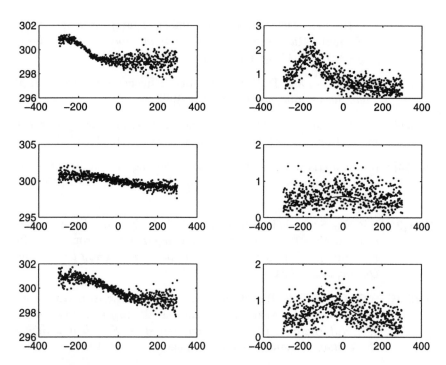

FIGURE 5. Frequency and amplitude estimates at three receivers

# 8   Conclusion

I have described above only a handful of problems which I posed and solved while working in Defence. I have chosen these as they show how the solution of fundamental problems such as the estimation of frequency may lead to the solution of important applied problems. There are many other important problems, particularly in the processing of array data, which are almost totally the preserve of Electrical Engineers. The area is obviously a fruitful one for Time Series Analysts.

# 9   References

[BH93]   R.F. Barrett and D.A. Holdsworth.   Frequency tracking using HMM models with amplitude and phase information. *IEEE Trans. on Signal Processing*, 41:2965–2976, 1993.

[Han73]   E.J. Hannan. The estimation of frequency. *J. App. Prob*, 10:510–519, 1973.

[Han79]   E.J. Hannan. The central limit theorem for time series regression. *Stoch. Proc. Appl*, 9:281–289, 1979.

[Han93]  E.J. Hannan. Determining the number of jumps in a spectrum. In T. Subba Rao, editor, *Developments in Time Series Analysis*, pages 127–138. Chapman and Hall, London, 1993.

[HH93]  E.J. Hannan and D. Huang. On-line frequency estimation. *J. of Time Series Analysis*, 14:147–161, 1993.

[HQ89]  E.J. Hannan and B.G. Quinn. The resolution of closely adjacent spectral lines. *J. of Time Series Analysis*, 10:13–31, 1989.

[KH94]  L. Kavalieris and E.J. Hannan. Determining the number of terms in a trigonometric regression. *J. of Time Series Analysis*, 15:613–625, 1994.

[LBQ94]  B.F. La Scala R.R. Bitmead and B.G. Quinn. An extended Kalman filter frequency tracker for high-noise environments. In *Proceedings of 7th SP Workshop on Statistical Array Processing, Quebec*, June 1994.

[QBS94]  B.G. Quinn R.F. Barrett and S.J. Searle. The estimation and HMM tracking of weak narrowband signals. In *Proceedings of the 1994 Internatonal Conference on Acoustics, Speech and Signal Processing, Adelaide*, volume IV, pages 341–344, 1994.

[QF91]  B.G. Quinn and J.M. Fernandes. A fast efficient technique for the estimation of frequency. *Biometrika*, 78:489–497, 1991.

[QP94]  B.G. Quinn and P.J. Kootsookos. Threshold behavior of the maximum likelihood estimator of frequency. *IEEE Trans. on Signal Processing*, 42:3291–3294, 1994.

[Qui94]  B.G. Quinn. Estimating frequency by interpolation using Fourier coefficients. *IEEE Trans. Signal Processing*, 42:1264–1268, 1994.

[Qui95]  B.G. Quinn. Doppler speed and range estimation using frequency and amplitude estimates. *J. Acoustical Society of America*, 97, 1995.

[RJ86]  L.R. Rabiner and B.H. Juang. An Introduction to Hidden Markov Models. *IEEE ASSP Magazine*, 4–16, January, 1986.

[RR88]  J.A. Rice and M. Rosenblatt. On frequency estimation. *Biometrika*, 75:477–484, 1988.

[SB90]  R.L. Streit and R.F. Barrett. Frequency line tracking using hidden Markov models. *IEEE Trans. ASSP*, 38:586–598, 1990.

[TV90]  B. Truong-Van. A new approach to frequency analysis with amplified harmonics. *J. Roy. Statist. Soc B*, 52:203–221, 1990.

Goldsmiths College, University of London

# Iterative Bandwidth Estimation for Nonparametric Regression with Long-range Dependent Errors

Bonnie K. Ray
Ruey S. Tsay

ABSTRACT We discuss the problem of bandwidth selection for a kernel regression trend estimator when the errors are long-range dependent. The iterative plug-in bandwidth selection method is investigated and modified to account for long memory in the errors. We compare the mean average-squared errors of the trend estimates using the bandwidth obtained from the modified procedure to that obtained assuming short-range dependent errors. For illustration, we apply the modified method to estimate a trend in a series of sea surface temperatures which exhibits both nonlinearity and a long-range dependent error structure.
**Keywords:** Kernel smoothing, Long-range dependence, Plug-in method

## 1  Introduction

In recent years, several automatic methods of nonparametrically estimating a trend in time series data have been proposed. These methods are important, in part, because they allow estimation of a nonlinear trend without requiring the specification of a particular form for the nonlinearity.

Assume that univariate data $x_1, \ldots, x_n$ are observed and that they follow the fixed-design setup

$$x_i = f(\frac{i - 1/2}{n}) + \epsilon_i, \quad i = 1, \ldots, n, \tag{1.1}$$

where $f$ is a smooth function defined on $[0, 1]$ and $\{\epsilon_i\}$ is a zero-mean, covariance stationary process. Our goal is to estimate the trend function $f$ nonparametrically. It has been shown that when the error terms $\epsilon_i$ are correlated and the correlation decays fast enough, *i.e.* the errors have short-range dependence, kernel estimates of the trend of the form

$$\hat{f}_h(x) = h^{-1} \sum_{i=1}^{n} x_i \int_{(i-1)/n}^{i/n} K(\frac{x - u}{h}) du, \tag{1.2}$$

converge to the true, unknown trend function at the same rate as in the case of independent errors (Hart, 1987; Hall and Hart, 1990). Here, $h > 0$ is called the bandwidth and $K$ is a kernel function with specified properties.

The main problem of nonparametric trend estimation involves the selection of the bandwidth $h$. Various automatic methods for such a selection have been proposed in the literature, although it has been shown that these methods fare poorly when the data are sufficiently positively correlated (Diggle and Hutchinson, 1989; Hart, 1991). Modifications of bandwidth selection procedures have been proposed to account for correlation in the errors when the errors are short-range dependent (e.g. Altman, 1991; Chiu, 1991; Herrmann, Gasser, and Kneip, 1992; Hart, 1994).

However, many real-world series have been found to have error terms with serial dependence which decays so slowly that the covariance function of the errors is not finitely summable (in the frequency domain, the spectrum may be infinite at frequency zero). These processes are termed *long-range dependent*. See Beran (1992) for a review of long-range dependent processes and their applications. When estimating a trend nonparametrically when the error terms exhibit long-range dependence, a different picture of kernel estimation emerges. Hall and Hart (1990) show that kernel estimates converge at a slower rate than in the case of independence or short-range dependence. Cheng and Robinson(1991) studied kernel density estimation for a purely stochastic process which is long-range dependent. In that case, the effect of the long-range dependence on convergence of the kernel estimate depends on the functional estimated. To the best of our knowledge, automatic bandwidth selection with long-range dependence has not been investigated in the fixed design case.

Nonparametric trend estimation with long-range dependent error terms is concerned with modeling both nonlinear behavior and long-range dependence, both of which have been observed in real-world time series, especially in the areas of oceanography and meteorology. See, for example, Haslett and Raftery (1989) and Lewis and Ray (1993). In this paper, we consider the iterative bandwidth estimation method (Herrmann, Gasser, and Kneip, 1992) in the case of correlated errors. We empirically investigate the performance of the method when long-range dependence is present, and propose a modification to the method to account for the long-range dependence. The usefulness of the modification is investigated via simulation. For illustration, we apply the modified method to estimate a trend in a series of sea surface temperatures which appears to have long-range dependent error structure.

# 2 Iterative Bandwidth Estimation

## 2.1 Optimal Bandwidth Estimation with Short-range Dependent Errors

When the residuals in Model (1.1) have serial correlations which are summable, Hall and Hart (1990) found that the bandwidth, $h_{opt}$, minimizing the mean integrated squared error

$$MISE(h) = \mathbf{E}[\int_0^1 v(x)\{f(x) - \hat{f}_h(x)\}^2 dx], \qquad (1.3)$$

where $v$ is a twice continuously differentiable function with support $[\delta, 1-\delta]$ and $v(x) > 0$ for all $x \in (\delta, 1-\delta)$ and some $\delta > 0$, is, asymptotically, given by

$$h_{opt} = (C_1 S_\epsilon(0)/[nC_2^2 \int_0^1 v(x)\{f''(x)\}^2 \, dx])^{1/5} \qquad (1.4)$$

for a kernel of order 2, where $S_\epsilon(0)$ denotes the spectral density of the error process $\{\epsilon_i\}$ evaluated at frequency zero, $C_1 = \int_0^1 v(t) \, dt \int \{K(x)\}^2 \, dx$ and $C_2 = \int x^2 K(x) \, dx$. Herrmann, Gasser, and Kneip (1992), hereafter referred to as HGK, proposed estimating the optimal bandwidth iteratively by estimating the unknown terms in $h_{opt}$. Thus estimates of $S_\epsilon(0)$ and $f''(x)$ are needed. They suggest estimating $S_\epsilon(0)$ using a function of the sample autocorrelations of the original data differenced an appropriate amount, or using the sample autocorrelations of the estimated residuals up to lag $m$, a prespecified positive integer. The function $f''(x)$ is also estimated, using a kernel estimator having kernel obeying special properties and a bandwidth which is a modification of the bandwidth estimated at the previous iteration.

This method fails asymptotically for long-range dependent errors, however, because $S_\epsilon(0)$ is infinite. To better understand the effects of long-range dependence on the iterative method in finite samples, we conduct a simulation study. We also propose a modified iterative method in the next subsection, and assess its performance via simulation. The simulation results are discussed in Section 2.5.

## 2.2 A Modified Plug-in Estimator

When the error terms in Model (1.1) have serial correlations which decay as $C_3 k^{-\alpha}$, where $0 < \alpha < 1$ and $C_3$ is a positive real number, Hall and Hart (1990) found that the bandwidth, $h'_{opt}$, minimizing the asymptotic MISE is

$$h'_{opt} = (C_3 \alpha C_4/[n^\alpha C_2^2 \int_0^1 v(x)\{f''(x)\}^2 \, dx])^{1/(4+\alpha)}, \qquad (1.5)$$

where $C_2$ is as in (1.4) and $C_4 = \int_0^1 v(t)\, dt \int \int |x - y|^{-\alpha} K(x) K(y)\, dx dy$. Thus to estimate the optimal bandwidth under strong correlation, we must estimate $h'_{opt}$. This requires estimating $\alpha$ and $C_3$, although a full parametric model for the errors is not necessarily required. Delgado and Robinson (1994) describe several "semi-parametric" methods of estimating $\alpha$ and $C_3$ in both the time domain and the frequency domain. We summarize two of these methods below and investigate their performance in estimating $\alpha$ and $C_3$ for the estimated residual process.

## 2.3   Estimation of Long Memory Parameters needed in Iterative Bandwidth Estimation

There are several approaches one may take to estimate the values of $\alpha$ and $C_3$ needed in the calculation of the bandwidth in Eq. (1.5).

The first is based on using estimated error terms. The error terms $\epsilon_i$ are unobservable, however they may be estimated in the following manner. Using a specified bandwidth, we can obtain an initial estimate of $f$. Then $\hat{\epsilon}_i = x_i - \hat{f}((i - 1/2)/n)$, $i = 1, ..., n$ can be used to estimate the amount of long-range dependence as characterized by $\alpha$. As Hart (1991) points out, using ordinary cross-validation to obtain an initial bandwidth will result in very poor estimates of $f$ and hence in very poor estimates of $\epsilon_i$. In the iterative bandwidth estimation procedure, however, the estimated error terms can be updated at each iteration, and updated estimates of $\alpha$ obtained.

Once we have an estimate of the error process, methods for estimation of $\alpha$ are available in both the time and frequency domain. Since we would like to be as "nonparametric" as possible, we mention only those that pertain to estimation of the long memory characteristics of $\epsilon_i$ as characterized by $\alpha$, without specification of any possible short-memory behavior of $\epsilon_i$.

In the time domain, Delgado and Robinson (1994) suggest two estimators based on exploitation of the form of the covariance function of a long memory process at large lags. Recall that

$$\gamma(k) \approx C_3 k^{-\alpha} \text{ as } k \to \infty. \tag{1.6}$$

Then

$$\ln \gamma(k) \approx \ln C_3 - \alpha \ln(k) \text{ as } k \to \infty. \tag{1.7}$$

This relation suggests estimating $\alpha$ and $C_3$ by substituting $\hat{\gamma}(k)$ for $\gamma(k)$ in Eq. (1.7) and using least-squares regression for $k = l, \cdots, l + M$, where $l$ is chosen to be suitably large. No asymptotic distributional properties of this estimator are known. However, Hosking (1984) shows that the estimated covariances of a long memory process with known mean term (in our case, we assume $\mu_\epsilon = 0$) are consistent and have the usual rate of convergence $o(n^{-1})$. A disadvantage of this estimator is that it is possible for $\hat{\gamma}(k)$ to take negative values, especially when $\gamma(k)$ is close to zero.

Because of this difficulty, Delgado and Robinson suggest an alternative procedure based on minimizing the squared distance between $\hat{\gamma}(k)$ and $C_3 k^{-\alpha}$ for large $k$, i.e.

$$(\hat{\alpha}, \hat{C}_3) = argmin_{C_3, \alpha} \sum_{k=l}^{l+M} (\hat{\gamma}(k) - C_3 k^{-\alpha})^2. \tag{1.8}$$

The sets over which minimization is carried out will typically be compact with respect to $\alpha$ and $C_3$. Again, no distributional properties of this estimator have been obtained.

For a long memory process $\{\epsilon_i\}$ the following relation holds for the spectrum of the process:

$$S_\epsilon(\omega) \approx c\omega^{\alpha-1} \tag{1.9}$$

for $\omega$ near zero. Estimates of $\alpha$ and $c$ can be obtained using the method proposed by Geweke and Porter-Hudak (1983) and modified by Robinson (1992). This method is based on estimating the spectrum of the process at low frequencies and regressing the logarithm of the estimated spectral values evaluated at the Fourier frequencies against the logarithm of the spectrum. A commonly used number of frequencies in the regression is $M = \sqrt{n}$.

There remains the problem of relating the constant $c$ to the necessary constant $C_3$. The constant $c$ may be thought of as the evaluation at $\omega = 0$ of the function $L(\omega)$, which describes the short memory behavior of the series and is slowly varying at low frequencies. For example, for a fractionally differenced ARMA (ARFIMA)$(p, d, q)$ process $(1 - \phi_1 B - \cdots - \phi_p B^p)(1 - B)^d X_t = (1 - \theta_1 B - \cdots - \theta_q B^q)a_t$, which has long memory when $0 < d < .5$ ($\alpha = 1 - 2d$), $c = \frac{\sigma_a^2}{2\pi} \left|\frac{\theta(1)}{\phi(1)}\right|^2$. The ARFIMA process has $\gamma(k) \approx \sigma_a^2 \left|\frac{\theta(1)}{\phi(1)}\right|^2 \frac{\Gamma(1-2d)}{\Gamma(d)\Gamma(1-d)} k^{2d-1} \approx C_3 k^{2d-1}$ for large $k$ (Hosking, 1981). Thus

$$C_3 = 2\pi c \frac{\Gamma(1 - 2d)}{\Gamma(d)\Gamma(1 - d)}. \tag{1.10}$$

Hart (1991) suggests estimating the correlation structure of the residuals without initial estimation of the trend function $f$. Since the original data $x_i$ is nonstationary, differencing the data twice (assuming $f$ is twice-differentiably continuous) will approximately eliminate the effect of $f$. Let $\delta_i = (1 - B)^2 x_i, j = 3, \ldots, n$. Since $f''$ is continuous, $\delta_i = n^{-2} f''(x_i) + d_i$, where $(i - .5)/n \leq x_i \leq (i + .5)/n$ and $d_i = (1 - B)^2 \epsilon_i$. For large $n$, $\delta_i \approx d_i$. The differenced data $\delta_i$ is then used to estimate $\alpha$.

In the frequency domain, the relation between the original data and the differenced data is quite simple. The spectrum of $\{\epsilon_i\}$, $S_\epsilon(\omega)$, is related to the spectrum of $\{d_i\}$, $S_d(\omega)$, through the relation

$$S_d(\omega) = |1 - e^{i\omega}|^4 S_\epsilon(\omega), \omega \in [-\pi, \pi]. \tag{1.11}$$

Thus $S_d(\omega) \sim c\omega^{4+\alpha-1}$ when $\omega$ near zero. The log periodogram regression method can again be used to estimate the coefficient $4+\alpha-1$. Tapering the data before estimating the spectrum is recommended (Hart, 1989; Hurvich and Ray, 1995).

## 2.4  Algorithm for Iterative Estimation of the Optimal Bandwidth under Long-range Dependence

The following steps are used to estimate the bandwidth minimizing MISE. Case 1: Assume $\alpha$ and $C_3$ known.

1. Let $h_0' = n^{-\alpha}$.

2. Set

$$h_i' = (C_3\alpha C_4/[n^\alpha C_2^2 \int_0^1 v(x)\{\hat{f}_2(x, h_{i-1}' n^{\alpha/2(4+\alpha)})^2\}] \, dx)^{1/(4+\alpha)}, i = 1, ..., i^*$$
$$(1.12)$$

3. Set $\hat{h}_{opt}' = h_{i^*}'$.

In order that the final estimated bandwidth be of the correct order, the number of iterations $i^*$ is fixed at $1 + [2(4+\alpha)/\alpha]$. The estimator $\hat{f}_2$ of $f''$ is given by a kernel estimator with bandwidth $h_2$ defined by

$$\hat{f}_2(t, h_2) = \sum_{i=1}^n x_i \int_{(i-1)/n}^{i/n} \frac{1}{h_2^3} K_2(\frac{t-u}{h_2}) du.$$
$$(1.13)$$

The bandwidth $h_2 = h_{i-1}' n^{\alpha/2(\alpha+1)}$, i.e. the bandwidth from the previous iteration is inflated in an asymptotically appropriate proportion with respect to the bandwidth to be estimated and is used to estimate $f''$. Case 2: Assume $\alpha$ and $C_3$ unknown.

1. Estimate an optimal bandwidth, $\hat{h}_{opt}$, using the short-range dependence method of HGK.

2. Let $h_0' = \hat{h}_{opt}$.

3. Estimate $f((i-1/2)/n)$ using $h_{i-1}'$ and let $\hat{\epsilon}_i = x_i - \hat{f}((i-1/2)/n)$. Estimate $\alpha$ and $C_3$ using $\hat{\epsilon}_i$. Evaluate $C_4$ using the estimated $\alpha$.

4. Set

$$h_i' = (\hat{C}_3\hat{\alpha}\hat{C}_4/[n^{\hat{\alpha}} C_2^2 \int_0^1 v(x)\{\hat{f}_2(x, h_{i-1}' n^{\hat{\alpha}/2(4+\hat{\alpha})})^2\}] \, dx)^{1/(4+\hat{\alpha})}, i = 1, ..., i^*$$
$$(1.14)$$

5. Repeat steps 3 and 4 until convergence is reached. The number of iterations required should generally be less than 10 because of the starting point.

6. Set $\hat{h}'_{opt} = h'_{i\cdot\cdot}$.

Alternatively, $\alpha$ and $C_3$ may be estimated initially using the twice differenced data as described above, and $h'_{opt}$ estimated as in the case $\alpha$ and $C_3$ known.

Following the proof of the Theorem given on p.789 of HGK, we may show that the bandwidth estimated using the above algorithm is asymptotically equal to the bandwidth minimizing the asymptotic MISE, provided $\alpha$ and $C_3$ are estimated consistently.

One method that may be used to determine if long-range dependence is present in the error terms is to estimate the optimal bandwidth assuming short-range dependence, with, $e.g.$, $m = [n^{.25}], [n^{.50}], [n^{.75}]$. If long-range dependence is present, the estimated optimal bandwidth should increase as $m$ increases for fixed $n$. Note that in finite samples, continuing to increase $m$ should result in very large bandwidths using the method of HGK, because $\hat{S}_\epsilon(0)$ will continue to increase for long-range dependent errors. If, however, a fixed number $m$ of lags is used, asymptotically, the bandwidth will be of the wrong order, $i.e.$ too small.

## 2.5   Simulation Results

We conducted a small simulation study to investigate the modified iterative method in practice. We generated series of length $n = 500$ and $n = 1000$ with trend functions

- $f_1(t) = 2 - 5t + 5\exp\{-100(t - .5)^2\}$ and

- $f_2(t) = 2\sin(4\pi t)$.

These are two of the trend functions used by HGK. To each trend function, we added four different ARFIMA$(p, d, 0)$ error processes, the first three having $p = 0, d = 0.2, 0.4, 0.45$ and the fourth having $p = 1, \phi = 0.5, d = 0.4$. The ARFIMA errors were generated using the algorithm of Hosking (1984). The variance of the errors was fixed at 1.5 for $f_1$ and 1.0 for $f_2$. The number of replications was 50. We used $K(x) = .75(1 - x^2), -1 < x < 1$ and $K_2(x) = \frac{15(3x^2-1)}{4}, -1 < x < 1$ as our kernel functions. For each series, five bandwidth estimates were considered:

1. $h_{sm1}$: The bandwidth resulting from using the HGK iterative procedure with $\hat{S}(0) = \hat{\gamma}(0) + 2\sum_{i=1}^{10}\hat{\gamma}(i)$ computed using the estimated residuals at each step.

2. $h_{sm2}$: The bandwidth resulting from using the HGK procedure with $\hat{S}(0) = \hat{\gamma}(0) + 2\sum_{i=1}^{m}\hat{\gamma}(i)$, where $m = 75$ when $n = 500$ and $m = 125$ when $n = 1000$. Thus $m \approx [n^{.7}]$.

3. $h_{lm1}$: The bandwidth resulting from using the modified iterative procedure assuming $\alpha$ and $C_3$ known.

4. $h_{lm2}$: The bandwidth resulting from using the modified iterative procedure with $\alpha$ and $C_3$ estimated. The log periodogram regression method with $m = \sqrt{n}$ periodogram ordinates was used to estimate $\alpha$ and $c$ from the estimated residuals at each iteration, and relation (1.10) used to obtain $C_3$. However, if $\hat{\alpha} \leq 0$, we set $\hat{\alpha} = 0.01$, and if $\hat{\alpha} \geq 1$, we set $\hat{\alpha} = 0.99$. The spectrum of the short-memory component at frequency zero was estimated from the fractionally differenced residuals and then used to estimate $c$.

5. $h_*$: the bandwidth minimizing $\sum_{i=1}^{n}\{f((i-.5)/n) - \hat{f}((i-.5)/n)\}^2$ over a grid of 100 equally spaced points between 0 and 0.5. This is the empirically optimal bandwidth, and is not necessarily equal to $h'_{opt}$.

The number of iterations was fixed at 11 for the HGK method. In the modified procedure, we found that setting $h_0 = n^{-\alpha}$ resulted in a starting point that was too large in finite samples. Thus we set $h_0 = h_{sm1}$ in practice. This modification also allows us to obtain reasonable estimates of the residuals at the first step when $\alpha$ must be estimated. We also found that 12 iterations were typically sufficient for convergence of the estimated bandwidth using the modified method.

Tables 1 and 2 show the simulation results. When $n = 500$, we see that for weak long memory ($d = 0.2$), the empirically optimal bandwidth is slightly overestimated for $f_1$ and slightly underestimated for $f_2$, using the HGK method with $m = 10$, but is overestimated when $m = 75$, more so for $f_1$ than for $f_2$. The bandwidth $h_{lm1}$ is very close to the empirical optimal for $f_1$ and slightly greater than the empirical optimal for $f_2$. The bandwidth $h_{lm2}$ is larger than the empirical optimal, with a larger positive bias for $f_2$. For stronger long-range dependence with no AR component ($d = 0.4, d = 0.45$), the estimator obtained using the modified method with $\alpha$ and $C_3$ known is smaller than the empirical optimal. This may be due to the fact that $h_{sm1}$ was used as a starting point and $h_{sm1}$ tends to be too small. Using $m \approx [n^{.7}]$ lags in the HGK method produces a pretty good estimate of the bandwidth, although it still tends to be too large for $f_1$. When $\alpha$ and $C_3$ are estimated, the resulting bandwidth tends to be slightly larger than the empirical optimal for $f_2$ and smaller than the empirical optimal for $f_1$. When an AR component is present, $h_{sm1}$ is extremely small, while the other estimated bandwidths are slightly smaller than the empirical optimal. When $n = 1000$, the results are similar, although the estimated bandwidths have a smaller standard deviation.

These results can be understood in light of the following results. Altman (1990) shows that for short-range dependent processes, the method of moments estimator of the correlation at lag $k$,

$$\hat{\rho}(k, h, n) = \frac{\sum_{i=[nh/2]}^{n+1-[nh/2]-k} \hat{\epsilon}(i, h, n)\hat{\epsilon}(i+k, h, n)}{\sum_{i=[nh/2]}^{n+1-[nh/2]-k} \hat{\epsilon}^2(i, h, n)},$$

has expectation

$$\mathbf{E}\rho(k) \approx \frac{\rho(k) + h^4(C_2/2)^2 \frac{\int \{f''(t)\}^2 dt}{\sigma^2} + \frac{S_\epsilon(0)}{nh}[C_1 - 2K(0)]}{1 + h^4(C_2/2)^2 \frac{\int \{f''(t)\}^2 dt}{\sigma^2} + \frac{S_\epsilon(0)}{nh}[C_1 - 2K(0)]} \qquad (1.15)$$

as $h \to 0$ and $nh \to \infty$. For long-range dependence, we have $\frac{\alpha C_3}{(nh)^\alpha}[C_4 - 2K(0)]$ in place of $\frac{S_\epsilon(0)}{nh}[C_1 - 2K(0)]$ in the above expression. Thus for kernels such that $C_4 < 2K(0)$, correlation produces a negative bias, which may be offset by a positive bias dependent on the signal to noise ratio of the underlying trend and noise, measured by $\int \{f''(t)\}^2 dt/\sigma^2$. Note that $C_4$ depends on $\alpha$. In our simulations, $C_4 > 2K(0)$ when $d = 0.2$ and $C_4 < 2K(0)$ when $d = 0.4, 0.45$. Additionally, the signal to noise ratio is larger for $f_1$ in our study, which can be evidenced by a smaller negative bias in the estimated bandwidths for $f_1$ than for $f_2$ when $d = 0.4, 0.45$.

The time domain method for estimating $\alpha$ and $C_3$ using the estimated residuals was also tried, but gave very different answers depending on the number of correlations used in the regression. To avoid the problem of bias in the estimated correlations using the estimated residuals, we tried using the twice-differenced series to initially estimate $\alpha$ and $C_3$. The method consistently resulted in overestimated $\alpha$ values.

Although the modified iterative procedure produced bandwidths closer to optimal in most cases, the different bandwidths did not produce significantly different Integrated Squared Errors (ISE) or Relative Integrated Squared Errors (RISE) on average for series of length $n = 500$. The greatest improvements in ISE and RISE were seen for series of length $n = 1000$ having strong long-range dependence.

# 3    Application of Method to Sea Surface Temperatures

We apply the iterative estimation method to a series of 20 years of daily sea surface temperatures measured off the California coast. Figure 1 shows the raw data (dashed line), along with estimated trend functions. There is an evident yearly cycle present, as well as a four- to five-year cycle, which has been attributed to El Niño episodes. Additional shorter term periodic effects are also present. Lewis and Ray (1993) detrended the data using a sine-cosine curve with a 1-year cycle. A plot of the log periodogram versus log frquency for the detrended data indicated that long-range dependence was present. They attempted to model the long-range dependent behavior of this series using an ARFIMA$(1, d, 0)$ model with $\hat{d} = 0.375$. However the model did not adequately capture the nonlinear behavior of the series, for example the sharp, downward swings in temperature which occur every spring. To incoporate nonlinear behavior, a threshold AR model was used, with long AR components included to try and capture the long-range dependence. We attempt to model the nonlinear behavior using a deterministic nonlinear trend estimated nonparametrically. As discussed in Lewis and Ray (1993), the variability of the temperatures appears to increase as the average temperature increases, thus we estimate the bandwidth for the log transformed temperatures.

Fixing $d = 0.375$ and estimating $C_3$ using the estimated spectrum at zero for the fractionally differenced residuals, a bandwidth of 663 days, or approximately 1.8 yrs, was obtained. Estimating $d$ and $C_3$, we obtained a bandwidth of 642 days, with an estimate of $\hat{d} = 0.39$. The trend estimate using $h = 642$ is shown in Figure 1, along with the trend resulting from using a bandwidth of 87 days, which was obtained by Altman (1990) for 12 years of raw sea surface data, assuming the errors followed an AR(2) process. We see that the trend estimate obtained using the result of the modified iterative method is much smoother than that obtained by Altman. Yearly effects and other small changes in the temperature are smoothed out, resulting in a much clearer picture of the long upwards trend in the data from 1971 to 1984.

# 4    Conclusion

We have presented a modified iterative method for estimating the bandwidth of a kernel density trend estimate. The modified method performs better in simulations than the unmodified method of HGK for series having long-range dependent errors. Future research will investigate alternative methods for estimating the parameters needed for the iterative method. Additionally, other techniques for bandwidth estimation, such as the Time Series Cross-Validation method of Hart (1994), will be investigated for series with long-range dependence when the goal is prediction.

# Acknowledgements

This research was supported by NSF under grant #DMS-9409273 (BKR) and Grant #DMS-9305045(RST). The authors wish to thank Peter Lewis of the Naval Postgradaute School for providing the sea surface temperature data.

# References

Altman, N.S. (1991) "Kernel smoothing of data with correlated errors." *J. Am. Statist. Ass.*, **85**, 749–759.

Beran, J. (1992) "Statistical methods for data with long-range dependence." *Statistical Science*, **7**, 402–427.

Cheng, B. and Robinson, P. (1991) "Density estimation in strongly dependent non-linear time series." *Statistica Sinica*, **1**, 335–359.

Chiu, S.T. (1989) "Bandwidth selection for kernel estimate with correlated noise." *Statist. Probab. Lett.*, **8**, 347–354.

Delgado, M.A. and Robinson, P.M. (1994) "New methods for the analysis of long-memory time series: Application to Spanish inflation", *Journal of Forecasting*, **13**, 97–107.

Diggle, P.J. and Hutchinson, M.F. (1989) "On spline smoothing with correlated errors." *Aust. J. Statist.*, **31**, 166–182.

Geweke, J. and Porter-Hudak, S. (1983) "The estimation and application of long memory time series models", *Journal of Time Series Analysis*, **4**, 221–238.

Hall, P. and Hart, J. (1990) "Nonparametric regression with long-range dependence," *Stoch. Processes Appl.*, **36**, 339–351.

Hart, J.D. (1987) "Kernel smoothing when the observations are correlated." *Technical Report 35*, Dept. of Statistics, Texas A&M University, College Station.

Hart, J.D. (1989) "Differencing as an approximate de-trending device," *Stoch. Processes Appl.*, **31**, 251–259.

Hart, J.D. (1991) "Kernel regression estimation with time series errors," *J. R. Statist. Soc.* B, **53**, 173–187.

Hart, J.D. (1994) "Automated kernel smoothing of dependent data by using time series cross-validation," *J. R. Statist. Soc.* B, **56**, 529–542.

Haslett, J. and Raftery, A. E. (1989) "Space-time modelling with long-memory dependence: assessing Ireland's wind power resource (with discussion)," *Appl. Statist.*, **38**, 1-50.

Herrmann, E., Gausser, T. and Kneip, A. (1992) "Choice of bandwidth for kernel regression when residuals are correlated", *Biometrika*, **79**, 783–795.

Hosking, J.R.M. (1981) "Fractional differencing", *Biometrika*, **68**, 165–176.

Hosking, J.R.M. (1984) "Modeling persistence in hydrological time series using fractional differencing", *Water Resources Research*, **20**, 1898-1908.

Hurvich, C. and Ray, B.K. (1995) "Estimation of the memory parameter for nonstationary or noninvertible fractionally differenced time series models", Journal of Time Series Analysis **16**, 17-42.

Lewis, P.A.W. and Ray, B.K. (1993) "Nonlinear modelling of multivariate and categorical time series using Multivariate Adaptive Regression Splines", in **Dimension Estimation and Models**, H. Tong, ed., World Scientific, Singapore.

Robinson, P. (1992) "Log-periodogram regression for time series with long-range dependence", Preprint.

TABLE 1.1. Average estimated bandwidths, Integrated Squared Error(ISE), and Relative Integrated Squared Error(RISE) for $f_1(t) = 2 - 5t + 5 \exp\{-100(t-.5)^2\}$

| | | | $n = 500$ | | | $n = 1000$ | |
|---|---|---|---|---|---|---|---|
| | Method | $h$ | ISE | RISE | $h$ | ISE | RISE |
| | SM1 | 0.071 | 0.112 | 1.066 | 0.063 | 0.090 | 1.067 |
| $d = 0.2$ | SM2 | 0.103 | 0.156 | 1.497 | 0.096 | 0.119 | 1.507 |
| | LM1 | 0.072 | 0.113 | 1.077 | 0.065 | 0.090 | 1.066 |
| | LM2 | 0.077 | 0.115 | 1.095 | 0.070 | 0.092 | 1.092 |
| | EOpt | 0.066 | 0.105 | – | 0.062 | 0.084 | – |
| | SM1 | 0.064 | 0.456 | 1.103 | 0.058 | 0.345 | 1.153 |
| $d = 0.4$ | SM2 | 0.098 | 0.469 | 1.129 | 0.093 | 0.337 | 1.110 |
| | LM1 | 0.074 | 0.453 | 1.079 | 0.070 | 0.335 | 1.106 |
| | LM2 | 0.086 | 0.481 | 1.163 | 0.074 | 0.336 | 1.111 |
| | EOpt | 0.079 | 0.431 | – | 0.082 | 0.313 | – |
| | SM1 | 0.060 | 0.603 | 1.063 | 0.054 | 0.520 | 1.157 |
| $d = 0.45$ | SM2 | 0.093 | 0.616 | 1.086 | 0.089 | 0.514 | 1.090 |
| | LM1 | 0.067 | 0.604 | 1.064 | 0.065 | 0.514 | 1.125 |
| | LM2 | 0.076 | 0.600 | 1.055 | 0.067 | 0.508 | 1.069 |
| | EOpt | 0.077 | 0.576 | – | 0.078 | 0.486 | – |
| | SM1 | 0.024 | 0.781 | 1.592 | 0.020 | 0.735 | 1.587 |
| $d = 0.4$ | SM2 | 0.091 | 0.596 | 1.088 | 0.089 | 0.569 | 1.048 |
| $\phi = 0.5$ | LM1 | 0.082 | 0.604 | 1.115 | 0.072 | 0.571 | 1.077 |
| | LM2 | 0.090 | 0.603 | 1.105 | 0.075 | 0.569 | 1.060 |
| | EOpt | 0.090 | 0.562 | – | 0.079 | 0.549 | – |

TABLE 1.2. Average Estimated bandwidth($h$), Integrated Squared Error(ISE), and Relative Integrated Squared Error(RISE) for $f_2(t) = 2\sin(4\pi t)$

|  | | $n = 500$ | | | $n = 1000$ | | |
|---|---|---|---|---|---|---|---|
|  | Method | $h$ | ISE | RISE | $h$ | ISE | RISE |
| | SM1 | 0.063 | 0.071 | 1.086 | 0.058 | 0.047 | 1.053 |
| $d = 0.2$ | SM2 | 0.083 | 0.073 | 1.093 | 0.076 | 0.049 | 1.105 |
| | LM1 | 0.070 | 0.070 | 1.062 | 0.065 | 0.046 | 1.042 |
| | LM2 | 0.084 | 0.074 | 1.114 | 0.071 | 0.047 | 1.059 |
| | EOpt | 0.070 | 0.066 | – | 0.063 | 0.045 | – |
| | SM1 | 0.061 | 0.308 | 1.053 | 0.055 | 0.268 | 1.200 |
| $d = 0.4$ | SM2 | 0.083 | 0.307 | 1.054 | 0.076 | 0.257 | 1.098 |
| | LM1 | 0.076 | 0.304 | 1.041 | 0.070 | 0.258 | 1.109 |
| | LM2 | 0.083 | 0.308 | 1.061 | 0.072 | 0.258 | 1.112 |
| | EOpt | 0.077 | 0.296 | – | 0.082 | 0.243 | – |
| | SM1 | 0.053 | 0.400 | 1.200 | 0.052 | 0.350 | 1.134 |
| $d = 0.45$ | SM2 | 0.074 | 0.384 | 1.095 | 0.074 | 0.337 | 1.046 |
| | LM1 | 0.063 | 0.393 | 1.157 | 0.068 | 0.342 | 1.084 |
| | LM2 | 0.076 | 0.384 | 1.087 | 0.067 | 0.338 | 1.050 |
| | EOpt | 0.082 | 0.368 | – | 0.080 | 0.327 | – |
| | SM1 | 0.021 | 0.535 | 1.878 | 0.018 | 0.497 | 1.614 |
| $d = 0.4$ | SM2 | 0.081 | 0.390 | 1.143 | 0.077 | 0.365 | 1.030 |
| $\phi = 0.5$ | LM1 | 0.080 | 0.393 | 1.160 | 0.072 | 0.370 | 1.053 |
| | LM2 | 0.084 | 0.390 | 1.145 | 0.073 | 0.368 | 1.038 |
| | EOpt | 0.093 | 0.364 | – | 0.084 | 0.357 | – |

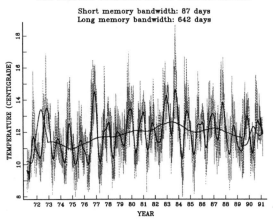

FIGURE 1. Raw and smoothed logged daily sea surface temperatures at Granite Canyon. Dashed line indicates raw data, while trend estimates are depicted using solid lines.

New Jersey Institute of Technology

University of Chicago

# The Likelihood of an Autoregressive Scheme

## M. Rosenblatt

Author address:

UNIVERSITY OF CALIFORNIA, SAN DIEGO
*E-mail address*: mrosenblatt@ucsd.edu

# The Likelihood of an Autoregressive Scheme

## 1. Gaussian Sequences

Ted Hannan among others spent some time investigating the asymptotic properties of maximum likelihood estimates of the parameters of autoregressive moving average (ARMA) Gaussian schemes. The asymptotics implicitly assume that the estimate is close enough to the actual parameter value so that a linear approximation is sufficiently accurate. Of course, this is often not true in actual computation and for that reason it is of interest to see whether one can get a reasonable approximation to the likelihood surface. Let $x_t, t = \ldots, -1, 0, 1, \ldots$ be the stationary solution of the system of equations

$$(1) \qquad \sum_{k=0}^{p} a_k x_{t-k} = \sum_{j=0}^{q} b_j \xi_{t-j}, \quad a_0 = b_0 = 1, \quad a_p, b_q \neq 0,$$

where the $\xi_t$ random variables are independent $N(0, \sigma^2)$, $\sigma^2 > 0$, and the polynomials

$$a(z) = \sum_{k=0}^{p} a_k z^k,$$

$$b(z) = \sum_{k=0}^{q} b_k z^k$$

have all their zeros outside the unit disc in the complex plane $\{z, |z| \leq 1\}$ and have no zeros in common. The solution $x_t$ is an ARMA scheme and the problem is that of estimating the parameters $\underline{\nu} = (\underline{\beta}, \sigma) = (\beta_1, \ldots, \beta_{p+q}, \sigma) = (a_1, \ldots, a_p, b_1, \ldots, b_p, \sigma)$. Assume that one observes $\underline{x} = (x_1, \ldots, x_n)$. The maximum likelihood estimate $(\hat{\underline{\beta}}, \hat{\sigma})$ of the true value $(\underline{\beta}_0, \sigma_0)$ maximizes $2/n$ times the loglikelihood

$$(2) \quad -\log(2\pi) - \frac{1}{n}\log|R| - \frac{1}{n}\underline{x}R^{-1}\underline{x}'$$

$$= -\log(2\pi) - 2\log\sigma - \frac{1}{n}\log|G| - \frac{1}{n\sigma^2}\underline{x}G^{-1}\underline{x}'$$

as a function of $\beta$, where $R = R(\underline{\beta}_0) = \sigma^2 G(\underline{\beta}_0)$ is the covariance matrix of $\underline{x}$ and $|R|$ is the determinant of $R$. The covariance matrix $R = (r_{j-k}; j, k = 1, \ldots, n)$ is a Toeplitz matrix with $r_j = E(x_s x_{s+j})$. The spectral density of the ARMA scheme $x_t$ is

$$f(\lambda) = \frac{\sigma^2}{2\pi}|a(e^{-i\lambda})|^2 / |b(e^{-i\lambda})|^2 = \frac{\sigma^2}{2\pi}g(\lambda) = \frac{\sigma^2}{2\pi}g(\lambda;\underline{\beta}).$$

Let $\tilde{\sigma}$ be the value of $\sigma$ maximizing (2) with $\beta$ values fixed. On inserting this value in (2) one obtains aside from the constant the expression

(3) $$-n^{-1}\log(\underline{x}G^{-1}\underline{x}'/n) - n^{-1}\log|G|$$

since

$$\tilde{\sigma}^2 = \underline{x}G^{-1}\underline{x}'/n.$$

The maximum likelihood estimate $\hat{\underline{\beta}}$ maximizes (3) as a function of $\underline{\beta}$. The first result of Szegö on Toeplitz matrices (see [4]) implies that

(4) $$n^{-1}\log G(\underline{\beta}) \to \frac{1}{2\pi}\int_{-\pi}^{\pi}\log g(\lambda;\underline{\beta})\,d\lambda$$

and $\int_{-\pi}^{\pi}\log g(\lambda;\beta)\,d\lambda \equiv 0$ since $a_0 = b_0 = 1$ and the zeros of $a(z)$, $b(z)$ lie outside the unit disc in the complex plane. The second result of Szegö on Toeplitz matrices (see [4]) tells us at what rate (4) approaches zero as $n \to \infty$. Specifically

(5) $$n\{n^{-1}\log G(\underline{\beta})\} \to \frac{1}{\pi}\iint\left|\frac{h'(z)}{h(z)}\right|^2 dz$$

where

$$h(z) = \frac{a(z)}{b(z)}$$

and the integral on the right of (5) is extended over the unit circle $|z| \le 1$. This suggests that a good approximation to $\hat{\underline{\beta}}$ is obtained just by minimizing

(6) $$n^{-1}\underline{x}G^{-1}(\beta)\underline{x}'$$

as a function of $\beta$. Peter Whittle proposed the approximation in which $G^{-1}(\beta)$ is replaced by the $n \times n$ Toeplitz matrix with $(j,k)$th element given by

$$\int_{-\pi}^{\pi}e^{i(j-k)\lambda}g(\lambda)^{-1}\,d\lambda.$$

The expression (6) is then replaced by

(7) $$\int_{-\pi}^{\pi}I_n(\lambda)/g(\lambda;\underline{\beta})\,d\lambda$$

where $I_n(\lambda)$ is the periodogram

$$I_n(\lambda) = \frac{1}{2\pi n}\left|\sum_{t=1}^{n}x_t e^{it\lambda}\right|^2.$$

The basic result on the asymptotic distribution of the parameter estimates states that the maximum likelihood estimates $(\hat\beta, \hat\sigma)$ are asymptotically normally distributed with $\hat\beta$ and $\hat\sigma$ asymptotically independent. $\hat\beta$ is asymptotically $N(\underline\beta_0, n^{-1}W^{-1}(\underline\beta_0))$ where

$$W(\beta_0) = \frac{1}{4\pi} \int_{-\pi}^{\pi} \left[\frac{\partial \log g(\lambda, \underline\beta_0)}{\partial\beta}\right] \left[\frac{\partial \log g(\lambda, \underline\beta_0)}{\partial\beta}\right]' d\lambda$$

and $\hat\sigma$ is asymptotically $N(\sigma_0, n^{-1}\sigma_0^2/2)$ with $(\underline\beta_0, \sigma_0)$ the correct parameter specification. The mean value of (7) tends to

$$\frac{\sigma_0^2}{2\pi} \int_{-\pi}^{\pi} \frac{g(\lambda, \underline\beta_0)}{g(\lambda, \beta)} d\lambda$$

as $n \to \infty$. Notice

$$(2\pi)^{-1} \int_{-\pi}^{\pi} \frac{g(\lambda; \underline\beta_0)}{g(\lambda; \beta)} d\lambda > 1$$

for all $\beta \neq \underline\beta_0$ on the closure of the set $C = \{\beta : a(z)b(z) \neq 0 \text{ for } |z| \leq 1, a_p \neq 0, b_q \neq 0, \text{ with } a(z), b(z) \text{ having no common zeros}\}$ (see [2]). It is this property that leads to the consistency of the maximum likelihood estimates as $n \to \infty$. Let $\eta_j, \mu_k, j = 1, \ldots, p, k = 1, \ldots, q$ be the zeros of the polynomials $a(z), b(z)$ respectively. For any fixed $\varepsilon > (\sqrt5 - 1)/2$ let

(8)  $C_\varepsilon = \{\underline\eta, \underline\mu : \text{zeros of } a(z), b(z) \text{lie in the closed annulus}$

$$1 + \varepsilon \leq |\eta_j|, |\mu_k| \leq \varepsilon^{-1}, \quad j = 1, \ldots, p, k = 1, \ldots, q\}.$$

Since the parameters $\underline\beta = \underline\beta(\underline\eta, \underline\mu)$ are continuous functions of the zeros $\underline\eta, \underline\mu$, we can alternatively consider $C_\varepsilon$ as the set of parameters $\beta$ that are functions of the zeros satisfying the conditions specified in definition of $C_\varepsilon$ as given in (8).

Consider the sequence

$$\mathcal{F}_n(\underline\beta, \sigma^2) = \frac{1}{n\sigma^2}(\underline{x}G_n^{-1}(\beta)\underline{x}' - E\{\underline{x}G_n^{-1}(\beta)\underline{x}'\}).$$

$\mathcal{F}_n(\cdot, \cdot)$ can be considered a sequence of probability measures on the space of continuous functions of $\beta, \sigma^2$ with $\beta \in C_\varepsilon$. M. Kramer [5] *showed that the sequence of measures of* $n^{\frac{1}{2}}\mathcal{F}_n(\underline\beta, \sigma^2)$ *converge weakly as* $n \to \infty$ *to a Gaussian measure* $\mathcal{F}(\cdot, \cdot)$ *on the continuous functions of* $\beta, \sigma^2$ *where* $\underline\beta_0 \in C_\varepsilon$ *with the property that for any finite* $\underline\beta_1, \ldots, \underline\beta_k \in C_\varepsilon$ $\mathcal{F}(\underline\beta_1, \sigma_1^2), \ldots, \mathcal{F}(\underline\beta_k, \sigma_k^2)$ *is a multivariate normal distribution with mean* 0 *and covariance function*

$$V(s, t) = \frac{\sigma_0^4}{\pi\sigma_s^2\sigma_t^2} \int_{-\pi}^{\pi} \frac{g^2(\lambda, \underline\beta_0)}{g(\lambda, \underline\beta_s)g(\lambda, \underline\beta_t)} d\lambda.$$

Actually if the process $x_t$ is a nonGaussian ARMA process (the $\xi$'s are nonGaussian), the expression (2) can still be used to obtain consistent estimates of the parameters $\beta$ by considering extremal values if the condition on the roots of the polynomial $a(z), b(z)$ having modulus greater than one (minimum phase condition) is satisfied. For in that case the process $x_t$ is causal and invertible, i.e. $x_t$ has a linear representation in terms of the present and past of the $\xi$ sequence

$$x_t = \sum_{k=0}^{\infty} h_k \xi_{t-k}$$

and the $\xi$ sequence has a linear representation in terms of the present and the past of the $x$ sequence

$$\xi_t = \sum_{k=0}^{\infty} \ell_k x_{t-k}.$$

The weights $h_k$ and $\ell_k$ in these representations decrease exponentially fast as $k \to \infty$. If the minimum phase condition is not satisfied but none of the zeros of the polynomials $a(z), b(z)$ are on the boundary of the unit circle $|z| = 1$ in the complex plane, there is a nonGaussian solution $x_t$ to the system (1) which now has a two-sided representation in terms of the $\xi$ sequence

$$x_t = \sum_{k=-\infty}^{\infty} h_k \xi_{t-k}.$$

We shall look at aspects of this nonminimum phase case for autoregressive schemes later on. However, one should note that Kramer actually obtained a more general result on convergence of measures than that cited above for the Gaussian minimum phase case.

## 2. NonGaussian Nonminimum Phase Sequences

At this point we wish to investigate some aspects relative to nonminimum phase nonGaussian autoregressive schemes. In this some results on random fields will be helpful and a bit of the background will be indicated. First let $x_t$ be the stationary solution of the system of equation

(9) $$\sum_{k=0}^{p} a_k x_{t-k} = \xi_t, \quad a_0 = 1,$$

where it is understood that the $\xi_t$'s are independent and identically distributed. Also $E\xi_t = 0$, $E\xi_t^2 = \sigma^2 > 0$, and the $\xi$'s have positive continuous density $f_\sigma(x) = (1/\sigma)f(x/\sigma)$. Factor $a(z) = a^+(z)a^*(z)$ where

(10) $\quad a^+(z) = 1 - \theta_1 z - \cdots - \theta_r z^r \neq 0$           for $|z| \leq 1$,

$\quad a^*(z) = 1 - \theta_{r+1} z - \cdots - \theta_p z^s \neq 0$      for $|z| \geq 1$, $r + s = p$.

Let

(11) $$U_t = a^*(B)x_t, \quad V_t = a^+(B)x_t$$

with $B$ the one-step backshift operator $(Bx_t = x_{t-1})$. The sequences $\{U_t\}$ and $\{V_t\}$ are causal and purely noncausal sequences respectively. Since the random variables $U_\ell$, $\ell \leq t$ are independent of $V_\ell$, $\ell > t - s + 1$ the joint probability density of $(U_1, \ldots, U_n, V_{n-s+1}, \ldots, V_n)$ can be written

$$h_U(U_1, \ldots, U_r) \left\{ \prod_{t=r+1}^{n} f_\sigma(U_t - \theta_1 U_{t-1} - \cdots - \theta_r U_{t-r}) \right\} h_V(V_{n-s+1}, \ldots, V_n)$$

with $h_U$ and $h_V$ the joint probability densities of $(U_1, \ldots, U_r)$ and $(V_{n-s+1}, \ldots, V_n)$ respectively. The linear transformation $T_n$ is determined by

$$
\begin{bmatrix}
U_1 \\ \vdots \\ U_s \\ U_{s+1} \\ \vdots \\ U_n \\ V_{n-s+1} \\ \vdots \\ V_n
\end{bmatrix}
=
\begin{bmatrix}
U_1 \\ \vdots \\ U_s \\ x_{s+1} - \theta_{r+1}x_s - \cdots - \theta_p x_1 \\ \vdots \\ x_n - \theta_{r+1}x_{n-1} - \cdots - \theta_p x_{n-s} \\ x_{n-s+1} - \theta_1 x_{n-s} - \cdots - \theta_r x_{n-s+1-r} \\ \vdots \\ x_n - \theta_1 x_{n-1} - \cdots - \theta_r x_{n-r}
\end{bmatrix}
= T_n
\begin{bmatrix}
U_1 \\ \vdots \\ U_s \\ x_1 \\ \vdots \\ x_n
\end{bmatrix}.
$$

A small argument (see [1,6]) shows that $x_t$ has a left-sided Markov property of order $p$. Let $\mathcal{B}_{[n,\infty[}$ be the sigma-field $\mathcal{B}(x_t, t \geq n)$ generated by $x_t$, $t \geq n$ and $\mathcal{B}_{[n-p,n[}$ the $\sigma$-field generated by $x_t$, $n-p \leq t < n$. The left-sided Markov property of order $p$ is the property that

$$m(A|\mathcal{B}_{]-\infty,n[}) = m(A|\mathcal{B}_{[n-p,n[})$$

$m-$ almost surely for all $n$ and for all $A \in \mathcal{B}_{[n,\infty[}$ where it is understood that $m$ is the measure of the process. The one step conditional probability density of the sequence $\{x_t\}$

(12) $\quad f_\sigma(x_n - a_1 x_{n-1} - \cdots - a_p x_{n-p}) \dfrac{h_V(a^+(B)x_{n-s+1}, \ldots, a^+(B)x_n)}{h_V(a^+(B)x_{n-s}, \ldots, a^+(B)x_{n-1})}$

$$\cdot \frac{|\det(T_n)|}{|\det(T_{n-1})|}$$

$$= g(x_n|x_k, k < n) = g(x_n|x_k, n - p \leq k < n)$$

and the fact that it depends on the past $x_k$, $k < n$ only through $x_{n-1}, \ldots$, $x_{n-p}$ implies that the sequence satisfies the left-sided Markov property of order $p$. The ordinary left-sided Markov property is what we would call the left-sided Markov property of order one. The right-sided Markov property of order $p$ is the left-side Markov property of order $p$ for the process with time reversed. Geogii [3] shows that the left-sided and right-sided Markov properties are equivalent and so they might just be called the one-sided Markov property. One can show that the right-sided and left-sided Markov properties of order $p$ are equivalent and so one can refer to them as the one-sided Markov property of order $p$. The two-sided Markov property of order $p$ is the property that

$$(13) \qquad m(A|\mathcal{T}_{]i,k[}) = m(A|\mathcal{B}_{[i-p+1,i]\cup[j,j+p-1]})$$

$m$-almost everywhere for all $i,j$ with $i+1 < k$ and all $A \in \mathcal{B}_{]i,k[}$ with $\mathcal{T}_{]i,k[}$ the $\sigma$-field generated by the random variables $x_t$ with $t \notin ]i,k[$. A measure satisfying condition (13) is a Markov field of order $p$. One can show that a sequence with the one-sided Markov property of order $p$ is a Markov field of order $p$. A possibly stronger property is the local Markov property of order $p$

$$(14) \qquad m(A|\mathcal{T}_V) = m(A|\mathcal{B}_{\partial_p V})$$

$m$-almost everywhere for all finite index sets $V$ and all $A \in \mathcal{B}_V$ where $\mathcal{T}_V$ is the $\sigma$-field generated by $x_t$, $t \notin V$, and $\partial_p V$ is the $p$-boundary of $V$, that is all indices within $p$ units of a point in $V$ but not in $V$ themselves. The autoregressive scheme of order $p$ with one step conditional probability density (12) can be shown to be local Markov of order $p$ with all conditional probabilities of type (14) expressible in terms of the conditional probability density (12) and its iterates and the invariant density function.

One way of studying random fields is to specify the conditional distribution of finite collections of random variables relative to $\sigma$-algebras of random variables external to the collections. The specification of structure is given by a family of conditional probabilities

$$\gamma_\Lambda(A|\mathcal{T}_\Lambda)$$

of events $A \in \mathcal{B}$ (the $\sigma$-algebra of the sequence) given $\mathcal{T}_\Lambda$ where $\Lambda$ is a finite index set. The collection of such conditional probabilities is specified for all finite index sets $\Lambda$. The consistency condition

$$(15) \qquad \gamma_\Delta \gamma_\Lambda(A|z) = \int \gamma_\Delta(dy|z)\gamma_\Lambda(A|y) = \gamma_\Delta(A|z)$$

for any two finite index sets $\Lambda$, $\Delta$ with $\Lambda \subset \Delta$ for any $A \in \mathcal{B}$ is required. One of the usual questions of interest is the existence of a probability measure $m$ (a process) consistent with the conditional probability specification spoken of above. Of course, without any of this additional notation we saw that there is a unique stationary sequence that is a solution of the system (9) and hence a translation invariant measure consistent with the conditional probability density (12). This measure $m$ together with the one step conditional probability density (12) determine a consistent specification of conditional probabilities $\{\gamma_\Lambda(A|\mathcal{T}_\Lambda),\ A \in \mathcal{B},\ \Lambda\ \text{finite}\}$. $m$ is the unique invariant probability measure consistent with this specification $\{\gamma_\Lambda\}$.

For each finite index set $\Lambda$ let $\mathcal{L}_\Lambda$ be the set of bounded $\mathcal{B}_\Lambda$ measurable functions. The collection $\mathcal{L} = \cup_{\Lambda \in \zeta}\mathcal{L}_\Lambda$, where $\zeta$ is the class of finite index sets, is the set of bounded local functions. The real-valued function $f$ is quasilocal if there is a sequence of local functions $f_n$ such that $\lim_{n\to\infty} \|f - f_n\| = 0$ where $\|\cdot\|$ is the supnorm. A specification $\{\gamma_\Lambda\}$ is quasilocal if for each finite $\Lambda$,

$$\gamma_\Lambda f(\cdot) = \int \gamma_\Lambda(dx|\cdot)f(x)$$

is quasilocal whenever $f$ is quasilocal. It is clear that if a specification $\{\gamma_\Lambda\}$ is such that $\gamma_\Lambda f$ is local whenever $f$ is local, then the specification is quasilocal. The specification determined by the autoregressive scheme has this property and hence is quasilocal.

Let $m_1$ and $m_2$ be two probability measures on the $\sigma$-field $\mathcal{B}$. Consider a sub-$\sigma$-algebra $\mathcal{A}$ of $\mathcal{B}$ and set

(16) $\qquad \mathcal{H}_\mathcal{A}(m_1|m_2) = \begin{cases} m_2(f_\mathcal{A} \log f_\mathcal{A}) & \text{if } m_1 << m_2 \text{ on } \mathcal{A} \\ \infty & \text{otherwise.} \end{cases}$

The function $f_\mathcal{A}$ is any Radon-Nikodym derivative of $m_1$ with respect to $m_2$ on $\mathcal{A}$. $\mathcal{H}_\mathcal{A}(m_1|m_2)$ is the relative entropy of $m_1$ with respect to $m_2$ on $\mathcal{A}$ and is often referred to as a Kullback-Leibler information. Given a finite index set $\Lambda$ let us understand by $\mathcal{H}_\Lambda(m_1|m_2)$ $\mathcal{H}_{\mathcal{B}_\Lambda}(m_1|m_2)$. We set $\Lambda_n = [1, 2, \ldots, n]$.

Our object in introducing the perspective via specifications is to state the following case of a theorem in Georgii [4].

THEOREM. *Let $\gamma$ be a quasilocal specification and $m_2$ a translation invariant probability measure consistent with $\gamma$. Let $m_1$ be a translation invariant probability measure such that*

$$\liminf_{n\to\infty} n^{-1}\mathcal{H}_{\Lambda_n}(m_1|m_2) = 0.$$

*Then $m_1$ is a measure consistent with $\gamma$.*

The invariant density $\alpha(x_1, \ldots, x_p)$ of the $p$th order Markov sequence $x_t$ is the solution of the integral equation

$$\int \alpha(x_1, \ldots, x_p) \prod_{s=p+1}^{2p} g(x_s|x_k, s-p \leq k < s)\, dx_1 \ldots dx_p = \alpha(x_{p+1}, \ldots, x_{2p})$$

with $g(\cdot|\cdot)$ given by (12). Of course, the invariant density $\alpha$ is the joint density of the random variables

$$x_t = \sum_{j=-\infty}^{\infty} \psi_j \xi_{t-j}, \quad t = 1, \ldots, p,$$

where

$$a(z)^{-1} = \sum_{j=-\infty}^{\infty} \psi_j z^j.$$

It is also clear from (10), (11) that

$$U_t = \sum_{j=0}^{\infty} \alpha_j \xi_{t-j}, \quad V_t = \sum_{j=s}^{\infty} \beta_j \xi_{t-j}$$

with

$$a^+(z)^{-1} = \sum_{j=0}^{\infty} \alpha_j z^j, \quad a^*(z)^{-1} = \sum_{j=s}^{\infty} \beta_j z^{-j}.$$

If observations are made on $x_1, \ldots, x_n$ the corresponding $\frac{1}{n} \times loglikelihood$ is

(17)
$$\frac{1}{n} \log \alpha(x_1, \ldots, x_p) + \frac{1}{n} \sum_{s=p+1}^{n} g(x_s | x_k, s - p \le k < s)$$

$$= \frac{1}{n} \log \alpha(x_1, \ldots, x_p) + \frac{1}{n} \sum_{s=p+1}^{n} \log f_\sigma(x_s - a_1 x_{s-1} - \cdots - a_p x_{s-p})$$

$$+ \frac{1}{n} \log h_V(a^+(B) x_{n-s+1,\ldots}, a^+(B) x_n)$$

$$- \frac{1}{n} \log h_V(a^+(B) x_{p+1-s,\ldots}, a^+(B) x_p)$$

$$+ \frac{1}{n} \log\{| \det(T_n) / \det(T_p)|\}.$$

Let $\underline{\nu}_0 = (a_{10}, \ldots, a_{po}, \sigma_0)$, $\underline{\nu} = (a_1, \ldots, a_p, \sigma) \in C_\varepsilon$ for some $\varepsilon > 0$ with $\underline{\nu}_0$ the true parameter specification. With an additional subscripting by 0 the true parameter specification is to be understood. The following proposition is an almost immediate consequence of the theorem.

PROPOSITION. *Assume that the following expectations are finite*

$$E_{\nu_0}|\log\alpha(x_1,\ldots,x_p)|, \quad E_{\nu_0}|\log\alpha_0(x_1,\ldots,x_p)|,$$

$$E_{\nu_0}|\log h_V(a^+(B)x_{n-s+1},\ldots,a^+(B)x_n)|,$$

(18) $\quad E_{\nu_0}|\log h_{V,0}(a^+(B)x_{n-s+1},\ldots,a^+(B)x_n)|,$

$$E_{\nu_0}|\log f_\sigma(x_s - a_1 x_{s-1} - \cdots - a_p x_{s-p})|,$$

$$E_{\nu_0}|\log f_{\sigma_0}(x_s - a_{10}x_{s-1} - \cdots - a_{p_0}x_{s-p})|.$$

*Then if* $\underline{\nu} \neq \underline{\nu}_0$

$$E_{\nu_0}\log f_{\sigma_0}(x_s - a_{10}x_{s-1} - \cdots - a_{p_0}x_{s-p}) + \log|\theta_{p0}|$$
$$> E_{\nu_0}\log f_\sigma(x_s - a_1 x_{s-1} - \cdots - a_p x_{s-p}) + \log|\theta_p|.$$

The specification determined by the transition probability density of the $p$th order Markov sequence (12) and the corresponding unique invariant probability density (16) with parameter $\underline{\nu}$ is quasilocal. Let the unique measure consistent with this specification be $m_2$. Call the measure corresponding to the specification determined by parameter $\nu_0$ $m_1$. A direct application of the theorem with the loglikelihoods as given by (17) leads to the desired conclusion.

Notice that if the expectations (18) are finite for all $\underline{\nu} \in C_\varepsilon(\varepsilon > (\sqrt{5} - 1)/2)$ we have the unique maximum in $C_\varepsilon$ (8) attained by $\underline{\nu} = \underline{\nu}_0$.

On heuristic grounds, under sufficiently strong additional conditions (moment conditions, smoothness conditions on $\log f$, etc.) one would expect to be able to show that

$$\frac{1}{\sqrt{n}}\{loglikelihood_\nu(x_1,\ldots,x_n) - E\,loglikelihood_\nu(x_1,\ldots,x_n)\}$$

as a process in the parameter $\nu \in C_\varepsilon$ converges weakly to a Gaussian process with mean zero and covariance function

$$cov_{\nu_0}\{\log f_\sigma(x_s - a_1 x_{s-1} - \cdots - a_p x_{s-p}), \log f_{\sigma'}(x_s - a_1' x_{s-1} - \cdots - a_p' x_{s-p})\}$$

where $\underline{\nu}_0$ is the true parameter specification and $\underline{\nu} = (a_1,\ldots,a_p,\sigma)$, $\underline{\nu}' = (a_1',\ldots,a_p',\sigma')$.

**Acknowledgment.** Th research for this paper was partially supported by ONR Grant N00014-90-J-1371 at the University of California, San Diego and NSF Grant DMS-9504596 at Colorado State University.

## 3. REFERENCES

[1] F. J. Breidt, R. A. Davis, K. S. Lii and M. Rosenblatt, "Maximum likelihood estimation for noncausal autoregressive processes," *J. Mult. Anal.*,**36**, pp. 175–198, 1991.

[2] P. Brockwell and R. Davis, *Time Series: Theory and Methods*, Springer-Verlag, 1991.

[3] H.-O. Georgii, *Gibbs Measures and Phase Transitions*, W. de Gruyter, 1988.

[4] U. Grenander and G. Szegö, *Toeplitz Forms and Their Applications*, University of California Press, 1958.

[5] M. Kramer, "The fluctuation of the Gaussian likelihood for stationary Gaussian sequences," *Ph.D. thesis*, University of California, San Diego, 1993.

[6] M. Rosenblatt, "Prediction and non-Gaussian autoregressive stationary sequences," *Ann. Appl. Prob.* **5**, p. 239–247, 1995.

# TESTING FOR SERIAL INDEPENDENCE USING MEASURES OF DISTANCE BETWEEN DENSITIES

By Hans Julius Skaug and Dag Tjøstheim

November 1995

## SUMMARY

A class of measures of the distance between densities is presented. The class contains the familiar Hellinger distance and the Kullback–Leibler information. A third special case is the weighted difference distance which was introduced in Skaug and Tjøstheim [14]. It is argued that the notion of the distance between densities forms a basis for testing the hypothesis of serial independence. The asymptotic null distributions of the test statistics are given. The normal approximation provided by the central limit theorem do not work well in practice. Instead randomization is used to ensure a correct level of the tests. In simulation experiments the power of the tests is compared to the power of the correlation test, two traditional rank tests and the BDS test. In these simulation the test based on the weighted difference distance performs very good, and the tests based on the Hellinger distance and the Kullback–Leibler information compete well with the BDS test. Also the van der Waerden test shows good power properties. Finally, the tests have been applied to a set of exchange rate data.

Keywords: Hellinger distance; Kullback–Leibler information; nonparametric test; serial independence.

Hans Julius Skaug
Norwegian Computing Center
P.O.Box 114 Blindern
N-0314 Oslo
NORWAY

Dag Tjøstheim
Department of Mathematics
University of Bergen
5007 Bergen
NORWAY

# 1  Introduction

Recently there has been a considerable research activity in testing of serial independence for a time series $\{X_t\}$. It is well known that correlation based tests fail in several examples of practical interest. Most of the recent work in this field, and in the related field of testing for independence between two random variables $X$ and $Y$, has been concerned with test functionals based on nonparametric estimates of the distribution or density functions. This is the case for the so-called BDS test (Brock et al. [2]) and for the tests considered by Blum et al. [1], Chan and Tran [4], Robinson [8], Rosenblatt [10], Rosenblatt and Wahlen [11] and Skaug and Tjøstheim [13, 14]. Many of these tests can be subsumed under a distance measure formalism described in Section 2. Asymptotic theory has been developed for most of the functionals, and this theory has been used to determine critical values for the tests. However, several authors have pointed out discrepancies between the real and nominal level of the tests constructed in this way. These differences persist for quite large sample sizes. Based on our development of the asymptotic theory we conjecture that traditional use of limit theory cannot be expected to work well in general, and that resampling is needed to obtain an approximately correct level.

# 2  Distance measures and test functionals

## 2.1  Measures of dependence

Let $\{X_t, t \geq 1\}$ be a stationary process. The hypothesis $H_0$ of serial independence states that $\{X_t\}$ is a collection of independent random variables with some unspecified distribution function $F$, say. For simplicity we will start out by discussing the less restrictive hypothesis $H_1$ that $X_{t-1}$ and $X_t$ are independent. Let $F_1$ be the distribution function of $(X_{t-1}, X_t)$. By definition $X_{t-1}$ and $X_t$ are independent if and only if $F_1(x,y) = F \otimes F(x,y)$ for all $x$ and $y$, where $F \otimes F(x,y) = F(x) \cdot F(y)$. The problem of measuring the dependence between $X_{t-1}$ and $X_t$ can be formulated as a problem of measuring the distance $\rho(F_1, F \otimes F)$ between $F_1$ and $F \otimes F$. We introduce two "natural" requirements. First, we require that

$$\rho(F_1, F \otimes F) \geq 0 \quad \text{and} \quad \{\rho(F_1, F \otimes F) = 0\} \Leftrightarrow \{F_1 = F \otimes F\}. \quad (1)$$

The hypothesis $H_1$ is then characterized by $\rho = 0$. Second, since the hypothesis problem of testing $H_1$ (and more generally $H_0$) is invariant under a continuous and strictly increasing transformation $h$ of the observations in the sense that $X_{t-1}$ and $X_t$ are independent if and only if $h(X_{t-1})$ and $h(X_t)$ are independent, it is natural to require that the distance measure $\rho$ should be invariant under the same transformation, that is

$$\rho(F_1^h, F^h \otimes F^h) = \rho(F_1, F \otimes F). \quad (2)$$

Here $F^h$ and $F_1^h$ are the distribution functions of $h(X_t)$ and $(h(X_{t-1}), h(X_t))$ respectively. The two most commonly used measures of dependence are probably the correlation and the rank correlation function. They can not be formulated in terms of distance measures satisfying (1) and (2). Neither can the distance measure associated with the BDS statistic.

## 2.2 Distance between densities

In the sequel we assume that densities exist, and base the test of serial independence on the distance between these. In this case (2) is required to hold for all continuous and strictly increasing transformations $h$ such that $h^{-1}$ is absolutely continuous. Let $f$ and $f_1$ be the densities of $X_t$ and $(X_{t-1}, X_t)$, respectively. We consider a class of distance measures

$$\rho = \int B\{f_1(x,y), f(x), f(y)\} \, dF_1(x,y), \tag{3}$$

obtained by varying the function $B$. If $B$ is of the form $B(z_1, z_2, z_3) = D(z_1/z_2 z_3)$ for some function $D$, we get

$$\rho = \int D\left\{\frac{f_1(x,y)}{f(x)f(y)}\right\} dF_1(x,y),$$

which by the change of variable formula for integrals is seen to have the property (2). Different choices of $D$ produce various well known distance measures. For instance, letting $D(u) = (1-u^{-1/2})^2$, we obtain the Hellinger distance

$$
\begin{aligned}
H &= \int \left\{\sqrt{f_1(x,y)} - \sqrt{f(x)f(y)}\right\}^2 dx\,dy \\
&= 2 - 2\int \sqrt{f(x)f(y)/f_1(x,y)} \, dF_1(x,y)
\end{aligned}
$$

between $f_1$ and $f \otimes f$. To our knowledge it has not been used before in independence testing. The Hellinger distance is a metric and thus satisfies (1). Another important choice of $D$ is $D(u) = \log(u)$, giving the Kullback–Leibler information

$$I = \int \log\left\{\frac{f_1(x,y)}{f(x)f(y)}\right\} dF_1(x,y),$$

which can be shown to satisfy (1) (see Robinson [8]). If we let $B(z_1, z_2, z_3) = z_1 - z_2 z_3$ in (3) we obtain

$$J = \int \{f_1(x,y) - f(x)f(y)\} \, dF_1(x,y), \tag{4}$$

considered by Skaug and Tjøstheim [14]. The distance measure $J$ has neither of the properties (1) or (2). However, for a large class of processes $J$ asymptotically has the property (1). Let $\{X_t\}$ be generated by

$$X_t = \alpha \cdot g(X_{t-1}) + e_t, \tag{5}$$

where $\alpha$ is a parameter, $\{e_t\}$ is a sequence of i.i.d. random variables with density $u$, and $g$ is a function such that $\{X_t\}$ is stationary. By series expansion of $f$ and $f_1$ in powers of $\alpha$ it can be shown, subject to mild regularity conditions on $u$ and $g$, that

$$J(\alpha) = \frac{1}{2}\alpha^2 \cdot C(g, u) + O(\alpha^3), \qquad \alpha \to 0,$$

where

$$C(g, u) = \int \left( [g(x) - E\{g(e_1)\}]^2 + \mathrm{Var}\{g(e_1)\} \right) u^2(x)\, dx \cdot \int \{u'(x)\}^2\, dx.$$

Here $u'$ denotes the derivative of $u$. Note that $C(g, u) \geq 0$, and that $C(g, u) = 0$ only if $g$ is constant, in which case $X_t$ and $X_{t-1}$ are independent, or if $u$ is the density of the uniform distribution.

The above measures of the dependence between $X_{t-1}$ and $X_t$ have obvious extensions to measuring the dependence between the components in $(X_{t-q}, \ldots, X_t)$, where $q \geq 1$. We may consider the distance between the simultaneous density of $(X_{t-q}, \ldots, X_t)$ and the corresponding product of marginals. However, estimating such a measure would involve estimating higher dimensional densities which is difficult due to the curse of dimensionality. We have decided instead to measure pairwise dependence at different lags. Let $f_m$ be the density of $(X_{t-m}, X_t)$. For a given distance measure $\rho$ define

$$\rho^{(q)} = \sum_{m=1}^{q} \rho_m,$$

where $\rho_m = \rho(f_m, f \otimes f)$. This approach has been shown to give good results in Skaug and Tjøstheim [13].

## 2.3 Estimated distance measures

Given estimators $\hat{f}$ and $\hat{f}_m$ of $f$ and $f_m$ based on the observations $\{X_t, 1 \leq t \leq n\}$, $\rho_m$ can be estimated by

$$
\begin{aligned}
\hat{\rho}_m &= \int B\left\{\hat{f}_m(x, y), \hat{f}(x), \hat{f}(y)\right\} w(x, y)\, d\hat{F}_m(x, y) \\
&= \frac{1}{n - m} \sum_{t=m+1}^{n} B\left\{\hat{f}_m(X_{t-m}, X_t), \hat{f}(X_{t-m}), \hat{f}(X_t)\right\} w(X_{t-m}, X_t),
\end{aligned}
$$

where $\hat{F}_m$ is the $m$-lag empirical distribution function. The weight function $w$ serves the double purpose of screening off extreme observations and simplifying the asymptotic analysis. We will take $\hat{f}$ and $\hat{f}_m$ to be kernel estimators, so that

$$\hat{f}(x) = \frac{1}{n} \sum_{t=1}^{n} \frac{1}{h_n} k\left(\frac{x - X_t}{h_n}\right) \tag{6}$$

and

$$\hat{f}_m(x,y) = \frac{1}{n-m} \sum_{t=m+1}^{n} \frac{1}{h_n} k\left(\frac{x - X_{t-m}}{h_n}\right) \frac{1}{h_n} k\left(\frac{y - X_t}{h_n}\right), \tag{7}$$

where $k$ is the kernel function and $h_n$ is the bandwidth. We have used the leave-one-out estimates in order to reduce bias. Further, to eliminate effects from different scalings for functionals not satisfying (2), $\{X_t\}$ has been normalized, i.e. it has been replaced by $\{X'_t = X_t/\widehat{SD}(X), 1 \leq t \leq n\}$, where $\widehat{SD}(X) = \{n^{-1} \sum_t (X_t - \overline{X})^2\}^{1/2}$. This introduces a slight dependence between the $X'_t$'s but it is a higher order effect and the results in the simulations are virtually indistinguishable from those obtained using the normalization $\{X''_t = X_t/SD(X), 1 \leq t \leq n\}$. Because of the normalization a natural choice of the bandwidth is $h_n = n^{-1/6}$ (cf. Silverman [12, p. 86] for optimal bandwidth selection for bivariate density estimation). Finally, in the simulations we have used $S = [-2, 2] \times [-2, 2]$. Clearly the choice of the design parameters $h_n$ and $w$ is open for discussion. A simple sensitivity experiment is included in Skaug and Tjøstheim [14].

In the rest of this paper we will limit ourselves to three functionals, namely the Hellinger distance statistic

$$\hat{H}_m = \frac{1}{n-m} \sum_{t=m+1}^{n} 2\left\{1 - \sqrt{\frac{\hat{f}(X_{t-m})\hat{f}(X_t)}{\hat{f}_m(X_{t-m}, X_t)}}\right\} w(X_{t-m}, X_t),$$

the Kullback–Leibler information statistic

$$\hat{I}_m = \frac{1}{n-m} \sum_{t=m+1}^{n} \log\left\{\frac{\hat{f}_m(X_{t-m}, X_t)}{\hat{f}(X_{t-m})\hat{f}(X_t)}\right\} w(X_{t-m}, X_t)$$

and the weighted difference statistic

$$\hat{J}_m = \frac{1}{n-m} \sum_{t=m+1}^{n} \left\{\hat{f}_m(X_{t-m}, X_t) - \hat{f}(X_{t-m})\hat{f}(X_t)\right\} w(X_{t-m}, X_t).$$

Summing over the different lags we obtain the portmanteau statistics

$$\hat{H}^{(q)} = \sum_{m=1}^{q} \hat{H}_m, \qquad \hat{I}^{(q)} = \sum_{m=1}^{q} \hat{I}_m, \qquad \hat{J}^{(q)} = \sum_{m=1}^{q} \hat{J}_m. \tag{8}$$

Robinson [8] has used a modified version of $\hat{I}_1$ as the test statistic of his test of serial independence.

# 3 Testing for serial independence

For a test statistic $\hat{\rho}$ based on a distance measure $\rho$ which satisfies (1) we reject the null hypothesis for large observed values of $\hat{\rho}$. The rejection criterion will be

$$\{\hat{\rho} \geq c_n(X_1, \ldots, X_n)\},$$

for a sequence of functions $c_n$. Two different ways of determining $c_n$ such that the test has an appropriate level are presented.

## 3.1 Test based on asymptotic theory

Assume that there exists a sequence $\{\mu_n(F), \sigma_n(F); n \geq 1\}$ such that in distribution

$$\frac{\hat{\rho} - \mu_n(F)}{\sigma_n(F)} \rightarrow N(0,1), \qquad n \rightarrow \infty.$$

Assuming that $\mu_n(F)$ and $\sigma_n(F)$ can be estimated from observations, say by estimators $\hat{\mu}_n$ and $\hat{\sigma}_n$, we use as a critical region

$$\{\hat{\rho} \geq \hat{\mu}_n + \hat{\sigma}_n u_\epsilon\},$$

where $u_\epsilon$ is the upper $\epsilon$-quantile in the standard normal distribution.

## 3.2 Permutation test

Let $X^{(\cdot)} = (X^{(1)}, \ldots, X^{(n)})$ be the order statistic. The fact that $X^{(\cdot)}$ is a sufficient statistic under $H_0$ (Lehmann [6, p. 133]) makes the construction of a permutation test possible. Conditioned on $X^{(\cdot)} = (x^{(1)}, \ldots, x^{(n)})$ the distribution of $(X_1, \ldots, X_n)$ is constructed by putting equal probability $1/n!$ on each of the $n!$ permutations of $x^{(1)}, \ldots, x^{(n)}$. The distribution of $\hat{\rho}$ conditioned on the observed value of $X^{(\cdot)}$, the permutation distribution, is then constructed by evaluating $\hat{\rho}$ at each of these $n!$ permutations. Define the $\epsilon$-quantile in the permutation distribution

$$c_\epsilon(X^{(\cdot)}) = \inf_c \left\{ c : \Pr\left(\hat{\rho} \geq c \mid X^{(\cdot)}\right) \leq \epsilon \right\}.$$

The permutation test has an exact level. In practice, for all but very small sample sizes it is impossible to compute $c_\epsilon(X^{(\cdot)})$ exactly, so Monte Carlo methods have to be used to obtain an approximation. This does not affect the level of the test, but has influence on the power properties of the test.

# 4 Asymptotic properties

The tests of Robinson [8], Rosenblatt and Wahlen [11], Blum et al. [1] and the BDS test (Brock et al. [2]) have all been constructed using asymptotic theory as outlined in Section 3.1. Is is natural to make an attempt along these lines also for the test statistics $\hat{H}^{(q)}$, $\hat{I}^{(q)}$ and $\hat{J}^{(q)}$. Below we give conditions which imply asymptotic normality.

## 4.1 Assumptions

i) The weight function $w$ is given as $w(x,y) = 1\{(x,y) \in S\}$, where $S = S_1 \times S_1$, with $S_1 = [a,b]$ for $a < b$.

ii) The marginal density $f$ is bounded and uniformly continuous on $R^1$.

iii) The kernel function $k$ is bounded, satisfies

$$\int uk(u)\,du = 0; \qquad \int u^2 k(u)\,du < \infty,$$

and has the representation

$$k(x) = \int \tilde{k}(\eta)e^{i\eta x}\,d\eta,$$

where $i = \sqrt{-1}$ and $\tilde{k}$ is a function of a real variable such that $\int |\tilde{k}(\eta)|\,d\eta < \infty$ (Robinson [9]).

iv) The bandwidth $h_n = cn^{-1/\beta}$ for some $c > 0$ and $4 < \beta < 8$.

v) The function $B$ is twice continuously differentiable with bounded second order partial derivatives.

Under an additional assumption of strong mixing with an exponentially decaying mixing coefficient, the test statistics $\hat{H}_m$, $\hat{I}_m$ and $\hat{J}_m$ are consistent estimators of their (weighted) population counterparts. Essentially this can be proved as in the consistency proofs of Skaug and Tjøstheim [13].

## 4.2 Asymptotic normality

In proving asymptotic normality of the test statistic $\hat{\rho}$ the following lemma is helpful.

**Lemma 1** *If $\{X_t, t \geq 1\}$ are i.i.d. random variables and the assumptions i)–v) above are satisfied, then*

$$\hat{\rho}_m =$$
$$\int \frac{\partial B}{\partial z_1}\{f_m(x,y), f(x), f(y)\}\, f_m(x,y)w(x,y)\,d\Big\{\hat{F}_m(x,y) - F_m(x,y)\Big\}$$
$$+ \int \Big[\int \frac{\partial B}{\partial z_2}\{f_m(x,y), f(x), f(y)\}\, f(y)w(y)\,dy\Big]\, f(x)w(x)\,d\Big\{\hat{F}(x) - F(x)\Big\}$$
$$+ \int \Big[\int \frac{\partial B}{\partial z_3}\{f_m(x,y), f(x), f(y)\}\, f(x)w(x)\,dx\Big]\, f(y)w(y)\,d\Big\{\hat{F}(y) - F(y)\Big\}$$
$$+ o_p(n^{-1/2}).$$

*Proof.* A proof of this result can be found in Skaug and Tjøstheim [15].
We then specialize to the test statistics $\hat{H}_m$, $\hat{I}_m$ and $\hat{J}_m$.

**Theorem 1** *Assume that $\{X_t, t \geq 1\}$ are i.i.d. random variables. Define*

$$\sigma_1^2 = \left[ \int f^3(x)w(x)\,dx - \left\{ \int f^2(x)w(x)\,dx \right\}^2 \right]^2.$$

*and*

$$\sigma_2^2 = \left\{ \int f(x)w(x)\,dx \right\}^2 \cdot \left\{ 1 - \int f(x)w(x)\,dx \right\}^2$$

*Then under the assumptions i)–iv) above we have in distribution as $n \to \infty$*

$$n^{1/2}\hat{J}^{(q)} \to N(0, q\sigma_1^2)$$

*if $\sigma_1^2 > 0$. If in addition for some $\gamma > 0$*

$$\inf_{x \in S_1} f(x) \geq \gamma,$$

*we have*

$$n^{1/2}\hat{H}^{(q)} \to N(0, q\sigma_2^2), \qquad n^{1/2}\hat{I}^{(q)} \to N(0, q\sigma_2^2)$$

*if $\sigma_2^2 > 0$.*

*Proof.* The result follows from Lemma 1 and the central limit theorem for $q$-dependent processes (Brockwell and Davis [3, p. 206]).

Observe that if we let $w \equiv 1$ we get $\sigma_2^2 = 0$. Thus the introduction of the weight function is essential in the argument leading to asymptotic normality of $\hat{H}^{(q)}$ and $\hat{I}^{(q)}$. For $\hat{J}^{(q)}$ on the other hand $w$ does not play the same important role in the asymptotic analysis.

## 4.3  The quality of the normal approximation

In Skaug and Tjøstheim [14] it is reported that the distribution of $\hat{J}^{(1)}$ has a shape which is fairly close to the shape of a normal distribution (with a slightly heavier tail to the right than to the left) for $n = 100$. The main problem with the asymptotic theory is that the asymptotic variances $\sigma_1^2$ and $\sigma_2^2$ in Theorem 1 give very poor approximation to the finite sample variances of the test statistics. ¿From the theorem we have that the limit distributions of $\sqrt{n}\hat{H}^{(1)}/\sigma_2$, $\sqrt{n}\hat{I}^{(1)}/\sigma_2$ and, $\sqrt{n}\hat{J}^{(1)}/\sigma_1$ are independent of how the sequence $\{h_n\}$ is chosen, as long as the assumption iv) is fulfilled. Clearly, for a finite sample size, the variance of the normalized statistics will depend on $h_n$. Indeed, simulation experiments in Skaug and Tjøstheim [14] show quite heavy dependence on $h_n$. Hence it is very problematic to use the results of Theorem 1 to construct critical values. One reason for the bad approximation is that the next order terms in a series expansion of the variance of the statistic are close in order to the leading term.

For an $n$-dependent support $S$ the leading term of the asymptotic expansion will in general depend on the smoothing bandwidth (cf. Robinson [8]), as is also the case for the functional of Rosenblatt and Wahlen [11] which has $S = R^2$. However, the fundamental obstacle of low accuracy of the leading term is not removed, as again the next order terms are close in magnitude. And it is a main point of this paper that application of asymptotic distribution results in tests of the type investigated here may be rather hazardous unless $n$ is very large. All of this points towards resampling or permutation techniques, as outlined in Section 3.1, as natural tools for obtaining an accurate level of the test.

# 5 The power of the tests

## 5.1 Other tests for serial independence

The $\hat{H}^{(q)}$, $\hat{I}^{(q)}$ and, $\hat{J}^{(q)}$ tests will be compared to the following standard portmanteau tests for serial independence: the correlation test, $\hat{C}_1^{(q)}$, Spearman's rank test, $\hat{C}_2^{(q)}$, the van der Waerden rank test, $\hat{C}_3^{(q)}$ and the BDS test. The theory of rank tests is surveyed in Hallin and Puri [5], while the BDS test is described in Brock et al. [2].

## 5.2 Alternatives to the hypothesis of independence

Since the BDS test has been an object of much interest in econometrics lately, we have chosen to compare the power of the tests against the time series alternatives considered by Brock et al. [2]. These are:
The ARCH(1), Autoregressive Conditional Heteroscedastic of order 1, process

$$X_t = e_t\sqrt{1 + 0.5X_{t-1}^2}. \tag{9}$$

The GARCH(1), Generalized Autoregressive Conditional Heteroscedastic of order 1, process

$$X_t = e_t\sqrt{h_t}, \qquad h_t = 1 + 0.1X_{t-1}^2 + 0.8h_{t-1}. \tag{10}$$

The NLMA, Nonlinear Moving Average, process

$$X_t = 0.5e_{t-1}e_{t-2} + e_t. \tag{11}$$

The ENLMA, Extended Nonlinear Moving Average, process

$$X_t = 0.8e_{t-1} \sum_{i=2}^{20} (0.8)^{i-2} e_{t-i} + e_t. \tag{12}$$

The TAR(1), Threshold Autoregressive of order 1, process

$$X_t = \begin{cases} -0.5X_{t-1} + e_t & X_{t-1} \leq 1 \\ 0.4X_{t-1} + e_t & X_{t-1} > 1 \end{cases}. \tag{13}$$

In all cases $\{e_t, t \geq 1\}$ is a sequence of i.i.d. $N(0,1)$ variables. Some other alternatives, among them a linear autoregressive process, are considered in Skaug and Tjøstheim [14].

## 5.3 Power of the tests

In addition to the hypothesis $H_0$ the more restrictive hypothesis

$$H_0^*: X_1, \ldots, X_n \text{ are i.i.d. } N(0,1) \text{ variables,}$$

is considered. For tests of $H_0^*$ the null distributions of the test statistics can be determined by Monte Carlo simulation, so there is no need for permutation tests.

The Tables 1–5 show the power of the tests against the alternatives (9)–(13) for $n = 250$ and $\epsilon = 0.05$. Both the critical values and the estimates of the power functions are found by Monte Carlo Simulation (16,000 replicas), except for the BDS test for which the results are taken from Brock et al. [2]. The "$*$" appearing in some of the entries indicates that we do not have result for the BDS test for this particular value of $q$.

It is seen from the tables that, with the exception of the TAR alternative (13), the $\hat{J}^{(q)}$ test has the highest power against all alternatives for small values of $q$. In particular, the $\hat{J}^{(q)}$ test performs better than the BDS test against all five alternatives. The $\hat{I}^{(q)}$ test, based on the Kullback–Leibler information, is best for the TAR alternative (13), and it and the Hellinger based test are comparable in quality to the BDS test for the other alternatives. Not unexpectedly the correlation test $\hat{C}_1^{(q)}$ performs poorly in these examples. Despite its nonparametric structure Spearman's test $\hat{C}_2^{(q)}$ performs even worse than the correlation test. The van der Waerden test $\hat{C}_3^{(q)}$, on the other hand performs very well against the alternatives (9)–(12), but it is inferior for the threshold example (13).

We see from the tables that the power of the tests in general depends heavily on the choice of $q$. As is natural, when the dependency is located on the first lag, the power decreases with $q$. Conversely, when the dependency is located mainly around a higher order lag, maximum power is obtained by chosing $q$ equal to this lag.

We also conducted an experiment with a smaller sample size and fewer simulation replicas, with the goal of assesing the power of the permutation tests. The power of the tests (for both $H_0$ and $H_0^*$) was compared against the same alternatives (9)–(12), but now only for $q = 1$. We used $n = 100$, $\epsilon = 0.05$, 500 permutation resamples and 2000 simulation replicas. It is observed from Table 6 that, although smaller, the power of the permutation tests is very close to the power of the corresponding non-permutation tests.

# 6   Testing the random walk hypothesis

Robinson [8] and Pinkse [7] have applied their independence test to exchange rate data to see if the random walk hypothesis can be rejected. If we let $Z_t$ denote the exchange rate between two currencies, the hypothesis is that the random variables $X_t = \log Z_t - \log Z_{t-1}$ are serially independent. Both Robinson [8] and Pinkse [7] reject the hypothesis for several currencies.

We will only consider the exchange rate between Pound Sterling and US Dollar. The data used are the daily recordings of the exchange rate in the period between October 1981 and July 1985, as recorded in the Bank of England Quarterly Bulletin, a total of 945 observations. The permutation tests based on $\hat{H}^{(q)}$, $\hat{I}^{(q)}$ and $\hat{J}^{(q)}$ reject the hypothesis very clearly on level $\epsilon = 0.05$, $q = 1, \ldots, 10$, while the tests based on $\hat{C}_1^{(q)}$, $\hat{C}_2^{(q)}$ and $\hat{C}_3^{(q)}$ do not give rejection.

We also wanted to see if the hypothesis could be rejected for a smaller number of observations. Table 7 shows the p-values for the tests when only the first 500 observations were used. For $\hat{H}^{(q)}$, $\hat{I}^{(q)}$, $\hat{J}^{(q)}$ and, $\hat{C}_1^{(q)}$ these are permutation $p$-values (calculated conditionally on the observed order statistic), but for $\hat{C}_2^{(q)}$ and $\hat{C}_3^{(q)}$ they are $p$-values in the usual sense. However, rejecting the hypothesis when the $p$-value is less than $\epsilon$ results in a test on level $\epsilon$ both for the conditional and the unconditional $p$-values. We used 1000 resampled realizations in approximating both the conditional and unconditional $p$-values.

For this real data example the van der Waerden test behaves in the same way as the ordinary and rank correlation test. They all fail to detect dependence in the data. It is interesting to notice that $\hat{I}^{(q)}$ and $\hat{J}^{(q)}$ do not reject for $q = 1$ (on level 0.05), but they do for $q \geq 5$.

# 7  Computer software

The routines for evaluation of $\hat{H}^{(q)}$, $\hat{I}^{(q)}$ and, $\hat{J}^{(q)}$, and the corresponding permutation tests, are available as S code. The source can be obtained by writing to the authors or by sending E-mail to "skaug@nr.no".

# 8  Acknowledgements

We are also grateful to Markku Rahiala, Peter Robinson and to two other readers for several helpful suggestions on an earlier version and to Joris Pinkse for providing us with the exchange rate data. Hans J. Skaug wants to thank the Norwegian Research Council (NFR) for financial support.

# 9  References

[1] J. R. Blum, J. Kiefer, and M. Rosenblatt. Distribution free tests of independence based on the sample distribution function. *Annals of Mathematical Statistics*, 32:485–498, 1961.

[2] W. A. Brock, W. D. Dechert, J. A. Scheinkman, and B. LeBaron. A test for independence based on the correlation dimension. Unpublished report, 1991.

[3] P. J. Brockwell and R. A. Davis. *Time Series: Theory and Methods*. Springer-Verlag, 1987.

[4] N. H. Chan and L. T. Tran. Nonparametric tests for serial dependence. *Journal of Time Series Analysis*, 13:19–28, 1992.

[5] M. Hallin and M. L. Puri. Rank tests for time series analysis: a survey. In Brillinger, D. et al., editor, *New Directions in Time Series Analysis*, pages 111–153. Springer-Verlag, 1992.

[6] E. L. Lehmann. *Testing Statistical Hypotheses*. John Wiley & Sons, Inc., 1959.

[7] C. A. P. Pinkse. A general characteristic function based measure applied to serial independence testing. Preprint, London School of Economics, 1993.

[8] P. M. Robinson. Consistent nonparametric entropy-based testing. *Review of Economic Studies*, 58:437–453, 1991.

[9] P. M. Robinson. Kernel estimation and interpolation for time series containing missing observations. *Annals of the Institute of Statistical Mathematics*, 36:401–412, 1984.

[10] M. Rosenblatt. A quadratic measure of deviation of two-dimensional density estimates and a test of independence. *Annals of Statistics*, 3:1–14, 1975.

[11] M. Rosenblatt and B. E. Wahlen. A nonparametric measure of independence under a hypothesis of independent components. *Statistics and Probability Letters*, 15:245–252, 1992.

[12] B. W. Silvermann. *Density Estimation for Statistics and Data Analysis*. Chapman and Hall, 1986.

[13] H. J. Skaug and D. Tjøstheim. A nonparametric test of serial independence based on the empirical distribution function. *Biometrika*, 80:591–602, 1993.

[14] H. J. Skaug and D. Tjøstheim. Nonparametric tests of serial independence. In T. Subba Rao, editor, *Developments in Time Series Analysis*, pages 207–229. Chapman and Hall, 1993.

[15] H. J. Skaug and D. Tjøstheim. Testing for serial independence using measures of distance between densities. Technical report, Department of Mathematics, University of Bergen, Norway, 1995.

## TABLES

TABLE 1. Power of the tests against the ARCH alternative (9) with $n = 250$ and $\epsilon = 0.05$.

| | $q = 1$ | $q = 2$ | $q = 3$ | $q = 4$ | $q = 5$ | $q = 6$ | $q = 8$ | $q = 10$ |
|---|---|---|---|---|---|---|---|---|
| $\hat{H}^{(q)}$ | 0.874 | 0.813 | 0.740 | 0.669 | 0.607 | 0.549 | 0.461 | 0.381 |
| $\hat{I}^{(q)}$ | 0.861 | 0.802 | 0.721 | 0.655 | 0.592 | 0.535 | 0.444 | 0.375 |
| $\hat{J}^{(q)}$ | 0.976 | 0.948 | 0.896 | 0.827 | 0.756 | 0.692 | 0.563 | 0.469 |
| $\hat{C}_1^{(q)}$ | 0.213 | 0.218 | 0.210 | 0.194 | 0.174 | 0.166 | 0.152 | 0.143 |
| $\hat{C}_2^{(q)}$ | 0.074 | 0.068 | 0.064 | 0.068 | 0.058 | 0.052 | 0.050 | 0.050 |
| $\hat{C}_3^{(q)}$ | 0.973 | 0.937 | 0.923 | 0.910 | 0.889 | 0.871 | 0.813 | 0.800 |
| BDS | 0.950 | 0.880 | 0.740 | * | * | * | * | * |

TABLE 2. Power of the tests against the GARCH alternative (10) with $n = 250$ and $\epsilon = 0.05$.

| | $q = 1$ | $q = 2$ | $q = 3$ | $q = 4$ | $q = 5$ | $q = 6$ | $q = 8$ | $q = 10$ | |
|---|---|---|---|---|---|---|---|---|---|
| $\hat{H}^{(q)}$ | 0.180 | 0.252 | 0.277 | 0.294 | 0.297 | 0.291 | 0.288 | 0.268 | 0.254 |
| $\hat{I}^{(q)}$ | 0.179 | 0.232 | 0.264 | 0.287 | 0.288 | 0.289 | 0.281 | 0.268 | 0.261 |
| $\hat{J}^{(q)}$ | 0.331 | 0.434 | 0.479 | 0.510 | 0.514 | 0.506 | 0.494 | 0.462 | 0.444 |
| $\hat{C}_1^{(q)}$ | 0.085 | 0.092 | 0.095 | 0.099 | 0.103 | 0.105 | 0.105 | 0.106 | 0.103 |
| $\hat{C}_2^{(q)}$ | 0.052 | 0.053 | 0.055 | 0.061 | 0.059 | 0.054 | 0.053 | 0.055 | 0.057 |
| $\hat{C}_3^{(q)}$ | 0.277 | 0.285 | 0.336 | 0.357 | 0.355 | 0.352 | 0.321 | 0.329 | 0.323 |
| BDS | 0.210 | 0.230 | 0.220 | * | * | * | * | * | |

TABLE 3. Power of the tests against the NLMA alternative (11) with $n = 250$ and $\epsilon = 0.05$.

| | $q = 1$ | $q = 2$ | $q = 3$ | $q = 4$ | $q = 5$ | $q = 6$ | $q = 8$ | $q = 10$ |
|---|---|---|---|---|---|---|---|---|
| $\hat{H}^{(q)}$ | 0.242 | 0.332 | 0.254 | 0.217 | 0.193 | 0.169 | 0.143 | 0.129 |
| $\hat{I}^{(q)}$ | 0.221 | 0.303 | 0.244 | 0.205 | 0.180 | 0.168 | 0.139 | 0.126 |
| $\hat{J}^{(q)}$ | 0.395 | 0.556 | 0.441 | 0.358 | 0.300 | 0.264 | 0.206 | 0.175 |
| $\hat{C}_1^{(q)}$ | 0.092 | 0.105 | 0.092 | 0.086 | 0.079 | 0.075 | 0.070 | 0.069 |
| $\hat{C}_2^{(q)}$ | 0.055 | 0.057 | 0.050 | 0.055 | 0.048 | 0.044 | 0.044 | 0.044 |
| $\hat{C}_3^{(q)}$ | 0.359 | 0.405 | 0.386 | 0.360 | 0.326 | 0.302 | 0.247 | 0.240 |
| BDS | 0.290 | 0.350 | 0.310 | * | * | * | * | * |

TABLE 4. Power of the tests against the ENLMA alternative (12) with $n = 250$ and $\epsilon = 0.05$.

| | $q = 1$ | $q = 2$ | $q = 3$ | $q = 4$ | $q = 5$ | $q = 6$ | $q = 8$ | $q = 10$ |
|---|---|---|---|---|---|---|---|---|
| $\hat{H}^{(q)}$ | 0.870 | 0.881 | 0.840 | 0.795 | 0.751 | 0.700 | 0.606 | 0.521 |
| $\hat{I}^{(q)}$ | 0.865 | 0.877 | 0.836 | 0.791 | 0.745 | 0.695 | 0.600 | 0.514 |
| $\hat{J}^{(q)}$ | 0.957 | 0.965 | 0.945 | 0.915 | 0.872 | 0.820 | 0.708 | 0.605 |
| $\hat{C}_1^{(q)}$ | 0.311 | 0.327 | 0.312 | 0.302 | 0.294 | 0.279 | 0.249 | 0.229 |
| $\hat{C}_2^{(q)}$ | 0.181 | 0.165 | 0.148 | 0.146 | 0.134 | 0.121 | 0.115 | 0.111 |
| $\hat{C}_3^{(q)}$ | 0.866 | 0.835 | 0.821 | 0.804 | 0.773 | 0.748 | 0.679 | 0.663 |
| BDS | 0.840 | 0.880 | 0.850 | * | * | * | * | * |

TABLE 5. Power of the tests against the TAR alternative (13) with $n = 250$ and $\epsilon = 0.05$.

| | $q = 1$ | $q = 2$ | $q = 3$ | $q = 4$ | $q = 5$ | $q = 6$ | $q = 8$ | $q = 10$ |
|---|---|---|---|---|---|---|---|---|
| $\hat{H}^{(q)}$ | 0.957 | 0.869 | 0.765 | 0.667 | 0.584 | 0.490 | 0.375 | 0.282 |
| $\hat{I}^{(q)}$ | 0.974 | 0.909 | 0.826 | 0.739 | 0.661 | 0.575 | 0.453 | 0.349 |
| $\hat{J}^{(q)}$ | 0.907 | 0.765 | 0.635 | 0.522 | 0.434 | 0.364 | 0.266 | 0.211 |
| $\hat{C}^{(q)}$ | 0.096 | 0.081 | 0.072 | 0.069 | 0.072 | 0.071 | 0.065 | 0.061 |
| $\hat{C}_2^{(q)}$ | 0.100 | 0.083 | 0.076 | 0.077 | 0.067 | 0.059 | 0.056 | 0.055 |
| $\hat{C}_3^{(q)}$ | 0.567 | 0.404 | 0.378 | 0.355 | 0.314 | 0.289 | 0.230 | 0.227 |
| BDS | 0.680 | 0.520 | 0.670 | * | * | * | * | * |

TABLE 6. Power comparison of permutation tests and non-permutation tests for $n = 100$ and $\epsilon = 0.05$.

| | $\hat{H}^{(q)}$ | | $\hat{I}^{(q)}$ | | $\hat{J}^{(q)}$ | |
|---|---|---|---|---|---|---|
| | $H_0$ | $H_0^*$ | $H_0$ | $H_0^*$ | $H_0$ | $H_0^*$ |
| Alternative (9) | 0.449 | 0.464 | 0.456 | 0.448 | 0.683 | 0.694 |
| Alternative (10) | 0.107 | 0.112 | 0.101 | 0.102 | 0.162 | 0.171 |
| Alternative (11) | 0.133 | 0.137 | 0.125 | 0.132 | 0.206 | 0.206 |
| Alternative (12) | 0.431 | 0.450 | 0.444 | 0.446 | 0.581 | 0.633 |
| Alternative (13) | 0.625 | 0.627 | 0.657 | 0.658 | 0.516 | 0.509 |

TABLE 7. p-values for the tests ran on the exchange rate data.

| | $q = 1$ | $q = 2$ | $q = 3$ | $q = 4$ | $q = 5$ | $q = 6$ | $q = 8$ | $q = 10$ |
|---|---|---|---|---|---|---|---|---|
| $\hat{H}^{(q)}$ | 0.043 | 0.047 | 0.015 | 0.069 | 0.028 | 0.004 | 0.003 | 0.005 |
| $\hat{I}^{(q)}$ | 0.069 | 0.069 | 0.014 | 0.066 | 0.029 | 0.004 | 0.003 | 0.006 |
| $\hat{J}^{(q)}$ | 0.086 | 0.014 | 0.004 | 0.006 | 0.002 | 0.000 | 0.000 | 0.000 |
| $\hat{C}_1^{(q)}$ | 0.351 | 0.381 | 0.321 | 0.400 | 0.508 | 0.632 | 0.559 | 0.590 |
| $\hat{C}_2^{(q)}$ | 0.416 | 0.477 | 0.149 | 0.176 | 0.218 | 0.257 | 0.307 | 0.374 |
| $\hat{C}_3^{(q)}$ | 0.326 | 0.344 | 0.266 | 0.342 | 0.430 | 0.554 | 0.510 | 0.536 |

# Regression in Long-Memory Time Series

## Richard L. Smith, Cambridge University and University of North Carolina
## Fan-Ling Chen, University of North Carolina

ABSTRACT

We consider the estimation of regression coefficients when the residuals form a stationary time series. Hannan proposed a method for doing this, via a form of Gaussian estimation, in the case when the time series has some parametric form such as ARMA, but we are interested in semiparametric cases incorporating long-memory dependence. After reviewing recent results by Robinson, on semiparametric estimation of long-memory dependence in the case where there are no covariates present, we propose an extension to the case where there are covariates. For this problem it appears that a direct extension of Robinson's method leads to inefficient estimates of the regression parameters, and an alternative is proposed. Our mathematical arguments are heuristic, but rough derivations of the main results are outlined. As an example, we discuss some issues related to climatological time series.

## 1 Introduction

Consider the model

$$y_t = \sum_{k=1}^{p} \beta_k x_{k,t} + u_t, \quad 1 \le t \le T, \tag{1}$$

where $\{y_t\}$ is an observed time series, $\{x_{k,t}, \ 1 \le k \le p\}$ is a vector of $p$ co-variates whose values at time $t$ are known, $\beta_1, ..., \beta_p$ are unknown regression coefficients, and $\{u_t\}$ is a stationary time series.

The problem of estimating $\beta_1, ..., \beta_p$ in this situation is well known, and fundamental results were known as long ago as Grenander (1954). In many cases, the asymptotic efficiency of the ordinary least squares estimators (OLSE), in comparison with the best linear unbiased estimators (BLUE), is very close to 1. Indeed, for many cases where the regressors are determined by simple analytic formulae, including the case of simple linear regression,

$$y_t = \beta_1 + \beta_2 t + u_t, \quad 1 \le t \le T, \tag{2}$$

Grenander showed that the asymptotic efficiency of the OLSE compared with the BLUE is 1. The assumptions on $\{u_t\}$, however, included continuity (and therefore boundedness) of the spectral density, which excludes the "long-memory" case we are interested in here.

Even in cases where the OLSE is exactly or approximately efficient, there is still the problem of how to obtain standard errors for the parameter estimates in a way which correctly allows for the autocorrelation of $\{u_t\}$. One solution to this problem is to use Gaussian estimation, as originally formulated by Whittle (1953) and extended to regression models by Hannan (1971, 1973). Hannan proved asymptotic normality for the estimators, and gave expressions for their asymptotic variances and covariances, but again under assumptions excluding long-memory.

Our interest in this problem was motivated, in part, by the problem of estimating trends in climatological time series, and of obtaining associated standard errors and significance tests. For discussion, see Bloomfield (1992) and Bloomfield and Nychka (1992). It is a well-known phenomenon that time series of global average temperatures exhibit a steady increasing trend over most of the last 150 years, but it is still a point of debate whether such trends are real evidence of an enhanced greenhouse effect due to human activity, or are simply long-term fluctuations in the natural climate. If we entertain the latter explanation as plausible, then it is natural to think of models for $\{u_t\}$ that would be consistent with it. One class of such models is *long-memory time series* for which the spectral density $f(\lambda)$, $0 < \lambda \leq \frac{\pi}{2}$ satisfies

$$f(\lambda) \sim c\lambda^{-2d}, \quad \lambda \downarrow 0, \tag{3}$$

where $c > 0$ and $d$ are constants. A model satisfying (3) is stationary and invertible provided $-\frac{1}{2} < d < \frac{1}{2}$, while all the standard time series models, such as stationary and invertible ARMA models, satisfy (3) with $d = 0$. The case $0 < d < \frac{1}{2}$ is of particular interest as this represents a stationary time series with long-range fluctuations.

For models whose spectral density satisfies (3), the asymptotic efficiency of OLSE compared with BLUE has been studied by Yajima (1988, 1991). In this case the asymptotic efficiency is no longer 1, though Yajima's results show that it still typically over 90% for linear regression models such as (2). Motivated by these results, Smith (1993) fitted the model defined by (2) and (3) to a number of climatic time series. Smith estimated $\beta_1$ and $\beta_2$ using the OLSEs, $\tilde{\beta}_1$ and $\tilde{\beta}_2$ say, and then $c$ and $d$ by a method which will be explained in Section 2. Treating the estimated $c$ and $d$ as the true values, he then used asymptotic expressions equivalent to those obtained by Yajima (1988) to estimate the standard error of $\tilde{\beta}_2$. For one series of global temperatures, he obtained a statistically significant value of $\beta_2$ (of about 0.4 $^\circ$C per century) but with a standard error (about 0.1) considerably greater than those produced by ignoring the time-series correlations or treating them as the realization of a low-order AR process.

This discussion raises the question of whether it is possible to estimate the parameters $\beta_1, \beta_2, c$ and $d$ jointly. The purpose of this paper is to propose such a method, and to outline preliminary results about its properties. The mathematical arguments are incomplete at the present time, but will be published elsewhere in due course.

## 2   Long-memory Time Series

In this section we outline some results on estimation of time series whose spectral density satisfies (3).

An important class of models are the *fractionally integrated* ARIMA $(p, d, q)$ models with non-integral $d$. These models were introduced independently by Granger and Joyeux (1980) and Hosking (1981), and have enjoyed wide application. Fox and Taqqu (1986) established asymptotic estimation results based on an approximate maximum likelihood procedure, and these were extended to exact maximum likelihood by Dahlhaus (1989). Haslett and Raftery (1989) discussed numerous applied aspects with respect to a problem on estimating wind speeds.

However, in practice we may not feel able, or may not wish, to specify a fully parametric model, and it is of interest to take a semiparametric approach in which we assume that $f$ satisfies (3), but without any further parametric assumptions on the stationary process. A further point is that when (2) and (3) hold, the asymptotic variance of the OLSE $\tilde{\beta}_2$ depends on $f$ just through the constants $c$ and $d$ (Smith 1993), so this provides further reason to concentrate on the lower tail of the spectral density. For most of our discussion we assume $0 < d < \frac{1}{2}$, since for reasons explained in Section 1, this is the main case of interest.

An ingenious development was due to Geweke and Porter-Hudak (1983) (henceforth GPH), who made a direct attempt to fit the model (3) using the sample periodogram. The periodogram of a time series $\{y_t, \ 1 \leq t \leq T\}$ is defined by

$$I_{y,T}(\lambda) = \frac{1}{2\pi T} \left| \sum_{t=1}^{T} y_t e^{i\lambda t} \right|^2 . \tag{4}$$

Usually the periodogram is evaluated just at the *Fourier frequencies* $\lambda_{j,T} = \frac{2\pi j}{T}$, $0 \leq j \leq \lfloor \frac{T}{2} \rfloor$ and it is a well-known result of time series analysis (e.g. Brockwell and Davis 1991, section 10.3) that the following "standard results" hold: *the periodogram ordinates at distinct Fourier frequencies are approximately independent exponentially distributed, with* $E\{I_{y,T}(\lambda_{j,T})\} \approx f(\lambda_{j,T})$. With these results in mind, they proposed (with a minor modification which need not concern us) that $-2d$ be estimated by regressing $\log I_{y,T}(\lambda_{j,T})$ on $\log \lambda_{j,T}$ for $1 \leq j \leq n_T$, where $n_T << T$. Asymptotic results for this estimator were conjectured by assuming that the just-mentioned "standard results" remain valid in this situation.

However a number of authors, beginning with Künsch (1986), cast doubt on this method by showing that in the extreme lower tail as $\lambda \downarrow 0$, the "standard results" do not hold for a long-memory process. In particular, Künsch calculated the asymptotic mean of $I_{y,T}(\lambda_{j,T})/f(\lambda_{j,T})$, as $T \to \infty$ for fixed $j$, and showed that it is not 1. Subsequently Hurvich and Beltrao (1993) extended the calculation by finding all the asymptotic covariances of the components of the discrete Fourier transform at frequencies $\lambda_{j,T}, \lambda_{k,T}$, for fixed $j$ and $k$ as $T \to \infty$, and showing that they do not correspond to the "standard results". For example, in this range the periodogram ordinates are no longer asymptotically independent. Nevertheless, it remained a conjecture that the GPH method could be validated with some suitable strengthening of the assumptions. This question remained open for a long time and was only answered in recent work by Robinson (1995a). Robinson introduced some restrictions which are needed for his method of proof, but may not be required for the validity of the result: he assumed that the process is Gaussian (rather than linear, as in the "standard results"), and also that the regression should be performed over Fourier frequencies $m_T + 1 \le j \le n_T$, where, as $T \to \infty$, $m_T \to \infty$, $m_T/n_T \to 0$, $n_T/T \to 0$ (with some additional restrictions which we will not spell out here). The GPH estimator corresponds to $m_T = 0$ which is excluded by Robinson's conditions. While it is still an open question whether these additional conditions are needed, the result itself was a considerable technical breakthrough.

However, Robinson (1995b) showed that an alternative method, essentially Gaussian estimation, yields results which are statistically superior as well as being simpler to derive. The standard method of Gaussian estimation is as follows: suppose we have a stationary process with spectral density $f(\lambda; \theta)$, parametrically determined by some finite-dimensional $\theta$. Define the periodogram as above, and choose $\theta$ to minimize

$$\sum_j \left\{ \log f(\lambda_{j,T}; \theta) + \frac{I_{y,T}(\lambda_{j,T})}{f(\lambda_{j,T}; \theta)} \right\}. \tag{5}$$

Robinson's proposal was an adaptation of (5): replace $f(\lambda; \theta)$ by its approximate form $c\lambda^{-2d}$, and minimize (5) where the sum is taken over just $1 \le j \le n_T$, for some $n_T$ with $n_T \to \infty$, $n_T/T \to 0$ as $T \to \infty$ (plus some additional conditions which we shall not specify here). It should be noted that this method may also be derived from the "standard results" on the periodogram, deriving a "likelihood function" based on the exponential distribution of periodogram ordinates, and this led Smith (1993) to call it the "maximum likelihood method". Robinson's "Gaussian estimation" terminology is certainly closer to the standard use of the phrase in time series analysis, though as we shall see later, thinking of the estimator as a form of maximum likelihood estimator also has its advantages.

Robinson (1995b) proved asymptotic normality for the Gaussian estimator under weaker conditions than he used for the GPH estimator. In this case he did not need to cut off the lower $m_T$ Fourier frequencies, nor did he need to assume a Gaussian process — linearity of the process, with innovations forming a martingale difference sequence, suffices for the result. The reason why it was possible to prove these results under simpler conditions than for GPH is a highly technical one: GPH involves linear combinations of *logarithms* of the periodogram ordinates, whereas Gaussian estimation is based on linear combinations of the periodogram ordinates themselves, and it is much easier to prove central limit theorems for the latter than for the former. Moreover it is also the case that the Gaussian estimation procedure yields an estimator with smaller asymptotic variance (by a factor $\frac{\pi^2}{6}$), so it seems better from every point of view. Henceforth, we do not consider GPH any further but concentrate on the Gaussian estimator, and its extensions to the regression case.

## 3   Gaussian estimation for a particular process

In order to explicate certain features of our derivation we concentrate on a very specific process: we assume $\{u_t\}$ in (1) is a Gaussian ARIMA(0,$d$,0) process with $0 < d < \frac{1}{2}$, i.e. one for which $f(\lambda) = c|1 - e^{i\lambda}|^{-2d}$. However the method of estimation is developed in a way which should allow extensions to models satisfying the much weaker assumption (3), and under weaker assumptions than Gaussianity.

For a given time series $\{y_t, \ 1 \le t \le T\}$, we define the discrete Fourier transform (DFT)

$$D_{y,T}(\lambda) = C_{y,T}(\lambda) + iS_{y,T}(\lambda), \tag{6}$$

where

$$C_{y,T}(\lambda) = \sqrt{\frac{1}{2\pi T}} \sum_{t=1}^{T} y_t \cos \lambda t \tag{7}$$

and

$$S_{y,T}(\lambda) = \sqrt{\frac{1}{2\pi T}} \sum_{t=1}^{T} y_t \sin \lambda t \tag{8}$$

are the cosine and sine transforms. Note that $I_{y,T}(\lambda) = |D_{y,T}(\lambda)|^2$. By taking the DFT of each side in (1), we have

$$D_{y,T}(\lambda) = \sum_{k=1}^{p} \beta_k D_{x_k,T}(\lambda) + D_{u,T}(\lambda) \tag{9}$$

and similarly for the cosine and sine transforms.

The "standard results" in this setting are as follows: for Fourier frequencies $\{\lambda_{j,T}\}$, the cosine and sine tranforms $C_{u,T}(\lambda_{j,T})$ and $S_{u,T}(\lambda_{j,T})$ are approximately mutually independent (of each other and of the values for all other Fourier frequencies) and normally distributed, each with mean 0 and variance $\frac{1}{2}f(\lambda_{j,T})$. Note, however, that one can in principle compute the exact variances and covariances of the DFT.

Fix some $n_T \ll \frac{T}{2}$ and let $\tilde{y}$ denote the vector $\{C_{y,T}(\lambda_{1,T}), S_{y,T}(\lambda_{1,T}),$ $C_{y,T}(\lambda_{2,T}), S_{y,T}(\lambda_{2,T}), ..., C_{y,T}(\lambda_{n_T,T}), S_{y,T}(\lambda_{n_T,T})\}$. In view of (9), we write $\mathrm{E}\{\tilde{y}\} = \tilde{X}\beta$ where the entries of $\tilde{X}$ are $C_{x_k,T}(\lambda_{j,T})$ and $S_{x_k,T}(\lambda_{j,T})$ for $1 \leq k \leq p$, $1 \leq j \leq n_T$. We also have that the covariance matrix of $\tilde{y}$ is $\Sigma(c,d)$ which, in view of our assumption of an exact ARIMA(0,d,0) process, is in principle exactly computable (it also depends on $T$ and $n_T$, as well as $c$ and $d$, but we suppress that in the notation).

We propose that the parameters $c, d, \beta_1, ..., \beta_p$ be chosen as those values $\hat{c}, \hat{d}, \hat{\beta}_1, ..., \hat{\beta}_p$ which minimize the expression

$$Q(c, d, \beta_1, ..., \beta_p) = \frac{1}{2} \log |\Sigma(c,d)| + \frac{1}{2}(\tilde{y} - \tilde{X}\beta)^T \Sigma(c,d)^{-1}(\tilde{y} - \tilde{X}\beta). \quad (10)$$

Note that this is the exact negative log likelihood based on the DFT evaluated at the first $n_T$ Fourier frequencies. If we were solely concerned with the parametric model there would be no reason to restrict ourselves in this way, but we are looking towards extensions to the semiparametric case.

One can compute the information matrix associated with (10). Write $\theta = (\theta_1, \theta_2) = (c, d)$. Using obvious vector notation, one easily verifies that

$$\mathrm{E}\left\{\frac{\partial^2 Q}{\partial \beta \partial \beta^T}\right\} = \tilde{X}^T \Sigma(\theta)^{-1} \tilde{X}, \quad (11)$$

$$\mathrm{E}\left\{\frac{\partial^2 Q}{\partial \theta \partial \beta^T}\right\} = 0, \quad (12)$$

$$\mathrm{E}\left\{\frac{\partial^2 Q}{\partial \theta \partial \theta^T}\right\} = \frac{1}{2}\frac{\partial^2}{\partial \theta \partial \theta^T}\left\{\log|\Sigma(\theta)|\right\} + \frac{1}{2}\mathrm{tr}\left\{\frac{\partial^2}{\partial \theta \partial \theta^T}\left(\Sigma(\theta)^{-1}\right)\Sigma(\theta)\right\}. \quad (13)$$

Assuming that the asymptotic properties of maximum likelihood are valid in this situation (this has not been checked, but there is no reason to disbelieve it), we deduce that (a) the asymptotic covariance matrix of $\hat{\theta}$ is the same as in the no-regression case, (b) the asymptotic covariance matrix of $\hat{\beta}$ is $\tilde{X}\Sigma(\theta)^{-1}\tilde{X}$, just as if $c$ and $d$ were known, (c) $\hat{\theta}$ and $\hat{\beta}$ are asymptotically independent. In other words, $\theta$ and $\beta$ are orthogonal parameters.

Now if we were to assume that the "standard results" hold in this situation, we could simplify the above considerably by replacing $\Sigma$ with the diagonal matrix with entries $\frac{c}{2}\lambda_{1,T}^{-2d}, \frac{c}{2}\lambda_{1,T}^{-2d}, \frac{c}{2}\lambda_{2,T}^{-2d}, \frac{c}{2}\lambda_{2,T}^{-2d}, ..., \frac{c}{2}\lambda_{n_T,T}^{-2d}, \frac{c}{2}\lambda_{n_T,T}^{-2d}$. This leads us to propose minimizing the objective function

$$Q'(c, d, \beta_1, ..., \beta_p) = \sum_{j=1}^{n_T}\left[\log c - 2d\log\lambda_{j,T} + \frac{\lambda_{j,T}^{2d}}{c}\left\{\left(C_{y,T}(\lambda_{j,T})\right.\right.\right.$$

$$-\sum_k \beta_k C_{x_k,T}(\lambda_{j,T})\Big)^2 + \Big(S_{y,T}(\lambda_{j,T}) - \sum_k \beta_k S_{x_k,T}(\lambda_{j,T})\Big)^2\Big\}\Big]. \quad (14)$$

Originally our intention was to propose (14) as our primary method of estimation. This would be a direct generalization of Robinson (1995b) to the regression problem, or, from a different point of view, of Hannan's (1973) method of estimating a regression model with time series errors. However, it would appear that this does not lead to asymptotically efficient estimates of $\beta$, for a reason which we shall now explain.

## 4   Theoretical properties

Our arguments in this section are preliminary and very rough, but they are intended to motivate the estimation procedure just described, and to indicate some plausible extensions.

We first concentrate on the estimation of $\beta$. One question which arises is the following: to what extent is information lost by using only the first $n = n_T$ components of the DFT? Note that Yajima (1988) established the asymptotic covariance matrix of the BLUE, in cases involving polynomial regression, which is equivalent to using all $\lfloor T/2 \rfloor$ components of the DFT.

A general argument is as follows. Suppose we run two experiments to estimate the same parameter $\beta$. The models are $y_1 = X_1\beta + \epsilon_1$, $y_2 = X_2\beta + \epsilon_2$, and the joint covariance matrix of $(\epsilon_1^T, \epsilon_2^T)$ is

$$A = \begin{pmatrix} A_{11} & A_{12} \\ A_{21} & A_{22} \end{pmatrix}.$$

Assuming joint normality, the information matrix for the combined experiment may be partitioned as

$$X_1^T A_{11}^{-1} X_1 + (X_2 - A_{21}A_{11}^{-1}X_1)^T (A_{22} - A_{21}A_{11}^{-1}A_{12})^{-1}(X_2 - A_{21}A_{11}^{-1}X_1)^T$$

where the first term is based on the distribution of $y_1$ and the second on the conditional distribution of $y_2$ given $y_1$. Thus the second term represents the additional information about $\beta$ contained in $y_2$, given that we already know $y_1$.

Let us apply this in the case where we identify $y_1$ with $\tilde{y}$, the DFT based on the first $n$ Fourier frequencies, and $y_2$ with either the cosine or the sine component of the $(n+1)$'st frequency. The results of Künsch (1986), Hurvich and Beltrao (1993) and Robinson (1995b) are all consistent with the statement that covariances between the $j$'th and $k$'th components of the DFT differ from the "standard results" by an expression of order $1/\sqrt{jk}$. For example, Lemma 3 of Robinson (1995b) states this explicitly in the case $j = k$. Let us assume it is true for all $j$ and $k$.

We consider only the case of simple linear regression, cf. (2), and center the regressor so that $x_t = t - \frac{T+1}{2}$, $t = 1, ..., T$. From the formula

$$\sum_{t=1}^{T} e^{i\lambda(t-\frac{T+1}{2})} = \frac{\sin T\lambda/2}{\sin \lambda/2}$$

and differentiating with respect to $\lambda$, we deduce that at Fourier frequencies $\lambda_{j,T}$,

$$\left| \sum_{1}^{T} \left( t - \frac{T+1}{2} \right) e^{i\lambda t} \right| = \frac{T}{2\sin \lambda/2} \sim \frac{T}{\lambda}.$$

Thus $C_{x,T}(\lambda_{j,T})$ and $S_{x,T}(\lambda_{j,T})$ are each of order $T^{3/2}/j$.

The magnitude of $X_2 - A_{21}A_{11}^{-1}X_1$ is determined primarily by that of $X_2$ itself, which is $O(T^{3/2}/n)$, while $A_{22} - A_{12}A_{11}^{-1}A_{21}$ is of the same order of magnitude as $A_{22}$ itself, which is $O(\lambda_{n+1,T}^{-2d}) = O((n/T)^{-2d})$. Thus the "information" in the $(n+1)$'st frequency of the DFT is of $O(T^{3-2d}n^{2d-2})$, and the total information in all the frequencies from $n+1$ onwards is of $O(T^{3-2d}n^{2d-1})$. Note that $2d < 1$ and therefore $\sum n^{2d-2} < \infty$. On the one hand, this implies that the total information in the data is of $O(T^{3-2d})$ and therefore that the variance of the BLUE is $O(T^{-3+2d})$, which turns out to be consistent with the results of Yajima (1988). On the other hand, the *relative* loss of information if we neglect all frequencies after the $n_T$'th is of $O(n_T^{2d-1})$, which tends to 0 provided only that $n_T \to \infty$.

This also explains, in intuitive terms, why estimation based on (14) will not be efficient. The $j$'th Fourier frequency (for fixed $j$, as $T \to \infty$) contributes a term of relative size $O(1)$ to the Fisher information matrix, so if this is wrong by a factor of $O(1)$, as is indeed the case if we assume the "standard results" (cf. Hurvich and Beltrao 1993), there will be a nonnegligible error in the final result.

In contrast, for the estimation of $c$ and $d$, similar calculations show that the contribution to the Fisher information from any single Fourier frequency is negligible and so, for this part of the problem, we can use the "standard results", as Robinson's (1995b) results made precise for the case without any regression terms.

A motivation for restricting attention to the first $n_T$ Fourier frequencies, as opposed to using all of them, is that the results should then be valid for a wide class of spectral densities satisfying (3), instead of just the exact ARIMA(0,$d$,0) process studied in Section 3. Again following results by Robinson (1995a, 1995b), we anticipate that it will be possible to make such a statement precise, though it remains to be determined exactly what conditions on $f$, and on the rate of growth of $n_T$, are needed to achieve this.

A final comment is that whereas many of the results in the area assume Gaussianity of the underlying process (e.g. Künsch 1986, the distributional results of Hurvich and Beltrao 1993, and Robinson 1995a), in fact this appears not to be necessary, and Robinson (1995b) obtained results under the assumption that the process is linear with the errors forming a martingale difference sequence. Again we anticipate that the procedure described here will be valid under that form of assumption.

## 5   An application in climatology

As an example of the methodology, we analyse the data plotted in Fig. 1. These are 108 annual values (1880–1987), in °C relative to an unspecified standard value, of the data series collected by Hansen and Lebedeff (1987, 1988), which purport to represent global temperature averages. Also shown on the plot is the least squares regression line, showing a clear increasing trend with an estimated slope of 0.55 °C per century.

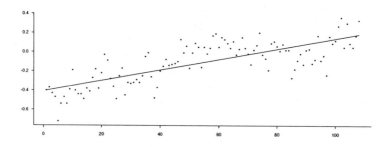

**Fig. 1.** Plot of 108 annual values from the Hansen-Lebedeff temperature series, with fitted least squares straight line.

A critical question in studies of this nature is to calculate the standard error of such a slope, making due allowance for time series correlations. In this connection, we investigate the possibility of long-memory dependence.

The residuals from the linear trend have been calculated, and their periodogram is displayed in Fig. 2. Also shown in Fig. 2 is a smoothed curve computed using the lowess function of Becker *et al.* (1987). The plot gives some evidence that the periodogram ordinates are rising at low Fourier frequencies, and this justifies pursuing the long-memory approach.

The model (3) may be fitted to the first $n$ periodogram ordinates by the Gaussian method advocated by Robinson (1995b). The value of $n$ is, of course, unspecified, but a plot of the estimates of $d$ and their approximate 95% confidence limits (Fig. 3) suggests that the estimates settle down after about $n = 12$. We have arbitrarily selected $n = 15$ for the subsequent study. In this case, we find $\hat{d} = 0.42$ with a standard error of 0.15. We also have $\hat{c} = .016$. Thus $d$ is significantly different from 0 and this provides substantial evidence that the series is indeed long-memory, though the large standard error means that this is not conclusive.

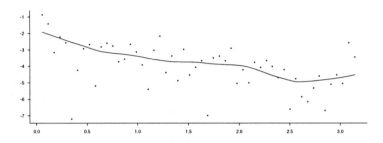

**Fig. 2.** Periodogram of the residuals from the straight-line fit in Fig. 1, with smoothed spectral density computed by lowess.

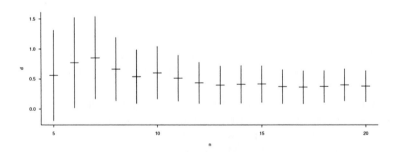

**Fig. 3.** Estimates of long-memory parameter $d$ by Gaussian method, based on periodogram ordinates in Fig. 2. The estimates are computed for a variety of values of $n$, along with approximate 95% confidence limits, represented by the vertical lines.

Based on these estimates of $c$ and $d$ and using the asymptotic formula (7.9) in Smith (1993), we deduce that the standard error of the slope of the least squares regression line is 0.37 ($^{o}$C per century). Since the estimated slope is 0.55, this leads us to doubt whether the effect is real.

As a comment in passing, we mention that there is another way to calculate this standard error which does not rely so much on asymptotics (though it still assumes (3)). If our estimate of the slope $\beta_2$ is of the form $\hat{\beta}_2 = \sum w_t y_t$ and if we define $W(\lambda) = \sum w_t e^{i\lambda t}$, then the variance of $\hat{\beta}_2$ is $2 \int_0^\pi |W(\lambda)|^2 f(\lambda) d\lambda$ (Bloomfield and Nychka, 1992) and this may be evaluated by numerical integration. In this case it leads to exactly the same standard error.

The standard error quoted here is larger than several values for the same data set quoted by Bloomfield (1992). For example, fitting an ARIMA(0,$d$, 0) process by maximum likelihood, Bloomfield obtained an estimated slope of 0.57 with standard error 0.11, and based on an ARIMA(0,$d$,1) process, the same point estimate with a standard error 0.10. The estimates of $d$ are respectively 0.42, 0.25.

Now let us try the new method. As shown in Section 3, we transform to frequency space and calculate the cosine and sine components of the DFT evaluated at the first $n$ Fourier frequencies. It is logical to estimate the mean and slope simultaneously, so we also use the DFT at frequency 0, that is, the sample mean. Thus $\tilde{y}$ is a vector of dimension $2n + 1$ whose mean is of the form $\tilde{X}\beta$ and whose covariance matrix is $\Sigma(c, d)$ depending on the unknown parameters of the spectral density, $c$ and $d$. The hardest part computationally is evaluating the entries of $\Sigma(c, d)$. If two of the components of $\tilde{y}$, say $\tilde{y}_k$ and $\tilde{y}_\ell$, are written as $\tilde{y}_k = \sum_s b_{ks} y_s$ and $\tilde{y}_\ell = \sum_t b_{\ell t} y_t$, then we may write

$$\text{Cov}(\tilde{y}_k, \tilde{y}_\ell) = \sum_s \sum_t b_{ks} b_{\ell t} \int_{-\pi}^\pi e^{i(s-t)\lambda} f(\lambda) d\lambda$$

$$= \int_{-\pi}^\pi B_k(\lambda) B_\ell^*(\lambda) f(\lambda) d\lambda \tag{15}$$

where $B_j(\lambda) = \sum_s b_{js} e^{i\lambda s}$ and the asterisk denotes complex conjugate. In calculating the likelihood function, (15) has been evaluated by direct numerical integration based on 1,000 sampling points.

A further point about this method of estimation is that the structure of the transformed model — a regression model whose covariance matrix depends on unknown parameters — is one which arises in other contexts, such as variance components estimation and geostatistics. In those contexts, it is often recommended that ordinary maximum likelihood should be replaced by "restricted maximum likelihood" or REML. The idea is that, rather than use the likelihood based on the full data set $\tilde{y}$, one first defines a maximal set of contrasts of $\tilde{y}$ to the columns of $X$, and then uses the joint density of the set of contrasts, which does not depend on $\beta$. The method leads to a slight theoretical reduction in efficiency, but this is often compensated for by a reduction in the bias for estimating the parameters of the covariance matrix. See, for example, Patterson and Thompson (1971) for the original idea and Cressie (1993) for the application to geostatistics. Since this method is computationally no more than a minor amendment to the MLE procedure, we have implemented it here alongside MLE.

Now we describe our results. If we base the results on $n = 15$, as in the ones quoted above when evaluating the standard error of the LSE, we find $\hat{d} = .307$ (standard error .129), $\hat{\beta}_2 = .553$ (.105). The REML results are $\hat{d} = .343$ (.137), $\hat{\beta}_2 = .554$ (.113).

It is natural to ask how sensitive the results are to the choice of $n$, so for comparison, $n = 10$ and $n = 20$ were also tried. The results were:

n=10, MLE: $\hat{d} = .447$ (.183), $\hat{\beta}_2 = .602$ (.124)
n=10, REML: $\hat{d} = .498$ (.199), $\hat{\beta}_2 = .610$ (.136)
n=20, MLE: $\hat{d} = .302$ (.110), $\hat{\beta}_2 = .556$ (.104)
n=20, REML: $\hat{d} = .336$ (.116), $\hat{\beta}_2 = .557$ (.113)

There is very little difference between the results for $n = 15$ and $n = 20$. Those for $n = 10$ are somewhat different, mainly in a larger estimated value for $d$ and corresponding increase in the standard error for $\beta_2$, but even in this case the results provide clear evidence for the statistical significance of $\beta_2$ which, in the context of the climatological interpretation, is probably the most important point. The results are also very consistent with Bloomfield's (1992) results for the same data set, which were based on parametric ARIMA$(p, d, q)$ models.

# 6 Conclusions

It seems worthwhile to combine the estimation of the long-memory and regression parameters into a single estimating equation, rather than keep them separate as in Smith (1993). Although the theoretical efficiency of the least squares estimate of $\beta$ is quite high (Yajima 1988), the example in Section 5 suggests that in practice, it may be possible to achieve quite a substantial reduction in standard error. The results may be contrasted with several previous methods for estimating regression parameters in the presence of correlated errors, including those introduced by Hannan for parametric ARMA models. In particular, the closest approach to ours in the previous literature is that of Bloomfield (1992), based on a parametric fractional ARIMA model, and analyzing the same data set, he obtained numerical results very similar to ours. However we believe there are a number of reasons why our approach may be superior, since there is very little loss of efficiency (for the estimation of $\beta$) in restricting attention to a small number of Fourier frequencies, and the method avoids having to specify a parametric model for $f$ over its entire range.

On the theoretical side, our results are largely conjectural and work is in progress on a more systematic development of the results.

# 7   Acknowledgements

RLS has been supported, in part, by NSF grant DMS/9205112 and by a grant from the EPSRC.

# References

Becker, R.A., Chambers, J.M. and Wilks, A.R. (1988), *The New S Language: A Programming Environment for Data Anaysis and Graphics.* Wadsworth & Brooks/Cole, Pacific Grove, CA.

Bloomfield, P. (1992), Trends in global temperature. *Climatic Change* 21, 1-16.

Bloomfield, P. and Nychka, D. (1992), Climate spectra and detecting climate change. *Climatic Change* 21, 275-287.

Cressie, N. (1993), *Statistics for Spatial Data.* Second edition, John Wiley, New York.

Dahlhaus, R. (1989), Efficient parameter estimation for self-similar processes. *Ann. Statist.* 17, 1749-1766.

Fox, R. and Taqqu, M. (1986), Large sample properties of parameter estimates for strongly dependent stationary Gaussian time series. *Ann. Statist.* 14, 517-532.

Geweke, J. and Porter-Hudak, S. (1983), The estimation and application of long-memory time series models. *J. Time Series Anal.* 4, 221-238.

Granger, C.W.J. and Joyeux, R. (1980), An introduction to long-memory time series models and fractional differencing. *J. Time Series Anal.* 1, 15-29.

Grenander, U. (1954), On estimation of regression coefficients in the case of an autocorrelated disturbance. *Ann. Math. Statist.* 25, 252-272.

Hannan, E.J. (1971), Non-linear time series regression. *J. Appl. Probab.* 8, 767-780.

Hannan, E.J. (1973), The asymptotic theory of linear time-series models. *J. Appl. Probab.* 10, 130-145 (correction on p. 913).

Hansen, J. and Lebedeff, S. (1987), Global trends of measured surface air temperatures. *J. Geophys. Research* D92, 13345-13372.

Hansen, J. and Lebedeff, S. (1988), Global surface air temperatures: update through 1987. *Geophys. Research Letters* 15, 323-326.

Haslett, J. and Raftery, A.E. (1989), Space-time modelling with long- memory dependence: assessing Ireland's wind power resource (with discussion). *Applied Statistics* 38, 1-50.

Hosking, J.R.M. (1981), Fractional differencing. *Biometrika* 68, 165-176.

Hurvich, C.M. and Beltrao, K.I. (1993), Asymptotics for the low frequency ordinates of the periodogram of a long-memory time series. *Journal of Time Series Analysis* 14, 455-472.

Künsch, H. (1986), Discrimination between monotonic trends and long-range dependence. *J. Appl. Probab.* 23, 1025–1030.

Patterson, H.D. and Thompson, R. (1971), Recovery of inter-block information when block sizes are unequal. *Biometrika* 58, 545–554.

Robinson, P.M. (1995a), Log-periodogram regression of time series with long range dependence. *Ann. Statist.* 23, 1048–1072.

Robinson, P.M. (1995b), Gaussian estimation of long range dependence. *Ann. Statist.* 23, 1630-1661.

Smith, R.L. (1993), Long-range dependence and global warming. In *Statistics for the Environment* (V. Barnett and F. Turkman, editors). John Wiley, Chichester, 141-161.

Whittle, P. (1953), Estimation and information in stationary time series. *Ark. Math.* 2, 423–434.

Yajima, Y. (1988), On estimation of a regression model with long-memory stationary errors. *Ann. Statist.* 16, 791-807.

Yajima, Y. (1991), Asymptotic properties of the LSE in a regression model with long-memory stationary errors. *Ann. Statist.* 19, 158-177.

# A frequency domain approach for estimating parameters in point process models

T. Subba Rao
R.E. Chandler

ABSTRACT A great deal of theoretical work has been done in the study of point processes; however, procedures for estimation and inference are much less well developed. This shortfall is partly due to the difficulty of finding likelihood functions for many of the models in use. In this work, collections of sample Fourier coefficients are considered, for which approximate likelihood functions can be derived. These functions lead to the use of the Whittle criterion, which is used extensively as an alternative to exact maximum likelihood estimation in time series analysis, to estimate parameters. The performance of the method, when applied to point processes, is investigated and the method is illustrated with reference to a data set arising in the field of astronomy. For this dataset, a generalization of the Neyman-Scott process is developed; second-order properties of this generalization are presented in the Appendix.

## 1  Introduction

There is a wide range of theoretical models available with which to describe point processes; for some examples, see [CI80], [Dig83]. However, the construction of many of these models (for example, the Neyman-Scott and Bartlett-Lewis clustering models) prohibits the easy formulation of a likelihood function. In practice, such models are of considerable importance as they attempt to represent the mechanics of physical processes; hence their parameters have a readily interpretable meaning.

The difficulty of obtaining likelihood functions has led to the use of various alternative techniques, most of which are effectively methods of moments. Perhaps the most sophisticated method to date is that described by [Dig83]; here, one has to choose a 'test function' which can be estimated from data (such as the spectral density or nearest-neighbour distribution). The parameters are then estimated so as to minimize the Mean Integrated Squared Error between the theoretical value of the chosen function and its estimate. Disadvantages of this method include its dependence on (arbitrarily-chosen) 'tuning constants' which govern, for example, the range of integration for the MISE; and the fact that the chosen test function may not completely characterize the process so the application of such techniques effectively results in a loss of data.

Our approach is to develop an approximate likelihood function based on collections of sample Fourier coefficients. For simplicity, we deal with processes on the real line, although extensions to more general spaces are straightforward.

## 2    Theoretical framework

Suppose that, on the interval $(0, T]$, we observe a point process $N(.)$ which satisfies the following criteria:

1. The process is stationary with mean rate $\lambda$.

2. The complete $k^{th}$-order cumulant density of the process,

$$c_k^c(r_1, \ldots, r_{k-1}) = \text{cum}\{dN(t), dN(t + r_1), \ldots, dN(t + r_{k-1})\},$$

   satisfies

$$\int_{-\infty}^{\infty} \cdots \int_{-\infty}^{\infty} |r_j c_k^c(r_1, \ldots, r_{k-1})| dr_1 \ldots dr_{k-1} < \infty \qquad (1.1)$$

   for all $j = 1, \ldots, k - 1$. This condition implies that there is no long-range dependence in the process. A particular consequence is that the second-order spectral density of the process exists; it is given by

$$f(\omega) = \frac{1}{2\pi} \int_{-\infty}^{\infty} c_2^c(r) e^{-i\omega r} dr = \frac{\lambda}{2\pi} + \frac{1}{2\pi} \int_{-\infty}^{\infty} c_2(r) \cos(\omega r) dr \quad (1.2)$$

   where $c_2(r)$ is the incomplete covariance density which is defined, for convenience, to be zero at the origin — see [Bar63a] for example.

If we now define the sample Fourier coefficients:

$$J_p = \frac{2}{T} \int_0^T e^{i\omega_p t} dN(t) \qquad p = 0, 1, 2, \ldots \qquad (1.3)$$

$$\text{where } \omega_p = \frac{2\pi p}{T} \qquad (1.4)$$

then, because the trigonometric functions form a basis over the range $(0, T]$ observation of the process $N(.)$ on $(0, T]$ is equivalent to observation of the infinite sequence $\{J_p\}$. For $p \neq 0$, we have $E\{J_p\} = 0, Var\{J_p\} \approx 8\pi f(\omega_p)/T$; and we have $E\{J_0\} = 2\lambda, Var\{J_0\} \approx 8\pi f(0)/T$.

Consider now a collection of integer-valued functions $s_\ell(T)$ ($\ell = 1, \ldots, L$) such that, for each $\ell$, $\omega_{s_\ell(T)}$ tends to a constant $\omega^{(\ell)}$ as $T \to \infty$. Then by Theorem 4.2 of [Bri71], the random variables $J_{s_1(T)}, \ldots, J_{s_L(T)}$ are asymptotically jointly Normally distributed and mutually independent. This enables us to write down an approximate likelihood function for the model parameter vector ($\Theta = (\theta_1, \ldots, \theta_K)'$, say), regarding the coefficients $J_{s_1(T)}, \ldots, J_{s_L(T)}$ as data, and suggests that if the sequence $\{J_p\}$ is computed up to a finite value of $p$, say $M$, then $\Theta$ may be estimated by maximizing the function

$$
\begin{aligned}
L(\Theta) \;=\; & \frac{1}{4\pi}\sqrt{\frac{T}{f(0, \Theta)}} \exp\left[\frac{-T(J_0 - 2\lambda(\Theta))^2)}{16\pi f(0, \Theta)}\right] \\
& \times \prod_{p=1}^{M} \frac{T}{8\pi^2 f(\omega_p, \Theta)} \exp\left[\frac{-T|J_p|^2)}{8\pi f(\omega_p, \Theta)}\right]
\end{aligned}
\tag{1.5}
$$

or alternatively, maximizing

$$
\begin{aligned}
\ln L(\Theta) \;=\; & \frac{1}{4\pi}\left(\frac{-T\left(\overline{N} - \lambda(\Theta)\right)^2)}{f(0, \Theta)} - \sum_{p=1}^{M} \frac{I(\omega_p)}{f(\omega_p, \Theta)}\right) \\
& -\left(\frac{1}{2}\ln f(0, \Theta) + \sum_{p=1}^{M} \ln f(\omega_p, \Theta)\right) + \text{constant}, \quad (1.6)
\end{aligned}
$$

where $I(\omega_p) = T|J_p|^2/2$ is the periodogram introduced by [Bar63a], [Bar63b], and $\overline{N} = T^{-1}\int_0^T dN(t)$ is the intuitive estimate of $\lambda$. These cannot be regarded strictly as likelihood functions, because of the restrictions under which asymptotic joint normality of the Fourier coefficients holds. However, asymptotic distribution theory for estimators deriving from (1.6) parallels the standard theory for likelihood estimators ([Ric79]); this enables us to use standard likelihood-based procedures in comparing models, and is one of the major advantages of this method of estimation.

The function (1.6) is the same as that used in Whittle's method for time series parameter estimation ([DY83, Section 5]), and is extensively used to approximate the exact log-likelihood when the data are Gaussian. Note that the motivation here was not to approximate the exact likelihood; rather, to combine the approximate likelihoods of small collections of the Fourier coefficients in a mathematically tractable, and intuitively reasonable, way. [AH73] did in fact apply Whittle's method in a point process context; our contribution is to justify its use and investigate its properties. Note also that [AH73] did not include a term at zero frequency — our experience indicates that it can be vital to incorporate this term, as it is often only here that information on the mean rate of the process can be incorporated.

In practice, maximization of (1.6) can often only be done numerically, which can be computationally expensive if $M$ is large. We now develop a modification for use with large samples, the aim being to reduce computation while preserving as much information as possible and maintaining a likelihood-based approach.

## 2.1 Improving the computational efficiency for large samples

The basic idea behind the proposed modification is that, because the process is assumed to be stationary, then if the time interval $(0, T]$ is subdivided into $S$ intervals of length $T/S$, the $S$ resulting 'subprocesses' are essentially replications of the same process on the interval $(0, T/S]$. These replications are, providing $T/S$ is large enough, approximately independent of each other (recall equation (1.1)). A convenient way to ensure that the replications tend to independence as $T \to \infty$ is to set

$$S = (\lambda T)^\alpha \tag{1.7}$$

where $0 < \alpha < 1$ (the factor $\lambda$ is included to ensure that $S$ remains invariant if the time axis is scaled). In practice, $\lambda$ is not known, and (1.7) is replaced by

$$S = [N^\alpha] . \tag{1.8}$$

If we now treat the $S$ subprocesses as being independent realizations of a process on $(0, T/S]$ then we can compute Fourier coefficients, and a periodogram, for each subprocess. The objective function, corresponding to (1.6), of this new collection of Fourier coefficients can be obtained quite straightforwardly using similar arguments: denote by $\overline{I}(\omega_{S_p})$ the mean of the $S$ periodograms at frequency

$$\omega_{S_p} = \frac{2\pi p S}{T} \qquad (p = 0, 1, 2, \ldots) \tag{1.9}$$

and denote by $N_s$ the number of events in the $s^{th}$ subinterval $(s = 1, \ldots, S)$. Then the required function is

$$
\ln L_S(\Theta) = \frac{-[T/S]}{4\pi f(0, \Theta)} \sum_{s=1}^{S} \left( \frac{N_s}{[T/S]} - \lambda(\Theta) \right)^2 - \frac{S}{4\pi} \sum_{p=1}^{M} \frac{\overline{I}(\omega_{S_p})}{f(\omega_{S_p}, \Theta)}
$$
$$
- S \left[ \frac{1}{2} \ln f(0) + \sum_{p=1}^{M} \ln f(\omega_{S_p}, \Theta) \right] + \text{constant}. \tag{1.10}
$$

Note that, because the individual replications are fully determined by their sequences of Fourier coefficients, we have lost very little information about the original process through this procedure. The only information which has been lost is the order in which the replications occurred; however, given that we are assuming them to be almost independent this need not be a cause for concern.

It is interesting to note that this modified procedure is in fact equivalent to smoothing the periodogram using Bartlett's window (see [Pri81, pp.439–40]) and working with a coarser grid of frequencies. This suggests the use of other smoothed spectral estimates in place of $\overline{I}(\omega_{S_p})$ in (1.10); however, this matter will not be pursued here. The use of $\overline{I}(\omega_{S_p})$ enables us to preserve as much information as possible while maintaining the likelihood-based interpretation of (1.10).

# 3   Performance of the Method

We now outline some of the large-sample properties of estimators obtained by this method. For estimators resulting from (1.6), the arguments involved are the same as those presented in [Ric79]; and for the 'large-sample' method (1.10), trivial modifications of these arguments are required (for details, see [Cha95]). For this reason, we do not go into detail regarding the results. The asymptotics hold if the estimation is carried out over all frequencies in a finite union of intervals. In our case, if the frequency range is $[0, \omega_{MAX}]$ then we must have $M = [\omega_{MAX}T/2\pi]$ for the upper limit in the sum (1.6).

In what follows, we will denote the true value of the parameter vector by $\Theta_0$, and its estimate using (1.6) by $\Theta_I$. Following [Ric79], we have

**Lemma 3.1** *Providing*

  1. *the set $\Omega$ of possible $\Theta$-values is compact;*

  2. *the spectral density $f(\omega, \Theta)$ is continuous with respect to $\Theta$ in $\Omega$, and is strictly greater than zero throughout $[0, \omega_{MAX}] \times \Omega$;*

  3. *for any $\Theta \neq \Theta_0$, $f(\omega, \Theta)$ and $f(\omega, \Theta_0)$ differ on a set of non-zero measure in $[0, \omega_{MAX}]$*

*then $\Theta_I$ is a consistent estimator of $\Theta_0$.*

If we now define $\phi^{(i)}(\Theta) = \partial \ln L_S(\Theta)/\partial\theta_i$ $(i = 1, \ldots, K)$, and define $\Phi(\Theta) = \left(\phi^{(1)}(\Theta), \ldots, \phi^{(K)}(\Theta)\right)'$, then this consistency result tells us that we can write

$$\Phi(\Theta) = \mathbf{A}\left(\Theta_I - \Theta_0\right) \qquad (1.11)$$

where $\mathbf{A}$ is a $K \times K$ matrix whose $(i, j)^{th}$ element is

$$A_{ij} = \left.\frac{\partial^2 \ln L_S(\Theta)}{\partial\theta_i\partial\theta_j}\right|_{\Theta=\Theta'} \qquad (1.12)$$

and $\Theta'$ is on the line segment joining $\Theta_I$ and $\Theta_0$. The expansion (1.11) holds with probability tending to 1 as $T \to \infty$, and enables us to examine the behaviour of the error

$$\Theta_I - \Theta_0 = \mathbf{A}^{-1}\Phi(\Theta)$$

by developing large-sample approximations for the elements of the matrix $\mathbf{A}$: for large $T$, $A_{ij}$ can be approximated by the matrix $T\alpha/2\pi$, where $\alpha$ has $(i,j)^{th}$ element

$$-\int_{\omega=0}^{\omega_{MAX}} \left(\frac{\partial}{\partial\theta_i}\ln f(\omega;\Theta)\right)\left(\frac{\partial}{\partial\theta_j}\ln f(\omega;\Theta)\right)d\omega \ . \tag{1.13}$$

We are now in a position to state the following theorem, which is effectively Theorem 3.1 of [Ric79]:

**Theorem 3.1** *Under the conditions of Lemma 3.1, and providing that:*

1. *the matrix $\alpha$ is nonsingular;*

2. *$f(., \Theta)$ is twice $\Theta$-differentiable in the neighbourhood of $\Theta_0$, and all its first and second $\Theta$-derivatives are bounded in this neighbourhood;*

3. *the first and second $\Theta$-derivatives of $\lambda(\Theta)$ exist and are bounded in the neighbourhood of $\Theta_0$;*

*then the estimation error $\Theta_I - \Theta_0$ is $O(T^{-1/2})$ with probability tending to 1 as $T \to \infty$. Furthermore, the quantity $T^{1/2}(\Theta_I - \Theta_0)'$ is approximately multivariate normally distributed, with mean tending to zero and variance-covariance matrix depending on the second- and fourth-order cumulant spectra of the process.*

An analogue of this theorem holds if the 'large-sample' method (**1.10**) is used. The main differences are: firstly that the error term is $O(S/T) + O(S^{-1/2})$, so that convergence of the estimator is slower in this case; and secondly that expressions for the variance-covariance matrix of the estimator are complicated by the smoothing of the periodograms in the objective function. These expressions arise from the approximation of the error by functions of the periodograms; and they can be deduced from the sampling theory of smoothed periodograms (see [Pri81, Section 6.2.3] for example).

# 4   Practical Considerations

In Section 1, the 'test function' method of estimating parameters was criticized on the grounds that it effectively results in a loss of data. It may be argued that, since the new method involves computation of only $M$ Fourier coefficients out of an infinite sequence, this also results in a loss of data. However, for the class of models which we have been considering, all of the 'information' in the spectral density is concentrated towards the origin ie. at low frequencies (in the sense that the spectral density approaches a limiting value as $\omega \to \infty$). Hence we would expect that our parameter estimates will vary very little once a 'threshold' value of $M$ is reached. Simulation studies of various models verify this; as a rough guide, the estimates seem to settle down into stable behaviour when $M$ is taken to be around 3 times the effective realization length ($T$ if (1.6) is used, $T/S$ if (1.10) is used). To determine what is an appropriate value for $M$ it is suggested that, for a given data set, the estimation procedure is repeated using several different values of $M$; it should then be straightforward to examine the results and determine when the estimates have stabilized.

Although Theorem 3.1 yields, in principle, explicit distributional results for the estimators, its range of applicability is limited by the dependence of the limiting distribution on the trispectrum of the model; for many point process models, the formula for the trispectrum is difficult to derive. This means that, in general, it is not possible to develop hypothesis tests and confidence intervals for the estimated parameters. However, the formulation of the problem in Maximum Likelihood terms allows the use of standard techniques such as likelihood ratio tests and the Akaike Information Criterion (AIC) to distinguish between models. In this work, we use the AIC for illustration purposes. This is defined as

$$AIC(\Theta) = -2\ln L(\Theta) + 2k \tag{1.14}$$

where $k$ is the number of parameters in $\Theta$. The best choice of model, from the point of view of AIC, is that which minimizes (1.14). In the present context, AIC measures the distance between the estimated joint probability density function of the sample Fourier coefficients and its true value — see [Aka74, Section IV]. Hence it is important that, if AIC is to be used to compare two models, the same value of $M$ should be used for both models and, if (1.10) is used, the same value of $\alpha$ in (1.8) should also be used. In practice therefore, the value of $M$ should be chosen so that the estimation procedure has stabilised for both models under consideration.

## 5    Analysis of data

The above results are now illustrated with reference to a data set from astronomy. The data consists of the arrival times (in seconds) of 329 X-ray pulses at a recorder; a graph of the cumulative process $N(t)$ is presented as Figure 1. A preliminary analysis of this data may be found in [SR92]. The first event occurred after 29890s, and the last after 215800s. The estimated intensity for this process is thus $329/215800 = 1.525 \times 10^{-3}$ events/sec. The resolution of the data was, in absolute terms, low; arrival times were presented in the form $a \times 10^b$, where $a \in [0, 1)$ and has 4 decimal places, and

$b$ is an integer. This meant that the resolution changed partway through the period of observation, with the first 133 arrival times being rounded off to the nearest ten seconds and the remainder to the nearest hundred. Because of this rounding, some pairs of events were recorded as being coincident — however, we will shortly see that the rounding errors are in fact small compared with the scale at which the process operates, so they are unlikely to cause too many problems. From this point on in the analysis, the time axis was scaled so that the mean intensity was 1, to facilitate computation.

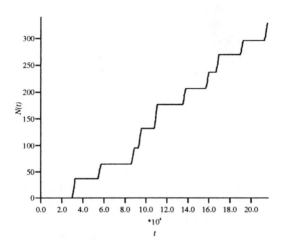

FIGURE 1. Plot of the cumulative process $N(t)$, for scaled X-ray pulse data

It is apparent from Figure 1 that the data consists of 10 compactly clustered groups widely separated from each other; hence a clustering model of some sort seems appropriate. To try and ascertain a suitable class of models, the spectral density of the process was estimated by smoothing the raw periodogram using the Tukey-Hanning spectral window ([Pri81, p.443]). The result, together with the unsmoothed periodogram for the data, is shown in Figure 2. The frequency range considered is $(0, \pi)$; although, in terms of the unscaled frequency units, this corresponds to only a tiny range, examination of the plot shows that there is little of interest at frequencies larger than $3\pi/8$ — thus, any structure in the process is at a very large scale. Another point is that the true spectral density of the process may have a maximum away from the origin, indicating some sort of 'quasi-periodicity' in the process.

Two different clustering models were fitted to the data set, using (1.6) (the number of events here is not so large that computational efficency is an issue; and besides, it is clear that in this case adjacent segments of the time axis are far from independent, which is a requirement if (1.10) is to be used). The first model was a Neyman-Scott model ([CI80, Section

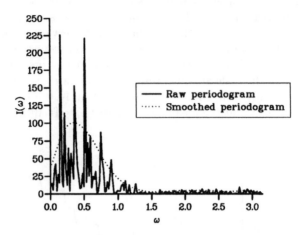

FIGURE 2. Scaled X-ray pulse data — raw and smoothed periodograms

3.4]), and the second was of generalized Neyman-Scott form, such that 'parent' events, instead of being Poisson, form a renewal process whose inter-event times have a Gamma distribution with parameters $(\nu, \ell)$ (such a choice of parent process should give rise to the kind of 'quasi-periodicity' in the resulting process which has been observed here). For both models, the number of offspring per parent was taken to be Poisson with mean $\mu$, the dispersal distribution of offspring about their parents was $N(0, \sigma^2)$ and parents were not included in the final process. The second-order properties of the generalized Neyman-Scott model are presented in the Appendix; for a more extensive discussion, see [Cha93].

For the basic Neyman-Scott model, the theoretical spectral density is given by

$$f(\omega) = \frac{\rho\mu}{2\pi} + \frac{\rho\mu^2}{2\pi}e^{-(\sigma\omega)^2} \tag{1.15}$$

where $\rho$ is the rate of the parent process. In this case, numerical methods must be used for the maximization of (1.6); it is straightforward to obtain first derivatives with respect to $\rho$, $\mu$ and $\sigma^2$, and the NAG routine E04KCF (maximization using first derivatives) was employed.

For the generalized model, the spectral density is given (see the Appendix) by

$$f(\omega) = \frac{\mu\ell}{2\pi\nu} + \frac{\mu^2 e^{-(\sigma\omega)^2}}{2\pi}\left[\frac{\ell}{\nu} + 2\pi f_P^*(\omega)\right] \tag{1.16}$$

where $f_P^*$ is the incomplete spectral density of a gamma $(\nu, \ell)$ renewal process ie. the real part of

$$\frac{\ell}{\pi\nu}\left[\left((1 - \frac{i\omega}{\ell})^\nu - 1\right)^{-1}\right] .$$

The NAG subroutine **EO4JAF** was used to maximize **(1.6)** in the absence of first derivatives.

The starting values for the numerical maximization procedures were obtained by inspection of the data. For both models, the estimates were calculated for $M = 700, 750, \ldots, 1000$, to determine whether the procedure had stabilized. For each value of $M$, the estimates obtained from the previous value were used as starting values in the optimization algorithm.

## 5.1   Results of the estimation procedure

For both models, it was apparent that the estimates had stabilized for the range of $M$-values considered here. For $M = 1000$, the estimates were as follows:

| Model | Parameter estimates | AIC value |
|---|---|---|
| Basic Neyman-Scott model | $\hat{\rho} = 0.0317$ <br> $\hat{\mu} = 33.1381$ <br> $\hat{\sigma}^2 = 2.3713$ | -4098.074 |
| Generalized clustering model | $\hat{\nu} = 7.3761$ <br> $\hat{\ell} = 0.1960$ <br> $\hat{\mu} = 36.5629$ <br> $\hat{\sigma}^2 = 2.4460$ | -4102.349 |

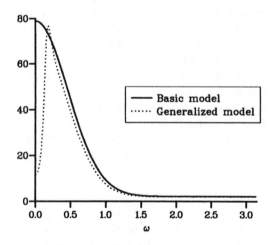

FIGURE 3. Theoretical spectra for both models fitted to the scaled X-ray pulse data

Comparison of the AIC values indicates that the generalized clustering model fits the data better than the Neyman-Scott model — this is perhaps to be expected, given that the generalized model was selected specifically to describe this data set. There are notable points of similarity between the parameter estimates for the two models. Firstly, the estimate of $\sigma^2$, the variance of the dispersal distribution, is very similar for the two models; and secondly, the estimated rates of the two models are approximately the same ($\hat{\rho}\hat{\mu} = 1.050$ for the first model, and $\hat{\mu}\hat{\ell}/\hat{\nu} = 1.051$ for the second).

For the purposes of illustration, graphs of the theoretical spectra for both of the models fitted to this data set are provided in Figure 3; the vertical scaling, by a factor of $4\pi$, enables comparison with Figure 2. It is clear from the Figures that the second model fits the data better than the first.

# 6　Discussion

The estimation procedure used here has been shown to produce consistent estimates; in addition, it is relatively simple to apply in practice. The parameters of any model can be estimated from data, providing the spectral density of that model is known. The main advantage of this method is that, because it is essentially likelihood-based, it allows different models to be compared using established, and objective, techniques. In addition, because the method uses all first- and second-order properties of the data, the data is used efficiently compared with, say, methods of moments.

It is of interest to compare the performance of this method with that of other methods of parameter estimation, in particular the 'test function' method. Simulations of that method have not been undertaken here because of the computational cost; however, [Dig83, Section 5.3.2] presents simulation results for such a method, estimating the parameters in a Neyman-Scott process on the plane (although his choice of test function effectively reduces the problem to one on the real line). We have simulated such a process on **R**, the construction being identical to that presented in [Dig83] ie. the number of offspring per parent has a Poisson distribution, the dispersal distribution of offspring about their parents is Normal with zero mean, and parents are not included in the final process. The results of our simulations are very similar to those reported by Diggle, so it is likely that the methods perform comparably.

*Acknowledgments:* RC was supported by the Science and Engineering Research Council during the period of this research. TSR is grateful to NATO for awarding a collaborative research grant to work on the applications of time series analysis to astronomy.

The authors are indebted to Julian Krolik (Department of Astronomy, John Hopkins University, Baltimore, Maryland, USA) for supplying the dataset used in section 5.

# Appendix: A Generalized Neyman-Scott Model

The Neyman-Scott and Bartlett-Lewis clustering models both have an intuitively appealing structure, but are restricted by the constraint that the parent process be Poisson (in particular, the spectral density of both models is always maximum at the origin). It is clear that if this restriction is removed, the applicability of the models will widen considerably. A few general theoretical results exist for such generalized processes — see, for example, [CI80, Section 3.4]. We here derive the second-order properties of a process where the Neyman-Scott clustering mechanism operates — for the sake of simplicity, we consider a process, on $\mathbf{R}$, where the parent events are *not included*. The process may be defined in the following manner:

1. There exists a point process, $P$ say, on $\mathbf{R}$, which we shall call the parent process. $P$ is assumed to be stationary with rate $\rho$, incomplete covariance density $\rho_2(.)$ and incomplete spectral density $f_P^*(.)$.

2. Associated with each event of $P$ is a cluster of 'offspring'. The number of offspring per parent is a random variable $Z$, realized independently and identically for each parent; in addition, each offspring is dispersed relative to its parent according to some continuous probability density $h(.)$ — again, independently and identically for each offspring.

3. The final process, $N$ say, is formed by superimposing all of the clusters arising from different parents. The parent events themselves are not included.

The general form of the probability generating functional for this process is known ([CI80, equation **(3.50)**]); however, it is easier to derive second-order properties directly rather than via the generating functionals. The rate of the process is clearly

$$\lambda = \rho E\{Z\} ; \tag{1.17}$$

for the incomplete covariance density we have, following standard arguments (the derivation may be found in full in [Cha93])

$$\begin{aligned} c_2(r) &= \rho h_2'(r) E\{Z(Z-1)\} + \\ &\quad (E\{Z\})^2 \int \int h(-p)h(r-(p+q))\rho_2(q)dpdq \end{aligned}$$

where $h_2'(r) = \int h(u)h(r+u)du$.

The use of $\rho_2(.)$ to denote the *incomplete* covariance density of the parent process eliminates the need to worry about the integral where $q = 0$. Notice also that when the parent process is Poisson, $\rho_2$ is identically zero and the above expression reduces to the form for a standard Neyman-Scott process without parents.

The incomplete spectral density for the process is now obtainable as

$$f_2^*(\omega) = \frac{|\phi_h(\omega)|^2}{2\pi} \left[ \rho E\{Z(Z-1)\} + (E\{Z\})^2 (2\pi f_P^*(\omega)) \right]$$

— note once again that $f_P^*(\omega)$ is the *incomplete* spectral density of the parent process.

The form of this spectral density indicates that, with a suitable choice of parent process $P$, we can incorporate the Neyman-Scott clustering mechanism into a model whose spectral density has a maximum away from the origin. In addition, the derivation of properties in terms of those of the parent process renders computation feasible.

# 7  References

[AH73]   L. Adamopoulos and A.G. Hawkes.  Cluster models for earth-quakes — regional comparisons. *ISI Conference, Vienna*, 1973.

[Aka74]  H. Akaike.  A new look at the statistical model identification. *IEEE Transactions on Automatic Control*, AC-19(6):716–723, 1974.

[Bar63a] M.S. Bartlett. The spectral analysis of point processes. *Journal of the Royal Statistical Society, Series B*, 25:264–296, 1963.

[Bar63b] M.S. Bartlett.  Statistical estimation of density functions. *Sankhyà, Series A*, 25:245–254, 1963.

[Bri71]  D.R. Brillinger. The spectral analysis of stationary interval functions. In *Proceedings, Sixth Berkeley Symposium*, pages 483–513, 1971.

[Cha93]  R.E. Chandler. *The Application of Spectral Methods to the Analysis of Point Process Data*. PhD thesis, University of Manchester Institute of Science and Technology, 1993.

[Cha95]  R.E Chandler.  A spectral method for estimating parameters in rainfall models.  Technical Report 142, Department of Statistical Science, University College London, 1995. http://www.ucl.ac.uk/Stats/research/abstracts.html .

[CI80]   D.R. Cox and V. Isham. *Point Processes*. Chapman and Hall, London, 1980.

[Dig83]  P. J. Diggle. *Statistical Analysis of Spatial Point Patterns*. Academic Press, 1983.

[DY83]   K.O. Dzhaparidze and A.M. Yaglom. Spectrum parameter esti-
         mation in time series analysis. In P.R. Krishnaiah, editor, *De-
         velopments in Statistics, Volume 4*, pages 1–96. Academic Press,
         1983.

[Pri81]  M.B. Priestley. *Spectral Analysis and Time Series*. Academic
         Press, 1981.

[Ric79]  J. Rice. On the estimation of the parameters of a power spectrum.
         *Journal of Multivariate Analysis*, 9:378–392, 1979.

[SR92]   T. Subba Rao. Discussion on papers by J.H. Krolik, P. Hertz
         and J.P. Norris. In E. Feigelson and G.J. Babu, editors, *Statis-
         tical Challenges in Modern Astronomy*, pages 398–409. Springer-
         Verlag, New York, 1992.

University of Manchester Institute of Science and
Technology

University College, London

# HIGHER ORDER ASYMPTOTIC THEORY FOR TESTS AND STUDENTIZED STATISTICS IN TIME SERIES

Masanobu Taniguchi

Osaka University

ABSTRACT  Let $\{X_t\}$ be a Gaussian ARMA process with spectral density $f_\theta(\lambda)$, where, $\theta$ is an unknown parameter. We consider to test a simple hypothesis $H : \theta = \theta_0$ against the alternative $A : \theta \neq \theta_0$. For this testing problem we introduce a class of tests $S$ , which contains many famous tests. Then it is shown that if $T \in S$ is modified to be second-order asymptotically unbiased , it is second-order asymptotically most powerful . Furthermore we derive the third-order asymptotic expansion of the distribution of $T \in S$ under a sequence of local alternative. Using this result we elucidate various third-order asymptotic properties of $T \in S$ ( e.g., Bartlett's adjustment, third-order asymptotically most powerful properties ). Some numerical results are given to confirm the theoretical results. We also discuss the higher order asymptotics of studentized statistics. The second-order Edgeworth expansions of studentized statistics are compared with those of non-studentized ones. Then a duality of them is illuminated in terms of dual connection.

## 1  Introduction

In multivariate analysis, the asymptotic expansions of the distributions of various test statistics have been investigated in detail ( e.g., Peer[Pe71], Hayakawa and Puri[HP85]). In time series analysis, Whittle[W51] seems to be the first to suggest the use of the Edgeworth expansion of distribution of statistics, though he did not conduct concrete analysis. Phillips[Ph77] gave the Edgeworth expansion of the t-statistic for tests of the coefficient of an AR(1) process. For that process, Tanaka[Tana82] gave the asymptotic expansion of the distributions of the likelihood ratio test, Wald test and Lagrange multiplier test under the null as well as alternative hypotheses.

In Section 2 we consider a Gaussian ARMA process with spectral density $f_\theta(\lambda)$, $\theta \in \Theta \subset \mathbf{R}^1$. The problem considered is that of testing a simple hypothesis $H : \theta = \theta_0$ against the alternative $A : \theta \neq \theta_0$. For this problem we propose a class of tests $S$ which contains the likelihood ratio (LR), Wald (W), modified Wald (MW) and Rao (R) tests as special cases. The second-order $\chi^2$ type asymptotic expansion for $T \in S$ under the sequence of alternatives $A_n : \theta = \theta_0 + \epsilon/\sqrt{n}, \ \epsilon > 0$, is given. Using this asymptotic expansion we can compare the second-order local powers of the LR, W, MW and R tests. Then it is shown that none of them is uniformly superior. However, if we modify them to be second-order asymptotically unbiased we can show that their second-order local powers are identical. Some numerical results are given to confirm the theoretical results.

In Section 3, for a general stochastic process, we derive the third-order asymptotic expansion of the distribution of $T \in S$ under $A_n$. Based on this result we elucidate various third-order asymptotics of $T \in S$ ( e.g., Bartlett's adjustments, third-order asymptotically most powerful properties ). These results are very general, and can be applied to the non-i.i.d. case, multivariate analysis and time series analysis. A numerical study for Bartlett's adjustments is given.

In Section 4 we discuss the second-order asymptotics of studentized statistics. The second-order Edgeworth expansions of studentized statistics are compared with those of non-studentized ones. Then a duality of them is illuminated in terms of dual connection. For an ARMA process we show an interesting duality between the second-order Edgeworth expansion for the studentized MLE of the AR-parameter and that for the non-studentized MLE of the MA-parameter. It seems that the higher order asymptotic theory for studentization is related to dual differential geometry.

# 2 Second-order asymptotic theory for testing problems

Suppose that $\{X_t; \ t = 0, \pm 1, \pm 2, ...\}$ is a Gaussian ARMA process with mean 0 and spectral density $f_\theta(\lambda)$ ($\theta \in \Theta \subset \mathbf{R}^1$) which is assumed differentiable with respect to $\theta$ up to necessary order. For the observation vector $\mathbf{X}_n = (X_1, ..., X_n)'$, the likelihood function is

$$L_n(\theta) = (2\pi)^{-n/2} |\Sigma_n|^{-1/2} exp\{-(1/2)\mathbf{X}'_n \Sigma_n^{-1} \mathbf{X}_n\},$$

where $\Sigma_n$ is the covariance matrix of $\mathbf{X}_n$. We denote the maximum likelihood estimator of $\theta$ by $\hat{\theta}_{ML}$.

We consider here the test of the parameter $\theta$ of the ARMA process $\{X_t\}$. The problem considered is that of testing a simple hypothesis $H : \ \theta = \theta_0$ against the alternative $A : \ \theta \neq \theta_0$. Let

$$Z_1 = Z_1(\theta) = \frac{1}{\sqrt{n}} \frac{\partial}{\partial \theta} \log L_n(\theta),$$

$$Z_2 = Z_2(\theta) = \frac{1}{\sqrt{n}} \{ \frac{\partial^2}{\partial \theta^2} \log L_n(\theta) - E_\theta \frac{\partial^2}{\partial \theta^2} \log L_n(\theta) \},$$

$$g(\theta) = \frac{1}{4\pi} \int_{-\pi}^{\pi} \{ \frac{\partial}{\partial \theta} \log f_\theta(\lambda) \}^2 d\lambda.$$

The following test statistics

$$LR = 2[\log L_n(\hat{\theta}_{ML}) - \log L_n(\theta_0)],$$

$$W = n(\hat{\theta}_{ML} - \theta_0)^2 g(\hat{\theta}_{ML}),$$

$$MW = n(\hat{\theta}_{ML} - \theta_0)^2 g(\theta_0),$$

$$R = Z_1(\theta_0)^2 g(\theta_0)^{-1},$$

are called the likelihood ratio (LR) test, the Wald (W) test, the modified Wald (MW) test and the Rao (R) test, respectively. For our test problem we here introduce the following class of test statistics which includes LR, W, MW and R ;

$$S = \{T| \; T = Z_1^2/g(\theta_0) + \frac{1}{\sqrt{n}}(\frac{a_1 Z_1^2 Z_2}{g(\theta_0)} + \frac{a_2 Z_1^3}{g(\theta_0)^{3/2}}) + o_p(n^{-1/2}),$$

under $H$, where $a_1$ and $a_2$ are nonrandom constants $\}$.

The next theorem enables us to evaluate the second-order local power of $T \in S$.

**Theorem 2.1**(Taniguchi[Tani91a]).
Under the local alternative hypothesis $A_n : \; \theta = \theta_0 + \epsilon/\sqrt{n}$ $(\epsilon > 0)$, the distribution function of $T \in S$ has the asymptotic expansion

$$P^n_{\theta_0 + \epsilon/\sqrt{n}}[T \le x] = P[\chi_1^2(\delta) \le x] + \frac{1}{\sqrt{n}} \sum_{j=0}^{3} B_j^{(T)} P[\chi_{1+2j}^2(\delta) \le x] + o(n^{-1/2}),$$

where the $B_j^{(T)}$ are expressed explicitly in $f_\theta(\lambda), a_1, a_2$ , $\epsilon$ and $\delta^2 = g(\theta_0)\epsilon^2/2$, and $\chi_j^2(\delta)$ is a noncentral $\chi^2$ random variable with j degrees of freedom and noncentrality parameter $\delta^2$.

From this theorem we can compare the second-order difference between the local powers of $T_1, T_2 \in S$ in terms of

$$\lim_{n \to \infty} \; \sqrt{n}[ \; P^n_{\theta_0 + \epsilon/\sqrt{n}}\{T_1 > x\} - P^n_{\theta_0 + \epsilon/\sqrt{n}}\{T_2 > x\}]$$

$$= \sum_{j=0}^{3}\{B_j^{(T_1)} - B_j^{(T_2)}\}P[\chi_{1+2j}^2(\delta) > x]$$

$$= r(T_1, T_2), \quad \text{(say)}.$$

But no test can be uniformly more powerful in $S$. To see this we give an example of the comparison between LR and R.

**Example 2.1.** Suppose that $\{X_t\}$ has the spectral density of ARMA(1,1) type ;

$$f_\theta(\lambda) = \frac{\sigma^2}{2\pi} \frac{|1 - \psi e^{i\lambda}|^2}{|1 - \rho e^{i\lambda}|^2}.$$

(1) In the case of $\theta_0 = \psi$, if $\psi > 0$, then $r(LR, R) > 0$, which implies that LR is more powerful than R ,and if $\psi \leq 0$, R is more powerful than LR.
(2) In the case of $\theta_0 = \rho$, if $\rho > 0$, then $r(LR, R) < 0$, which implies that R is more powerful than LR ,and if $\rho \leq 0$, LR is more powerful than R.

Next we show that an appropriate modification of $T \in S$ leads to a unified result. If $T \in S$ satisfies

$$\frac{\partial}{\partial \epsilon} P^n_{\theta_0 + \epsilon/\sqrt{n}}[T > x]|_{\epsilon=0} = o(n^{-1/2})$$

it is said to be second-order asymptotically unbiased.

**Theorem 2.2** (Taniguchi[Tani91a]).
(1) For $T \in S$, we can choose a smooth function $m(\theta)$ so that $T^* = m(\hat{\theta}_{ML})T \in S$ is second-order asymptotically unbiased.
(2) If $T \in S$ is second-order asymptotically unbiased, then it is most powerful up to order $n^{-1/2}$ in the class of second-order unbiased tests.

We give a numerical result.

**Example 2.2.** Consider the following AR(1)-model,

$$X_t = \theta X_{t-1} + \epsilon_t, \quad |\theta| < 1, \tag{2.1}$$

where $\{\epsilon_t\}$ is a sequence of i.i.d.N(0,1) random variables. The testing problem is $H : \theta = \theta_0$ against $A : \theta \neq \theta_0$. In this case LR is second-order asymptotically unbiased, but R is not so. From the modification procedure given in Theorem 2.2, it follows that

$$R^* = \{1 + (\hat{\theta}_{ML} - \theta_0)\frac{-2\theta_0}{1 - \theta_0^2}\}R$$

is second-order unbiased. Suppose that a stretch $(X_1, ..., X_{200})'$ is observed from (2.1) under the sequence of alternative $A_n : \theta = \theta_0 + \epsilon/\sqrt{g(\theta_0)n}$. For each $\epsilon = 2.5, 3.0, 3.2$, we computed the probabilities ;

$$PR = P^n_{\theta_0 + \epsilon/\sqrt{g(\theta_0)n}} [R > x_{0.05}],$$

$$PR^* = P^n_{\theta_0 + \epsilon/\sqrt{g(\theta_0)n}} [R^* > x_{0.05}],$$

$$PLR = P^n_{\theta_0 + \epsilon/\sqrt{g(\theta_0)n}} [LR > x_{0.05}],$$

by 1000 trials simulation for $\theta_0 = -0.9 \ (0.2) - 0.3$, where $x_{0.05}$ is the level 0.05 point of $\chi^2$. Tables 2.1 -2.3 give the values of them for $\epsilon = 2.5, 3.0, 3.2$, respectively.

Table 2.1.

$\epsilon = 2.5$

| $\theta_0$ | $PR$ | $PR^*$ | $PLR$ |
|------|------|------|------|
| $-0.9$ | 0.189 | 0.532 | 0.562 |
| $-0.7$ | 0.475 | 0.624 | 0.614 |
| $-0.5$ | 0.562 | 0.633 | 0.632 |
| $-0.3$ | 0.626 | 0.658 | 0.661 |

Table 2.2.

$\epsilon = 3.0$

| $\theta_0$ | $PR$ | $PR^*$ | $PLR$ |
|------|------|------|------|
| $-0.9$ | 0.295 | 0.717 | 0.698 |
| $-0.7$ | 0.641 | 0.781 | 0.774 |
| $-0.5$ | 0.745 | 0.811 | 0.814 |
| $-0.3$ | 0.788 | 0.817 | 0.815 |

Table 2.3.

$\epsilon = 3.2$

| $\theta_0$ | $PR$ | $PR^*$ | $PLR$ |
|------|------|------|------|
| $-0.9$ | 0.345 | 0.778 | 0.754 |
| $-0.7$ | 0.706 | 0.825 | 0.825 |
| $-0.5$ | 0.805 | 0.853 | 0.854 |
| $-0.3$ | 0.838 | 0.863 | 0.866 |

Tables 2.1 - 2.3 show that LR is more powerful than R if $\theta_0 < 0$, and that the modified test $R^*$ is as good as LR. Thus the results confirm Theorem 2.2, and show that the higher order asymptotic theory works very effectively.

# 3 Third-order asymptotic theory for testing problems

In this section we extend the results in the previous section to the case where (i) the n-consecutive observations are from a general vector-valued stochastic process and (ii) the approximation of the distribution is given by using the third-order asymptotic expansion. Therefore our results can be applied to regression analysis , multivariate analysis and time series analysis .

Suppose that $\mathbf{X}_n = (X_1, ..., X_n)$ is a collection of m-dimensional random vectors forming a stochastic process. Let $p_{n,\theta}(\mathbf{x}_n)$, $\mathbf{x}_n \in \mathbf{R}^{mn}$, be the probability density function of $\mathbf{X}_n$ depending on $\theta \in \Theta$, where $\Theta$ is an open set of

$\mathbf{R}^1$. We require the following assumption.

**Assumption 3.1.**

(1) $p_{n,\theta}(\mathbf{x}_n)$ is continuously differentiable with respect to $\theta \in \Theta$ up to necessary order.

(2) The partial derivative $\partial/\partial\theta$ and the expectation $E_\theta$ with respect to $p_{n,\theta}(\mathbf{x}_n)$ are interchangeable.

(3) For an appropriate sequence $\{c_n\}$ satisfying $c_n \to \infty$ as $n \to \infty$, the asymptotic cumulants of

$$Z_i \ = \ c_n^{-1}\{\frac{\partial^i}{\partial\theta^i}\log p_{n,\theta}(\mathbf{X}_n) \ - \ E_\theta\frac{\partial^i}{\partial\theta^i}\log p_{n,\theta}(\mathbf{X}_n)\}, \quad (i=1,2,3),$$

possess asymptotic expansions of the form

$$cum_\theta\{Z_i, Z_j\} \ = \ \kappa_{ij}^{(1)} \ + \ c_n^{-2}\kappa_{ij}^{(2)} \ + \ o(c_n^{-2}),$$

$$cum_\theta\{Z_i, Z_j, Z_k\} \ = \ c_n^{-1}\kappa_{ijk}^{(1)} \ + \ o(c_n^{-2}),$$

$$cum_\theta\{Z_i, Z_j, Z_k, Z_l\} \ = \ c_n^{-2}\kappa_{ijkl}^{(1)} \ + \ o(c_n^{-2}),$$

i,j,k,l = 1,2,3, and the $J$th-order $(J \geq 5)$ cumulants satisfy

$$cum_\theta^{(J)}\{Z_{i_1}, ..., Z_{i_J}\} \ = \ O(c_n^{-J+2}),$$

where $i_1, ..., i_J \in \{1, 2, 3\}$.

Henceforth we use the following notations ; $g = \kappa_{11}^{(1)}$, $J = \kappa_{12}^{(1)}$, $L = \kappa_{13}^{(1)}$, $M = \kappa_{22}^{(1)}$, $K = \kappa_{111}^{(1)}$, $N = \kappa_{112}^{(1)}$, $H = \kappa_{1111}^{(1)}$ and $\Delta = \kappa_{11}^{(2)}$, because the resulting formula in the asymptotic expansion can be expressed solely in terms of them.

Consider the transformation

$$W_1 \ = \ Z_1/\sqrt{g},$$

$$W_2 \ = \ Z_2 \ - \ Jg^{-1}Z_1,$$

$$W_3 \ = \ Z_3 \ - \ Lg^{-1}Z_1.$$

For the testing problem $H : \theta = \theta_0$ against $A : \theta \neq \theta_0$, we introduce the following class of tests:

$$\mathcal{S}' \ = \ \{T|\ T = W_1^2 \ + \ c_n^{-1}(a_1W_1^2W_2 + a_2W_1^3)$$

$$+ \ c_n^{-2}(b_1W_1^2 + b_2W_1^2W_2^2 + b_3W_1^4 + b_4W_1^3W_2 + b_5W_1^3W_3) \ + \ o_p(c_n^{-2}),$$

under H, where $a_i(i = 1, 2)$ and $b_i(i = 1, ..., 5)$ are nonrandom constants $\}$.

This class $S'$ is very natural because the four famous tests LR, W, MW and R belong to $S'$ (see Taniguchi(1991b)).

**Theorem 3.1** (Taniguchi[Tani91b]).
The distribution function of $T \in S'$ under a sequence of local alternative $\theta = \theta_0 + \epsilon/c_n$ has the asymptotic expansion

$$P_{n,\theta_0+\epsilon/\sqrt{n}}[\, T \leq x \,]$$

$$= P[\chi_1^2(\delta) \leq x\,] + c_n^{-1} \sum_{j=0}^{3} B_j^{(T)} P[\chi_{1+2j}^2(\delta) \leq x\,]$$

$$+ c_n^{-2} \sum_{j=0}^{6} C_j^{(T)} P[\, \chi_{1+2j}^2(\delta) \leq x\,] + o(c_n^{-2}),$$

where $\delta^2 = g\epsilon^2/2$, and the coefficients $B_j^{(T)}$ and $C_j^{(T)}$ are expressed explicitly in $\epsilon, a_i, b_i, g, J, L, M, K, N, H$ and $\Delta$.

Next we explain Bartlett's adjustment in our situation. Under the null hypothesis H it is easy to see that the expectation of $T \in S'$ can be written as

$$E(T) = 1 - \rho/c_n^2 + o(c_n^{-2}),$$

and that

$$T/E(T) = (1 + \frac{\rho}{c_n^2})T + o_p(c_n^{-2}).$$

Henceforth $\rho$ is called the Bartlett adjustment factor. If the terms of order $c_n^{-2}$ in the asymptotic expansion of the distribution of $T^* = (1 + \rho/c_n^2)T$ vanish ( i.e., $P_{n,\theta_0}[T^* \leq x] = P[\chi_1^2 \leq x] + o(c_n^{-2})$), we say T is adjustable in the sense of Bartlett ( B-adjustable for short ). Except for LR, $T \in S'$ is not generally B-adjustable. Thus we consider modification of T to $T^* = h(\hat\theta_{ML})T$ so that $T^*$ is B-adjustable, where $h(\theta)$ is a smooth function and $\hat\theta_{ML}$ is the MLE of $\theta$. Then we can state,

**Theorem 3.2**( Taniguchi[Tani91b]).
Suppose that $h(\theta)$ is continuously three times differentiable.
(1) For $T \in S'$, the modified test $T^* = h(\hat\theta_{ML})T$ is B-adjustable if $h = h(\theta_0)$, $h' = h'(\theta_0)$ and $h" = h"(\theta_0)$ satisfy
(i)   $h = 1$,
(ii)  $h' = -(3a_2 g^{3/2} + K)/3g$,
(iii) $h" = -g\tilde{M}a_1^2/2 - \tilde{N}a_1 - 2gb_3 + (2g^{3/2}a_2 - J - K)$

$$\times(3a_2 g^{3/2} + K)/3g^2 - (gH - 3K^2)/6g^2,$$

where $\tilde{M} = M - J^2/g$ and $\tilde{N} = N - JK/g$.
(2) If h satisfies (i), (ii) and (iii), the Bartlett adjustment factor $\rho^*$ of $T^*$ is given by

$$\rho^* = -\{12g^2\Delta + 12g^3 b_1 + 12g^3 \tilde{M} b_2 - 9g^3 \tilde{M} a_1^2 - 6g^2 \tilde{N} a_1 - 3gH + 5K^2\}/12g^3.$$

**Theorem 3.3**( Taniguchi[Tani91b]).
Suppose that $h(\theta)$ satisfies (i) - (iii) in Theorem 3.2. Then, for $T \in S'$, the distribution function of the modified test $T^{**} = (1 + c_n^{-2}\rho^*)h(\hat{\theta}_{ML})T$ under a sequence of local alternatives $\theta = \theta_0 + \epsilon/c_n$, has the third-order asymptotic expansion

$$P_{n,\theta_0+\epsilon/c_n}[\, T^{**} \leq x \,] = P[\, \chi_1^2(\delta) \leq x \,]$$

$$+ c_n^{-1} \sum_{j=0}^{2} B_j^{(T^{**})} P[\, \chi_{1+2j}^2(\delta) \leq x \,] + c_n^{-2} \sum_{j=0}^{4} C_j^{(T^{**})} P[\, \chi_{1+2j}^2(\delta) \leq x \,] + o(c_n^{-2}),$$

where the coefficients $B_j^{(\cdot)}$ are independent of $T \in S'$, and the coefficients $C_j^{(\cdot)}$ depend on $T \in S'$ unless $\tilde{M} = 0$.

Theorem 3.3 implies that all the powers of the modified tests $T^{**}$ are identical up to second-order, and that, in general, there is no test which is third-order uniformly most powerful in $S'$ unless $\tilde{M} = 0$. Here we note that $\gamma_\theta = \tilde{M}^{1/2}/g$ is a counterpart of Efron's statistical curvature in our situation. Therefore the results above agree with those of Kumon and Amari[KA83] and Amari[A85] which elucidate higher order asymptotics of tests for a curved exponential family.

Here we give a numerical example.

**Example 3.1.** Suppose that a stretch $(X_1, ..., X_{100})$ is observed from the AR(1) model

$$X_t = \theta X_{t-1} + \epsilon_t, \qquad |\theta| < 1,$$

where $\{\epsilon_t\}$ is a sequence of i.i.d.$N(0,1)$ random variables. Let LR be the likelihood ratio test for the testing problem $H : \theta = \theta_0$ against $A : \theta \neq \theta_0$. From Theorem 3.1, setting $\epsilon = 0$, we can get the asymptotic expansion of LR under H,

$$P_{n,\theta_0}[LR \leq x] = P[\chi_1^2 \leq x] + \frac{1}{n}[P\{\chi_1^2 \leq x\} - P\{\chi_3^2 \leq x\}] + o(n^{-1}).$$

It can be shown that Bartlett's adjustment factor for LR is 2. Hence the adjusted LR test is $LR^* = (1 + \frac{2}{n})LR$. It is natural to expect that $\chi_1^2$-approximation of $LR^*$ is better than that of LR. Now we verify this numerically. Let $x_\alpha$ be the level $\alpha$ point of $\chi_1^2$. We computed the following probabilities ;

$$P(0.05) = P_{n,\theta_0}[LR > x_{0.05}], \quad P(0.025) = P_{n,\theta_0}[LR > x_{0.025}],$$

$$P(0.01) = P_{n,\theta_0}[LR > x_{0.01}], \quad BP(0.05) = P_{n,\theta_0}[LR^* > x_{0.05}],$$

$$BP(0.025) = P_{n,\theta_0}[LR^* > x_{0.025}], \quad BP(0.01) = P_{n,\theta_0}[LR^* > x_{0.01}],$$

by 1000 trials simulation for $\theta_0 = -0.9(0.3)0.9$. These values are given by the table below.

Table 3.1.

| $\theta_0$ | $P(0.05)$ | $P(0.025)$ | $P(0.01)$ | $BP(0.05)$ | $BP(0.025)$ | $BP(0.01)$ |
|------|-----------|------------|-----------|------------|-------------|------------|
| $-0.9$ | 0.051 | 0.021 | 0.006 | 0.051 | 0.024 | 0.008 |
| $-0.6$ | 0.041 | 0.027 | 0.011 | 0.044 | 0.028 | 0.012 |
| $-0.3$ | 0.037 | 0.020 | 0.009 | 0.042 | 0.021 | 0.009 |
| $0.0$ | 0.039 | 0.017 | 0.008 | 0.042 | 0.018 | 0.009 |
| $0.3$ | 0.037 | 0.021 | 0.007 | 0.040 | 0.021 | 0.007 |
| $0.6$ | 0.043 | 0.022 | 0.013 | 0.048 | 0.022 | 0.014 |
| $0.9$ | 0.057 | 0.029 | 0.008 | 0.060 | 0.032 | 0.011 |

Table 3.1 shows that Bartlett's adjustment is effective to attain the level probability. Since Bartlett's adjustment is a sort of third-order correction in the sense of higher order asymptotic theory, Example 3.1 shows this is also useful.

## 4 Higher order asymptotic theory for studentized statistics

In this section, higher order asymptotics of studentized statistics will be illuminated. We consider the following time series regression model;

$$\begin{cases} X_t = \beta_0 + u_t, & t = 1, ..., n, \\ u_t = \theta u_{t-1} + \epsilon_t, & |\theta| < 1, \end{cases} \tag{4.1}$$

where $\{\epsilon_t\}$ is a sequence of i.i.d. random variables with $E\{\epsilon_t\} = 0$, $Var\{\epsilon_t\} = 1$, $E\{\epsilon_t^8\} < \infty$ and probability density function $g(x) > 0$ on $\mathbf{R}^1$. Let $\hat{\beta}_n = n^{-1} \sum_{t=1}^{n} X_t$. It is known that

$$\lim_{n \to \infty} Var[\sqrt{n}(\hat{\beta}_n - \beta_0)] = (1 - \theta)^{-2}.$$

Thus it is natural to define the studentized statistic

$$\tilde{t} = \sqrt{n}(1 - \hat{\theta}_n)(\hat{\beta}_n - \beta_0),$$

where $\hat{\theta}_n = \{\sum_{t=1}^{n-1} \hat{u}_t \hat{u}_{t+1}\} / \{\sum_{t=1}^{n} \hat{u}_t^2\}$, with $\hat{u}_t = X_t - \hat{\beta}_n$.

**Theorem 4.1**(Taniguchi-Puri[TP85]).

$$P_{\beta_0}^n[\tilde{t} < y] = \int_{-\infty}^y \varphi(x)\{1 + \frac{\kappa_3}{6\sqrt{n}}(x^3 - 3x)\}dx + o(n^{-1/2}),$$

where $\varphi(x) = \frac{1}{\sqrt{2\pi}}\exp\{-(1/2)x^2\}$, and $\kappa_3$ is the third-order cumulant of $\epsilon_t$.

Implication of this theorem seems interesting. The second-order Edgeworth approximation of $\tilde{t}$ is independent of $\theta$ ( i.e., the dependent structure of the residual process ). Also if $\kappa_3 = 0$, $\tilde{t}$ becomes the second-order normalizing transformation. Hence we are led to develop the higher order asymptotic theory for studentized statistics in general models as in Section 3.

Let $\mathbf{X}_n = (X_1, X_2, ..., X_n)'$ be a collection of m-dimensional random vectors $X_i$ which are not necessarily i.i.d.. Let $p_n(\mathbf{x}_n; \theta)$ denote the probability density function of $\mathbf{X}_n$ with respect to a carrier measure, where $\mathbf{x}_n \in \mathbf{R}^{mn}$ and $\theta = (\theta^1, ..., \theta^p)' \in \Theta \subset \mathbf{R}^p$. We also write $\theta = (u, v)$ where $u = (u^1, ..., u^a, ..., u^q)'$ and $v = (v^{q+1}, ..., v^\kappa, ..., v^p)'$. Henceforth we use a,b,c,... for indices of u-coordinates and related quantities ; use $\kappa, \mu, \lambda, ...$ for indices of v-coordinates and related quantities ; use i, j, k,... for indices of $\theta$-coordinates and related quantities. It is assumed that $p_n(.; \theta)$ is differentiable with respect to $\theta$ up to necessary order. Define

$$Z_i = n^{-1/2}\partial_i l_n(\theta),$$
$$Z_{ij} = n^{-1/2}\{\partial_i\partial_j l_n(\theta) - E_\theta[\partial_i\partial_j l_n(\theta)]\},$$

where $i, j = 1, ..., p$, $l_n(\theta) = \log p_n(\mathbf{X}_n; \theta)$, and $\partial/\partial\theta^i$ is abbreviated to $\partial_i$. We make the following assumption, which is very reasonable even in the non-i.i.d.case because it is satisfied by many regular statistical models.

**Assumption 4.1.** The asymptotic moments ( cumulants ) of $Z_i$ and $Z_{ij}$ are evaluated as follows :

$$E(Z_iZ_j) = g_{ij} + O(n^{-1}),$$

$$E(Z_iZ_{jk}) = J_{ijk} + O(n^{-1}),$$
$$E(Z_iZ_jZ_k) = n^{-1/2}K_{ijk} + O(n^{-1}),$$

and Jth-order cumulants of $Z_i$ and $Z_{ij}$ are all $O(n^{-J/2+1})$.

To avoid many regularity conditions for the stochastic expansion of MLE, we also make the following assumption.

**Assumption 4.2.** The $i$th component of the MLE $\hat{\theta} = (\hat{u}, \hat{v})$ of $\theta = (u, v)$ has the stochastic expansion

$$\hat{\theta}^i = \theta^i + \frac{1}{\sqrt{n}}g^{ij}Z_j + \frac{1}{n}g^{ij}g^{km}Z_{jk}Z_m$$

$$- \frac{1}{2n}(K_{jkm} + J_{jkm} + J_{kmj} + J_{mjk})g^{ij}g^{kk'}g^{mm'}Z_{k'}Z_{m'} + o_p(n^{-1}),$$

where $g^{ij}$ denotes the (i,j)-component of the inverse matrix of $\{g_{ij}\}$. Here we adopt the Einstein summation convention.

Many regular statistical models satisfy this assumption. We further make the orthogonal condition.

**Assumption 4.3.**

$$g_{a\kappa} = 0, \quad \text{for } a = 1, ..., q, \ \kappa = q + 1, ..., p.$$

Now we are interested in the parameter u. Letting $G = \{g_{ab}; a, b = 1, ..., q\}$, we denote the (a,b)-components of $G^{-1}$ and $G^{1/2}$ by $g^{ab}$ and $g_{ab}^{(1/2)}(= g_{ab}^{(1/2)}(u, v))$, respectively. Consider the studentized statistics of $\hat{u}$;

$$\bar{u}_a^{(S)} = \sqrt{n} g_{ab}^{(1/2)}(\hat{u}, \hat{v})(\hat{u}^b - u^b), \quad a = 1, ..., q.$$

First, we derive the stochastic expansion of $\bar{u}_a^{(S)}$ in terms of $Z_i$ and $Z_{ij}$. Second, we evaluate the asymptotic cumulants of $\bar{u}_a^{(S)}$. Then we obtain,

**Theorem 4.2.**

$$P_\theta^n[\bar{u}_1^{(S)} < y_1, ..., \bar{u}_q^{(S)} < y_q]$$
$$= \int_{-\infty}^{y_1} \cdots \int_{-\infty}^{y_q} N(y; I_q)[1 + \frac{1}{\sqrt{n}} c_a^{(S)} H^a(y) + \frac{1}{6\sqrt{n}} c_{aa'a''}^{(S)} H^{aa'a''}(y)]dy + o(n^{-1/2}), \quad (4.2)$$

where $y = (y_1, ..., y_q)'$, $N(y; I_q) = (2\pi)^{-q/2} \exp(-y'y/2)$,

$$H^{j_1\cdots j_s}(y) = \frac{(-1)^s}{N(y; I_q)} \frac{\partial^s}{\partial y_{j_1} ... \partial y_{j_s}} N(y; I_q),$$

and

$$c_a^{(S)} = \frac{1}{2}[2g^{bl}(\partial_l g_{ab}^{(1/2)}) - g_{ab}^{(1/2)} g^{bb'} g^{b''b'''}(K_{b'b''b'''} + J_{b'b''b'''}) - g_{ab}^{(1/2)} g^{bb'} g^{\kappa\kappa'}(K_{b'\kappa\kappa'} + J_{b'\kappa\kappa'})],$$

$$c_{aa'a''}^{(S)} = -2g_{ab}^{(1/2)} g_{a'c}^{(1/2)} g_{a''d}^{(1/2)} g^{bb'} g^{cc'} g^{dd'} K_{b'c'd'} + \ll -g_{ab}^{(1/2)} g_{a'c}^{(1/2)} g_{a''d}^{(1/2)} g^{bb'} g^{cb''} g^{db'''} J_{b'b''b'''}$$
$$+ g^{bc} g^{ld}(\partial_l g_{ab}^{(1/2)}) g_{a'c}^{(1/2)} g_{a''d}^{(1/2)} + g^{bd} g^{lc}(\partial_l g_{ab}^{(1/2)}) g_{a'c}^{(1/2)} g_{a''d}^{(1/2)} \gg_{aa'a''},$$

with $\ll A_{aa'a''} \gg_{aa'a''} = A_{aa'a''} + A_{a'a''a} + A_{a''aa'}$.

Although the expression (4.2) is very general, it is not so informative. Hence we consider the case where u and v are scalar (i.e., q = 1, p = 2 ). Then

$$\bar{u}^{(S)} = \sqrt{n} g_{11}^{1/2}(\hat{u}, \hat{v})(\hat{u} - u),$$
$$\bar{u}^{(NS)} = \sqrt{n} g_{11}^{1/2}(u, v)(\hat{u} - u),$$

are, respectively, the studentized statistic and the non-studentized statistic of u.

**Theorem 4.3.** Suppose that u and v are scalar, and that $K_{122} + J_{122} = 0$. Then

(1) $\quad E(\bar{u}^{(NS)}) = -\dfrac{\Gamma^*}{2\sqrt{n}g_{11}^{3/2}} + o(n^{-1/2}), \quad cum^{(3)}(\bar{u}^{(NS)}) = -\dfrac{2\Gamma^* + \Gamma}{\sqrt{n}g_{11}^{3/2}} + o(n^{-1/2}),$

(2) $\quad E(\bar{u}^{(S)}) = \dfrac{\Gamma}{2\sqrt{n}g_{11}^{3/2}} + o(n^{-1/2}), \quad cum^{(3)}(\bar{u}^{(S)}) = \dfrac{\Gamma^* + 2\Gamma}{\sqrt{n}g_{11}^{3/2}} + o(n^{-1/2}),$

where $\Gamma = J_{111}$ and $\Gamma^* = J_{111} + K_{111}$, which correspond to the 1-connection and the dual connection ( see Amari[A85]). Hence the second-order Edgeworth expansions $Edg(NS; \Gamma; \Gamma^*)$ and $Edg(S; \Gamma; \Gamma^*)$ for $\bar{u}^{(NS)}$ and $\bar{u}^{(S)}$, respectively, have the following duality ;

$$Edg(NS; \Gamma; \Gamma^*) = Edg(S; -\Gamma^*; -\Gamma).$$

This theorem shows that the higher order asymptotic theory for studentization is related to dual differential geometry. We next discuss a concrete example. Let $\{X_t\}$ be a Gaussian ARMA(1,1) process generated by

$$X_t - \alpha X_{t-1} = \epsilon_t - \beta \epsilon_{t-1}, \qquad |\alpha|, |\beta| < 1, \tag{4.3}$$

where $\{\epsilon_t\}$ is a sequence of i.i.d. $N(0, \sigma^2)$ random variables.

**Proposition 4.1.** In the model (4.3), suppose that $u = \alpha$ (AR- parameter ), $v = \sigma^2$ ( innovation variance ). Then, the distribution of the non-studentized statistic of $\hat{u}$ has the asymptotic expansion ;

$$
\begin{aligned}
P_\theta^n[\bar{u}^{(NS)} < y] &= \Phi(y) - \frac{1}{\sqrt{n}}\varphi(y)\{\frac{-u}{\sqrt{1-u^2}}(y^2+1)\} + o(n^{-1/2}), \\
&= Edg(NS; AR; y; u) + o(n^{-1/2}), \quad (say),
\end{aligned}
$$

and that of the studentized statistic of $\hat{u}$ has ;

$$
\begin{aligned}
P_\theta^n[\bar{u}^{(S)} < y] &= \Phi(y) - \frac{1}{\sqrt{n}}\varphi(y)\{\frac{-u}{\sqrt{1-u^2}}\} + o(n^{-1/2}), \\
&= Edg(S; AR; y; u) + o(n^{-1/2}), \quad (say),
\end{aligned}
$$

where $\Phi(y) = \int_{-\infty}^y \varphi(x)dx$.

This proposition means that, for the AR-parameter $\alpha$, studentization gives uniformly better approximation to N(0,1) than non-studentization.

**Proposition 4.2.** In the model (4.3), suppose that $u = \beta$(MA-parameter), $v = \sigma^2$ ( innovation variance ). Then the distribution of the non-studentized statistic of $\hat{u}$ has the asymptotic expansion ;

$$P_\theta^n[\bar{u}^{(NS)} < y] = \Phi(y) - \frac{1}{\sqrt{n}}\varphi(y)\{\frac{u}{\sqrt{1-u^2}}\} + o(n^{-1/2}),$$

$$= Edg(NS; MA; y; u) + o(n^{-1/2}), \quad (say),$$

and that of the studentized statistic of $\hat{u}$ has ;

$$P_\theta^n[\bar{u}^{(S)} < y] = \Phi(y) - \frac{1}{\sqrt{n}}\varphi(y)\{\frac{u}{\sqrt{1-u^2}}(y^2+1)\} + o(n^{-1/2}),$$

$$= Edg(S; MA; y; u) + o(n^{-1/2}), \quad (say).$$

This proposition means that, for the MA-parameter $\beta$, studentization gives uniformly worse approximation to $N(0,1)$ than non-studentization. From Propositions 4.1 and 4.2 we can see the following dual relations;

$$Edg(NS; AR; y; u) = Edg(S; MA; y; -u),$$

$$Edg(S; AR; y; u) = Edg(NS; MA; y; -u),$$

which are due to Theorem 4.3 and the fact that the 1-connection $\Gamma_{AR}$ for AR-parameter is equal to the dual connection $\Gamma_{MA}^*$ for MA- parameter. Regarding the duality of AR-model and MA-model and that of studentization and non-studentization, further investigation will be required.

# References

[A85]  S.I.Amari.  *Differential Geometrical Methods in Statistics.* Springer Lecture Notes in Statistics. Vol.28, Springer-Verlag, Heidelberg ,1985.

[HP85]  T. Hayakawa and M.L.Puri. Asymptotic expansions of distributions of some test statistics. *Ann. Inst. Stat. Math.* 37, 95-108,1985.

[KA83]  M. Kumon and S.I.Amari. Geometrical theory of higher-order asymptotics of test, interval estimator and conditional inference. *Proc. Roy. Soc. London Ser. A* , 387, 429-58,1983.

[Pe71]  H.W. Peers. Likelihood ratio and associated test criteria. *Biometrika* 58, 577-87,1971.

[Ph77]      P.C.B. Phillips.  Approximations to some finite sample distributions associated with a first-order stochastic difference equations. *Econometrica*  45,  463-86,1977.

[Tana82]    K. Tanaka.  Chi-square approximations to the distributions of the Wald, likelihood ratio and Lagrange multiplier test statistics in time series regression.  Tech. Rep. 82,  Kanazawa University,1982.

[Tani91a]   M. Taniguchi.  *Higher Order Asymptotic Theory for Time Series Analysis.*  Springer Lecture Notes in Statistics, Vol.68,  Springer-Verlag, Heidelberg,1991.

[Tani91b]   M. Taniguchi.  Third-order asymptotic properties of a class of test statistics under a local alternative.  *J. Multivariate Anal.* 37, 223-38,1991.

[TP95]      M. Taniguchi and M.L. Puri.  Valid Edgeworth expansions of M-estimators in regression models with weakly dependent residuals.  To appear in *Econometric Theory* ,1995.

[W51]       P. Whittle.  *Hypothesis Testing in Time Series Analysis.* Almqvist and Wiksells, Uppsala,1951.

# Semi-parametric graphical estimation techniques for long-memory data.

Murad S. Taqqu
Vadim Teverovsky [1] [2] [3]

ABSTRACT This paper reviews several periodogram-based methods for estimating the long-memory parameter $H$ in time series and suggests a way to robustify them. The high frequencies tend to bias the estimates. Using only low frequencies eliminates the bias but increases the variance. We hence suggest plotting the estimates of $H$ as a function of a parameter which balances bias versus variance and, if the plot flattens in a central region, to use the flat part for estimating $H$. We apply this technique to the periodogram regression method, the Whittle approximation to maximum likelihood and to the local Whittle method. We investigate its effectiveness on several simulated fractional ARIMA series and also apply it to estimate the long-memory parameter $H$ in computer network traffic.

## 1 Introduction

Time series with long memory have been considered in many fields including hydrology, biology and computer networks. Unfortunately, estimating the long memory (long-range dependence) parameter $H$ in a given data series is a difficult task. Some of the more accurate estimation methods, e.g. the Whittle estimator, are parametric in that the exact form of the spectral density must be known *a priori*. If the assumed spectral density function is not the correct one, then the estimated $H$ may be biased. In this paper we review several periodogram-based methods for estimating $H$ and propose a graphical extension in order to reduce the bias.

The methods we will try to modify are the periodogram regression method originally introduced by Geweke and Porter-Hudak [4], the Whittle approximate likelihood method (see Fox and Taqqu [3], Beran [1]) and the "local Whittle" method developed by Robinson [18]. An empirical study of

---

[1] Research partially supported by NSF grants NCR-9404931 and DMS-9404093 at Boston University.

[2] *AMS Subject classification:* 60G18, 62-09.

[3] *Keywords:* Long-range dependence, long memory, periodogram, Whittle.

these methods can be found in Taqqu, Teverovsky and Willinger [23] and in Taqqu and Teverovsky [22].

We will plot the estimates of $H$ as a function of a parameter which balances bias versus variance. For the periodogram regression and for the local Whittle methods we vary the fraction of frequencies which is used in the estimation of $H$. For the Whittle method we aggregate the given data series and then apply the method assuming a Fractional Gaussian Noise (FGN) model. The goal is to achieve a "flattening" of the plot of the estimates of $H$ for a certain range of values of the varying parameter. We will then use the value of $\widehat{H}$ given by the flat part of the plot as an estimate of the long-memory parameter.

We test our methods on simulated Gaussian FARIMA$(1, d, 1)$ series and also apply them to two very long data series of Ethernet traffic, which are believed to exhibit long memory, i.e. have $H \in (1/2, 1)$. A FARIMA$(1, d, 1)$ series $(H = d + 1/2)$ is defined by:

$$(1 - \phi B)(1 - B)^d X_i = (1 - \theta B) Z_i, \tag{1.1}$$

where $Z_i$ are a Gaussian white noise sequence, $\phi$, $\theta$ and $d$ are the parameters of the process and $B$ is the backward shift operator, $BX_i = X_{i-1}$. For more information about FGN and FARIMA see Samorodnitsky and Taqqu [20] and Beran [1].

We consider the regression on the periodogram method in Section 2, the "aggregated Whittle" in Section 3 and the "local Whittle" in Section 4. Our conclusions will be presented in Section 5.

## 2   Regression on the Periodogram

This method is based on the periodogram of the time series, defined by

$$I(\nu) = \frac{1}{2\pi N} \left| \sum_{j=1}^{N} X_j e^{ij\nu} \right|^2, \tag{1.2}$$

where $\nu$ is the frequency, N is the number of terms in the series, and $X_j$ are the data. A series with long-range dependence has a spectral density proportional to $|\nu|^{1-2H}$ close to the origin. Since $I(\nu)$ is an estimator of the spectral density, a regression of the logarithm of the periodogram vs. the logarithm of the frequency $\nu$ should give a coefficient of $1 - 2H$.

This method was first developed by Geweke and Porter-Hudak [4] in a slightly different version which is referred to as the GPH estimator. Their estimator regresses on $\log|2\sin(\nu/2)|$ instead of $\log|\nu|$. Clearly, at very small frequencies the two are equivalent. A FARIMA$(0, d, 0)$ model has the spectral density $|2\sin(\nu/2)|^{1-2H}$, whereas a Fractional Gaussian Noise (FGN) model has a spectral density closer, although not equal, to $|\nu|^{1-2H}$.

The function $C|\nu|^{1-2H}$ lies approximately half-way between these two theoretical spectral densities. For plots see [22].

Whatever the independent variable, not all of the periodogram frequencies are used in the regression, since the proportionality to $|\nu|^{1-2H}$ or $|2\sin(\nu/2)|^{1-2H}$ holds only for $\nu$ close to the origin. Finding the right cut-off is very important. Geweke and Porter-Hudak [4] suggested using a cut-off of $M = \sqrt{N}$ where $N$ is the series length, and $M$ is the number of frequencies used for the regression.

In our work we regress the logarithm of the periodogram on $\log|\nu|$, as a compromise between the FARIMA$(0, d, 0)$ and FGN models (see Fig. 2 [22]). In [23] we used this method on simulated series of both FGN and FARIMA$(0, d, 0)$, focusing on Fourier frequencies $\nu_j = 2\pi j/N$, $j = 1, \ldots, M$, with $N = 10,000$ and $M \simeq 500$. For those types of series the method works quite well. On the other hand, when series such as FARIMA$(1, d, 1)$ are used, the periodogram can be biased because of the short-range dependence structure.

There have been many attempts to justify the periodogram regression method. Yajima [26] showed that the discrete Fourier transforms of long memory series become independently normal as the sample size $N$ approaches infinity, but his work applies only to a fixed finite number of frequencies that does not change with the sample size. Künsch [13] showed that the normalized periodogram is asymptotically i.i.d. $\chi^2$ if there are a fixed finite number of varying frequencies that tend to zero with $N$ but such that the smallest frequency multiplied by $N^{1/2}$ approaches infinity. He also proposed a variation on the GPH estimator by ignoring some portion $L$ of the smallest frequencies. The variation was motivated by a desire to distinguish between monotonic trends and long-range dependence. Robinson [19] has established the consistency and asymptotic normality of a similar modification of the GPH estimator. He uses $\log|\nu|$ as the regressor, and does not fix the number of frequencies used in the regression. Robinson shows that, if $M = o(N^{4/5})$ and

$$\frac{L}{M} + \frac{M^{1/2}\log M}{L} + \frac{(\log N)^2}{M} \to 0,$$

then the periodogram for frequencies between $2\pi L/N$ and $2\pi M/N$ is asymptotically i.i.d. $\chi^2$, and the $H$ estimated from the regression is consistent and asymptotically normal. Robinson has also suggested using an estimator based on the averaged (or integrated) periodogram [17], and derived criteria for choosing $M$ [16], but they depend, unfortunately, on $H$. The estimator, moreover, is biased downward, especially for $H$ close to 1.

Hurvich and Beltrao [9] have shown that the periodogram is, in fact, asymptotically neither i.i.d. nor $\chi^2$ distributed at the very low frequencies of order $1/N$. This does not contradict the results of Künsch and Robinson, since they were dealing with frequencies that tend to zero much more slowly. It implies a bias for the GPH estimator which is noticeable at the very low frequencies only. In the same paper, Hurvich and Beltrao propose tapering the periodogram as a way to reduce the bias. They provide in [10] some criteria for choosing $M$ and $L$, based on frequency domain cross-validation and a minimization procedure over $M$ and $L$. Simulation results are presented to show that their method is slightly better than the GPH estimator, and better, as well, than the variation proposed by Künsch and Robinson. Beran, in his recent book [1], presents several of the above results.

In a recent paper, Hurvich, Deo and Brodsky [11] have established some asymptotic properties of the original (GPH) periodogram estimator. They show that if the spectral density is $f(\nu) = |2\sin(\nu/2)|^{1-2H} f^*(\nu)$, then the bias behaves asymptotically like

$$-\frac{2\pi^2}{9} \frac{f^{*''}(0)}{f^*(0)} \frac{M^2}{N^2}.$$

Here $f^*$ is an even, positive, continuous function on $[-\pi, \pi]$, with certain other regularity properties that ensure that it posseses only short-range dependence.

Hurvich, Deo and Brodsky also show that the variance of the GPH estimator is equal to $\pi^2/(24M)$ and thus obtain a theoretical expression for the mean squared error (MSE) of the estimator. They obtain a formula for the $M$ which minimizes the MSE of the GPH estimator. This $M^{OPT}$ is proportional to $N^{4/5}$, but the proportionality constant again depends on $f^*$. Even if the asymptotic relations were to hold exactly, $M$ can not be chosen *a priori*.

Instead of picking *a priori* a value of $M$ (whether blindly or using the $M^{OPT}$ above) and then finding an estimate of $H$, we suggest letting $M$ vary through a fairly wide range. The various estimates of $H$ are then plotted as a time series. Starting at large values of $M$ (small values of $N/M$), we would expect to find a range where the estimates of $H$ are incorrect because of the short-range effects. Then, as $M$ decreases ($N/M$ increases), the short-range effects should disappear and the value of $H$ obtained should represent the true long-memory dependence. Thus there should be a period of relative stability, where the estimates of $H$ are approximately constant. Then, if we move to smaller $M$'s, we will get into a region where the estimates of $H$ are very scattered and unreliable because there are not enough frequencies left to have an accurate regression. Thus, we should expect to see a flat region somewhere in the middle of the plot of the estimates of $H$ and we can estimate an overall $H$ from that region. In practice, of course, the region will not be completely flat.

We have applied this method to 10 simulated series of FARIMA$(1, d, 1)$ of length $N = 10,000$, with parameters $\phi = 0.3$, $\theta = 0.7$, $d = 0.3$ $(H = 0.8)$. We used the above method with $N/M = 4(4)200$. For each series, a line was drawn connecting the estimates at the various values of $M$. The plot is given

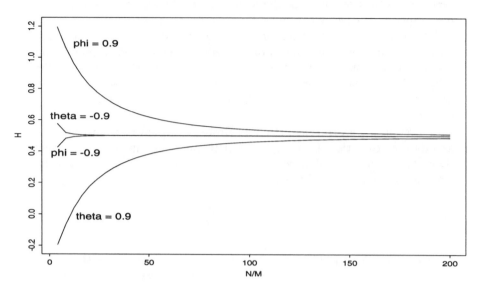

FIGURE 1. Periodogram method applied to the theoretical spectral density of AR(1) processes with $\phi = \pm 0.9$ and MA(1) processes with $\theta = \pm - 0.9$.

in Figure 2. In addition the plot contains some "theoretical" results related to this method: we used the theoretical spectral density of the FARIMA process in place of the periodogram in the regression, and generated the "estimates" of $H$ at the same values of $M$ as above. This is illustrated by the thickest curve in the plot, which starts at approximately 0.5 and levels off at 0.8, which is the nominal value of $H$. The other two curves in the plot denote the approximate 95% confidence intervals ($\pm 1.96\sigma$) around the "theoretical" result, where $\sigma = \pi/\sqrt{24M}$ is the asymptotic standard deviation (see [19], [11]). It can be seen that the simulated results stay within these bounds almost all of the time, and that most of the lines flatten out appreciably at higher $N/M$. On the other hand, the bounds are large enough that the estimates of $H$ thus obtained can be fairly scattered.

We also include the "theoretical" curves for several other series, namely MA(1) with $\theta = \pm 0.9$ and AR(1) with $\phi = \pm 0.9$ (see Figure 1). These are extreme choices, that include both "good" and "bad" case scenarios. In accordance with the results of Hurvich, Deo and Brodsky the magnitude of the bias is the same in the cases $\theta = 0.9$ and for $\phi = 0.9$. The magnitude of the bias is also the same, but much smaller, in the cases $\theta = -0.9$ and $\phi = -0.9$. All the "theoretical" curves, however, are different, illustrating the fact that care has to be exercized when using a periodogram estimator.

In real life one does not have the luxury of either examining ten replicates of a time series or using their theoretical spectral density. One has only one actual time series to analyze. We applied the graphical extension to the periodogram regression method to two very long series of Ethernet traffic, considered by Leland, Taqqu, Willinger and Wilson [14], [25]. The first Ethernet data set gives the number of bytes per 10 milliseconds passing through a monitoring system during a "normal traffic hour" in August 1989. The second gives the number of packets sent per 10 milliseconds during a "normal traffic hour" in August 1989. Thus, each of the series is 360,000 long. The authors obtain estimates of approximately $H = 0.8$ for the *byte* data, and $H = 0.9$ for the *packet* data, by using several different methods (see also Teverovsky and Taqqu [24]). The results of using the graphical periodogram regression method on these series are shown in Figure 5 ($N/M = 10(10)1000$). The curves do flatten out eventually, and $H$ can be estimated to be approximately 0.70 to 0.75 for the *byte* data and approximately 0.90 to 0.95 for the *packet* data.

## 3 Aggregated Whittle Estimator

The Whittle estimator involves the function

$$Q(\eta) := \int_{-\pi}^{\pi} \frac{I(\nu)}{f(\nu;\eta)} d\nu, \qquad (1.3)$$

where $I(\nu)$ is the periodogram (see (1.2)) and $f(\nu;\eta)$ is the spectral density at frequency $\nu$, where $\eta$ denotes the vector of unknown parameters. The Whittle estimator is the value of $\eta$ which minimizes the function Q. When dealing with fractional Gaussian noise or fractional ARIMA$(0, d, 0)$, $\eta$ is simply the parameter $H$ or $d$. If the series is assumed to be FARIMA$(p, d, q)$ then $\eta$ also includes the unknown coefficients in the autoregressive and moving average parts.

Ted Hannan contributed in an important way to the development of the (classical) Whittle's method. For a stationary Gaussian time series with spectral density $f(\nu, \beta)$, $-\pi < \lambda < \pi$, Whittle's method, which provides an estimate of $\beta$, requires replacing the inverse covariance matrix that appears in the Gaussian likelihood by a Toeplitz (covariance) matrix with spectral density $1/f$ and then maximizing the quadratic form. Hannan, in the seminal paper [6], applied Whittle's method to finite variance ARMA time series. He proved that the estimator is consistent and asymptotically normal. An ARMA time series, however, has short-range dependence because the correlations decrease exponentially fast. Fox and Taqqu [3] extended this result to Gaussian time series with long-range dependence such as fractional Gaussian noise or fractional ARIMA by appealing to a central limit theorem for weighted quadratic forms whose weights are chosen in such a way as to compensate for the long-range dependence. Fox and Taqqu's result, which was later generalized to the full maximum likelihood by Dahlhaus [2], is the basis of one of the most commonly used techniques for estimating the intensity of long-range dependence in Gaussian time series (see Beran [1]). Giraitis and Surgailis [5] extended Fox and Taqqu's result to finite variance innovations without Gaussian assumptions and Heyde and Gay [8] to random fields. Mikosch, Gadrich, Klüppelberg and Adler [15] went back to the original Hannan's ARMA result and investigated what happens if the Whittle estimator is used on ARMA time series with infinite variance; Kokoszka and Taqqu [12] then provided the extension to FARIMA sequences with long-range dependence and infinite variance.

The Whittle method is parametric. If the assumed spectral density function is not the correct one, then the estimated $H$ obtained by the Whittle method may be incorrect. To alleviate this problem, one may try to aggregate the series, namely consider

$$X_k^{(m)} := \frac{1}{m} \sum_{i=(k-1)m+1}^{km} X_i$$

where $X_i$ is the original time series and $m$ is the aggregation level. As the aggregation level becomes large, $X^{(m)}$ approaches FGN if $X_i$ is Gaussian with long-range dependence (see Taqqu [21]). Thus if we aggregate the series at a large enough level, we can use the Whittle estimator with the FGN spectral density model to estimate $H$. Since the appropriate $m$ is not known *a priori*, we again estimate $H$ at different levels of aggregation, plot the results, and try to find a region where the graph is approximately flat. This idea was first used in Leland, Taqqu, Willinger and Wilson [14], in the context of Ethernet series.

We applied this method to the same simulated series of FARIMA$(1, d, 1)$ considered above with $d = 0.3$ ($H = 0.8$). We chose $N/m = 4(4)100$. For each series, a line was drawn connecting the estimates at the various values of $m$. The plot is given in Figure 3. As $m$ grows the graph flattens to $\hat{H} = 0.8$. Although individual estimates might range from 0.7 to 0.9, most of the estimates seem fairly close to 0.8.

We have also produced the aggregated Whittle plots for the two Ethernet series introduced in Section 2. We used $N/m = 10(10)1000$ (Figure 6). After initial periods of rise or fall, the plots for both of the series flatten out. The estimate of $H$ seems to be about 0.9 for the packet series and about 0.75 to 0.8 for the byte series. (Refer also the Ethernet papers [14], [25]).

# 4   Local Whittle Estimator

This method shares features with both of the previous methods. Like the regression on the periodogram, local Whittle only assumes the behavior of the spectral density close to the origin and just like the Whittle method it involves minimizing a modification of the function Q in (1.3). It was recently introduced by Robinson [18] who derived its asymptotic properties. Taqqu and Teverovsky [22] studied its robustness. Assume

$$f(\nu) \sim G(H)|\nu|^{1-2H} \quad \text{as} \quad \nu \to 0.$$

For such a spectral density, the discrete analogue of the function $Q$ in (1.3) is

$$Q(G, H) := \frac{1}{M} \sum_{j=1}^{M} \left( \frac{I(\nu_j)}{G\nu_j^{1-2H}} + \log G\nu_j^{1-2H} \right) \qquad (1.4)$$

where $\nu_j = 2\pi j/N$. Note that: (a) the frequencies are integrated (summed) only up to $2\pi M/N$; (b) the spectral density $f$ is replaced by its asymptotic form. M is assumed to satisfy:

$$\frac{1}{M} + \frac{M}{N} \to 0 \quad \text{as} \quad N \to \infty.$$

Replacing the constant $G$ by its estimate $\widehat{G} = M^{-1} \sum_{j=1}^{M} I(\nu_j)/\nu_j^{1-2H}$, one minimizes $R(H) := Q(\widehat{G}, H) - 1$, that is

$$R(H) = \log \left( \frac{1}{M} \sum_{j=1}^{M} \frac{I(\nu_j)}{\nu_j^{1-2H}} \right) - (2H - 1) \frac{1}{M} \sum_{j=1}^{M} \log \nu_j.$$

Robinson [18] shows that under some additional technical conditions, $\widehat{H} = \arg\min R(H)$ converges in probability to the actual value $H_0$, as $N \to \infty$. Moreover, under slightly stronger assumptions on $f$ and $M$,

$$M^{1/2}(\widehat{H} - H_0) \to_d \text{Normal}\,(0, 1/4), \quad \text{as} \quad N \to \infty. \qquad (1.5)$$

This last relation yields asymptotic values for the standard deviations, confidence intervals, etc. The asymptotic variance is smaller than for the regression on the periodogram method.

Note that the larger the value of $M$, the smaller the variance of $\widehat{H}$. On the other hand, if the series is not "ideal", e.g. if it is FARIMA$(1, d, 1)$, then $M$ has to be small in order to avoid bias. In such a case we only want to use the frequencies which are close to zero, since at larger frequencies the short-range behavior of the series will affect the form of the spectral density. In a previous paper [22] we have examined several FARIMA series at several discrete values of $M$. In particular, we found that $M$ had to be no more than approximately $N/32$ when $N = 10,000$ in order to achieve reasonable accuracy for the series we used.

Here, we introduce the continuous version of the above method, where we will calculate $H$ for many values of $M$ and plot the resulting estimates. If the range of values of $M$ is chosen correctly then we should again see a fairly constant region when the estimates of $H$ converge to the correct long-memory parameter.

We apply this method, with $N/M = 4(4)200$, to the same simulated FARIMA series used for the two previous estimators. As in the regression on the periodogram method, we plot the "theoretical" values of the estimates and the approximate confidence intervals in Figure 4. The "theoretical" values are obtained by using the spectral density of the known process in place of the periodogram in the estimator. The confidence intervals correspond to the standard deviations (1.5) around these "theoretical" values. The simulated results stay almost always within these bounds. Henry and Robinson [7] have recently derived results for the bias of this estimator. These are very similar to the results obtained for the regression on the periodogram.

We have used this estimator to estimate $H$ for the two Ethernet series discussed above using $N/M = 10(10)1000$. The results are shown in Figure 7 and are very similar to the results for the regression on the periodogram. Based on the local Whittle method, $\widehat{H}$ is 0.75 for the *byte* series and 0.9 for the *packet* series. The estimates for all three methods are hence fairly consistent.

We should also mention that there is a very large downward spike for the packet data on the left edge of the plot for all three methods. This is due to a large spike in the periodogram of the data series occuring at a frequency of approximately $2\pi/64$. Thus, when this frequency is included in the periodogram regression or the function Q for the Whittle estimators it causes a spike in the estimates of $H$. Once $M$ is small enough for this frequency not to be included, the estimates of $H$ tend to be more consistent.

## 5 Conclusion

The original Whittle method for estimating the long memory parameter $H$ is parametric: it is highly accurate if the underlying model is correct. Other methods, such as the regression on the periodogram and the local Whittle use a cut-off parameter to ensure robustness. It is difficult, unfortunately, to decide *a priori* on the best choice of values for these parameters.

We have introduced several modifications to these methods, in order to increase robustness with respect to short memory effects and applied them to two real-life series of Ethernet traffic of length $N = 360,000$. The graphical technique used for the aggregated Whittle estimator in [14], [25], was extended to the regression on the periodogram and the local Whittle methods. In the aggregated Whittle method, the data is aggregated at various levels and the estimates of $H$ are computed for the aggregated series. In the other two methods, the values of $H$ are computed for various cut-off values. The results in all cases are displayed in a plot which should first display variability and then become approximately flat. Since at that point, the short-range dependencies are no longer affecting the estimates to

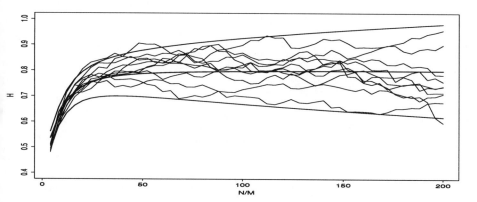

FIGURE 2. Periodogram regression estimates for 10 series of FARIMA(.3, .3, .7).

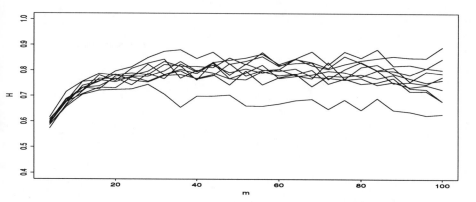

FIGURE 3. Aggregated Whittle estimates for 10 series of FARIMA(.3, .3, .7).

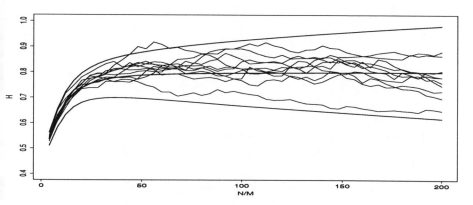

FIGURE 4. Local Whittle estimates for 10 series of FARIMA(.3, .3, .7).

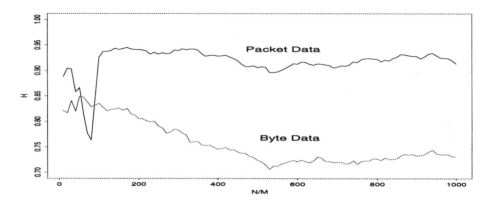

FIGURE 5. Periodogram regression estimates for Ethernet series.

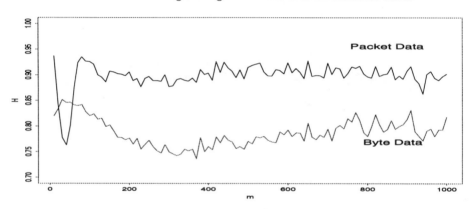

FIGURE 6. Aggregated Whittle estimates for Ethernet series.

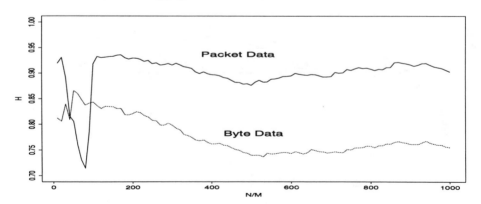

FIGURE 7. Local Whittle estimates for Ethernet series.

431

any noticeable degree, the resulting value of $H$ should be fairly unbiased.

It is difficult to evaluate these methods in a quantifiable way since the estimates of $H$ are, of necessity, somewhat subjective. This is why we presented several plots of the results using simulated series of length $N = 10,000$. We also presented corresponding "theoretical" graphs for several other types of series. The results seem to confirm that the estimators do behave as expected and that it is preferable to use these plots of $\widehat{H}$ rather than the $\widehat{H}$ corresponding to a fixed *a priori* value of $N/M$.

# 6 References

[1] J. Beran. *Statistics for Long-Memory Processes.* Chapman & Hall, New York, 1994.

[2] R. Dahlhaus. Efficient parameter estimation for self similar processes. *The Annals of Statistics*, 17(4):1749–1766, 1989.

[3] R. Fox and M. S. Taqqu. Large-sample properties of parameter estimates for strongly dependent stationary Gaussian time series. *The Annals of Statistics*, 14:517–532, 1986.

[4] J. Geweke and S. Porter-Hudak. The estimation and application of long memory time series models. *Journal of Time Series Analysis*, 4:221–238, 1983.

[5] L. Giraitis and D. Surgailis. A central limit theorem for quadratic forms in strongly dependent linear variables and application to asymptotical normality of Whittle's estimate. *Probability Theory and Related Fields*, 86:87–104, 1990.

[6] E. J. Hannan. The asymptotic theory of linear time series models. *Journal of Applied Probability*, 10:130–145, 1973.

[7] M. Henry and P. M. Robinson. Bandwidth choice in Gaussian semiparametric estimation of long range dependence. In *Proceedings of the Athens Conference on Applied Probability and Time Series Analysis*, Springer-Verlag, New York, 1996. Time series volume in honour of E. J. Hannan. Appears in this volume.

[8] C. C. Heyde and R. Gay. Smoothed periodogram asymptotics and estimation for processes and fields with possible long-range dependence. *Stochastic Processes and their Applications*, 45:169–182, 1993.

[9] C. M. Hurvich and K. I. Beltrao. Asymptotics for the low-frequency ordinates of the periodogram of a long-memory time series. *Journal of Time Series Analysis*, 14:455–472, 1993.

[10] C. M. Hurvich and K. I. Beltrao. Automatic semiparametric estimation of the memory parameter of a long memory time series. *Journal of Time Series Analysis*, 15:285–302, 1994.

[11] C.M. Hurvich, R. Deo, and J. Brodsky. The mean squared error of Geweke and Porter-Hudak's estimator of the memory parameter of a long memory time series. Preprint, 1995.

432

[12] P. S. Kokoszka and M. S. Taqqu. Parameter estimation for infinite variance fractional ARIMA. To appear in *The Annals of Statistics*, 1995.

[13] H. Künsch. Discrimination between monotonic trends and long-range dependence. *Journal of Applied Probability*, 23:1025–1030, 1986.

[14] W. E. Leland, M. S. Taqqu, W. Willinger, and D. V. Wilson. On the self-similar nature of Ethernet traffic (Extended version). *IEEE/ACM Transactions on Networking*, 2:1–15, 1994.

[15] T. Mikosch, T. Gadrich, C. Klüppelberg, and R. J. Adler. Parameter estimation for ARMA models with infinite variance innovations. *The Annals of Statistics*, 23:305–326, 1995.

[16] P. M. Robinson. Rates of convergence and optimal bandwidth in spectral analysis of processes with long range dependence. *Probability Theory and Related Fields*, 99:443–473, 1994.

[17] P. M. Robinson. Semiparametric analysis of long-memory time series. *The Annals of Statistics*, 22:515–539, 1994.

[18] P. M. Robinson. Gaussian semiparametric estimation of long range dependence. *The Annals of Statistics*, 1995. To appear.

[19] P. M. Robinson. Log-periodogram regression of time series with long range dependence. *The Annals of Statistics*, 23:1048–1072, 1995.

[20] G. Samorodnitsky and M. S. Taqqu. *Stable Non-Gaussian Processes: Stochastic Models with Infinite Variance*. Chapman and Hall, New York, London, 1994.

[21] M. S. Taqqu. Weak convergence to fractional Brownian motion and to the Rosenblatt process. *Zeitschrift für Wahrscheinlichkeitstheorie und verwandte Gebiete*, 31:287–302, 1975.

[22] M. S. Taqqu and V. Teverovsky. Robustness of Whittle-type estimates for time series with long-range dependence. Preprint, 1995.

[23] M. S. Taqqu, V. Teverovsky, and W. Willinger. Estimators for long-range dependence: an empirical study. *Fractals*, 3:785–798, 1995.

[24] V. Teverovsky and M. S. Taqqu. Testing for long-range dependence in the presence of shifting means or a slowly declining trend using a variance-type estimator. Preprint, 1995.

[25] W. Willinger, M. S. Taqqu, W. E. Leland, and V. Wilson. Self-similarity in high-speed packet traffic: analysis and modeling of Ethernet traffic measurements. *Statistical Science*, 10:67–85, 1995.

[26] Y. Yajima. A central limit theorem of Fourier transforms of strongly dependent stationary processes. *Journal of Time Series Analysis*, 10:375–383, 1989.

Murad S. Taqqu and Vadim Teverovsky
Boston University
Department of Mathematics
111 Cummington Street
Boston, MA 02215-2411, USA
*Email:* murad@math.bu.edu, vt@math.bu.edu

# Lecture Notes in Statistics

For information about Volumes 1 to 33
please contact Springer-Verlag